THE
SCIENCE
OF LIFE

CONTRIBUTORS

Robert Day Allen
Dartmouth College

Anthony C. Breuer
Harvard University

William D. Cohen
Hunter College

Fred Diehl
University of Virginia

Marvin Druger
Syracuse University

Frank Fisher
Rice University

Carl George
Union College

Donald Mandell
Bronxville, New York

Daniel McKinley
State University of New York—Albany

Raymond Rappaport
Union College

Lionel I. Rebhun
University of Virginia

Walter Rosen
University of Massachusetts—Boston

Michael Rosenzweig
University of Arizona

John W. Saunders, Jr.
State University of New York—Albany

Susan Schlee
Editorial Assistant

THE SCIENCE OF LIFE

Robert Day Allen
Dartmouth College

and Contributors

Harper & Row, Publishers
New York / Hagerstown / San Francisco / London

Special Projects Editor: Marlene Ellin
Project Editor: Karla B. Philip
Designer: Emily Harste
Production Supervisor: Kewal K. Sharma
Compositor: Ruttle, Shaw & Wetherill, Inc.
Printer: The Murray Printing Company
Binder: Halliday Lithograph Corporation
Art Studio: J & R Technical Services Inc.
Cover Photos: Emily Harste and Robert Day Allen

THE SCIENCE OF LIFE

Library of Congress Cataloging in Publication Data

Allen, Robert D.
  The science of life.

  Includes bibliographies and index.
  1. Biology. I. Title.
QH308.2.A42   574   76-40004
ISBN 0-06-040207-5

To our families, whose sacrifices made this book possible

Contents

Contents in Detail

DEVELOPMENT: FROM CELLS TO ORGANISMS

13. SEXUAL REPRODUCTION AND DEVELOPMENT IN ANIMALS 91

Author, Raymond Rappaport
Union College
Editor, John W. Saunders, Jr.
State University of New York—Albany

14. PRINCIPLES AND PROCESSES OF DEVELOPMENT 104

Author, Raymond Rappaport
Union College
Editor, John W. Saunders, Jr.
State University of New York—Albany

III   GENETICS:
HEREDITY,
VARIATION,
AND CONTROL
OF DEVELOPMENT

15. MENDELIAN GENETICS 113

Author, Marvin Druger
Syracuse University

16. CYTOGENETICS: THE CHROMOSOMAL BASIS OF HEREDITY 118

Author, Marvin Druger
Syracuse University

17. MOLECULAR GENETICS 125

Author, Marvin Druger
Syracuse University

18. HUMAN HEREDITY 134

Author, Marvin Druger
Syracuse University

IV ORGANISMS: BIOLOGICAL DIVERSITY

19. BIOSYSTEMATICS: AN INTERPRETATION OF THE ORIGINS AND RELATIONSHIPS OF LIVING THINGS 147

Author, Robert Day Allen
Dartmouth College

31. BEHAVIOR 266

Author, Robert Day Allen
Dartmouth University

MAN

32. SUPPORT AND LOCOMOTION 279

Author, Anthony C. Breuer
Harvard University
Editor, Robert Day Allen
Dartmouth College

33. INTEGRATION AND REGULATION: THE NEURON 286

Author, Anthony C. Breuer
Harvard University
Editor, Robert Day Allen
Dartmouth College

VI EVOLUTION

43. THE THEORY OF EVOLUTION BY NATURAL SELECTION 359

Author, Michael Rosenzweig
University of Arizona

44. THE ROLE OF MUTATION IN EVOLUTION 364

Author, Michael Rosenzweig
University of Arizona

45. ADAPTATION: THE MAIN PRODUCT OF EVOLUTION 368

Author, Michael Rosenzweig
University of Arizona

46. THE EVOLUTION OF DIVERSITY 373

Author, Michael Rosenzweig
University of Arizona

VII ECOLOGY

Preface

The Science of Life is a new kind of textbook for one-semester introductory courses at the college level. The informational content of this book is similar to that of many other texts, but both the philosophy and the organization are significantly different.

The underlying philosophy has been that future biologists as well as future nonscientist citizens need the same kind of textbook—one that provides an even, organized coverage of contemporary biology. This does not necessarily mean a recital of the latest research results and controversies in all fields, but rather a serious effort to incorporate new findings into a balanced, up-to-date perspective on biology—the most socially relevant and popular of the natural sciences.

The organization of this book is based on an analysis of the main objectives of introductory courses in biology. The most important purpose of such a course is to enable students to determine whether they are sufficiently interested in biology to engage in further study or to entertain the possibility of a career in biology, medicine, or a related field. The second aim is to inform students of the scope and organization of the field of biology so that they can make intelligent choices of courses for future study. Third, one hopes that even a superficial knowledge of biology will enrich the lives of young people by enhancing their appreciation of nature and the good things in life, and by helping them understand their bodies and minds through an appreciation of the relationship of humans to their animal relatives and their environment.

It is becoming clear that these goals are achievable in a one-semester course, and that the second semester of a two-course sequence might better be spent on higher level, specialized material.

The organization of this book is unique in that it is cyclic. It begins and ends with the unifying theme of energy viewed in cellular and ecological perspectives. The text can be entered at any point, and each part can stand alone. There is, however, a clear rationale for following the numerical order of the chapters. Part I, The Cellular Basis of Life, presents the structure and function of cells. The chemistry that is introduced is the absolute minimum required for an elementary understanding of fundamental cellular processes. The chemistry of the cellular processes is, in turn, useful background knowledge for Parts II, III, and V, Development, Genetics, and Human Biology, respectively.

An excellent alternate starting point is Part IV, Organismal Diversity and Behavior, which contains an evolutionary theme based on comparative anatomy, physiology, and behavior of selected representative organisms. The concluding chapter, Chapter 31 on Behavior, is a pivotal one because it not only discusses the science of behavior and its problems and challenges, but also traces some of the elements of behavior as they have evolved. The chapter then summarizes the behavioral information in Part IV and sets the stage for an in-depth consideration of the physiology of human behavior in Chapters 34 and 35.

Although physiology is central to the cell biology discussion (Part I) and is a recurrent theme in the diversity presentation (Part IV), it is emphasized mainly in Part V, Human Biology, in which physiology is closely integrated with anatomy.

The last two parts of the book, Part VI, Evolution, and Part VII, Ecology, can be read independently of the earlier portions of the book; however, the perspective they offer is more rewarding after the areas of organismal diversity and genetics have been mastered.

The Science of Life has several features that make it easy for both instructors and students to use. The chapters are relatively short, and are introduced by a short list of "Questions to Keep in Mind" that focus on the broader aspects of the material to be discussed. The chapters are divided into sections and each section of every chapter is introduced by a topic sentence. Students should spend a few seconds reading these both before and especially after reading each chapter because they serve as an excellent summary of the major concepts introduced. New terms appear in boldface type when they are defined or explained, and most technical terms also appear in the extensive glossary and index.

Another unique feature of this book is a collection of "Bioepicurean Delights" which appear throughout. In addition to providing a little light relief, they demonstrate that biologists are, on the whole, interesting and fun-loving people who use their knowledge to get the most out of life.

Throughout the preparation of this book, it has been a pleasure to work with the consultants who prepared much of the original manuscript and with Susan Schlee, who did much to make the final version concise and readable.

Robert Day Allen

THE
SCIENCE
OF LIFE

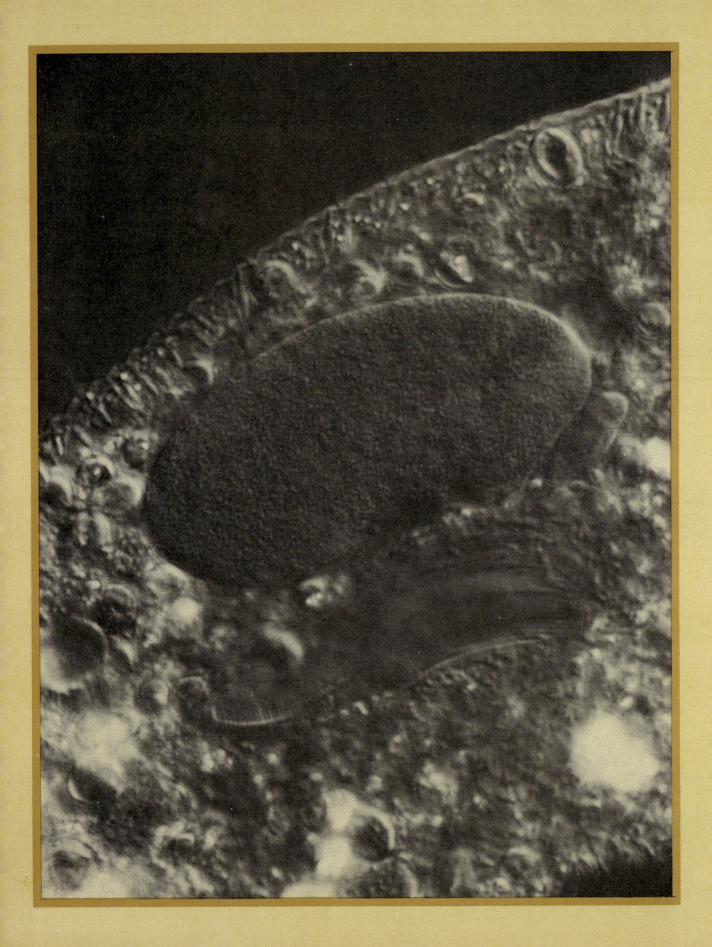

THE CELLULAR BASIS OF LIFE

There are several reasons why the cell is an ideal "point of entry" into the study of biology. First, the vast majority of living things exist either as single cells or as aggregates of cells. For this reason, the nature of cells and cellular activities is basic to the study of the entire biological world, from bacteria to humans. Second, reproduction and development (Part II) and heredity (Part III) have a cellular basis and depend on the specialized biochemical activities of cells. Finally, the study of cells is one of the most exciting frontiers of research in biology and medicine. This field, perhaps more than any other, promises the solutions to many of man's health problems. In this field one can never be sure which new research finding will lead to a more effective antibiotic or drug, a safer contraceptive, a deeper understanding of the aging process, or a potential cure for cancer.

1

Matter, Energy, and the Cell

QUESTIONS TO KEEP IN MIND

What is matter? What is special about *living* matter?

What is energy? In what forms is it found?

How is energy captured by plants? by animals?

Into what form do cells convert the energy they capture?

In what ways and for what purposes do cells use energy?

How does genetic information influence the use of energy by cells?

There is something very special and very difficult about defining the living state. One of our first tasks is to try to distinguish living things from those that are nonliving and those that were formerly alive. To do this, we must first examine the properties and interactions of matter and energy.

1.1 All things living, nonliving, and dead are composed of matter.

Everything we are familiar with is made of matter: our parents, our food, our possessions, the water we drink, and the air we breathe. Only in a "perfect" vacuum and in the void of interstellar space is matter in such short supply that we can say with some authority that it is lacking.

Given a chunk of matter, such as a log, we know that is is here to stay. Even if the log is burned, the wood is transformed to other matter—carbon diox-

ide and water—by oxidation, but matter remains. This observation has been expressed as the law of conservation of matter, which states that matter can be transformed, but can be neither created nor destroyed. It applies to all reactions in which living organisms take part.

Thanks to the science of chemistry, we have a pretty good idea of what matter is. It consists ultimately of over a hundred elements, which may be combined in an almost infinite number of substances. Living matter is very ordinary indeed in its elemental composition. What makes man, worm, leaf, bacterium, and their 2 million relatives special is their relationship to energy: *Living matter can capture and manipulate energy.*

The smallest unit of living matter that is organized enough to manipulate energy to its own advantage is the **cell.** The first chapters of this book will serve to explain the interrelationships between matter and energy that take place in the cell. The starting point must be energy itself, for without it matter cannot be transformed in any way.

1.2 Energy is the capacity to do work, and may exist in many forms.

Although energy exists in thousands of different forms—a candy bar, a gallon of gasoline, a ray of light, a burning log, a current of electricity, or a boulder perched on a precipice—all of it may be divided into two categories, kinetic and potential. **Kinetic energy** is associated with motion. It may occur in the form of light (the movement of photons, or light particles), an electric current (the movement of electrons from atom to atom along a wire), heat (the random movement of atoms and molecules), or a chemical reaction (the movement of atoms within or between molecules).

The other kind of energy, **potential energy**, does not involve movement. It represents stored energy. A boulder ready to roll down a hill and an unstable chemical compound ready to break down are both examples of potential energy. If the boulder does come tumbling down the hill, its potential energy is converted into kinetic energy; to endow it once more with potential energy, an outside source of energy would have to be called in to haul it back up the hill.

Transformations may occur not only between kinetic and potential energy but also between forms of energy: heat and light, chemical and electrical energy, and so forth. The study of these energy transformations, the science of **thermodynamics**, has made many important contributions to biology, including the useful generalizations known as the laws of thermodynamics.

The first law of thermodynamics states that during any energy transformation, energy is neither gained nor destroyed. The total energy involved remains the same before and after the transformation.

The **second law of thermodynamics** states that during any energy transformation there occurs a decrease in the amount of *useful* energy. Since according to the first law, energy cannot be destroyed, every decrease in useful energy must be accompanied by an increase in some other kind of energy —"use*less* energy." Obviously, the kind of energy that cells, humans, and factories want is the useful variety called **free energy**. Free energy can perform useful work, and as will become clear later, the cell (or factory) using the energy becomes increasingly better organized.

"Useless energy" at first seems to be a contradiction in terms, since energy is the capacity to do work. However, in an energy transformation not all the energy is harnessed as work. Some always escapes and causes disorder. The tendency toward disorganization of a system is called **entropy**. As an analogy, a small child running loose in a well-ordered house quickly increases the household's entropy. There will be an increasingly random arrangement of various items as the child strews them over the floor, and an expenditure of free energy on someone's part will be required to restore the original degree of order. A cell is a highly organized system in which an orderly arrangement of parts is necessary for continued existence. To remain alive,

a cell must keep its house in order, which is to say it must obtain free energy and perform work.

1.3 The energy used by cells comes largely from the sun.

Every cell stores a certain amount of energy in various forms. If the useful energy were not replenished, as one form of energy was transformed into another the cell would become increasingly disorganized, and it would finally die. Not only must a cell have an outside source of free energy, it must also export its useless energy to the environment as excretory products. Since both energy and matter are able to move back and forth across the cell membrane, a cell is considered an **open system**. If such a system is to continue to function, the loss of matter and energy to the environment must be balanced by the entrance of new matter and energy into the cell. Such a balance is called a **steady state**.

Where does all this energy used by cells come from? Most animals catch their energy as food, in the form of other cells or cell products. The energy they obtain in this way is potential energy stored as chemical bonds in the molecules of the food they have eaten.

Obviously, all energy cannot come from cells or the supply would soon be devoured. Plant cells solve this dilemma for all of the earth's inhabitants with their unique ability to trap energy from sunlight. Plant cells and their products are eaten by animal cells, which in turn fall prey to other animal cells, and so on. This, in essence, is the **food chain**, and it will be discussed in Part III. The preponderance of energy upon which the existence of all cells depends, then, is the energy of sunlight trapped by plant cells.

1.4 Cells must convert the energy they trap into the potential energy of chemical bonds.

Even when energy has been obtained by a plant or animal cell, the energy must be converted to usable form. For a cell this is the potential energy of chemical bonds that hold atoms together in molecules. Thus a plant cell must convert the trapped kinetic energy of sunlight into the potential energy of chemical bonds, and an animal cell must convert food into the same energy form.

To accomplish this conversion, the plant cell is provided with an **energy transducer**—a device which converts one form of energy into another. This is the **chloroplast**, a small green body within plant cells. Within this structure the energy of streams of **photons** (particles of light) is used to synthesize large, energy-rich molecules from smaller ones containing less energy. The process is called **photosynthesis.**

For free-living animal cells, the situation is somewhat different. Having trapped their energy in the form of food, they must break that food down into molecules that are relatively small, but that contain plenty of chemical bonds. (The smaller the molecule the fewer the chemical bonds and the less energy it contains.) This breakdown process, known as **digestion**, is carried out by all free-living animal cells (protozoa). It is also accomplished by the cells comprising the digestive systems of multicellular animals.

Animal cells cannot rely on digestion alone for all their needs, however. Like plant cells, they must synthesize molecules. **Synthesis**, the making of new chemical compounds, is one of the most important processes that takes place within a cell. What, in brief, does it entail?

Cells themselves are composed of molecules, most of which are water molecules, while the rest comprise the various structures of the cell. A cell synthesizes large molecules from smaller ones and from atoms. Molecules of cellular substances such as **proteins, nucleic acids,** and **polysaccharides** are the largest found in cells and are called **macromolecules.** The molecules of substances such as **amino acids, nucleotides,** and **sugars** are smaller and are the building blocks of macromolecules.

Synthesis increases the degree of order present in a cell. There is, of course, a corresponding decrease in entropy. Clearly, a small number of large molecules constitutes a more orderly system than a large number of smaller ones. The synthesis of molecules requires free energy, as do other cell functions which will be discussed later.

1.5 Cells require energy to reproduce and perpetuate the genetic information stored within them.

Certainly one of the most important functions of a cell is to reproduce, and however this is accomplished it requires energy. In the process, some form of **genetic information** will be passed on to the next generation, so that the new cells will resemble the ones that produced them. "Like produces like" is a rule that applies to single cells as well as to multicellular organisms. Parent cells have a kind of "tape recording" of genetic information specifying their characteristics, copies of which they pass to their daughter cells.

In the term *genetic information*, the word *genetic* refers to the genesis, or production, of characteristics. Just as the functional unit of written information is the sentence, the functional unit of genetic information is the **gene.** As you will see, genetic information is a concrete message expressed in physical terms, just as our more familiar forms of information require ink on paper, grooves in a phonograph record, or magnetic spots on a recording tape. In making a new cell, the original cell follows the directions which it contains in the form of genetic information, just as one might produce a cake according to a recipe in a cookbook.

A cell depends upon its genetic information for its own reproductive success and must preserve it carefully, copy it faithfully, and distribute the original and copy to daughter cells. Gross errors in either the copying process, called **replication** (making a replica), or the distribution of genetic information to daughter cells, called **mitosis,** would prove fatal. This is not to say that cells with new characteristics —that is, new genetic information—can never be produced. In a certain phase of sexual reproduction certain specialized cells deliberately produce cells that lead to offspring with new combinations of genetic information. By a process known as **meiosis,** these cells produce either egg or sperm cells (gametes), which contain half the amount of genetic information of other cells. The original amount is restored by the union of sperm and egg at the time of fertilization. When the egg and sperm which unite arise from different individuals, new combinations of genetic information are produced. This event is the fundamental reason for the existence of all sexual processes, from those of bacteria to those of man.

1.6 Cells also need energy to move and to be aware of their surroundings.

Motility, the ability to move, is a fundamental characteristic of a great many living organisms, many of which are single cells; this implies that motility is beneficial, that it helps cells meet some fundamental problems.

There are many forms of cellular motility. The whole cell may move, or movement may occur within the cell. The movement of cellular contents serves to distribute materials within the cell. When the entire cell moves, it is often to capture food. This behavior is therefore more prevalent among animal cells than plant cells. Although some plant cells glide along surfaces or propel themselves through liquids, the majority of plant cells have fairly rigid cell walls that tend to impede motility.

In general, motility enables a cell to move from a less favorable environment—for example, from a place that is too hot, too cold, too dark, or that contains too little food or oxygen—to a more favorable one. In addition, there are more specialized kinds of cellular motility, such as the movement of male gametes (sperm cells) which results in their union with an egg at the time of fertilization. Cellular motility, regardless of its type, is achieved only through the conversion of the potential energy of chemical bonds into the kinetic energy of motion.

In order to make effective use of its motility in search of food or a more favorable environment, a cell must be aware of its surroundings. An animal cell, for example, must recognize the food it captures as digestible or poisonous. A plant cell like *Euglena* must detect light from the sun in order to find and remain in a place where it can engage in photosynthesis. Information of this sort may come to the cell in the form of energy, such as the light which starts a plant photosynthesizing, or in the form of molecules or larger particles, such as food. However it arrives, such information has survival value; cells must have and maintain the equipment both to acquire information and to translate it into action. All of this requires energy.

1.7 For a cell, communal living solves many problems.

For the most part, a cell that is part of a multicellular (many-celled) organism has the same basic problems and the same energy requirements as does a single cell. But there are differences.

Take, for example, the problem of trapping energy in a plant. There are many cells in a multicellular plant which do not have the ability to trap light energy. Instead they obtain it from food molecules that other cells in the plant send them. These other cells, which do trap energy from sunlight, can feed a host of dependents if they receive certain commodities or services in return. For example, leaf cells can produce food for root and stem cells

because the roots supply them with water and various minerals and the stems enable the energy-trapping leaves to find a place where sunlight will fall on them.

In multicellular animals exactly the same situation exists. Each individual cell no longer needs to hunt food. Instead, certain cells in the group are specialized to form organs which can capture and take in food. It is then digested by other cells, and the food molecules obtained are sent to all cells via the circulatory system, which is itself constructed of cells.

In a community of cells, each of the many functions which must all be carried out within a single free-living cell is now performed by a specialized group of cells. Such problems as trapping energy and converting it into a useful form, keeping informed, moving about, and even producing offspring are all solved by such specialization. The result is greater efficiency, less work for all, more cooperation, less competition, and the possibility of doing things which would be impossible for one cell to do alone.

We have, then, the situation in which not all cells perform all functions. In fact, they must *not* perform

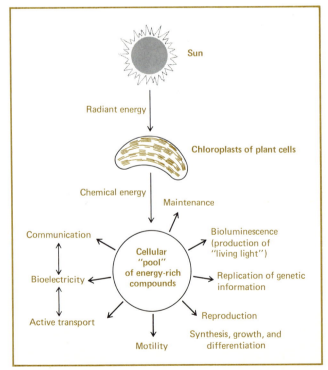

1-1 The radiant energy of sunlight captured by chloroplasts of plant cells finds its way into an imaginary "pool" of energy-rich chemical compounds in cells. The diagram shows the principal ways this energy is utilized by cells.

them all, for then the great advantages achieved by this association would be lost. Cells in communal life must have some mechanism for becoming and remaining specialists rather than generalists. The process by which cells, all of which descend from a single egg and all of which contain the same genetic information, become specialized is known as **differentiation,** and it is still a great mystery. Somehow a cell can disregard much of the genetic information it contains and focus its attention on the rest.

Obviously in a multicellular community, communication is of the utmost importance if the functions of the various parts are to be coordinated. Intercellular communication is effected by the use of certain kinds of molecules called **hormones,** which are produced by cells that specialize in their synthesis. The hormones carry messages from place to place via the circulatory system. In addition, multicellular animals have developed a more rapid system of communication, the nervous system. This system is made up of nerve cells, specialists in receiving and transmitting messages.

The ways in which cells utilize energy are shown in Figure 1-1. How they interact and specialize is truly amazing, and associations of cells are capable of functioning in ways that no individual cell could ever manage. In one well-known organ, cells can even work together to think about themselves, as yours are doing now.

2

Cells and Their Organelles

QUESTIONS TO KEEP IN MIND

When and how were cells discovered?

Why is the cell theory considered a milestone in the history of biology?

What is the structure of a "typical" cell?

How do the cells of animals, plants, and bacteria differ?

The ability to capture and manipulate energy characterizes the living state. This ability is attributable to the activities of the individual cell, existing either as a unicellular (one-celled) organism or as part of a multicellular animal or plant.

2.1 The invention of the microscope in the seventeenth century made possible the discovery of cells and the development of the cell theory.

The discovery of cells in 1664 by the British microscopist Robert Hooke was one of the most important milestones in the history of biology. It was a discovery for which the microscope was prerequisite, and this instrument was invented in its several forms in the early 1600s. Before this time, physicians, naturalists, and amateurs in many parts of the world had already learned a considerable amount about the structure and function of living organisms. For centuries they had faithfully recorded what they had seen, although by modern standards they were not able to see much. Limitations of the human eye restricted their investigations until the 1600s, when with the aid of the microscope, observers began to make acquaintance with the invisible inhabitants of the world.

For Hooke, who was then the curator of experiments for the Royal Society of London, one of the most striking characteristics of this newly discovered world was its apparent organization. In 1665, while continuing experiments he had begun the year before on several kinds of wood with a microscope of his own design (Figure 2-1), he examined a thin slice of cork. In his book, *Micrographia*, he describes what he found:

I could exceedingly plainly perceive it to be all perforated and porous, much like a Honeycomb, but that the pores of it were not regular . . . these pores, or cells, were not very deep, but consisted of a great many little boxes. . . .

Thus Hooke is credited with the discovery and naming of cells, those "little boxes" he observed in cork and other kinds of plants.

Another pioneer of microscopy was Anton van Leeuwenhoek of Delft, Holland. In 1673 Leeuwenhoek began a series of microscopic observations which he continued for 50 years. Using simple equipment and with fine lenses of his own making (Figure 2-1, *right*), he discovered protozoan, bacterial, sperm, and blood cells (although he did not call them *cells*), and was the first to see any of the structural details in cells.

In the following years so many other investigators observed and described the cells in plants and animals that it gradually became clear that the cell was the basic unit of life. It was not until 1838, however, that this idea was clearly set forth, in the writings of two German scientists, Matthias Schlei-

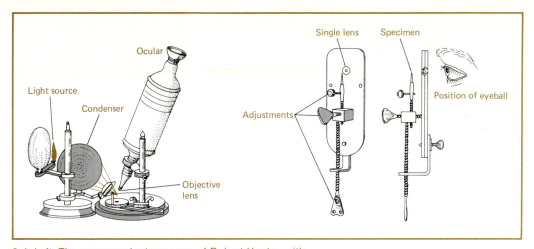

2-1 *Left:* The compound microscope of Robert Hooke, with which the "cells" of cork were first described. This microscope was similar in design to microscopes used today. *Right:* Single-lens microscope of the type designed, constructed, and used by Anton van Leeuwenhoek. It was held directly in front of the eye. Specimens were placed on the pointed specimen holder, which was moved by a delicate adjustment of thumbscrews.

Labels on figure: Ocular, Light source, Condenser, Objective lens, Single lens, Specimen, Adjustments, Position of eyeball

den, a botanist, and Theodor Schwann, a zoologist. Schleiden, working exclusively on plants, and Schwann, on animals, proposed that *all living systems are composed of cells* and substances produced by cells, a generalization known as the **cell theory.** They further demonstrated that the cell, the basic unit of life, is to some extent an independent unit capable of performing the functions necessary for life, yet subordinate to the organism as a whole.

In spite of their fine work, Schleiden and Schwann had a number of misconceptions, particularly about the origin of cells. This question was not settled until 1855, when a German physician, Rudolph Virchow, could state with good evidence that cells arise only from pre-existing cells—*omnis cellula e cellula*. This was an important addendum to the cell theory.

In substantiating the cell theory, scientists expected to find that any sexually reproducing organism, be it an onion, worm, or man, was made of cells and that all the cells in each came from a common ancestor—the fertilized egg. This expectation was confirmed many times in the century and a half following Schleiden and Schwann's pronouncement. Today, however, it is known that while the theory holds true for the vast majority of plants and animals, there are also tissues, organs, and organisms that are not compartmentalized into cells. These are sometimes called **syncytial** or **acellular** (without cells). These exceptions, however, do not diminish the great significance of the cell theory.

2.2 The cell theory led to the development of cellular physiology and to many advances in medicine.

Almost as soon as it was proposed, the cell theory focused attention on the possibility that vital functions such as respiration, movement, and excitation could be studied in single cells, populations of similar cells, and tissues, as well as in the organism as a whole. In fact, in many instances, research on the cellular level seemed the most promising approach. Thus in the mid-nineteenth century the foundations were laid for the science of cellular physiology. The enormous body of information acquired since then about cellular processes has had a major impact on the practice of medicine, and physicians are now able to understand many aspects of health and disease in terms of normal and abnormal cellular function.

The cell theory had other effects on medical practice that were less elemental but still important. For example, in order to find out how cells function, cell physiologists needed a ready supply of living cells. Techniques for growing cells and tissues outside the body in a cell culture were therefore developed. At first some cells were extremely difficult to culture, but now nearly any plant or animal cell, tissue, or organ can be cultured for extended periods. Modifications of these procedures have enabled medical scientists to preserve blood cells for transfusions and corneas for sight-restoring transplants, and also to maintain even larger organs such as kidneys, hearts, and livers over periods of hours or days for transplantation. None of these life-saving techniques would have been possible without basic research in cell physiology.

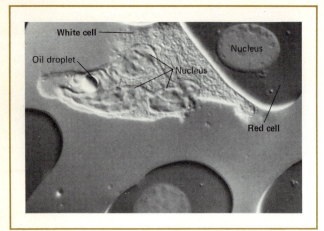

2-2 A photomicrograph of one white blood cell (a neutrophil) and portions of several red blood cells of the newt *Tarica granulosa*. Note the spherical oil droplet, multilobed nucleus, and large number of cytoplasmic particles in the neutrophil; compare the red blood cells of this species, which have an unlobed nucleus and very few cytoplasmic particles. (Courtesy of Robert Hard.)

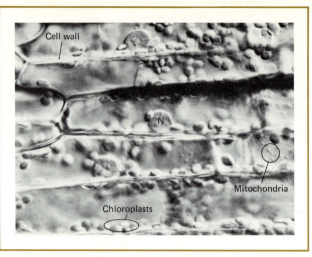

2-3 Cells in a leaf of the aquatic plant *Elodea canadensis*. Each cell is surrounded by a cell wall and contains a large nucleus (*N*), many chloroplasts (medium-sized particles), and much smaller mitochondria. The empty space in each cell is the large central vacuole. (Photograph by the author.)

2.3 The general structure of cells can be seen through a light microscope.

A look at Figures 2-2 and 2-3 will show you that cells come in a wide range of sizes and shapes, even if they come from a single organism. Yet they are all recognized as cells and clearly have many structural features in common. Figure 2-4 represents a schematic reconstruction of a generalized animal cell as seen through a light microscope. Such a cell appears to be entirely enclosed within a limiting **plasma membrane,** so thin that we can detect only its presence, not its structure. Near the center of the cell is a spherical structure, the **nucleolus.** The surrounding **nucleus,** with its **chromosomes,** is enclosed within the **nuclear envelope.**

The portion of the cell outside the nucleus is known as the **cytoplasm,** and several structures are embedded in it. Just outside the nucleus are two tiny dots, the **centrioles,** and elsewhere in the cytoplasm are oval or elongated structures, the **mitochondria** (*mitochodrion* in the singular) and denser bodies called **lysosomes.** Occasionally a **vacuole** or **vesicle**—a fluid-filled space enclosed by a membrane—may be observed. Occasionally **Golgi bodies** can be seen in some cells.

Had this book been written 20 or 25 years ago, we might already have neared the end of the generalized description of cell structure. But in recent years knowledge of cell structure has increased tremendously with the use of the electron microscope.

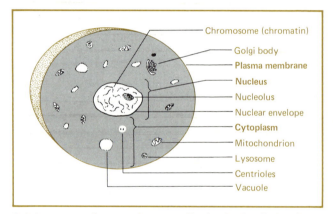

2-4 A cutaway diagram of a generalized animal cell showing sizes and positional relationships of the various cellular organelles visible in the light microscope.

2.4 In the past 25 years, several kinds of electron microscopes and many special methods for using them have revolutionized our knowledge of cell structure.

The light microscope can resolve objects as close as 0.2 micrometers apart; electron microscopes can resolve objects 1,000 times closer together. The most common of these instruments is the **transmission electron microscope** or **TEM,** in which electrons create an image of a specimen almost exactly as light forms an image in an optical microscope. The greater resolving power of the TEM is due to the fact that the wavelengths of electrons are much shorter than those of light. An electron beam can travel only in a vacuum. Therefore, the main limitations in transmission electron microscopy are that only dead specimens can be studied and they must be very thin. Because a specimen must be specially treated to be viewed by this microscope, the question arises as to whether all the structures observed were actually present in the living cell or whether they are "artifacts" which resulted from the preparatory treatment. Sometimes cell biologists prepare replicas of the surfaces of frozen-fractured specimens. These are "deliberate artifacts" that reveal aspects of membrane surfaces usually not seen in thin sections (Figure 2-5).

A newer instrument, the **high-voltage electron microscope** or **HVEM**, is similar except that the electron voltage has been increased considerably. This makes it possible to penetrate thicker specimens and obtain the three-dimensional relationships among cell components. It is also possible to view very thin living cells in special wet chambers. However, since living cells would die if treated with substances which heighten the image contrast in electron micrographs, it is difficult to observe their internal details with this procedure.

In the past several years a new class of instrument, the **scanning electron microscope (SEM)**, has found many uses in cell biology. Cells are dried in a special way that prevents them from becoming distorted, and their surfaces are coated thinly with a metal which scatters electrons. The SEM provides

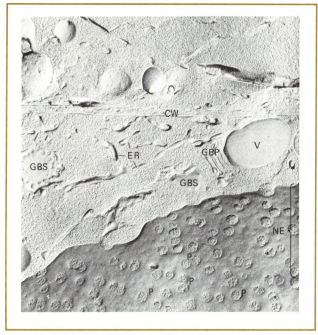

2-5 A transmission electron micrograph of an onion root tip cell that was frozen and then fractured. The micrograph shows the fractured surface. Features that can be seen include the nuclear envelope (*NE*) with its pores (*P*); endoplasmic reticulum (*ER*); cell wall (*CW*); Golgi bodies in surface and profile views (*GBS, GBP*); and vesicles (*V*). (Courtesy of D. Branton.)

2-6 A surface view of a culture of myoblasts (developing muscle cells) of a lizard (*Anolis carolinensis*) obtained with a scanning electron microscope. Cells that have recently divided have these characteristic blebs on their surfaces and interconnecting "filament-feet." (Courtesy of Ellen Kahn Bayne.)

BIOLOGICAL INSIGHTS

Cellular Measurements

Cells and their parts are measured in micrometers (formerly microns) in the light microscope, and in nanometers in the electron microscope. One meter (m) = 39.37 inches (in) = 10^2 centimeters (cm) = 10^3 millimeters (mm) = 10^6 micrometers (μm) = 10^9 nanometers (nm) = 10^{10} Ångstrom units (Å). Chemists measure molecular size and interatomic distances in Ångstrom units. It is convenient to remember that the resolution of (i.e., smallest observable separation in) the light microscope is about 0.2 μm. For the electron microscope it is about 0.2 nm.

a three-dimensional view of the surface of objects as small as a cell (Figure 2-6), or even a virus, and as large as a housefly. The resolving power of the SEM is intermediate between the light microscope and the TEM. It is not suitable for examining the interior of cells.

2.5 The structure of cells as seen through an electron microscope is vastly more complicated than is apparent through a light microscope.

If a thin section of the same animal cell shown in Figure 2-4 were observed with an electron microscope, it would appear to have all the complexities illustrated in Figure 2-7. Here the inner cytoplasm is filled with an extensive network of flat membranes called the **endoplasmic reticulum** (*endo*: internal; *plasmic*, referring to the cytoplasm; *reticulum*: a network). In some places the endoplasmic reticulum (ER) is studded with small particles called **ribosomes** (*ribo*, referring to the molecules of **ribonucleic acid** found in these bodies; *some*: body). These areas of the ER are referred to as

rough ER, in contrast to portions of it that are less flat, lack ribosomes, and are called **smooth ER.**

Ribosome clusters are called **polyribosomes** or simply **polysomes.** Polysomes and free ribosomes can be found free in the cytoplasm as well as spread on the surface of the rough ER.

Another membranous system is the **Golgi bodies,** which were actually discovered with a light microscope and is named after its discoverer. The Golgi bodies are composed of piles of flattened sacs (**cisternae**) and adjacent small spherical sacs (vesicles) used to store substances to be secreted. Smaller cytoplasmic components also bounded by membranes include the mitochondria and the **lysosomes,** bodies similar in size to the mitochondria but different in texture, once confused with them. The particles bounded by membranes divide the volume of the cell into compartments in which many different processes take place under a variety of chemical conditions.

In most animal cells there are two centrioles that lie almost at right angles to each other. Near them and elsewhere in the cell are a number of extremely thin tubular structures, some 27 nanometers (nm)

2-7 A cutaway diagram of a generalized animal cell showing the three-dimensional shapes of organelles that are seen as a rule only in thin sections with an electron microscope.

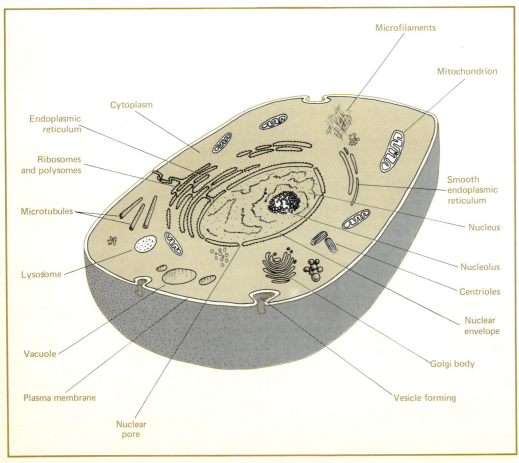

Microfilaments

Mitochondrion

Cytoplasm

Endoplasmic reticulum

Ribosomes and polysomes

Microtubules

Smooth endoplasmic reticulum

Nucleus

Lysosome

Nucleolus

Centrioles

Nuclear envelope

Vacuole

Plasma membrane

Golgi body

Vesicle forming

Nuclear pore

in diameter, called **microtubules.** Thinner still are **microfilaments,** 6 to 7 nm in diameter, bundles of which occur in most cells, especially those that move.

Closer examination of the nucleus shows that the nuclear envelope consists of two membranes and is perforated by channels or "pores." These pores may allow some passage of materials between the cytoplasm and the **nucleoplasm,** the contents within the nuclear membrane. The nucleoplasm houses a nucleolus and some very thin threads of **chromatin.** Before the cell reproduces, these chromatin threads condense into thicker bodies called **chromosomes.** As a final detail, we can see that ribosomes are readily visible on the cytoplasmic side of the nuclear envelope, but not on the inside.

All these features have been represented in an idealized manner in the diagram of a generalized animal cell (Figure 2-7). Many of the same structures are shown more realistically in Figure 2-8, an electron micrograph of a rat liver cell.

A generalized plant cell, shown in diagram in Figure 2-9, contains some things that the animal cell does not, namely a **cell wall,** a large vacuole,

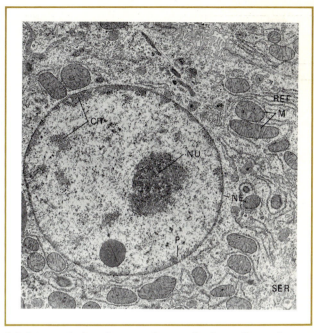

2-8 A transmission electron micrograph of a rat liver cell. Features visible include the nuclear envelope (*NE*) with a pore (*P*), the nucleolus (*NU*), chromatin (*CH*), mitochondria (*M*), and both rough and smooth endoplasmic reticula (*RER* and *SER*).

2-9 A diagram of a generalized plant cell showing the structures usually seen in electron micrographs.

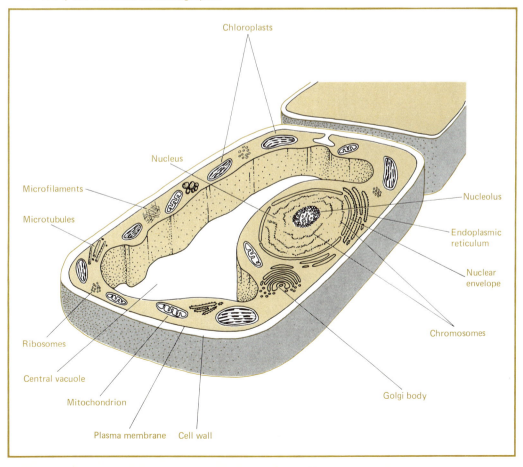

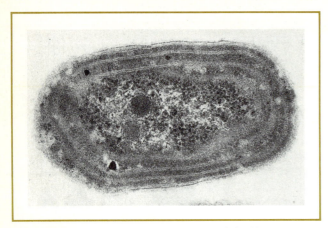

2-10 A transmission electron micrograph of the blue-green alga *Coccochloris pencocystis*. (Courtesy of Mary M. Allen.)

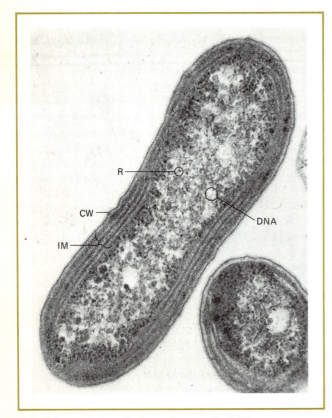

2-11 A transmission electron micrograph of *Nitrosococcus*, a marine bacterium that oxidizes ammonia to nitrite to obtain energy. Structures visible include the cell wall (*CW*), intracellular membranes (*IM*) where the oxidative enzymes are localized, thin filaments of deoxyribonucleic acid (*DNA*), and bacteria ribosomes (*R*). (Courtesy of Stanley Watson.)

and many chloroplasts; it lacks structures that most animal cells contain, namely centrioles. (Some plant cells do have centrioles, but most of the cells in multicellular plants do not.)

The structures found in both plant and animal cells are usually divided into two categories, **organelles** ("little organs") and **inclusions.** Organelles are the relatively permanent parts of the cell's structure and serve specialized functions in somewhat the same way that organs do in a multicellular organism. They include the nucleus, mitochondria, chloroplasts, Golgi bodies, and other structures illustrated in Figures 2-7 and 2-8. The inclusions, on the other hand, are not permanent working parts of the cell. Fat droplets, starch granules, and secretion granules (granules produced by a cell for export) are all inclusions. They are not shown on the generalized diagrams.

Up to now we have been describing the structure of complex **eucaryotic** cells (cells having a true nucleus). There is a second, much simpler type of cell—the **procaryotic** cell. Only blue-green algae and bacteria—both single-celled organisms—have procaryotic cells. These two types of organisms are shown in Figures 2-10 and 2-11. There is no nuclear envelope surrounding the very simple chromosomes in these cells. Procaryotes are said to contain a **nucleoid** rather than a nucleus. Their relative simplicity becomes even more apparent when you consider that they lack mitochondria, Golgi bodies, chloroplasts, endoplasmic reticulum, centrioles, lyosomes, microtubules and micofilaments.

2.6 Viruses are alive only when present in cells. Viruses are not cells.

Unlike cells, many viruses can be crystallized and kept in a frozen dehydrated state for years, to "come alive again" when allowed to infect living cells. Most scientists agree that viruses are not alive, because they can only reproduce inside living cells, using the synthetic machinery of those cells to produce more viruses. Nevertheless, when viruses do infect cells they are part of a fascinating living system—one which has taught us a great deal about cellular inheritance, as we shall see in Chapter 17.

3

The Molecules of the Cell

What are the principal molecules present in cells? Of what elements are they composed? How are the atoms of these elements held together?

What building blocks are used to construct the large molecules of the cells?

What is the fundamental way in which monomers are strung together into long chain polymers? How are polymers taken apart?

What are the characteristics of polysaccharides, proteins, nucleic acids, and lipids? Of what building blocks are they made?

To understand what goes on in a cell it is essential to know what it is made of. An enormous amount of research has been devoted to the task of analyzing the chemical components of cells and determining how they interact. Two things have emerged: a catalog (still incomplete) of the molecules of which cells are made, and the realization that the structure of molecules and their location and function within the cell can provide the key to understanding many of the properties of living organisms.

3.1 The small molecules in cells are composed mostly of atoms of carbon, oxygen, hydrogen, and nitrogen, held together with covalent bonds.

The forces that hold one atom to another are known as **chemical bonds.** When two or more atoms, either of the same element or of different elements, are held together by chemical bonds, they constitute a **molecule.** The kind of molecule formed depends upon what atoms are bonded together and in what arrangement. Molecules in living matter vary greatly in size and properties. An oxygen molecule, one of the simplest, has just two identical atoms; protein molecules consist of literally thousands of atoms of five different types. Not surprisingly, the properties of two such different molecules are quite different. All molecules of any one substance have the same number and kind of atoms.

Let us consider the ways in which molecules, large and small, are represented. One method is the use of **molecular formulas** that name the types of atoms present and give the numbers of constituent atoms. The molecular formula for water is H_2O; the formula for glucose is $C_6H_{12}O_6$. The formula for water indicates that each molecule of water contains two atoms of hydrogen (atomic symbol H) and one of oxygen (O). The formula for glucose indicates that every molecule has 6 carbon atoms, 12 hydrogen atoms and 6 oxygen atoms.

A molecular formula indicates nothing about the arrangement of atoms in a molecule. A **structural formula,** however, presents a two-dimensional picture of a molecule, by showing the relative positions of the atoms and also the bonds holding them together. The structural formula for water is H—O—H.

In a structural formula, the lines that connect atoms represent **covalent bonds,** strong chemical bonds that form when each of two atoms shares one of its electrons with the other atom. Hydrogen, for example, is much more stable as the molecule H—H, in which two atoms share two electrons (a covalent bond), than as H, a single atom with a single electron. Not all of the electrons possessed by an atom are available for sharing, and different kinds of atoms have different numbers of sharable electrons. If we use a dot to designate each electron available for sharing, then the number of dots around a given atom equals the number of covalent bonds that it usually forms. The atoms of the elements most commonly found in cells may be represented as follows:

H·   ·O·   ·N·   ·Ċ·   ·S·   :P·

Here, H stands for hydrogen, which forms one covalent bond; O for oxygen, which forms two; N for nitrogen, three; C for carbon, four; S for sulfur, two; and P for phosphorus, five. You can remember the first four easily, since in order, they spell HONC.

Now we are ready to construct some small molecules. One carbon atom will form four covalent bonds. If we add to it four hydrogen atoms, the molecule CH_4 is formed. This substance is methane, or swamp gas. For simplicity, we draw this as the structural formula in which lines are used to represent the covalent bonds.

$$\cdot \dot{C} \cdot + 4H \cdot \rightarrow \; H:\overset{H}{\underset{H}{\overset{\cdot\cdot}{C}}}:H \quad \text{or} \quad H-\overset{\displaystyle H}{\underset{\displaystyle H}{\overset{|}{\underset{|}{C}}}}-H$$

Each covalent bond consists of two shared electrons, one from each of the atoms bonded together.

Several common small molecules are presented in Table 1. You will note that every atom has made its expected number of covalent bonds. This will hold true for all the structural formulas presented in this book. When there are double and triple bonds between atoms, each shared pair of electrons is represented by a line.

Certain organic molecules in a watery environment (such as that inside a cell) may undergo ionization and lose some of their hydrogen. An atom of hydrogen consists of a positively charged proton, or atomic nucleus, with a negatively charged electron spinning around it. In the case described above, hydrogen leaves its electron behind with the main part of the molecule and is lost in the form of **hydrogen ions**—each of which has a positive electric charge of +1. Such an ion is represented as H^+. Molecules which lose H^+ in water are called **acids**; they become negatively charged because each electron they are left with has a

negative charge of −1. In contrast, some molecules, called **bases**, tend to pick up H^+ from the environment. When a neutral molecule picks up a hydrogen ion it becomes positively charged.

Molecules such as those in Table 1 contain carbon and are known as organic molecules. Carbon, nitrogen, oxygen, hydrogen, phosphorus, and sulfur are the major constituents of large biological molecules.

TABLE 1 STRUCTURAL FORMULAS OF SEVERAL SMALL MOLECULES

Molecule	Structural formula
Hydrogen gas	H—H
Oxygen gas	O=O
Nitrogen gas	N≡N
Water	H—O—H
Ammonia (in cleaning agents)	H—N—H with H below
Carbon dioxide gas	O=C=O
Methane (swamp gas)	H—C—H with H above and H below
Ethyl alcohol (in alcoholic beverages)	H—C—C—O—H with H's above and below
Acetaldehyde	H—C—C—H with H below and O double bonded above
Acetic acid (an organic acid in vinegar)	H—C—C—O—H with H below and O double bonded above
Glycine (an amino acid)	N—C—C—O—H with H's attached
Cysteine (a sulfhydryl amino acid)	H—S—C—C—C—O—H with H, N, O attached

3.2 Giant molecules are generally built up and broken down by dehydration synthesis and hydrolysis, respectively.

The giant molecules, or **macromolecules,** such as proteins found in cells, belong to a class of compounds known as **polymers** ("many segments"). Polymers are long, chainlike molecules which consist of large numbers of smaller molecules bonded together, that may be thought of as links in a chain. These single segments are called **monomers.** They are the building blocks from which the large molecules found in cells are formed. Joining two monomers results in the formation of a **dimer.**

Polymers are very much a part of everyday life; our clothes are made either of natural polymers (silk and wool are proteins) or of synthetic polymers such as nylon, rayon, and orlon. The polymers of natural fibers are formed from their particular monomers by **dehydration synthesis,** which is the usual way in which all biological macromolecules are synthesized. As the name suggests, this process involves the loss of water.

The reverse of this process, **hydrolysis** ("water-splitting"), is the common way of breaking down macromolecules and is one of the key steps in digestion. It is the process by which cells and organisms break down the macromolecules of the prey organism (food) so that the components may be resynthesized into the kinds of macromolecules characteristic of the predator. If, for example, you were stranded on a desert island and limited to a diet of raw oysters, you would not become an oyster, despite the saying "You are what you eat." Rather, the oyster's macromolecules would be broken down by hydrolysis into monomers, these would be selectively absorbed by your cells, and the monomers would be recombined into the same kind of "human macromolecules" that you would synthesize from a more normal diet.

It is clear, then, that cells must be able to take apart macromolecules and synthesize them as well. How is this done?

When two or more monomers join together, only the atom or atoms at one site on each molecule participate in linking with other monomers: one hydrogen atom in one location, and one hydrogen plus one oxygen atom in another. This is shown schematically in Figure 3-1. The two small molecules to be joined are depicted so that the atoms involved in the reaction protrude from the rest of the molecule. In the reaction step one bond on each is broken. This releases a water molecule, and the monomers bond together. Each of the latter is now called a **residue,** because it has lost an atom or two. This molecule has the same two reactive groups as a monomer and can react similarly with another monomer to release another water molecule and add another residue to the growing chain. The process can be repeated hundreds of times. The general outcome is a macromolecule.

A macromolecule is taken apart essentially by reversing the dehydration synthesis. In other words, a water molecule is split, and the atoms which were removed from the monomers are put back where

3-1 A diagram showing the generalized reaction of dehydration synthesis, by which all biological macromolecules are made.

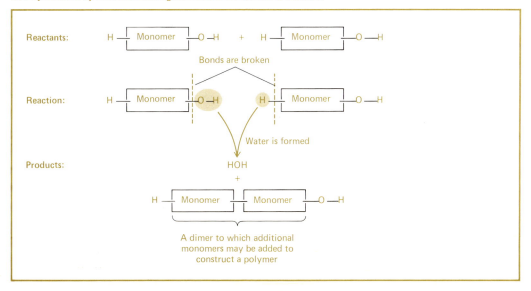

Reactants: H — | Monomer | — O — H   +   H — | Monomer | — O — H

Bonds are broken

Reaction: H — | Monomer | — O — H   H — | Monomer | — O — H

Water is formed

Products: HOH
+
H — | Monomer — Monomer | — O — H

A dimer to which additional monomers may be added to construct a polymer they were in the first place. It is precisely this process of hydrolysis that occurs in digestion.

Both dehydration synthesis and hydrolysis are somewhat more complex than indicated in this simplified overview. Enzymes and other small molecules (especially organic phosphates) are involved as well. Their roles will be discussed in Chapter 6.

3.3 Polysaccharides are one of the major categories of macromolecules.

Polysaccharide means "many sugars." Polysaccharides are polymers that consist of up to hundreds of similar glucose sugar monomers linked together by dehydration synthesis. Two examples are **cellulose,** a structural component of the cell walls of most plants, and **starch,** a common substance which plants usually form to store much of the captured energy of sunlight. Although glucose monomers compose both starch and cellulose, they are bonded together differently in these two polymers (Figure 3-2, bottom). Still another polysaccharide composed of glucose molecules is **glycogen,** a macromolecule found in many animal cells and some plant cells. In glycogen the glucose residues are arranged in

highly branched rather than in predominantly straight chains.

Sugars and polysaccharides are examples of the class of organic molecules known as **carbohydrates.** This name arose because their general formula contains one water molecule for each carbon atom — that is, $C_n(H_2O)_n$. Thus, glucose can be written $C_6H_{12}O_6$ or $C_6(H_2O)_6$. A glucose molecule can exist either as a straight chain or as a hexagonal ring (Figure 3-2, top). In solution, glucose folds into this ring configuration, and polymers of glucose are usually represented as chains of these rings. Carbohydrates constitute the major source of energy in the human diet.

3.4 Proteins comprise a second category of macromolecules.

As you will note in Table 2, proteins are polymers of many different **amino acids.** The name *amino acid* has a logical basis, as do the names given to most organic molecules. If we find out what *amino acid* means, the construction of proteins will be a little easier to understand.

There are several characteristic groups of atoms that appear in the structural formulas of many different organic molecules. These are called **func-**

3-2 Top: Four different ways of representing a molecule of the sugar glucose. Bottom: Glucose shown as the repeating unit in both starch and cellulose.

Glucose: straight chain form

Glucose: formation of ring form in solution

Glucose: Haworth ring formula

Shorthand form

Starch

Cellulose

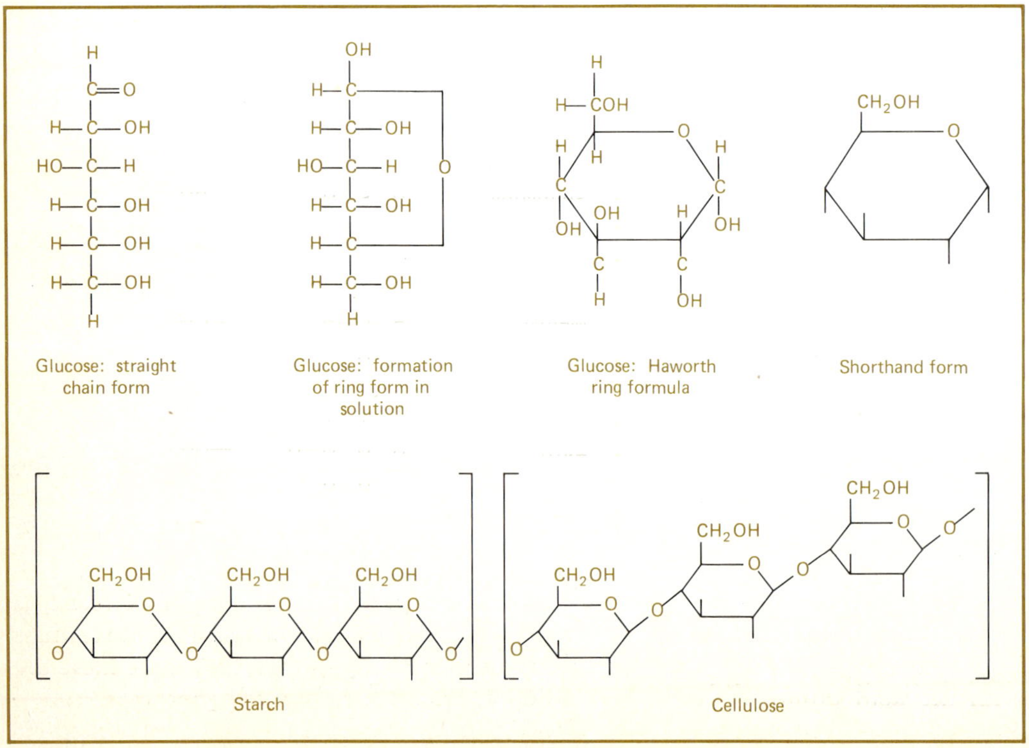

TABLE 2 MACROMOLECULES

Category of macro- molecule	Constituent monomers	Number of different kinds of monomers per macro- molecule	Total number of monomers per macromolecule
Polysaccharides	Sugars or sugar derivatives	Sometimes many, often only one kind	Sometimes over 1000
Proteins	Amino acids	About 20 different kinds of amino acids	Usually 100 to 1000
Nucleic acids (DNA, RNA)	Nucleotides	Four different kinds of nucleotides in each type of nucleic acid	Often over 3000; sometimes over 10 million

tional groups. When a particular group, say the acid group, makes up a part of several different kinds of molecules, each will share certain common characteristics. Table 3 lists some of the most common functional groups. In the table, R stands for the residue or rest of the molecule, the part other than the functional group.

A molecule may contain more than one functional group, as illustrated by amino acids. An amino acid molecule contains one amino group ($-NH_2$) and one acid group ($-COOH$); almost all of the roughly 20 different natural amino acids have the following general structural formula:

Amino group →
$$H-N-C-C-O-H$$
← Acid group

(with H R O above and H below)

As you can see, the amino and acid groups are both bonded to the same alpha (α) carbon atom. This α-C atom (adjacent to the acid or $-COOH$ group) is also bonded to one hydrogen atom and R, the rest of the molecule. The 20 different types of R groups are responsible for the 20 or so different kinds of amino acids that make up naturally occurring proteins.

Like all macromolecules, proteins are assembled from their building blocks by dehydration synthesis. Although the details are more complicated, we can understand in principle how a protein is formed from a study of how two amino acids are joined. Figure 3-3 shows how this is done. Of the two reacting amino acids, the amino group of one amino acid combines with the acid group of the other. A water molecule is split out, and the two

TABLE 3 FUNCTIONAL GROUPS IN ORGANIC MOLECULES

Name of group	General structural formula	Abbreviated group formula
(Carboxyl organic acid)	$R-C(=O)-O-H$	$-COOH$
Amino	$R-N(H)(H)$	$-NH_2$
Phosphate	$R-P(=O)(O-H)-O-H$	$-PO_3H_2$ (or $-\text{\textcircled{P}}$)
Sulfhydryl	$R-S-H$	$-SH$
Hydroxyl	$R-O-H$	$-OH$
Methyl	$R-C(H)(H)-H$	$-CH_3$
Ethyl	$R-C(H)(H)-C(H)(H)-H$	$-CH_2CH_3$ or $-C_2H_5$

residues bond together. The bond that forms is known as a **peptide bond.** The amino acid residues are themselves called **peptides,** and since there are two of them linked together, the product is called a dipeptide (di, "two"). By adding on more and

3-3 The formation of a dipeptide by dehydration synthesis from two amino acids. Water resulting from the reaction is derived from joining the amino (—NH₂) group of one amino acid with the acid (—COOH) group of the other to form a peptide bond. Up to several hundred different amino acids can be joined in this way to form a polypeptide chain or protein molecule.

more peptides we can make tri- ("three"); tetra- ("four"); and finally, polypeptides (poly, "many").

The simplest proteins are single polypeptide chains containing a large number of amino acid residues (peptides). The more complex proteins contain several interlaced polypeptide chains, each composed of thousands of amino acid residues. Between these two extremes exist a truly incredible variety of proteins. This great diversity results from combinations of the 20 kinds of amino acids in different amounts, sequences, and combinations. Even if no protein were larger than a tripeptide, 8000 different molecules would be possible. Since most proteins are polypeptides, the number of possible combinations is practically unlimited.

One protein with which you may be familiar is insulin. Each insulin molecule consists of 51 amino acid residues arranged in a specific sequence. Another is hemoglobin, which contains four polypeptide chains totaling 574 amino acid residues.

Polypeptide chains tend to coil and fold in a conformation (shape) that is highly specific for each polypeptide. Some segments of a polypeptide chain are held in a springlike alpha-helical configuration (Figure 3-4) by hydrogen bonds, while other parts assume a random coil configuration.

The sequence of amino acids in a protein determines the precise position of the interacting atoms or groups in the entire polypeptide chain and in that way also determines the conformation of the molecule. Because the interactions that produce this shape are delicate, protein conformation is sensitive to many factors in the environment and may

be totally distorted if the molecule's surroundings are altered, for example by heat. The molecule is then said to be denatured and loses all of its biological activity. Under appropriate conditions some proteins can become renatured and return to their original shape.

Each kind of protein molecule has a particular size and shape. Like precut building materials, many proteins lend themselves to the construction of specific cellular structures. Proteins with this function are known as structural proteins. Others, however, have another function. Their characteristic shapes enable them to make temporary interactions with other molecules, causing the latter to break apart or react with other molecules. Protein molecules with this function are called enzymes. All enzymes are protein molecules, but not all protein molecules are enzymes. Some protein molecules serve both as part of a structure and as an enzyme (e.g., the muscle protein, myosin; see Chapter 8).

3.5 Chemical reactions in cells require enzymes as well as energy.

In a typical chemical reaction, two molecules, A and B, react to form two products, C and D; a simple example being the reaction of sucrose and water to form glucose and fructose.

Sucrose + water → fructose + glucose

In all such reactions we are interested in two things: To what extent does the reaction reach completion, and at what rate does it proceed?

Many chemical reactions in cells proceed at much faster rates than could occur in a test tube if purified reactants A and B were simply put together in solution at room temperature. Such rapid chemical reactions are possible in cells because of the presence of enzymes, which are biological catalysts, substances that hasten chemical reactions.

Enzymes do not affect the extent to which the reaction will proceed. They only hasten the achievement of equilibrium, the point at which as much sucrose and water are reacting to form glucose and fructose as fructose and glucose are reacting to form sucrose and water. The point at which equilibrium is reached is determined by the change in the free energy of the reaction, the difference between the amount of energy still remaining in the bonds of the products and the original free energy of the reactants.

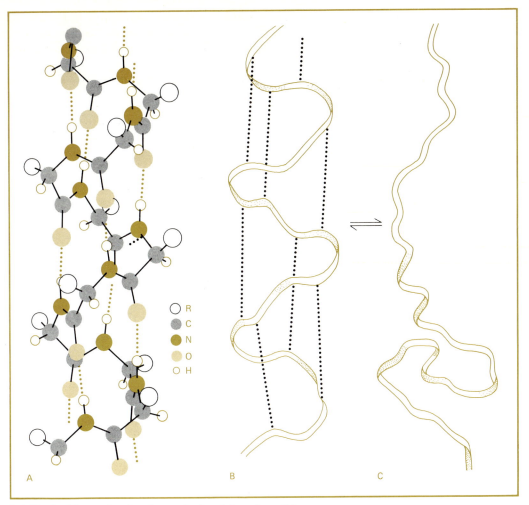

3-4 The backbone structure (———) of protein polypeptide chains can assume either an α-helical (A and B) or random coil (C) conformation or shape. The α-helical conformation is maintained by hydrogen bonds (·······) between the N—H in one part of the chain and the double-bonded O in another part of the chain. In this figure only the atoms composing the backbone and those involved in hydrogen bonding are shown, for simplicity.

○ R
● C
● N
○ O
○ H

The problem that reactions encounter outside the cell is one of insufficient energy. In order for molecules to react, they must collide with a certain amount of energy. This is called the **activation energy.** Outside the cell this activation energy can be provided in the form of thermal energy by heating reactants in a test tube to give the molecules enough energy to react. Enzymes also cause the reaction, but in a very different way. Instead of supplying activation energy to the reactants, enzymes lower the required amount of activation energy and so allow the reaction to proceed with less—much less—than the usual activation energy. (Sucrose and water will react 3 trillion times faster in the pres-

ence of the enzyme invertase than if left alone in solution in a test tube.)

Even with enzymes, however, chemical reactions are reversible, and all eventually reach a point where the reaction $C + D \rightarrow A + B$ is going on as fast as the forward one, $A + B \rightarrow C + D$. There they remain, with equal numbers of molecules reacting in both directions: $A + B \rightleftharpoons C + D$. Because the amounts of A, B, C, and D are no longer changing, the reaction is said to have reached equilibrium. If products C and D are removed from the site of the reaction—usually by becoming involved in a second set of reactions—then A and B, the original reactants, will continue to react to replace the C and D removed. Such a removal of products is almost the rule in cells, and products such as respiratory gases, excretory products, and **metabolites** (sugars, amino acids, organic acids, and other small molecules), are continually being carried away or used up as reactants in other enzymatically catalyzed reactions. Under these conditions, where the products are being removed as fast as they are formed, reaction is said to be in a **steady state.** It is proceeding at a certain rate dependent upon other reactions in the same cell or organelle.

How do enzymes get these reactions started? The reactants temporarily become attached to the enzyme. In this situation the molecules involved in the reaction are spoken of as the **substrate** of the enzyme. To enable the enzyme to attach to the substrate, the enzyme has an **active site** on its surface which "fits" the **binding site** on the substrate; the two are thought to go together more or less the way two pieces of a puzzle do. The activation energy of the reaction is lowered because the molecules of substrate are held in the grasp of the enzyme so that the desired bonding points are brought close together. The molecules of substrate usually react together while attached to the enzyme, then products and enzyme separate. The enzyme, unchanged, accepts new molecules of the substrate.

In some cases enzymes alone are insufficient to produce a reaction and work in cooperation with smaller molecules called coenzymes. Coenzymes are neither substrates nor proteins. Many are vitamins and will be discussed in later chapters.

Since enzymes are proteins, and since it is the specific shape of an enzyme that enables it to interact with its substrate, anything which spoils the natural shape of a protein molecule also destroys its enzyme activity. Hence denaturation renders any enzyme worthless.

3.6 Genetic information is carried by a third category of macromolecules, nucleic acids.

Nucleic acids, macromolecules often found in the nucleus or nucleoid of cells, were discovered in 1869 by the Swiss scientist F. Miescher. From white blood cells brought to him as pus on bandages, he extracted impure samples of the nucleic acid now called **deoxyribonucleic acid (DNA),** one of the two main types of nucleic acid. The other is the closely

TABLE 4 THE MOLECULAR CONSTITUENTS OF DNA AND RNA

related but significantly different **ribonucleic acid, (RNA).**

The building blocks of which DNA and RNA are composed are called **nucleotides,** and each of these monomers is a three-part molecule made up of one phosphate group (PO_4), one sugar residue, and one nitrogen-base residue. It is the phosphate group which makes nucleic acids acidic.

In both RNA and DNA molecules there are four different nucleotides. In each, these four differ only in the base portion: each nucleotide contains a different nitrogen base. The RNA bases are adenine, guanine, cytosine, and uracil. The phosphate group and the sugar residue of the nucleotide are always the same. In RNA, the sugar is ribose, a five-carbon sugar with the formula $C_5H_{10}O_5$. The four nucleotides that make up an RNA molecule are:

adenine-ribose-phosphate
adenosine
   monophosphate (AMP)

guanine-ribose-phosphate
guanosine
   monophosphate (GMP)

cytosine-ribose-phosphate
cytidine
   monophosphate (CMP)

uracil-ribose-phosphate
uridine
   monophosphate (UMP)

Any two of these nucleotides can be joined together by dehydration synthesis to form a dinucleotide; a third nucleotide can be added to form a trinucleotide; and so on, until a polynucleotide chain containing some thousands of nucleotides is constructed. As the nucleotides link, a ribose group of one joins to the phosphate of another. This forms a chain composed of alternating ribose and phosphate backbone with the nitrogen bases extending from it.

In a cell, the nucleotide residues are put together with the aid of a specific enzyme called **RNA polymerase** (the -*ase* suffix indicates an enzyme).

As with the amino acid residues in protein, the number and sequence of nucleotide residues vary from one RNA molecule to the next, and the number of different kinds of RNA molecules is tremendous (Table 2).

DNA differs from RNA in several respects. First, the sugar residue of its four nucleotides is **deoxyribose,** $C_5H_{10}O_4$, called such because it has one less oxygen than ribose, $C_5H_{10}O_5$. Second, one of the four bases in the nucleotides is different: uracil is replaced by the closely related thymine (Table 4). Thousands of these slightly different nucleotides are joined together by dehydration synthesis into a chain or strand, just as in RNA. The conformation of the DNA molecule, however, differs considerably from that of RNA. RNA usually consists of a single strand, but DNA is almost always found with two strands wound together in a configuration called a **double helix.** The two strands of the double helix are complementary; that is, they are constructed in such a way as to fit together snugly. This "fit" results from **hydrogen bonds** which can form only between specific bases. They form only between adenine and thymine or between guanine and cytosine. In Figure 3-5 the relatively weak hydrogen bonds between the bases are dotted to distinguish them from covalent bonds.

3.7 The double-stranded form of DNA is the key to its reproduction and to the replication of genetic information.

DNA can carry genetic information because its nitrogen bases are strung together in a specific sequence. **Genes** are merely sequences of nitrogen bases. Because the two strands of DNA are complementary, each contains the information needed to make the other.

Figure 3-6 outlines the replication of a double-stranded DNA molecule. Replication starts as the bonds holding the strands together begin to break at one end. In the next step, new nucleotides position themselves along the separated strands only at the places where their base is complementary to those in the existing strand. Once these nucleotides are aligned along the strand, they are joined together by dehydration synthesis. Thus one parent DNA molecule gives rise to two daughter molecules, each identical to the parent and each, in fact, composed of one of the two original parental strands. The original parental molecule is not conserved; instead, half of it—one strand—is conserved in each daughter molecule. This method of reproduction is called **semiconservative replication.**

Thus DNA meets one of the major requirements for a carrier of genetic information; it can be replicated in a dividing cell so that each of the daughter cells receives an exact copy of parental DNA. Just how genetic information is coded within the molecule and then translated into cellular activity will be discussed in Chapters 7 and 18.

An enzyme called **DNA polymerase** is required for DNA molecules to replicate. A test tube mixture containing this enzyme, the four nucleotides, some "primer" DNA as a template or model, and several other ingredients, can produce DNA identical to the primer DNA molecules.

In 1953 James Watson and Francis Crick dis-

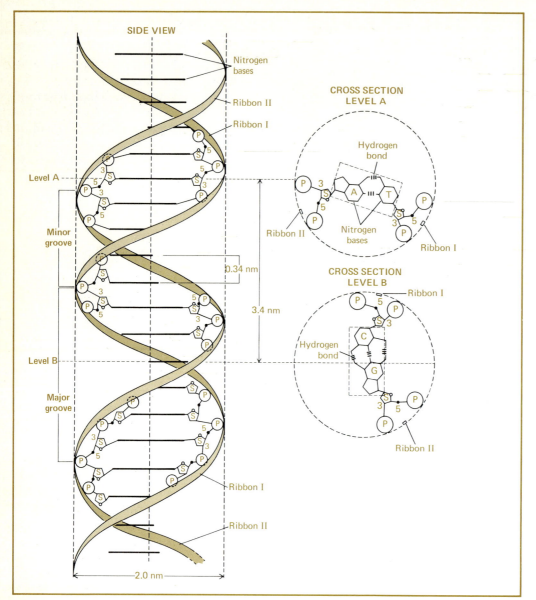

3-5 A diagram of a portion of the double-helical DNA molecule, as seen in side (*left*) and two cross-sectional levels. The backbone of each helix is composed of alternating phosphate (*P*) and deoxyribose sugar (*S*) groups, the phosphate lying at the periphery and the sugars about one-quarter of the distance to the center (*dotted line*). The complementary nitrogen bases (*A, T* or *G, C*) stretch across from one helical strand to the other and are held together by hydrogen bonds as shown.

covered the double-helical structure of DNA and realized its importance in the chemistry of inheritance. This was probably the most important contribution to biology in this century. It led to a period of rapid advance in nearly every field of biology and to enormous progress in many areas of medical science.

3.8 Lipids are an organism's most highly concentrated source of energy.

Lipids, a large and diverse group of compounds composed of carbon, hydrogen, oxygen, and sometimes phosphorus, store energy for the cell and are used as structural components as well. Fats, oils, waxes, and steroids are common lipids.

Like the other molecules we have considered, lipids are composed of building blocks, but far fewer join together in a single molecule than is the case with carbohydrates, proteins, and nucleic acids. A molecule of fat, for example, is formed from

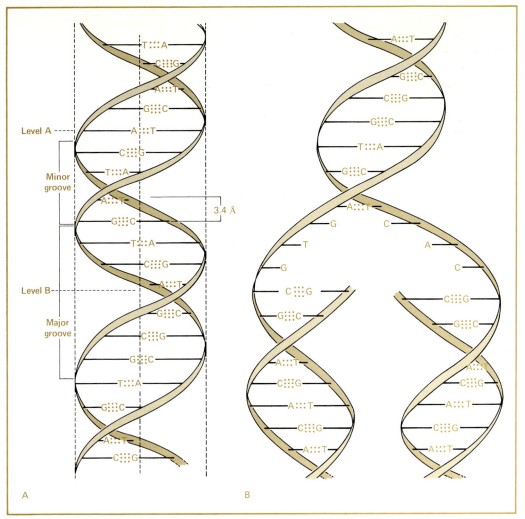

3-6 A diagram showing how a DNA double helix (*A*) replicates. Breakage of hydrogen bonds (*B*) causes the complementary strands to separate, allowing free nucleotides to bind to each strand, forming replicates.

only four molecules—a single molecule of glycerol and three fatty acids. The fatty acids in a single lipid molecule may be the same or different. Fat stores about twice the energy of an equal amount of carbohydrate or protein. When a cell needs the

3-7 The dehydration synthesis of one molecule of a fat from three fatty acid molecules and one molecule of glycerol. During digestion of fats the same reaction runs in reverse and is called hydrolysis.

3-8 Some other lipids: A. The phospholipid lecithin. (Note that one of the fatty acids is unsaturated and that the molecules contains a phosphate group and a nitrogen base.) B. The steroids cholesterol, testosterone (a male hormone), and progesterone (a female hormone). C. Vitamin A.

energy stored in fat, the fat molecules are broken down by hydrolysis into glycerol and fatty acids (Figure 3-7).

In a molecule of the animal fat tristearin, all three fatty acids are stearic acid. Like other fatty acids, each molecule of stearic acid is composed of carboxylic acid and a long chain of carbon atoms with hydrogens attached. If the hydrocarbon tail contains no double bonds, it is said to be **saturated.** A hydrocarbon tail with double bonds that can accept additional hydrogen molecules under certain circumstances is **unsaturated.** Because the long hydrocarbon chains have no electrically charged functional groups, most fats and oils are not soluble in water, an attribute which gives them an important role in the composition of cell membranes. (This is explained more fully in Chapter 4).

An important class of lipids found in all membranes contain phosphate and are called **phospholipids.** These contain glycerol and one or more fatty acids, but in addition have a phosphate group and a nitrogen base such as ethanolamine (HO—CH_2—CH_2—NH_2).

There are many biologically important lipids besides fats. Some examples are the **plant pigments** carotene, chlorophyll, and xanthophyll, and vitamins D, E, and K. A group of compounds called **steroids** are also lipids. These include animal sex hormones (androgens and estrogens), bile salts (important in digestion), cholesterol (a substance implicated in hardening of the arteries with age), and cortical steroids, a family of adrenal hormones of enormous medical importance. Steroids share a common fused ring structure made up of 17 carbon atoms. They differ in the placement and types of functional groups attached to this common structure (Figure 3-8).

4

The Cell Surface

QUESTIONS TO KEEP IN MIND

What are the molecules of the plasma membrane, and how do they interact?

How are the following concepts related: lipid bilayer, bimolecular leaflet, unit membrane?

What are the Davson-Danielli model and the fluid-mosaic models of plasma membrane organization?

How do diffusion and osmosis differ?

What differentiates passive from active transport?

How are macromolecules and food particles transported across the plasma membrane?

It has long been known that special properties of the cell surface determine to what degree substances can enter and leave and the rates at which they will do so. Because of its importance, the cell surface has been investigated at least as intensively as any other part of the cell. This chapter attempts to present not only the insights gained but also some idea of how they were achieved.

4.1 The general properties of the cell surface are due to the presence of lipids and proteins in the plasma membrane.

The presence of lipids in the **plasma membrane,** the outer cell membrane, was first suggested by the fact that cells break down in lipid solvents such as ethyl ether, chloroform, and acetone. In a quantitative study of the lipid content of red blood cell membranes, it was found that enough lipid was present per red blood cell to cover the cell with a layer of lipid two molecules thick. This suggested at first that the plasma membrane might consist simply of a lipid bilayer. Soon this idea had to be modified, for measurements of the tension at the surface of the cell showed that the cohesive forces at the cell surface were too great to be accounted for by lipids alone. The examination of cell membranes under polarized light also suggested the presence of some other kind of molecules—in all likelihood, proteins.

Our present concept of the molecular architecture of the plasma membrane is derived from the model (theoretical concept) of membrane structure proposed about 1940 by H. Davson and J. F. Danielli. According to their model the plasma membrane consists of a bilayer of lipid molecules surrounded on both sides by protein layers—in other words, a **bimolecular leaflet** (Figure 4-1).

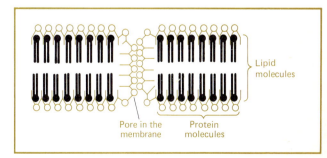

4-1 The Davson-Danielli model of membrane structure assumed that a lipid bilayer (*color*) was surrounded on both sides by protein molecules. In this model proteins lined pores through which ions could pass.

To understand how this bimolecular leaflet is constructed, it is first necessary to discuss the electrical charges of lipid molecules and the effect that these have on the molecules' compatibility with water. Because water molecules are polar (that is, have oppositely charged ends) they are attracted to each other (Figure 4-2) and to other charged molecules. In water, weak hydrogen bonds hold water molecules more or less in formation, in spite of constant rearrangement due to molecular motion. (At lower temperatures there is less molecular motion, and the structured form of water becomes more obvious, as in a snowflake or ice crystal.)

The fluid within cells contains polysaccharides, proteins, and nucleic acids, as well as water. Because these macromolecules contain many polar functional groups, such as the hydroxyl, amino, carboxyl, and phosphate groups (Table 3), water molecules are attracted to these macromolecules as well. Because of this attraction, these substances are soluble in water and are said to be hydrophilic, which literally means "water-loving."

There are also molecules that contain no (or almost no) polar groups and are not soluble in water and are therefore called hydrophobic, meaning "water-fearing." Lipids are largely hydrophobic. As you remember, in most natural fats and oils the three fatty acid molecules (all hydrophobic) that are attached to the glycerol are not usually all the same kind. When one fatty acid is replaced by a hydrophilic or water-loving group (Figure 4-3) the resulting lipid molecule is partly hydrophobic and partly hydrophilic. Phospholipids possess a phosphate containing a hydrophilic (charged) group bound to glycerol. Their structure may be simply represented with a hydrophilic "head" and two hydrophobic "sticks" or "tails." It is molecules such as these that are important in the construction of cell membranes.

Given its dual nature, what would happen to such a molecule in water? The two hydrophobic chains, staying close together to eliminate as much water as possible from their immediate vicinity, orient themselves so that they stick out of the water. The hydrophilic group, on the other hand, tends to bend down into the water, away from the hydrophobic end, as in Figure 4-4. There it attracts water molecules.

This is precisely what the phospholipid molecules made by cells can do. Because phospholipids carry electrical charges on their "head" portion, they are able to associate with other charged molecules such as proteins. Such an association with proteins forms lipoprotein. If phospholipids are put

4-2 A schematic diagram showing the mutual attraction of water molecules due to their polarity, and the hydrogen bonds that may form. Many of the unusual properties of water are due to these hydrogen bonds.

4-3 The structural formula for the phospholipid lecithin is related to this simplified shape used to show the "water-loving" and "water-fearing" portions of the molecule.

Hydrophobic chains

Hydrophilic end

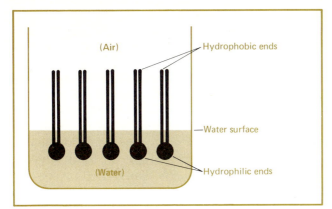

4-4 The behavior of phospholipids in water. The hydrophilic portions point down into the water, while the hydrophobic portions escape at the water surface.

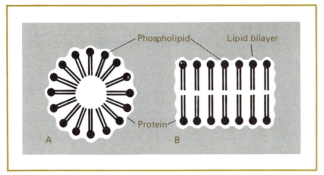

4-5 Lipoprotein bimolecular leaflets (*B*) form when charged proteins are added to an aqueous suspension of phospholipids (*A*). Note the similarity to the Davson-Danielli model for membrane structure (Figure 4-1).

into water they tend to form spontaneously many lipid bilayers (Figure 4-5). When electrically charged protein molecules are dissolved in this mixture, they attach themselves to the charged heads of the phospholipid molecules to form bimolecular leaflets. The numerous membranes of the cell were, until recently, believed to be composed of structures similar to lipoprotein bimolecular leaflet complexes.

4.2 Cellular membranes are similar in structure but not in function, and are apparently fluid.

Soon after thin sections of cells could be prepared for electron microscopy (section 2.4), J. D. Robertson observed the plasma membranes of enough cells to generalize that all cell surfaces showed the presence of two evenly spaced lines. This applied both to surfaces separating the cell from its environ-

ment and to those within the cell separating organelles from the rest of the cytoplasm. This double structure he called **unit membrane,** and suggested that his pictures confirmed the Davson-Danielli bimolecular leaflet model. He proposed that the two dark lines he photographed were probably the rows of protein molecules and that the lipids comprised the space between them. Before Robertson's work it had seemed doubtful that all cell membranes were structurally similar. The presence in all cells of the same unit membrane structure as seen with the electron microscope, however, called attention to a basic structural similarity, even though different membranes in cells are now known to function differently.

Even more recently, some interesting new discoveries have indicated that membranes have a fluid nature. Apparently both protein and lipid molecules are able to move about in the plane of the membrane. Some protein molecules penetrate the lipid bilayer and are free to move in the plane of the membrane, as suggested in Figure 4-6. S. J. Singer has called this particular view of membrane structure the **fluid-mosaic model.** D. M. Branton and others have observed the middle and outer surfaces of a cleaved membrane by electron microscopy (Figure 4-7). These studies show some protein molecules to be mobile and able to aggregate under certain conditions.

Depending on the kind of membrane, the proteins in it have different properties. Hence membranes can perform different functions.

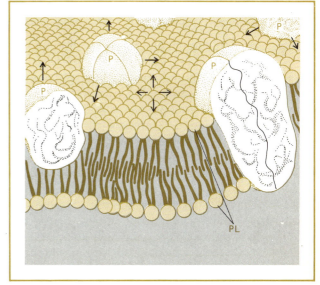

4-6 The "fluid-mosaic" model of membrane structure, in which both phospholipid (*PL*) and protein (*P*) molecules are free to move in the plane of the membrane.

4-7 Electron micrograph of a freeze-fracture replica of a red-blood-cell membrane showing (A) particles (probably protein molecules) exposed when the fracture plane splits the phospholipid double layer and (B) on the membrane outer surface. (Courtesy of K. Fisher.)

4.3 The plasma membrane is selectively permeable because of its lipids, "pores," and electrical charges.

The selectively permeable nature of the plasma membrane was discovered in experiments carried out at about the turn of the century, in which cells were immersed in solutions of known content and concentration. For example, red blood cells (erythrocytes) were placed in a variety of solutions, which caused the cells to swell and lose their red pigment if the dissolved substance, the solute, could penetrate the cells. The time required for a suspension of cells to break down was taken as a relative measure of the rate of penetration of the solute into the cells. As expected, the membrane of the red blood cell permitted the passage of different substances at different rates, while totally excluding some. The cell's membrane was therefore said to be selectively permeable.

Experiments of this kind, including experiments on plant cells, established the main facts regarding the permeability of the plasma membrane to dissolved substances. At first it was suspected that the molecular weight of the solute (more or less an indication of its size) would be the most important factor governing a molecule's ability to pass through the plasma membrane. It was found, however, that the solute's solubility in lipid solvents (compared to that in water) was more important. Molecules that are relatively more soluble in lipid solvents enter the cell more readily than those that are less so. For both groups, molecular size is still important; smaller molecules apparently pass more readily through some kinds of "pores" in the membrane. The third factor determining the penetration of solutes is their charge. All else being equal, charged and polar molecules enter more slowly than do molecules without polarity.

4.4 Diffusion is the net movement of molecules from a region of higher concentration to a region of lower concentration.

Molecules are constantly in motion, a fact that may be inferred from Brownian motion, a phenomenon in which particles of microscopic dimensions (up to a few micrometers in diameter) are jostled by the more rapid movement of submicroscopic molecules. Because, in a water solution, molecules of water and dissolved substances are constantly in motion, any local differences in concentration tend to be evened out in time. This process is called diffusion, and it occurs naturally as it is a direct result of random molecular motion (heat energy).

As an example of diffusion, imagine what happens to a lump of sugar dropped into a cup of coffee and not stirred. First it dissolves to produce a pool of sugary syrup at the bottom of the cup. Already some of the sugar molecules have begun to be propelled around the cup, because the thermal energy of the hot coffee causes the molecules to move faster. The higher the temperature of the coffee, the more rapidly the sugar diffuses. When the sugar concentration is similar in all parts of the cup, equilibrium has been established. Once the solution is in equilibrium the molecules of sugar and water, both of which have been moving, keep right on bumping about but there is no net movement in any direction. Over the distances found in a coffee cup, diffusion takes too long to mix the sugar and coffee effectively; therefore we stir. At the level of all but the largest cells, however, diffusion to equilibrium can occur in a few seconds or less.

Because diffusion tends to occur in cells wherever and whenever concentration differences exist, the plasma membrane must be selective. If it were not, diffusion would cause the molecules inside a cell to be present in exactly the same concentrations as outside the cell. If particular kinds of molecules were synthesized inside a cell and were not present outside it, away they would go. If unwanted molecules were outside and not inside, in they would come. The selectively permeable membrane prevents this.

4.5 Osmosis is the diffusion of water through a semipermeable membrane in response to a difference in water concentration across the membrane.

Water diffuses into or out of cells in response to a difference in water concentration. Like salts, sugars, or other solutes, water moves from a region of greater concentration to one of lesser concentration. In the case of the sugar lump in the coffee, as sugar diffused upward, water diffused downward into the sugary region where there were fewer water molecules. This same example can be used to illustrate osmosis, but only if the sugary syrup is enclosed in a semipermeable membrane (one that is impermeable to sugar molecules but permeable to water molecules) and then immersed in the cup of coffee. Now the sugar molecules are unable to diffuse out, and the water alone diffuses into the membranous bag, until the solution inside is so dilute that it resembles the water outside or until the bag bursts.

At times a cell may resemble a bag of sugar syrup in a cup of coffee. For example, consider what happens when an egg cell from a sea animal like a starfish or clam is removed from sea water and placed in distilled water. Such a cell contains a tremendous number of dissolved substances. The water concentration of the cell's contents is much lower than that of its surroundings. The large difference in water concentrations inside and outside the cell creates a steep gradient in water concentration across the plasma membrane. Water molecules pass through the membrane in both directions, but the net movement, or diffusion, of water is toward the lower concentration of water inside. Consequently, the cell swells until its membrane bursts.

All plasma membranes are not equally permeable to water. Water can move in and out of a red blood cell, for example, about 10,000 times faster than into and out of either a salmon egg or some species of amoebae. Plant cells have an additional safeguard against taking in too much water by osmosis: a rigid cell wall. As water moves into a plant cell, the cell expands and pushes against the cell wall to create a pressure known as turgor pressure. The wall (which itself is no barrier to water) holds firm by exerting wall pressure.

What if a plant cell is placed in a watery environment containing a greater concentration of solutes? Say, for example, that a leaf of lettuce is placed in a beaker of sea water. Then, by osmosis, water will leave the plant cells, causing the cytoplasm of each cell to shrink away from the cell wall. This shrinking due to osmosis is plasmolysis, illustrated in Figure 4-8.

The passage of water or dissolved substances by diffusion requires no energy expenditure by the cell. Consequently, the two processes are referred to as passive transport.

BIOEPICUREAN DELIGHTS

The Secret of a Good Salad

The making of a good salad depends on an understanding of cellular water relations in plants. Everyone prefers a salad that is crisp to one in which the leaves are limp and discolored, but few cooks understand the secret and apply it successfully. Most salad materials can profit from soaking for 10 minutes or more in cold fresh water while the best leaves are selected. By that time nearly all of the cells in the leaves of lettuce, spinach, watercress, and the like are turgid (section 4.5). This turgidity is preserved if the leaves are rapidly dried by shaking and blotting with a clean towel and then immediately coated with salad oil. After the oil is applied, vinegar, salt, herbs (especially dill, thyme, oregano, and tarragon) can be added without the leaves losing turgor. Salt and pepper are added last, seconds before serving.

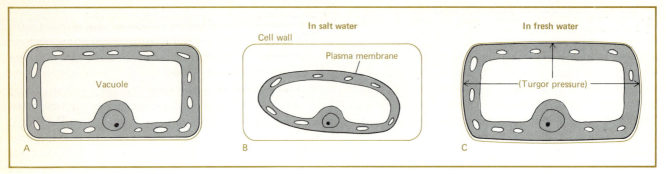

4-8 A plant cell (*A*) changes in appearance in different environments. If placed in salt water (*B*), the cell shrinks away from the cell wall as water leaves the cell. When placed in fresh water (*C*), the cell swells until its membranes press against the rigid cell walls and cause them to bow outward.

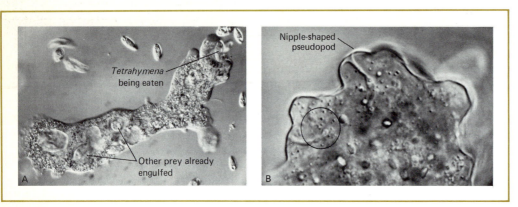

4-9 *A*. An *Amoeba proteus* phagocytizing (engulfing) several *Tetrahymena* (small ciliated protists) at once. Several are visible inside food vacuoles. *B*. Another specimen of *A. proteus*, showing a prominent nipple-shaped projection (a pseudopod) with pinocytosis channel leading into the cytoplasm. Another channel at the left is breaking off into pinocytic vesicles, or pinosomes. (Photograph by the author.)

4.6 All cells actively transport substances in and out.

When a cell moves material in or out against a concentration gradient—in the direction opposite from that dictated by diffusion—it is said to be engaged in **active transport**. Active transport requires energy, and this is usually furnished by the breakdown of the energy-rich substance ATP.

There are many examples of active transport. Plant root cells actively accumulate certain minerals. The cells lining an animal's intestine can actively transport amino acids and sugars obtained by digestion into the bloodstream, even when the bloodstream already contains a higher concentration of these molecules. The kidney filters all kinds of small, useless molecules out of the bloodstream but then actively transports useful molecules back into the blood. At its peak, active transport by cells in the tubules of the kidney requires as great an expenditure of energy as does the pumping of blood by the contracting muscle cells of the heart. Some organisms in fresh water environments are able to transport water out of their cells by means of a contractile vacuole. This is a special kind of active transport.

The usual cases of active transport may be thought of as involving specific enzymelike carrier molecules in the plasma membrane which act as "pumps." The exact mechanism is not clear, but it seems that the carrier molecules in the cell membrane can combine with an ion or molecule of the substance that the cell wishes to import or export and temporarily form a complex that can move within the lipid of the membrane (section 4.2). The complex moves across the membrane (from the inside to the outside, or vice versa) and is then split apart, the whole process utilizing energy in an unknown manner. The carrier molecule returns to its original position in the membrane, ready to repeat the process.

4.7 Some cells may take in large particles by phagocytosis ("cell eating") and pinocytosis ("cell drinking").

The methods of active and passive transport that we have discussed allow cells to import and export relatively small molecules and ions. Large macromolecules and still larger food particles cannot get into a cell in this way. Therefore, some cells use phagocytosis or pinocytosis to engulf large particles.

In **phagocytosis** (Figure 4-9A), parts of the cell actually flow out and around a particle and enclose it. The particle ends up inside a vesicle, the membrane of which was originally a portion of the plasma membrane. This means of getting things into a cell is employed primarily by amoebae, and by amoebalike cells such as white blood cells (leukocytes). These white blood cells are called **phagocytes,** and they move about the bloodstream and other tissues of multicellular animals, such as ourselves, engulfing bacterial cells that might otherwise cause diseases.

Pinocytosis (Figure 4-9B) usually involves the taking in of droplets rather than particles; hence the name, which means "cell drinking." Desirable molecules (such as dissolved proteins) may exist in the fluid surrounding the cell. To obtain these the cell apparently anchors one point on its plasma membrane to a gel region in the cytoplasm with an all-but-invisible filament, then advances to form a nipple-shaped pseudopool all around this stationary point. The result is the formation of a narrow channel, which of course contains a bit of the surrounding solution because it is open at one end. Next, a portion of the channel is pinched off at the closed end to form a vesicle or **pinosome,** which moves around within the cell. Pinosomes fuse with lysosomes (section 2.5), which contain enzymes capable of breaking down macromolecules in the pinosome.

5

Photosynthesis: Life from Light

QUESTIONS TO KEEP IN MIND

What are the raw materials and energy supply for photosynthesis? What are the products?

What is light energy and how is it transduced to electrical and, finally, chemical energy by plant cells?

How do some cells regulate the amount of light they absorb?

What are the "light reactions" and "dark reactions" of photosynthesis?

What are the pathways of electron flow in the light reactions of photosynthesis?

How do we know precisely what happens to atoms in water and carbon dioxide in photosynthesis?

What are redox reactions, and why are they important in photosynthesis?

The process by which plants transduce solar energy into the chemical energy needed by the plants themselves, by animals great and small, and by the expanding community of man is called **photosynthesis.** For a long time, it has been the most important process in the world.

Without photosynthesis there would be almost no life on earth, nor would there be coal, oil, natural gas, or the other fossil fuels that were formed from the bodies of plants and animals that lived eons ago. Today, it is estimated that plants incorporate 50 billion tons of carbon into photosynthetic products each year. Much of this is eaten, directly or indirectly, by all the animals of the world, and the rest is eventually decomposed by bacteria and fungi.

5.1 Photosynthesis is the utilization of solar energy in the manufacture of sugars and other organic compounds from water and carbon dioxide.

Plants take carbon dioxide and water from their surroundings and synthesize energy-rich compounds; as part of the process they produce oxygen gas. A generalized formula for photosynthesis is

$$6CO_2 + 12H_2O + \text{energy (sunlight)} \rightarrow$$
$$C_6H_{12}O_6 \text{ (sugar)} + 6O_2 + 6H_2O.$$

This is an **endergonic** process, which means that it requires energy, as all synthetic reactions do.

When the plants (or the animals that eat them) need the energy stored in the plant's sugars, they oxidize them in a series of **exergonic** (energy-releasing) reactions. In doing so, they perform a number of processes that are superficially the exact opposite of photosynthesis. For example, during photosynthesis, plants take in carbon dioxide and give off oxygen. Both animals and plants do the reverse when they oxidize sugars. All these transformations, endergonic and exergonic, are part of **metabolism,** the totality of chemical reactions in cells. In both plants and animals there would be no metabolism and no life without the chemical energy supplied by photosynthesis in plants.

5.2 Research in the eighteenth and nineteenth centuries laid the foundation for our present understanding of photosynthesis.

In 1772 a British chemist, Joseph Priestly, observed that oxygen was given off by green plants. Several years later, a Dutch contemporary, Jan Ingenhousz, added the knowledge that this occurred only in the presence of light and with an uptake of CO_2. That chlorophyll, the green pigment in leaves and stems is necessary for photosynthesis, was not discovered until about 1837. Shortly afterward the German chemist Justus von Liebig deduced that all organic compounds in plants must be derived from carbon dioxide. He also formulated his "minimum law," which states in essence that when any chemical necessary for photosynthesis is in short supply, the process stops.

The experimental tools for digging deeper into the mysteries of photosynthesis were not available until the twentieth century. By then it was known that plants need carbon dioxide, water, sunlight, chlorophyll, and minerals. With these, they can produce organic compounds and release oxygen. But how?

5.3 Plants use principally the blue and red regions of the light spectrum for photosynthesis.

An essential step toward understanding photosynthesis is a brief examination of the properties of light. In different situations it may be useful to think of light as rays, waves, or particles of energy. According to the ray theory, light travels in straight lines, but it is not specified just what constitutes a ray. The wave theory considers that light behaves as transverse waves vibrating at right angles to the direction of propagation of the light ray. Since the velocity of light in a vacuum is constant (about 3×10^{10} cm/sec), the wavelength, measured in centimeters (cm), and the frequency (waves per second) are inversely proportional. Sunlight, which we perceive to be white, is really a mixture of many waves of different lengths, each of which we can see when it is reflected separately, since the waves correspond to the colors we recognize. Waves that are too short or too long for the eye to detect are called ultraviolet and infrared, respectively. These, together with the visible wavelengths, comprise only a narrow portion of the complete spectrum of electromagnetic radiation, which includes the very short, powerful X-rays and the very long radio waves.

The quantum theory of light considers light to be streams of energetic particles, called photons, each an indivisible "package" of energy. The higher the frequency (or the shorter the wavelength), the larger the quantum (amount) of energy in each photon.

Because leaves appear to our eyes as mostly green, we know that they reflect or transmit (and therefore do not absorb) green light. Figure 5-1, a simplified absorption spectrum for chlorophyll, shows the wavelengths (colors) of light that are most readily absorbed. The plant can use only light that it absorbs. This light is mostly in the blue and orange-red portions of the spectrum. However, many plants can capture some of the energy of the green wavelengths because they contain accessory pigments which absorb them. These pigments can pass some of the energy they absorb on to chlorophyll molecules close by as shown by the action spectrum in Figure 5-1, which shows the rate of photosynthesis as a function of the wavelength of light.

5.4 Chlorophyll is the energy-transducing molecule of the plant cell, but only when built into a chloroplast membrane.

Most molecules do not absorb visible light and are therefore colorless. Chlorophyll and other biological pigments do absorb certain visible wavelengths, due to their molecular structure, and therefore have a characteristic color. The chlorophyll molecule has two sections, a long hydrophobic carbon chain and a porphyrin group, a cluster of ring-shaped groups of atoms in the center of which is a single magnesium atom (Figure 5-1). The porphyrin group contains many pairs of atoms with double bonds between them, which often alternate with single bonds (—C=C—N=C—, etc.). The structures of most pigments contain such conjugated bonds, and it is the number and arrangement of these that determines which wavelengths of light the molecules will absorb.

When photons of light strike the molecules of such pigments, their energy is transferred to certain electrons within the molecules. These electrons are said to be in an "excited state," and they assume a position with a higher energy content within the molecule. The movement of these electrons from one position to another represents a form of electrical energy. The chlorophyll molecule, be-

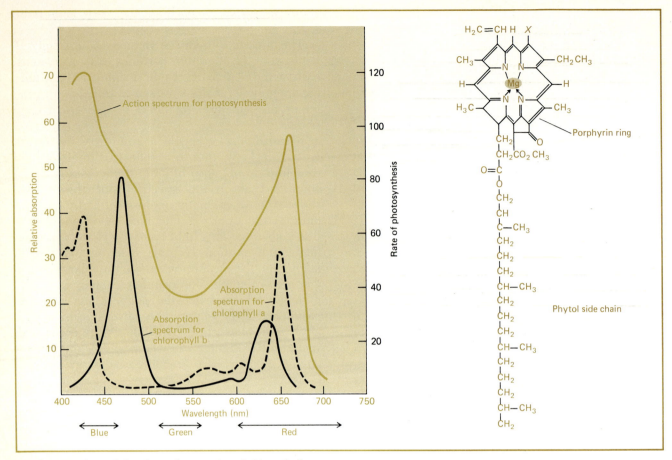

5-1 A comparison of the absorption spectra of chlorophylls a and b with the action spectrum of photosynthesis (*left*). Also shown (*right*) is the structural formula for chlorophyll. In chlorophyll a, *X* is —CH_3, in chlorophyll b, *X* is —CHO.

cause of its ability to transform light energy to electrical energy, is called a **biological photoelectric energy-transducing molecule.**

Chlorophyll only functions as an energy-transducing molecule for the cell when it is an integral part of the chloroplast membrane. This membrane puts chlorophyll in contact with other molecules to which it can transfer this electrical energy. In the living chloroplast, the chlorophyll molecules are built into membranes along with certain other substances that allow the electrical energy to be used for the synthesis of stable chemical compounds.

The chloroplast is bounded by a double membrane, and in many cells looks like an oval green capsule under the light microscope. Thin sections of chloroplasts of higher plants examined in the electron microscope show that these chloroplasts are packed with membranes, called **lamellae** ("thin plates"). These membranes are folded over in places to form stacks of disk-shaped **thylakoids** (Fig-

BIOEPICUREAN DELIGHTS

Creamed Chloroplasts

Spinach leaves are so rich in chloroplasts that they have been used as a standard source of isolated chloroplasts for studies of photosynthetic reactions. A slight modification of the procedure results in European-style creamed spinach. Steam well-soaked spinach leaves (a liter serves four) for 3 to 5 minutes, then drain and homogenize in a blender for about a minute with one-half cup of sour cream, one-half cup of cream of chicken soup, and one-half teaspoon of nutmeg. Add salt to taste.

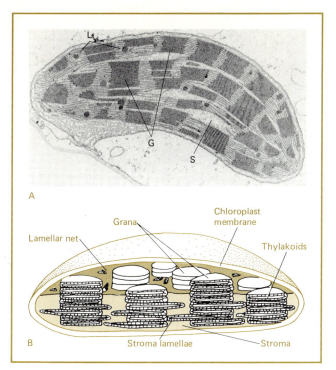

A

Grana

Chloroplast membrane

Lamellar net

Thylakoids

B

Stroma lamellae   Stroma

5-2 *A*. Transmission electron micrograph of a chloroplast showing grana (*G*), which are stacks of folded membranes or lamellae (*L*) suspended in the stroma (*S*) of the chloroplast. (Courtesy of L. K. Shumway.) *B*. A diagrammatic interpretation of a portion of a chloroplast.

ure 5-2). The stacks are visible in the light microscope as green flecks known as **grana**. The nonmembranous background material is called the **stroma**. Some of the lamellae extend across the stroma, connecting the grana together. They are called **stroma lamellae**. Those that do not extend to the next granum are called **lamellar frets**.

The structural details of these complex membranes are incompletely understood, but apparently they are built according to the same principles as other cell membranes and in fact are seen as unit membranes. Chloroplast membranes contain not only chlorophyll and accessory pigments, such as the carotenoids, but also the phospholipids, other lipids, and proteins found in other membranes. The accessory pigments lie in such close proximity to chlorophyll that the light energy they absorb can be transferred to neighboring chlorophyll molecules.

Physiological experiments have suggested that chlorophyll molecules associate in clusters of about 230 molecules each, along with accessory pigments. Freeze-fracture electron micrographs of cleaved chloroplast membranes have revealed regularly packed oval bodies named **quantasomes**, which are believed by some scientists to represent such clusters.

Before going on to discuss the reactions involved in photosynthesis, we will examine how plants can control to a certain degree the amount of light falling upon, or absorbed by, their chloroplasts.

5.5 Some plants can regulate the amount of light their cells absorb, either by leaf and stem movements or by moving or reorienting their chloroplasts.

Certain plants and plant cells have evolved simple kinds of "behavior" that regulate the amount of light they absorb. Anyone who has raised house plants knows that plants exhibit slow, light-mediated growth responses that permit them to move leaves into the most favorable positions for capturing light. Many plants also have rhythms of leaf movement that recur daily and probably serve to regulate light absorption. These rhythms are endogenous, or internally regulated, and persist to some degree in constant light or dark.

Certain plant cells have evolved a "behavioral mechanism" on the subcellular level which serves to position their chloroplasts so that they will absorb light maximally under dim illumination or minimally when the light is too bright. An example of this process, called **photodinesis,** is shown in Figure 5-3. Cytoplasmic motility appears to be involved in the transport or rotation of chloroplasts within individual cells. These movements are in response to signals received by pigments in the cell other than chlorophyll.

5.6 Photosynthesis consists of two sets of processes, the light and dark reactions.

Early in this century it was shown that if plants are illuminated by the same amount of light (the same number of quanta), administered as flashes rather than as continuous illumination, the efficiency of photosynthesis increases. It was reasoned that the reactions which require light, called **light reactions,** are faster than the subsequent dark reactions. The **dark reactions,** which do not require light themselves, do require the products of the light reactions. It was also shown by flashing at different intervals that the dark reactions take an appreciable amount of time to complete. When plants are con-

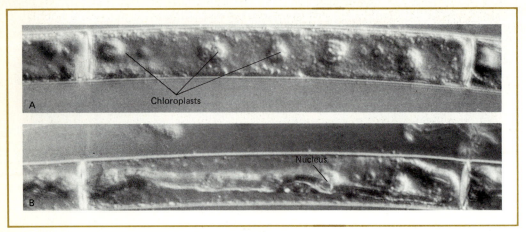

5-3 Photomicrographs illustrating chloroplast movement (photodinesis) in plant cells in response to light. Shown is a *Mougeotia* cell in which the plank-shaped chloroplast rotates from the "face" position (*A*) to the "profile" position (*B*). (Courtesy of N. S. Allen.)

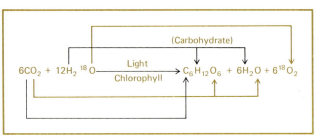

5-4 The overall reaction of photosynthesis, showing the fate of the isotope ¹⁸O, when incorporated into water and introduced into the photosynthetic reaction.

tinuously illuminated at high intensities, much of the energy is lost, simply because the dark reactions create a bottleneck.

5.7 Experiments have established what happens to the raw materials of photosynthesis during the light and dark reactions.

The work of C. B. Van Niel on a sulfur bacterium showed that its photosynthesis followed the following reaction:

$$6CO_2 + 6H_2S + \frac{light}{energy} \rightarrow C_6H_{12}O_6 + 6S + 6\tfrac{1}{2}O_2$$

The demonstration that the splitting of H_2S in bacterial photosynthesis suggested, by analogy, that in green plants, perhaps water was split and became the source for the oxygen evolved in photosynthesis.

This idea was substantiated in 1938, when R. Hill showed that isolated spinach chloroplasts could both produce oxygen and reduce certain added compounds. The Hill experiment was one of the earliest physiological experiments on an isolated organelle and was important in first demonstrating what Van Niel had hypothesized: the **splitting of water, called photolysis** in this instance, because the process utilizes light energy. Although scientists are not certain exactly how water is split in the chloroplast, it is clear that it must lose both an electron (e^-) and a proton (H^+) in the process.

In 1941, S. Ruben, M. Randall, M. D. Kamen, and

J. L. Hyde performed one of the first isotope experiments on intact plant cells. They succeeded not only in proving rigorously that water molecules are split in intact cells, but also in establishing the fates of the atoms of oxygen in carbon dioxide (CO_2) and water (H_2O), the raw materials for photosynthesis. In separate experiments they carried out photosynthesis, first with isotopic oxygen 18 (¹⁸O) incorporated into H_2O to form $H_2^{18}O$, then into CO_2 to form $C^{18}O_2$. Tracing the isotope, they found that the oxygen from the water was recovered entirely as evolved oxygen; while that from the carbon dioxide wound up in carbohydrates and water. From these experiments we can deduce the fates of all of the atoms in the raw materials for photosynthesis (Figure 5-4).

5.8 Many of the light reactions of photosynthesis are oxidation-reduction, or "redox," reactions.

In the **oxidation-reduction**, or **"redox" reaction**, electrons are transferred from one atom or molecule to another. As one reactant *loses electrons and is oxidized*, the other *gains these same electrons and*

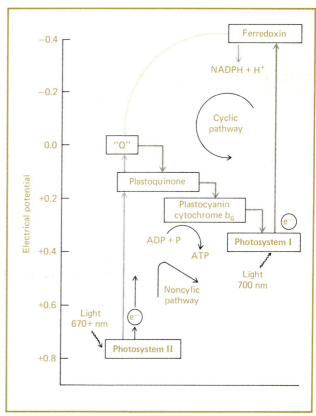

5-5 The redox reaction involving plastoquinone.

is **reduced.** Perhaps the most familiar example of oxidation is the rusting of iron. Metallic iron, with a valence of 0 (Fe^0), becomes oxidized when electons are removed from it, resulting in rust, or ferric oxide ($Fe_2^{3+}O_3^{2-}$). In this reaction each iron atom loses three electrons. As rust is formed, the oxygen gains electrons from the iron and is reduced. Generally speaking, when a substance is reduced it gains free energy and when oxidized it gives up free energy.

Redox reactions proceed in one direction to an equilibrium, because of a difference in the **redox potentials** of the **electron acceptor** and **donor.** This redox potential may be thought of as an "electron pressure" and is measurable. The maximum free energy obtainable from a redox reaction depends on the redox potential difference.

In biological oxidations entire hydrogen atoms, as well as electrons, may sometimes be transferred. An example relevant to photosynthesis is plastoquinone, which is reduced by gaining two hydrogens and two electrons to form plastoquinone·H_2 (Figure 5-5). The reaction is reversible.

Many biological redox reactions involve the release of significant amounts of free energy that can be captured as highly reactive, energy-rich compounds like ATP (adenosine triphosphate). Such compounds are used as "ready cash" for all kinds of cellular functions.

5.9 The light reactions consist mostly of the transfer, by redox reactions, of excited electrons from chlorophyll to a sequence of electron carriers.

The transfer of excited electrons in photosynthesis involves some redox reactions that release enough free energy to cause a **phosphorylation reaction**— the addition of a high-energy phosphate group

to ADP to form ATP. Since this process requires light, it is called **photophosphorylation.**

An equally important result of the light reactions for photosynthesis is the production of the reducing agent NADPH + H^+ (reduced nicotinamide adenine dinucleotide phosphate plus a proton), which is later used during the dark reactions for the synthesis of carbohydrate. Although the pathways of electron transport in chloroplasts are incompletely known, it is believed that there are at least two basic reaction schemes: **cyclic** and **noncyclic photophosphorylation** (Figure 5-6).

5-6 The cyclic (*color*) and noncyclic pathways of electron transport during the light reactions of photosynthesis. The broad arrows represent electron flow.

In cyclic photophosphorylation, only ATP is generated that can be used subsequently for carbohydrate synthesis in the dark reactions. The same ATP can of course be used for any other cellular activity. The pathway is cyclic because the electron that leaves illuminated chlorophyll returns to chlorophyll via a chain of reactions involving the electron carriers ferridoxin, flavine mononucleotide (FMN), and iron-containing intracellular **cytochrome** pigments similar to those that participate in the trapping of energy during oxidative phosphorylation (Chapter 6).

In noncyclic phosphorylation, the electrons from excited chlorophyll molecules are transferred to NADP to form the reducing agent NADPH + H[+]. Chlorophyll is resupplied with electrons from the photolysis of water. The clue that led to unraveling part of the mystery of how energy is actually captured by a transfer of electrons in the noncyclic pathway came from an observation by R. Emerson and C. M. Lewis in 1943. They found unexpectedly that plants illuminated with both red and far-red light simultaneously photosynthesize more carbohydrate than the sum of the carbohydrates photosynthesized by each wavelength independently. This intriguing observation led to the finding that there are two separate pathways of electron flow emanating from two different types—or associations—of chlorophyll molecules. These pathways (chlorophyll molecules and their associated electron carriers) are called **photosystems I** and **II** and they turned out to have different light requirements.

Photosystem II absorbs light maximally in the red range (680 nm) and loses electrons to a series of electron carriers, consisting of a substance "Q," the chemical nature of which is not known; plasto-

5-7 A simplified diagram of the dark reactions of the Calvin cycle. The reactants shown in color are supplied by the light reactions.

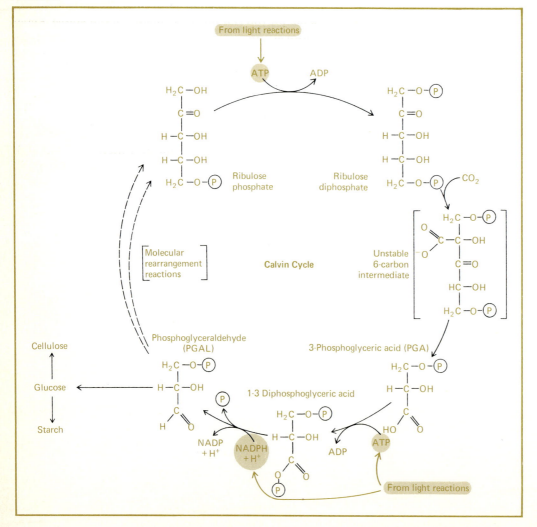

quinone; plastocyanin; and finally two cytochrome pigments. As the excited electron is transferred sequentially to several electron carriers, energy is lost, sometimes in sufficient quantity to generate ATP. The remaining electron energy is donated to a special kind of chlorophyll molecule, called P_{700} because it absorbs light maximally at 700 nm wavelength. This chlorophyll needs only the smaller 700 nm quantum to excite it, because it has already received a boost from photosystem II.

The chlorophyll P_{700} is also associated with a different group of electron transport substances, an association referred to as photosystem I. It is the cooperation between photosystems I and II that accounts for photosynthetic enhancement in the Emerson and Lewis experiments described earlier in this section.

5.10 The dark reactions of photosynthesis produce carbohydrates and are dependent upon the supplies of ATP and NADPH + H$^+$.

During the dark reactions, carbon dioxide is incorporated into organic molecules, a process called **carbon fixation**. This process can continue without light as long as a supply of free energy (such as ATP) and a reducing agent (such as NADPH + H$^+$) is available. The reactions of carbon fixation are not unique to plants, but can also occur in animals under appropriate conditions.

In the 1950s, M. Calvin and his colleagues used radioactive ^{14}C (in $^{14}CO_2$ supplied to the alga *Chlorella*) to trace the path of carbon in the synthesis of sugar, from its entry into a cell through the various intermediate chemical compounds. It was not until then that the complexity of the dark reactions was appreciated. In what is now called the **Calvin cycle** (Figure 5-7), it was found that CO_2 first combines with a phosphorylated sugar, ribulose diphosphate, to form an unstable six-carbon molecule. This breaks down into two three-carbon molecules, each of which is 3-phosphoglyceric acid (PGA). This compound is in turn reduced to an energy-rich triose (three-carbon sugar) called phosphoglyceraldehyde (PGAL). ATP and NADPH + H$^+$ (from the light reactions) are needed to provide both free energy and reducing power for these reactions. Both ADP and NADP can be "recharged" in the light reactions to participate in these reactions over and over again.

As the PGAL molecules are formed, some are converted to sugars such as glucose and fructose, whereas the majority remain in the Calvin cycle to become the phosphorylated sugar, ribulose diphosphate, that accepts the incoming CO_2 molecules.

There is at least one additional chemical reaction cycle for the synthesis of sugars. In the **Hatch-Slack pathway**, four-carbon organic acids are the main metabolites of the synthesis. This cycle is found mainly in grasses and related plants.

In all cases, the production of sugars is not necessarily an end in itself. Sugar molecules are the most readily available source of energy for plant cells, but they are also important building blocks for the construction of polysaccharides, including the storage product starch, stored in special plastids called **amyloplasts** and as cellulose, the material laid down in plant cell walls. Many plants synthesize considerable quantities of other carbohydrates (e.g., pectins) and lipids (corn, olive, and linseed oil, etc.).

6

Energy Conversion

QUESTIONS TO KEEP IN MIND

What is the ultimate source of energy for plants and animals?

What is meant by the "common pathway" for carbohydrate metabolism? What processes occur along the pathway?

What conditions are necessary for fermentation?

What are the roles of hydrogen acceptor molecules in fermentation and in anaerobic and aerobic glycolysis?

Where in the cell do the reactions of the common pathway, the Krebs cycle, and oxidative phosphorylation occur?

What processes occur in the Krebs cycle and in the electron transport system?

How is energy trapped for later use by the cell?

Both plants and animals must convert the carbohydrates that come directly or indirectly from photosynthesis into ready energy in the form of ATP. Virtually all the energy at the disposal of animals comes from green plants. Some forms of bacteria can obtain energy and material entirely from certain inorganic nutrients not manufactured by photosynthesis, but all other forms of life require the products of photosynthesis. Whether they receive these as simple sugars or as apples or lamb chops, the problem is the same—how can the compound or the food be broken down so that its stored energy becomes available as ATP, the ready-energy coin of the realm? For many organisms the process begins with digestion—or more precisely, the breakdown

of macromolecules. In this process (discussed in Chapter 39) compounds of many varieties are broken down into their building blocks and then passed along a few "metabolic pathways"—series of chemical reactions in which the molecules are broken down and release energy. Of all the building blocks, starch and glycogen are the most readily available sources of energy. To use them, the cell must first split them (in reactions catalyzed by enzymes) into simple sugars. Glucose is the most important of these.

6.1 The first steps of all three types of sugar breakdown occur along a "common pathway" of metabolism without the involvement of molecular oxygen.

There are three principal kinds of **glycolysis,** or sugar breakdown, all with the same first steps and each with a different end product (Figure 6-1). They are **alcoholic fermentation** (for example, in yeasts), which occurs in the absence of oxygen and yields alcohol; other forms of **anaerobic glycolysis,** which occur in the absence of oxygen and which produce lactate in animal tissues and a variety of other products in bacteria; and **aerobic glycolysis** which occurs with oxygen and which breaks the glucose all the way down to water and carbon dioxide. All of these reactions are catalyzed by enzymes.

The common pathway along which these three processes proceed is called the **Embden-Meyerhof common pathway** in honor of the scientists who worked out its steps. The reactions that occur are

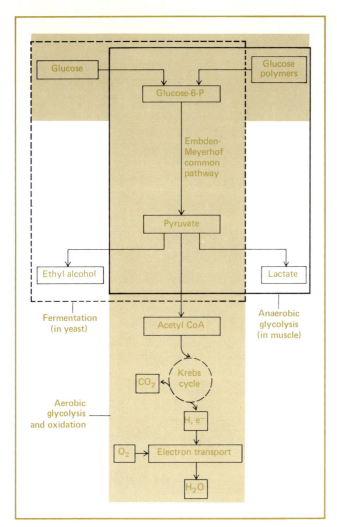

6-1 A diagram to show the relationships among the principal processes of energy metabolism: anaerobic and aerobic glycolysis, fermentation, and oxidation. Also shown are the roles of the Krebs tricarboxylic acid cycle and electron transport in oxidation.

again invested and fructose-1-6-diphosphate formed. This six-carbon sugar diphosphate is then cleaved into two three-carbon molecules (step 4), dihydroxyacetone phosphate and 3-phosphoglyceraldehyde. These three-carbon compounds are interconvertible and exist in an equilibrium mixture, but, as the arrows in step 5 indicate, the pathway essentially continues via 3-phosphoglyceraldehyde. As the latter is used up, the dihydroxyacetone phosphate changes into 3-phosphoglyceraldehyde.

In step 6 the aldehyde group is oxidized by removal of hydrogen atoms to an organic acid group —COOH—which is phosphorylated by inorganic phosphate ions always present in the cell. The hydrogen acceptor NAD (nicotinamide adenine dinucleotide) accepts the two hydrogens as NADH + H$^+$ and may then be thought of as "reducing power" held in reserve, ready to donate hydrogen atoms to one of several possible reactions.

In steps 7 through 10, the two reactive phosphates are sequentially removed from the three-carbon molecule, yielding one ATP at each point. Pyruvate[1] can be thought of as a key compound at the branch point in the three metabolic pathways shown in Figure 6-1.

The reactions along the common pathway have netted the cell two molecules of ATP for each molecule of glucose or three ATP molecules for each glucose monomer derived from glycogen. These reactions take place in the fluid **ground cytoplasm** of the cell, rather than in an organelle.

6.2 In the absence of free oxygen, glycolysis will proceed to the formation of either alcohol or lactate.

The formation of pyruvate is the last step which the three forms of glycolysis have in common. In some organisms, such as yeasts, fermentation then occurs (Figure 6-3). Carbon dioxide is removed from pyruvate to form acetaldehyde (step 11a). This compound is reduced by the NADH + H$^+$ generated in step 6, and ethyl alcohol is produced. (The NAD is again free to accept hydrogen from 3-phosphoglyceraldehyde.)

[1] Pyruvate is the anion, or negatively charged ion, formed by the dissociation of pyruvic acid. For the sake of simplicity, Figures 6-2, 6-3, and 6-4 present the structural formulas of the metabolic organic acids (R—COOH), even though in the living cell they are largely dissociated into their respective anions (R—COO$^-$). For correctness the text refers to the anion alone (*pyruvate* rather than *pyruvic acid*, *lactate* rather than *lactic acid*, etc.).

actually small molecular rearrangements which release part of the energy of glucose in small steps.

Figure 6-2 illustrates this common pathway. In step 1 the cell invests one molecule of ATP to change a glucose molecule into the more reactive form, glucose-6-phosphate. (The 6 indicates that the phosphate is on the sixth carbon away from the aldehyde (CHO) group of carbon 1.) If the glucose entering the pathway is derived from glycogen, however, the ATP has already been invested and no further expenditure is necessary. Consequently, a cell with a ready supply of glycogen can use its energy more efficiently.

Step 2 consists of a molecular rearrangement and produces fructose-6-phosphate. In step 3, ATP is

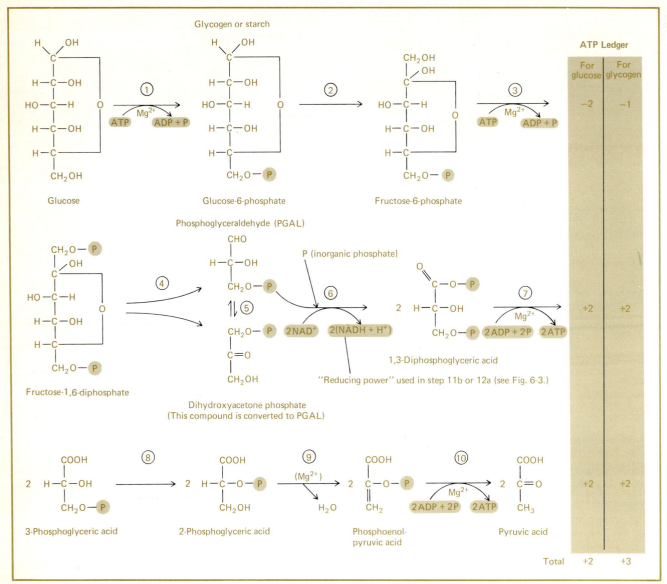

6-2 Steps in the Embden-Meyerhof common pathway from glucose and its derivatives to pyruvic acid. The reaction numbers correspond to those in the text. Each step is catalyzed by a different enzyme.

When animal cells are deprived of oxygen, anaerobic glycolysis (Figure 6-3) occurs and pyruvate is immediately reduced to lactate. This happens, for example, when people exert themselves to the point where they are required to pant for a minute or so afterward. The circulatory system can bring only a limited quantity of oxygen to the working muscles—less than they need when pushed to extremes. Consequently, not enough oxygen reaches the cell to oxidize the NADH + H$^+$ formed in step 6. Instead the pyruvate oxidizes the NADH + H$^+$ to form NAD and lactate (step 11b) in the muscle.

When a runner stops running but continues to pant, the lactate, which has been transported to the liver, is oxidized back to pyruvate. Runners are said to incur an **oxygen debt** when anaerobic glycolysis occurs in the muscles. They repay the debt by panting.

6.3 The oxidative reactions of aerobic glycolysis involve the Krebs tricarboxylic acid cycle and occur mainly in mitochondria.

The third kind of glycolysis and the most important for plants and animals, including ourselves, is aerobic glycolysis, in which pyruvate is broken

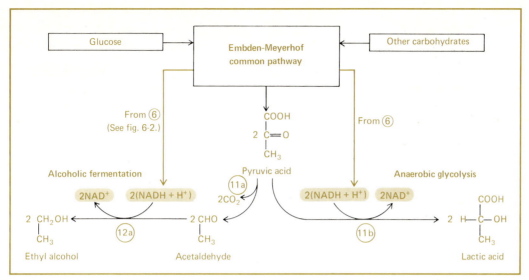

6-3 The chemical reactions in alcoholic fermentation and the formation of lactic acid in anaerobic glycolysis.

down into CO_2 and H_2O. Most of the oxidative reactions of this process occur after the formation of pyruvate and take place in the mitochondria. These organelles contain all the enzymes required.

The Krebs tricarboxylic acid cycle (Figure 6-4) is a metabolic pathway of enzymatically catalyzed reactions. This cycle fulfills two main functions: (1) It removes hydrogen atoms and electrons from the breakdown products of glucose and passes them to an electron transport system, where the hydrogens are oxidized to protons (H^+) and combined oxygen (OH^-) to form water; and (2) it forms CO_2 as a waste product. The oxidation of hydrogen furnishes the major part of the total energy available in aerobic glycolysis.

There are three processes occurring in the Krebs cycle which together break down pyruvate into CO_2 and H. These are **decarboxylation,** the removal of CO_2 (Figure 6-4, steps 1, 5, and 6); **dehydrogenation,** the removal of hydrogen, usually to a hydrogen acceptor, either NAD or FAD (steps 1, 4, 6, 8, and 10); and **hydration,** the incorporation of a water molecule (step 9), the hydrogens of which are later removed (step 10).

You will see in the same figure that coenzyme A (CoA), a sulfur-containing derivative of the B vitamin pantothenic acid, enters the cycle twice, in steps 1 and 7. The reaction in step 7 is the only reaction of the Krebs cycle releasing enough energy to synthesize a high-energy bond as ATP. The bulk of the ATP is made by the subsequent oxidation of hydrogen removed in the cycle.

6.4 The hydrogen removed in the reaction of the Krebs cycle is oxidized by an electron transport system, coupled to a phosphorylation system that efficiently captures the energy released by oxidation.

Hydrogen atoms removed from substrates such as pyruvate and isocitrate in the dehydrogenation reactions of the Krebs cycle are held by the hydrogen acceptors NAD or FAD as $NADH + H^+$ or $FADH_2$. In addition, electrons from the same substrates are transferred first to NAD or FAD and then to a series of electron carriers called **cytochromes** (Figure 6-5). The cytochromes, pinkish pigment molecules built into the inner membrane of the mitochondria, are proteins containing a porphyrin group similar to that of chlorophyll, except that cytochrome porphyrins contain an atom of iron instead of manganese. The iron atom can accept or donate an electron by undergoing the reaction $Fe^{2+} \rightleftharpoons Fe^{3+} + e^-$.

When an electron passes along the electron transport chain from the substrate to oxygen (Figure 6-5), its electrical potential drops. If the drop in potential is sufficiently large, that reaction can supply energy for the synthesis of ATP from ADP and inorganic phosphate. As the pairs of electrons are transferred by the cytochrome system to the oxygen at the end of the line, hydrogen (from $FADH_2$ or $NADH + H^+$) combines with oxygen to form water. Thus, the hydrogen is oxidized, and the oxygen reduced. The whole process—the oxidation of hydrogen and the synthesis of ATP—is called **oxidative phosphorylation.** These reactions occur in the folded inner membranes of mitochondria called **cristae** (Figure

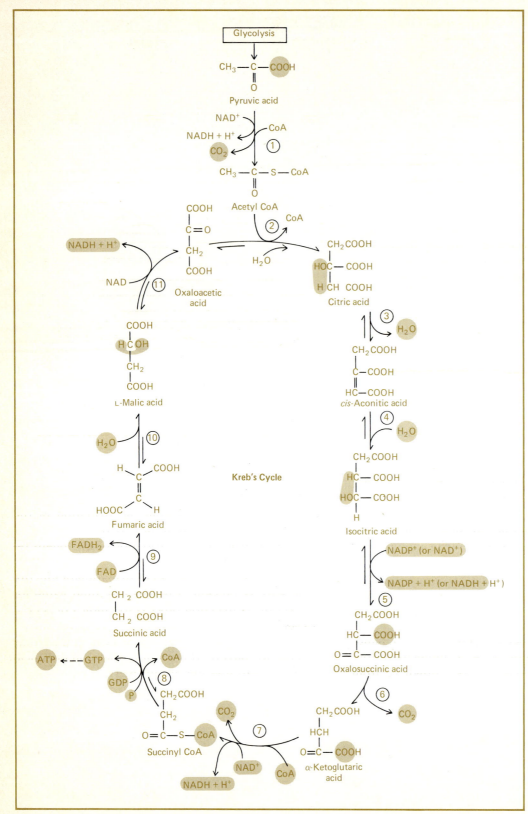

6-4 The chemical reactions of the Krebs tricarboxylic acid
cycle. Each reaction is catalyzed by a different enzyme.

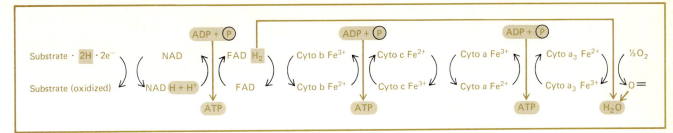

6-5 The electron transport system of the mitochondrial inner membrane. On the average, three molecules of ATP are generated for every pair of hydrogens oxidized and electrons transported. For exceptions, see reactions 8 and 9 of Figure 6-4.

6-6). The cytochrome molecules are an integral part of the inner membrane structure, as is the enzyme ATPase, which catalyzes the formation (or under certain conditions, the breakdown) of ATP.

6.5 Compared to anaerobic glycolysis, aerobic glycolysis yields considerably more energy.

In section 6.2 it was seen that the anaerobic reactions leading to the production of alcohol or lactate result in only two molecules of ATP from each molecule of glucose. The chemical energy in a gram-molecular weight, or **mole,** of glucose is about 686 Calories (Cal). (One kilogram calorie, or *calorie* with a capital *C*, is the amount of heat energy required to raise one liter of water one degree centigrade.) Since a mole of one substance contains the same number of molecules as a mole of another substance, the anaerobic breakdown of 1 mole of glucose leads to 2 moles of ATP. A mole of ATP has only 8 Calories of chemical energy stored in the bond to the terminal phosphate. The efficiency, therefore, of the anaerobic reactions is only $2 \times \frac{8}{686}$, or approximately 2.3 percent. On the other hand, aerobic glycolysis yields 38 ATP molecules from each molecule of glucose broken down (2 from

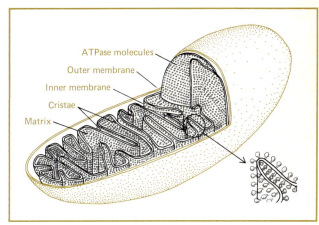

6-6 A schematic diagram of a mitochondrion showing the outer membrane and the inner membrane of which the cristae are composed. It is this inner membrane that contains the electron transport system and ATPase with which the mitochondrion captures the energy released in cellular oxidations.

anaerobic glycolysis, 6 from the oxidation of the hydrogens removed from step 6 of the common pathway, and 30 from the oxidation of hydrogens removed in the Krebs cycle).

The efficiency of the overall oxidation of glucose is at least 44 percent, a remarkably high value for any energy conversion process (38 moles of ATP × 8 Cal/mole = 304 Cal; 304/686 = 44 percent). By comparison, a well-designed gasoline engine may be able to convert 25 percent of the total chemical energy of gasoline into useful work.

7

The Utilization of Energy
for Macromolecular Synthesis

QUESTIONS TO KEEP IN MIND

How is the cell's "genetic blueprint" replicated?

How is the genetic information of DNA transcribed into a form used to synthesize proteins?

What are the three important roles of RNA molecules in protein synthesis?

How is energy utilized for macromolecular synthesis?

What is the triplet code? Where is it used?

How is a protein synthesized?

How are biological structures made from protein building blocks?

Chapters 5 and 6 have shown how plants capture the energy of sunlight and store much of it as glucose or polymers of glucose, and how both plants and animals engage in glycolysis and oxidation to recover the energy stored in glucose in the form of ATP. One of the most important functions of this "ready-cash" molecule is providing energy for the synthesis of three types of macromolecules—DNA, RNA, and proteins. In this chapter we will examine how these essential macromolecules are synthesized, how they are dependent upon one another, and how their synthesis requires ATP.

7.1 DNA molecules are made using other DNA molecules as templates, or models, in the process known as replication.

Because the synthesis of DNA, the carrier of the genetic code, is an essential step in the reproduc-

tion of cells (and of some organelles as well), it is important to understand how the process works. In Chapter 3 a general description was given of the four kinds of building blocks or nucleotides that go into the construction of DNA. We also saw the way in which the bases of a single parent strand of DNA pair in a complementary manner with the nitrogenous bases of a growing strand so that an exact copy is made of the original double-helical DNA molecule (Figure 3-6). Before each building block can be added to the growing strand, however, it must first pick up two high-energy phosphates from ATP (Figure 7-1). In this enzymatic reaction, the energy of ATP is not released, but only transferred to the nucleotide. This energy is then used when the building block is added to the new strand in a reaction catalyzed by the enzyme DNA polymerase.

7.2 DNA also serves as a template for the synthesis of RNA, but the genetic message is transcribed into a slightly different language.

The synthesis of RNA is similar to that of DNA, especially as regards the use of ATP. Again, a strand of DNA serves as the template or model, but there are two differences. First, building blocks of RNA, not DNA, are used. RNA nucleotides differ from DNA nucleotides in the sugar component and in one base. RNA uses uracil (U), instead of thymine (T). Secondly, only one strand of DNA is used as a template to form RNA. In both processes the bases in the original and new molecules line up in a complementary manner. In RNA synthesis U aligns with A, G with C, and so on, as is shown in Figure 7-2. The sequence of building blocks in the DNA mole-

48

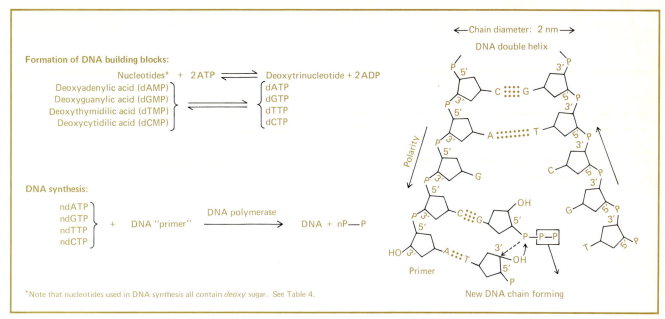

7-1 ATP generated by metabolism is used to prepare the building blocks for DNA synthesis. These building blocks only form DNA in the presence of DNA primer and the enzyme DNA polymerase. New subunits are inserted as shown. Note that the two primer DNA strands have opposite polarity.

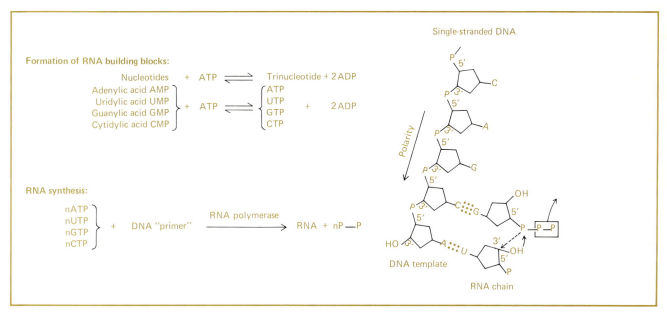

7-2 A mixture of trinucleotides (containing ribose) and single-stranded DNA primer reacts to form RNA and pyrophosphate.

cule will not be replicated or reproduced exactly, but will be transcribed into a different but closely related language. This process is known as transcription.

The method of synthesis is by no means the only difference between RNA and DNA. DNA is a fairly stable molecule and is found primarily in the nucleus, where it exists as very long strands of DNA-protein called chromatin. (These strands condense into chromosomes in preparation for cell division, a process that will be described in Chapter 9.) On the other hand, the cell synthesizes three kinds of RNA in the nucleus, and these are found in several different locations within the cell.

7.3 There are three different kinds of RNA — messenger RNA, transfer RNA, and ribosomal RNA — and they are all involved in protein synthesis.

Messenger RNA (mRNA) is a relatively large, uncoiled molecule composed of 900 to 1500 nucleotide building blocks (and perhaps even as many as 12,000). It is usually associated with ribosomes, which are the site of protein synthesis. As the name implies, messenger RNA has the function of carrying information from a source, DNA in the nucleus, to a destination, the ribosome. The message that mRNA carries is the "blueprint" for the specific sequence of amino acids in a new protein molecule. The blueprint is actually the mRNA's own sequence of nucleotides, which we have seen is acquired by transcription from a DNA molecule. The mRNA moves from the nucleus and attaches itself to the surface of a ribosome. Before protein synthesis is described further, we must know something of the other forms of RNA.

Transfer RNA (tRNA) is a much smaller molecule (about 80 nucleotide building blocks) in which the strand coils back on itself into a double-helical structure. Transfer RNA is found free in the cell and at the ribosomes, and its function is to guide the amino acid building blocks into position so that they can link together in proper sequence to form the desired protein. Transfer RNA does this first by attaching itself to an amino acid (Figure 7-3). For each of the 20 kinds of amino acids there is at least one kind of tRNA.

It is believed that if the tRNA–amino acid combination could be examined closely it would appear as in Figure 7-3B, with the amino acid at one end and three of the tRNA's numerous nitrogenous bases protruding at the other, due to the configuration of the molecule. These three bases enable the tRNA with its attached amino acid, like a key ranging along a row of locks, to find the position where it fits with a complementary set of three bases on the mRNA strand. Each set of three bases, or **triplet**, in the mRNA is known as a **codon**, and the complementary triplets in tRNA are known as **anticodons**. The bases can be thought of as "letters" in a code.

A genetic code using base triplets can code for the 20 different amino acids and several commands, such as "stop synthesis." RNA has only four nitrogenous bases. These taken one at a time could indi-

7-3 *A*. The cloverleaf model for the structure of alanine tRNA as suggested by R. W. Holley. Note the helical regions held in place by hydrogen bonds and the anticodon and amino acid attachment sites at opposite ends of the molecule. *B*. The spatial relationships between the ribosomal subunits, mRNA and four tRNA molecules. The triplet codons of the mRNA strand determine which kind of tRNA may attach. This in turn determines the sequence of amino acids in the growing polypeptide chain.

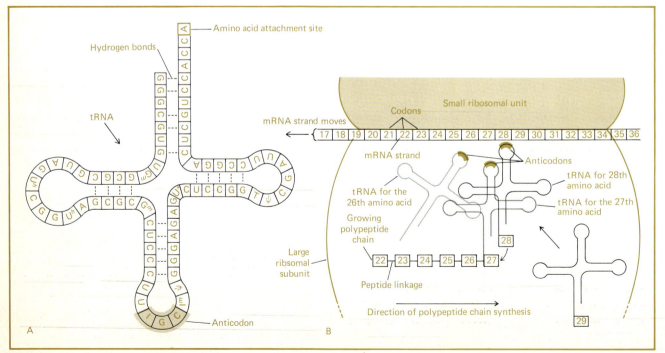

cate only 4 different amino acids; a two-base code could accommodate only 16; but a triplet code combines the four bases to provide 64 codons, more than enough. Each triplet specifies either a particular amino acid or a particular kind of punctuation. Cells use all 64 codons, so most amino acids are coded by more than one codon. Amazingly enough, these 64 codons have the same meaning for all cells —whether amoebae, algae, or human cells. The code is thus said to be universal.

Ribosomal RNA (rRNA) is the third type of RNA and makes up part of the structure of ribosomes. The latter, as seen in the intact eucaryotic cell, are composed of two subunits, one larger (60S) and one smaller (40S). (The S stands for *Svedberg unit*, a unit of sedimentation velocity.) Each of the subunits contains both rRNA and protein. The rRNA molecules are relatively large and partially coiled, and both subunits contain several thousand nucleotide building blocks. The function of the ribosome in protein synthesis has not yet been worked out, but it is clear that the ribosome serves as a kind of "workbench" for this process.

7.4 Protein synthesis is dependent on the genetic information that comes in mRNA's triplet code.

With the function and character of DNA and the RNAs in mind, we can now examine in a general way the process of protein synthesis. To put it another way, we can see how the genetic code is *translated* into a specific amino acid sequence. First a molecule of mRNA is transcribed from DNA in the nucleus and exiled to the cytoplasm, where it attaches itself to a ribosome. The nitrogenous bases of the mRNA are then in an easily accessible position (Figure 7-3B). Next, tRNAs with their amino acids attached cluster all around the mRNA until individual tRNA molecules find the three-base code word (codon) on the mRNA that is complementary to their anticodon. Once in place, their amino acids are added onto the growing peptide chain.

Actually the process is more complicated. (See Figure 7-4.) When the mRNA is in place, a ribosome moves along it, and mRNA binds to the small subunit of the ribosome, while the complementary tRNA (with its amino acid in tow) binds to the large subunit. Once the proper tRNA with its attached amino acid is in place, it is well positioned to trans-

fer its amino acid to the growing polypeptide chain. Then this uncharged tRNA is released from the ribosome. The next codon in mRNA then determines the binding of the proper tRNA with its attached amino acid. Now this attached amino acid is positioned perfectly for bonding to the amino acid just added to the chain, and an enzyme on the large subunit catalyzes this reaction.

The dipeptide thus formed is shifted (or the ribosome moves) so that one pair of bonding sites on the ribosomal subunits is free to accept the third triplet of the mRNA and a third tRNA molecule. The mRNA moves along the group of ribosomes, "reading" the genetic message and, with the help of tRNA, translating it into the amino acid sequence of a polypeptide that grows as each amino acid is added on. One of the triplets on the mRNA signals the end of synthesis, and the last point of attachment between polypeptide-ribosome-mRNA is broken. The new protein that has been coiling up in its characteristic way is now ready to function as an enzyme, structural member, or secretory product. The entire synthesis of an average protein molecule by a cell requires about one minute.

7.5 Ribosomes can occur singly or grouped with mRNA as polysomes, many of which are attached to the endoplasmic reticulum.

It has been known for decades that a cell which is growing rapidly or which synthesizes protein for export (i.e., for secretion) contains much more RNA in its nucleolus and cytoplasm than do other cells. In some of the early electron microscopic studies utilizing thin sections of such cells, K. R. Porter described a system of membranous sacs dotted with ribosomes, which he called the **rough endoplasmic reticulum** or **RER** (section 2.5). Much later it was observed that ribosomes are often arranged in groups called **polysomes,** which appear as either straight or curved lines because they are associated with strands of mRNA. In a polysome, many ribosomes are utilizing the same mRNA strand at the same time, with new protein molecules in various stages of completion (Figure 7-4). The RER serves to "package" these proteins into membrane-enclosed sacs for storage or secretion. Not all polysomes are associated with membranes, however. Those that are free in the ground cytoplasm are responsible for synthesizing the structural and enzymatic proteins that remain in the cell.

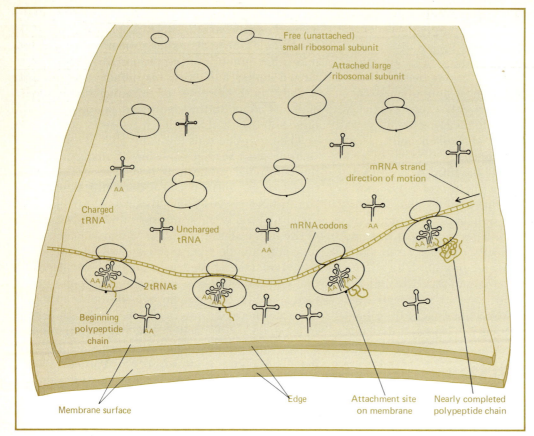

Free (unattached)
small ribosomal subunit

Attached large
ribosomal subunit

mRNA strand
direction of motion

Charged
tRNA

Uncharged
tRNA

mRNA codons

AA

2 tRNAs

Beginning
polypeptide
chain

Membrane surface

Edge

Attachment site
on membrane

Nearly completed
polypeptide chain

7-4 A "mind's eye view" of a single leaflet of rough endoplasmic reticulum, to which are attached the large ribosomal
subunits. A strand of mRNA winds its way along a path of
ribosomes, its codons "selecting" the appropriate tRNA, which
specifies the next amino acid (AA) to be added to the growing
polypeptide chain.

7.6 Proteins may polymerize or assemble into structures that can perform various functions.

Many polypeptides do not become biologically functional when released from a polysome. For example, many enzyme molecules are constructed of more than one polypeptide subunit. Some proteins polymerize into filaments and/or self-assemble to form entities with unique biological functions. Two examples will serve to illustrate the principles involved.

Many, if not all, eucaryotic cells contain a protein called **actin** that is, as its name implies, involved with action. It was long thought to occur only in muscle. Whether actin molecules are identical in different cells is still to be decided, but the properties of muscle actin can serve as an example of the properties of this protein in general.

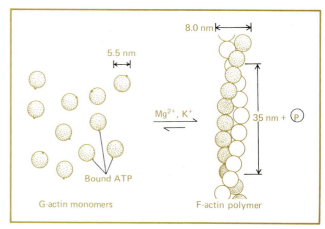

8.0 nm

5.5 nm

Mg^{2+}, K^+

35 nm + (P)

Bound ATP

G-actin monomers

F-actin polymer

7-5 The polymerization of G-actin monomers into F-actin filaments.

Muscle actin can exist in two forms: G- (globular) actin and F- (fibrous) actin. G-actin is observable in the electron microscope as a slightly ellipsoidal body 6 to 7 nm in largest diameter, and it has a molecular weight of about 47,000. Associated with each molecule of G-actin is a nucleotide molecule.

G-actin can polymerize — that is, join with molecules of its own kind — to form F-actin, which is a double chain of G-actin subunits, twisted into a helix with a crossover repeat distance of 37 nm (Figure 7-5).

F-actin forms the backbone of the so-called **thin filaments** of muscle and probably that of many of the microfilaments which have been discovered in most kinds of plant and animal cells. The fact that microfilaments in a variety of cells resemble muscle actin suggest that they play a role in cell motility and also in the transport of substances within cells.

8

The Utilization of Energy in Movement

QUESTIONS TO KEEP IN MIND

What is the source of energy for various kinds of movement?

What are the principal molecular constituents of muscle, and what are their functions?

What is a mechanochemical transducer?

How do calcium ions initiate muscle contractions? What releases the calcium ions?

How does a pseudopodium of an amoeba extend? What are the molecular processes involved? How are they controlled?

What is meant by cytoplasmic transport? In what cells does it occur? What is axoplasmic transport?

What are the structural components of cilia and flagella? How do these function according to the sliding filament theory?

How do lower plant cells glide?

Movement is one of the fundamental properties of life. It is also one of the principal ways in which cells use energy from metabolism. Nearly all living things move. They either move relative to their environment or move materials about within themselves. To do either, the organism must transduce chemical energy to mechanical energy. In all eucaryotic cells the energy for doing mechanical work is derived from ATP and other high-energy phosphate compounds. The actual machinery for moving, however, is not the same in all cells or organisms; there are many variations upon a few themes. This chapter will show the five funda-

mental mechanisms of movement and a few of their many variations. Other aspects of movement, including the evolution of the methods of locomotion and internal transport that abound in the living world today, will be discussed in Chapter 20 and in Chapters 24 to 31.

8.1 Muscular contraction is the most highly developed mechanochemical transduction system.

The muscles of higher animals, especially vertebrates and insects, possess the most highly specialized kinds of cells for the transduction of chemical energy to mechanical work. Muscular contraction is by far the most precisely understood mechanism of movement.

A muscle of a rabbit, for example, is considered to be an organ because it consists of several kinds of cooperating tissues and cells: Muscle fibers do the contracting, connective tissue holds the muscle together, blood tissue provides nutrients and oxygen, and nerve cells carry in the impulse to contract.

Let us examine the structure of a single muscle fiber as it would look if we carefully teased it out from a piece of muscle with a pair of thin needles (Figure 8-1). The muscle fiber is a multinucleated cell. It is surrounded by its **sarcolemma** ("muscle sheath"), a plasma membrane covered with fibers of the protein, **collagen.** Inside the cell are contractile fibrils, called **myofibrils,** stacked like logs, parallel to the long axis of the cell. Near the membrane are several nuclei.

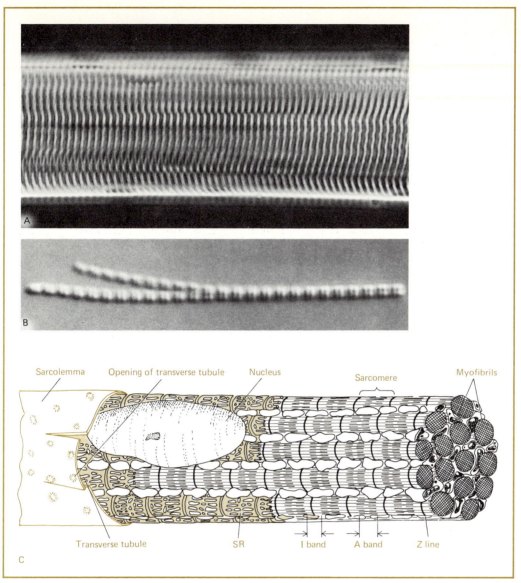

8-1 *A.* Photomicrograph of a living rabbit skeletal muscle fiber in phase contrast. Note the cross striations and myofibrils. *B.* Photomicrograph of a single isolated myofibril. *C.* A diagram of a short segment of a muscle fiber showing the relationships of the internal structures, especially the myofibrils, the transverse tubules, and the sarcoplasmic reticulum. (Photomicrographs courtesy of D. L. Taylor.)

To every muscle fiber in the rabbit's skeletal muscle a nerve cell **process** (here meaning *extension*) or **axon** is attached at a **motor end-plate.** The fiber does not contract unless it is stimulated by the passage of one or more electrical disturbances called **action potentials.** These are nerve impulses that move down the axon to the motor end-plate (section 33.3). The motor end-plate then releases a **transmitter substance** that causes a similar elec-

trical disturbance to sweep over the entire sarcolemma of the muscle fiber.

The surface of the sarcolemma is dotted with tiny invaginations that lead into a system of **transverse tubules,** which are in turn connected to membranous sacs collectively called the **sarcoplasmic reticulum or SR.** The SR contains calcium ions in a concentration much higher than that of the surrounding **sarcoplasm.** The SR also surrounds a substantial portion of each myofibril.

When the action potential sweeps over the muscle fiber, it transmits an impulse across the transverse tubules to the SR, causing the release of calcium into the neighborhood of the myofibrils.

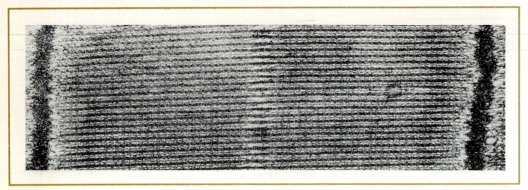

8-2 An electron micrograph of a single sarcomere of a myo-fibril, showing interdigitating thick (myosin) and thin (actin plus tropomyosin and troponin) filaments. (Courtesy of M. Reedy.)

8.2 The myofibrils contract when calcium causes two proteins, actin and myosin, to interact.

Myofibrils of skeletal muscle have a repeating unit of structure (Figure 8-1) which is itself striped. This repeating unit of structure is called the **sarcomere** ("muscle segment"). Each sarcomere is bounded by two **Z lines.**

With the electron microscope it can be seen that the sarcomere contains two kinds of filaments, one thick and one thin (Figure 8-2). The orderly arrangement of these into A and I bands gives the sarcomere its stripes. The thin filaments which comprise the **I bands** are about 7 nm in diameter. They extend in both directions from the Z lines. These thin filaments are not quite half the length of the sarcomere. They extend into and interdigitate with the more dense **A band,** comprised of thick filaments. The thick filaments are some 18 nm in diameter.

The thin filaments are made mostly of F-actin, a helically twisted double strand of G-actin monomers (section 7.6). However, two other proteins,

tropomyosin and **troponin,** are found in the groove formed by the actin double helix. The thick filament consists mostly of **myosin** molecules (section 8.3). It is the interaction of thick and thin filaments, initiated by calcium ions, that is ultimately responsible for muscle contraction.

8.3 Myosin is both a mechanochemical transducer molecule and an ATPase enzyme.

The myosin molecule produces movement by undergoing a change in shape at the expense of the energy in ATP. To understand how this is possible, it is necessary to look at the molecule more closely. Purified myosin molecules have been found to have a molecular weight of about 500,000 and a shape a little like that of a two-headed sperm (Figure 8-3). The molecule functions both as an enzyme that splits ATP and as a structural protein.

As structural proteins, myosin molecules have the capacity for self-assembling into myosin aggregates which resemble thick filaments. (Thick filaments also contain other proteins not discussed here.) In the assembly process the first myosin molecules begin to go together in an antiparallel

8-3 The generally accepted model for the structure of myosin from vertebrate striated muscle.

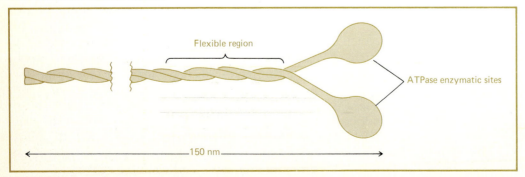

Flexible region

ATPase enzymatic sites

150 nm

manner—that is, tail to tail, in something like a bow-tie figure. The heads of the myosin fan out to either side of the aggregate, and because the heads are oriented in opposite directions the aggregate is structurally polarized in two directions, or bipolar.

8.4 Calcium ions initiate contraction by freeing actin to react with myosin.

When a muscle fiber is stimulated to contract, the release of calcium ions from the SR causes a change in shape of the control proteins troponin and tropomyosin (located in the groove of the helical F-actin filament) so that they free the actin to react with myosin. Put in more structural terms, the calcium counteracts the inhibitory effect of the control proteins and allows the thin filaments to react with— actually slide relative to—the thick filaments. After the contraction the SR membrane actively transports calcium back into the spaces of the SR, and the control proteins again restrict contraction.

The reaction or sliding between the two types of filaments depends on temporary attachments that the myosin heads make with the thin filaments. F-actin filaments have evenly spaced binding sites for these attachments. In intact cells the myosin heads protruding from the thick filaments attach to the thin filaments (Figure 8-4). In the presence of ATP and calcium ions these heads swing out and establish cross-bridges to the actin filaments. Because of the way the myosin molecules are packed in the thick filament, and the way the thin filaments are packed hexagonally around them, there is an actin binding site within reach of every myosin. Since both the actin filaments and the myosin aggregates are polarized, moreover, all the attachments are made at the same angle. (Before the polarization of these filaments was discovered, it was impossible to understand how the muscle filaments "knew" which way to slide. Sliding in both directions at once would result in no net contraction at all.)

As ATP molecules are split by the myosin ATPase at each cross-bridge, the "neck" region of the myosin molecule undergoes a change in shape which results in a small displacement of the thick and thin filaments relative to one another. The myosin head then binds to the next actin binding site within reach. This cycle is repeated, effecting a constant rowing motion. Each cycle requires the breakdown of ATP. In most muscles these cycles are not in phase (i.e., do not occur simultaneously in all myosin aggregates). The action of the cross-bridge causes the thin filaments to slide in toward the middle of the sarcomere. Thus the sarcomere and eventually the whole muscle contracts. The filaments themselves do not shorten. The sliding filament theory was proposed in 1953 simultaneously but independently by H. E. Huxley and A. F. Huxley (not related) and their coworkers.

ATP is required for three processes in muscle contraction: (1) the contraction as cross-bridges attach, disconnect, and reattach; (2) the relaxation;

8-4 *A.* A diagram showing the spatial relationships and polarities of the thick and thin filaments in a single sarcomere of a vertebrate striated muscle myofibril. *B.* Expanded detail of one thin filament, showing the relationships of F-actin, tropomyosin, and troponin. *C.* Expanded detail of the region of filament overlap, showing one myosin molecule extending toward the thin filament. The arrow shows the direction in which the myosin must flex to cause the sarcomere to shorten.

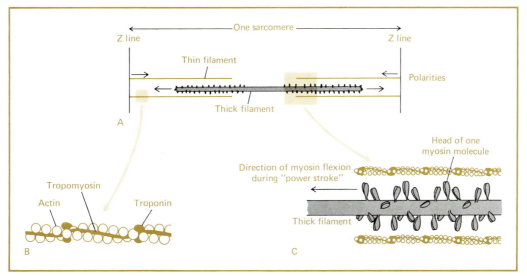

and (3) the transport of calcium ions back into the SR. If ATP is in short supply the muscle passes into a rigid state known as **rigor,** as a result of extensive cross-bonding between the actin and myosin filaments. The drop in ATP concentration following death is what leads to **rigor mortis.** By the addition of ATP, properly prepared dead fibers can be made to pass from rigor into a relaxed state.

8.5 Different kinds of muscle differ in structure and in speed and strength of contraction.

Muscles are usually divided into three categories— **striated** (for body movements), **smooth** (visceral organs), and **cardiac** (heart)—but this is an oversimplification. Each type actually varies widely in structure and function even within the same animal. For example, the two large claws of the Atlantic coast lobster are slightly different in function, one specialized for cutting, the other for crushing. Both claws have fast and slow skeletal muscle fibers, but the crushing claw contains more of the slow fibers, which are more powerful and tire less quickly.

In the visceral organs of most animals and in vertebrate arteries and arterioles, **smooth muscle** is found. Smooth muscle cells are smaller than striated muscle cells, have a single nucleus, and are not striated. Although they contain actin and myosin filaments, these are not lined up to produce striations. Smooth muscle function will be discussed in Chapters 36 and 39.

The hearts of some animals contain a third type of muscle: **cardiac muscle.** Cardiac muscle is striated and contains contractile proteins similar to those of striated and smooth muscle, but is organized somewhat differently. The most important feature of cardiac muscle is its ability to initiate and conduct impulses for contractions in such a way that the heart beats as a coordinated unit. This and other features of cardiac muscle will be discussed in Chapter 36.

8.6 Amoeboid movement is a process of widespread occurrence that, like muscle contraction, depends on the interaction of actin and myosin.

The most favorable cells for the study of amoeboid movement are the large free-living amoebae, such as *Chaos carolinensis* (Figure 8-5) and *Amoeba proteus* (section 20.8). Because they are large and easy to grow in the laboratory, we know much more

---

BIOEPICUREAN DELIGHTS

Sashimi: Japanese-Style Raw Fish

Of the many forms of muscle that can be consumed, one of the most delicious—and natural—is raw fish. In screening possible candidates for consumption as *sashimi*, it is important to choose only marine fish that do not migrate into freshwater streams (as do salmon and some kinds of trout), and it is important that they be freshly caught. Some of the best-tasting *sashimi* can be prepared from the following fishes available along North American coasts: tuna, swordfish, bluefish, striped bass, bonita, yellowtail, red snapper, flounder, fluke, and cod.

Filets of these fish are sliced into bite-size morsels and dipped briefly into a mixture of *shoyu* (Japanese soy sauce, made from fermented soy beans) and *wasabi* (a tangy Japanese horseradish). Wasabi can be purchased in powdered form in cans at most Oriental food stores. The mixture should be adjusted to taste, but half a teaspoon of *wasabi* to a quarter of a cup of *shoyu* should be about right; some may prefer to dilute this mixture with water.

In Japan and in a few specialty restaurants in the United States, raw fish appears in another delicacy called *sushi*. Long, thin slices of different kinds of fish, mollusks, sea urchin gonads, and whale meat* are arranged atop cakes of specially prepared rice cakes spread with a small amount of *wasabi* paste. Many gourmets consider that *sashimi* and *sushi* are the outstanding Japanese contribution to the fine foods of the world.

* Delicious, but frowned upon by conservationists, as whales are being hunted to the brink of extinction.

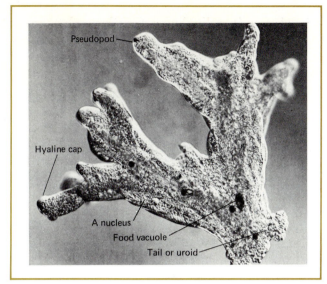

8-5 A photomicrograph of the giant amoeba *Chaos caroli-nensis*. This specimen is about 1 mm long and has many pseudopods. The cell also has many (50 to 200) nuclei; they are not easy to see. (Photo by the author.)

move in a manner similar to some species of amoeba.

Many other cells in multicellular animals exhibit amoeboid movement at some time during embryonic development. Cells of embryos change their shape and migrate, and in doing so exert forces on other cells to help bring about many developmental changes.

It begins to appear that many if not all of these amoeboid cells may have, in common with muscle, at least some of the molecules that constitute the mechanochemical transduction systems that we have just described. The supporting evidence is that the microfilaments found in many species of amoebae are actinlike, in that they will bind to fragments of muscle myosin. In the giant amoeba, *Chaos*, thick filaments seen in the electron microscope have been identified as aggregates of myosin similar to those that can be made to assemble from purified muscle myosin.

about them than about any other amoeboid cell. It must be remembered, however, that there are thousands of different amoeboid cells. These include not only the several hundred species of amoebae themselves, both free living and parasitic, that live in oceans, lakes, ditches, moist soils, intestines, and even in the spaces between human teeth; but also the amoeboid cells found as a natural part of most multicellular animals. For example, **amoebocytes** ("amoebalike cells") and **leukocytes** ("white cells"), found in blood and body fluids,

8.7 In large amoebae, cytoplasmic streaming and pseudopod formation are the result of localized contractions.

In a giant amoeba such as *Chaos*, the cytoplasm exists in two different states. The outer layer of **ectoplasm** ("outer juice") is a surprisingly rigid gel, while the **endoplasm** ("inner juice") is able to stream even though it is viscous and elastic. The difference in consistency between these layers is caused by different degrees of cross-bonding between myosin and actin filaments, similar to the thick and thin filaments of muscle.

Figure 8-6 illustrates the frontal contraction theory proposed in 1961 by R. D. Allen. According

8-6 Diagram showing the differentiated regions of an amoeba with one pseudopod and the processes underlying amoeboid movement according to the frontal contraction theory. *V* indicates the relative forward velocities, both of a particle (*P*) on the surface, and of internal organelles.

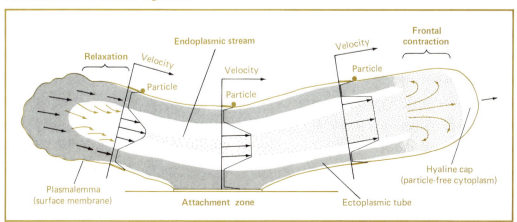

to this theory, the endoplasm is pulled toward the tips of pseudopodia (false feet) by a cytoplasmic contraction that is always localized close to the tip. The advancing rim of the ectoplasmic tube (an outer layer of gelated, contracted cytoplasm) is the point toward which the cytoplasm contracts as it becomes everted, like a cuff, at the tips of pseudopodia. It is possible to isolate an amoeba's cytoplasm (by removing its membrane) and make it contract and relax repeatedly and even form pseudopods under the conditions of ATP and calcium concentration found in muscle (section 8.2).

8.8 Cytoplasmic transport without cell movement takes place in a variety of plant and animal cells.

In many kinds of cells, especially large ones, transport systems exist in order to distribute organelles or inclusions to locations in which they or their products are needed. In other cells it is not yet known what function cytoplasmic transport serves.

In plant cells there are at least five types of cytoplasmic transport. **Saltation** ("jumping"), found in many cells, is the sudden displacement of single particles—either organelles or inclusions—to other parts of the cell. Other types involve different patterns of cytoplasmic streaming. In at least two of these types of streaming, the protein actin is involved.

The various processes of cytoplasmic transport in plants are affected by light received by the plant, but will continue in its absence. It is important to remember that chloroplasts move in many plants in response to light, a process referred to as photodinesis (discussed in section 5.5). High light intensities induce some forms of movement that probably seldom if ever occur in nature. Under the intense illumination required to observe *Euglena* in the higher powers of the microscope, for example, the cell carries on convulsive contractions called **euglenoid movement.**

Many unicellular animals or **protozoa** (Chapter 20) show some form of cytoplasmic streaming or transport of organelles. In *Paramecium* and other similar unicells, for example, the cytoplasm circulates around the cell, carrying food vacuoles and distributing digestive products.

Many animal cells also show saltations similar to that described above for plant cytoplasm. In egg cells and in some leukocytes there is a variation on this theme. The saltations are associated with micro-

tubules radiating out from one or more centrioles (section 2.3), and particles saltate along the paths of these microtubules. It is not yet known whether the microtubules exert the force that moves the particles.

8.9 Both particles and various substances are transported along nerve cells.

Nerve cells in large animals may grow to lengths of several meters, and although their principal function is to transmit nerve impulses (section 33.4), they also perform other functions. They are known in some instances to transport important substances and particles to nerve endings.

Transport in nerves is called **axoplasmic transport,** because it occurs in the **axoplasm** of nerve, a special name given to the cytoplasm of nerve cells, or **neurons.** Axoplasmic transport apparently includes several different processes, because nerves transport some substances rapidly and others slowly. A great deal of saltatory movement of axoplasmic particles takes place in neurons in a tissue culture. The particles move considerable distances at speeds up to half that of streaming in an amoeba.

Electron micrographs of nerve cells show them to be rich in microtubules and microfilaments. It is not yet clear whether either or both of these structures are involved in axoplasmic transport. The recent finding of actin and myosin in brains suggests the possibility that movement within neurons may have a molecular basis somewhat similar to muscle contraction and amoeboid movement.

Whatever the mechanism of axoplasmic transport, it plays an important role in human health. Several serious human neurological disorders are caused either by failure of the axoplasmic transport system or by the transport of viruses or other harmful entities within the nervous system.

8.10 Ciliary and flagellar motion are used both for cell locomotion and for moving fluids through the internal cavities of sessile organisms.

Cilia ("eyelashes") and **flagella** ("whips") are organelles that protrude from the surface of some cells and move in such a way that they apply a force to the surrounding fluid (Figure 8-7). Cilia tend to be short and to beat like an oarsman, with a power stroke and recovery stroke; while flagella undulate in a wavelike manner. Flagella may be single (as in

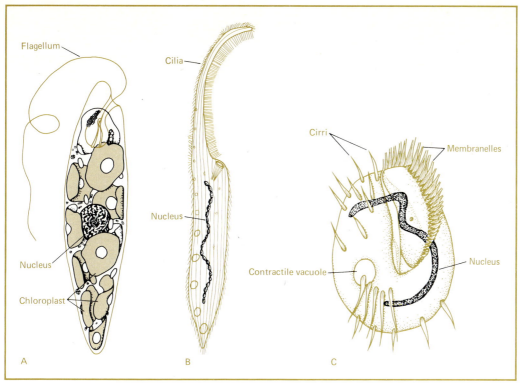

8-7 A diagram showing a flagellum, cilia, membranelles, and cirri in (A) *Euglena*, (B) *Dileptus*, and (C) *Euplotes*.

most animal sperm), paired (as in *Euglena*), or present in tens of thousands per cell (as in *Trichonympha*), whereas cilia tend to occur only in large numbers.

The **protists** (one-celled organisms; see Chapter 20) are classified according to their method of locomotion—cilia, flagella, or other organelles—and many thousands of different protists use each of these methods. Both cilia and flagella can "fuse" with one another and with membranes to form complex organelles which may create water currents (e.g., **membranelles**) or may serve as "walking legs" (**cirri** of *Euplotes*), as shown in Figure 8-7.

The sperm of multicellular animals generally move by means of a single flagellum. Flagella are also used by so called **collar cells** in sponges to pump water (with food particles) through the animal's internal cavity.

Ciliated cells perform many transport functions in multicellular organisms. For example, in humans the egg is transported along the oviduct by ciliary beating; the respiratory passages are cleared of debris by the beating of cilia in the nasal passages and trachea (section 38.1).

Cilia and flagella have a similar ultrastructure (Figure 8-8). Each has an outer covering of unit

membrane over an **axoneme** (axial bundle) consisting of nine **outer doublet microtubules** and two **central tubules**. The axoneme originates at the **basal body**, just beneath the cell surface from which the cilium or flagellum protrudes. Attached to each outer doublet are **side arms** that extend toward the adjacent doublet microtubules. The axoneme is held together by **spokes**, fibrils connecting the central tubules with the outer doublet tubules.

The outer doublet microtubules are constructed from two slightly different proteins, A- and **B-tubulin**. The microtubules are linked by the protein **nexin**, and the side arms are still another protein, **dynein**. Dynein was immediately implicated as the probable mechanochemical transducer molecule because it has ATPase enzymatic activity.

8.11 Cilia and flagella, like muscle, work on a sliding-filament principle, but the molecules are not the same as in muscle.

The presence of a probable mechanochemical transducer, dynein, between outer doublet pairs, suggested the possibility of tubule sliding. The sliding

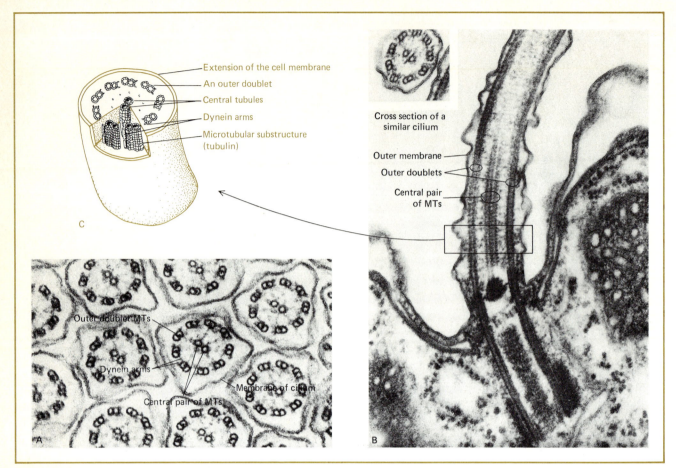

8-8 *A.* Cross section of mussel gill cilia showing the 9 + 2 pattern of microtubule (*MT*) arrangement. *B.* Longitudinal section of a cilium and its basal body from *Tetrahymena,* with inset cross section. *C.* An interpretation of the structure of the ciliary shaft based on thin sections. (Electron micrographs courtesy of P. Satir and B. Satir.)

filament model of ciliary and flagellar motion was proposed and tested successfully by P. Satir, who carefully sectioned beating cilia in clam gills for electron microscopy and showed that the outer doublet on the outside curve of a bent cilium does not extend as far into the tip of the cilium as the doublet on the inner surface of the bend. Since the two doublets are of equal length when the cilium is straight, the tubules must slide relative to each other.

The sliding filament theory has now been supported in the most elegant way by I. R. Gibbons and K. E. Summers, who isolated axonemes and selectively made the nexin soluble so that the outer doublets could move more extensively relative to one another. They then added ATP and showed that outer doublets could slide on one another's surfaces, constituting a telescoping system of microtubular doublets.

Flagellar locomotion also occurs in some procaryotic cells (bacteria), but here the flagella are entirely different. They are composed of a single protein, **flagellin,** and are not surrounded by a membrane. Bacterial flagella do not undulate but rotate by a "rotary motor" in their bases.

8.12 Some lower plant cells have a unique gliding mechanism that rescues them from burial after storms.

Diatoms, desmids, and blue-green algae have the ability to glide along solid surfaces. This process is essential to their dispersal and to their survival. After storms, these plant cells are often buried in mud or sand as the waters subside. Intertidal zones change color after a storm from muddy brown to brown with a green or blue-green tint as diatoms, desmids, and blue-green algae surface by a seemingly random back-and-forth gliding motion.

9

Cellular Reproduction

QUESTIONS TO KEEP IN MIND

How do cells reproduce?

Do all cells reproduce? Does the specialization of cells affect their reproduction?

What is the cell cycle, and how are its phases defined?

At what point in the cell cycle is the commitment made to divide?

How is DNA packaged? In what manner is it replicated?

What is the mitotic apparatus, and how is it formed?

The ability to reproduce is one of the characteristics of living cells, and to understand how reproduction occurs is fundamental to much of current biological and biomedical research. Reproduction is a beautifully ordered process, but complex and not easy to understand. Little progress was made until the cell theory, introduced in 1838 and 1839, led scientists to study reproduction on the cellular level. Even in the latter half of the nineteenth century, when chromosomes were discovered and their role during cell division elucidated, no penetrating insights were gained into the real nature of reproduction.

In the mid-twentieth century there occurred a true revolution in the biological sciences, and molecular biology was one of several new sciences to emerge. The progress made in this field enables us to look at cellular reproduction from a new vantage point. The sophisticated questions being asked today are not only intrinsically interesting but may well lead to a solution of one of the most distressing of human problems—cancer. In cancer, the re-straints on cellular reproduction are removed. Can reproduction be understood thoroughly enough to discover what these restraints are and why they fail?

9.1 Most cells divide periodically, either throughout the life of the cell or organism or until the cells take on a specialized function.

Relatively simple procaryotic cells, such as bacteria, can divide as frequently as every 20 minutes, given proper conditions of temperature and nutrition.

In cells of the more complex eucaryotes, cell division usually takes a much longer time—20 hours on the average—although there are exceptions such as fertilized egg cells, which, with their ample store of necessary molecules, can divide at intervals of less than an hour.

Free-living single eucaryotic cells, the protists, reproduce until they are eaten or otherwise die, but in multicellular organisms habits of cellular reproduction are varied, depending on the cells involved. For example, embryonic cells that will give rise to muscle and nerve cells divide for a time while they differentiate (i.e., take on special characteristics), but once a certain point is reached they do not divide again. If you run or exercise regularly you will add to the size and efficiency of your muscle cells, but you will not grow any new ones. Nor do brain cells divide in an adult human: Not only will you fail to acquire new brain cells, but you will lose millions and millions of the ones you

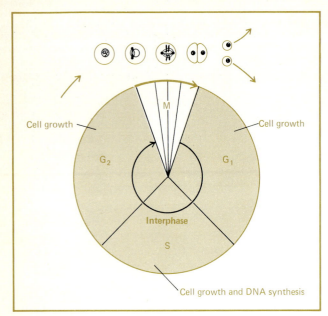

9-1 Mitosis (M) is a very short stage in the life cycle of a cell, during which it divides. The rest of the time it is said to be in interphase. Interphase is made up of two periods of cell growth (G₁ and G₂) separated by a period (S) during which DNA is synthesized.

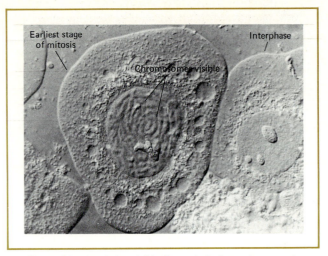

9-2 The cell in the right of this figure is in interphase, and no chromosomes can be seen in its nucleus. The cell on the left has begun mitosis, and is in early prophase stage. The DNA-protein fibers in its nucleus are beginning to "condense" (by folding in an intricate pattern) to form chromosomes. (Photo by author.)

now have through injury, poisoning (e.g., alcohol), and aging. The cells living in your intestine, however, will keep on reproducing throughout your lifetime, replacing those cells that are constantly sloughed off. Still other cells require a stimulus such as a wound or an infection to begin reproducing. Such is the case with the lymphocytes, one type of leukocyte (white blood cell) involved in immune responses to infection.

In higher plants, cell divisions occur almost exclusively near the tips of roots and shoots.

9.2 Cellular reproduction occurs in three main phases—the interphase, mitosis, and cytokinesis.

Most of the life of a "typical" eucaryotic cell is spent in the **interphase** (Figure 9-1), during which time the cell prepares for the next division by synthesizing proteins, DNA, and other constituents it will need. It must synthesize DNA and replicate its chromosomes so that a complete set of genetic directions can be passed on to the daughter cells. As far as DNA synthesis is concerned, the interphase is actually divided into three periods—G_1 (which stands for an apparent gap in activity), S (the period of DNA synthesis), and G_2 (another gap).

In the second phase of cellular reproduction,

called **mitosis**, the replicated chromosomes are separated and apportioned equally to the two daughter cells. The final phase is **cytokinesis**, in which the cytoplasm of the original cell is divided in two.

In most cells the three phases of reproduction follow each other smoothly. However some cells leave out an entire phase, and some vary the amount of time spent in each phase, depending on conditions such as temperature and the availability of nutrients.

9.3 DNA is packaged within each cell as part of its chromosomes.

The amount of DNA in a cell is a very small portion of the cell's weight. DNA molecules are extremely thin (≈ 2 nm) and amazingly long. The total length of DNA in any single human cell is about 2 meters, and if all the DNA from one person's 10^{10} cells were strung together, it would more than stretch to the moon and back.

The problem of folding two meters (2,000,000 μm) of DNA into a nucleus roughly 10 μm in diameter has been solved by packaging the DNA in chromosomes. The name means "colored bodies" and was given because the chromosomes were first seen in cells stained with dyes which had bound specifically to the DNA. Chromosomes form by "condensation," which in this case means the intricate folding of a DNA-protein fiber. They are visible in most cells only when the cells are about to divide. When the cell is not dividing, the DNA-protein fibers are so thin that optical sections of them appear to be granular (Figure 9-2). Chromosomal material in this uncondensed state was called **chromatin** ("colored stuff"), and the name has stuck.

9.4 Chromosomal DNA replication is semiconservative and takes place in different regions of the chromosome at different times.

It is while the DNA in the cell nucleus is in the form of chromatin threads that it replicates in order to form a second set of genetic directions for the new cell. This takes place in the S stage of the interphase. But how?

In the late 1950s it was shown that DNA replicates in a semiconservative manner (section 3.7). By this process each of the two daughter cells receives chromosomes composed of "hybrid" DNA: One of the strands of the double-stranded DNA is newly fabricated, and one is from the parent DNA.

In 1958, M. Meselson and F. W. Stahl completed a series of elegant experiments that proved the replication of DNA to be semiconservative in the nucleoid of certain bacteria (procaryotes). These bacteria were allowed to go through many cell cycles in a growth medium containing an isotope of nitrogen, namely ^{15}N, which has an extra neutron and is measurably heavier than common ^{14}N. That the bacteria did incorporate ^{15}N into their DNA strands could be seen by comparing the density of their DNA molecules with the density of DNA from bacteria growing in a normal growth medium. When these two DNA samples were put into a concentrated cesium chloride salt solution and subjected to high-speed centrifugation, the salt formed a density gradient, and the DNA molecules settled at two different levels matching their own different densities. (Figure 9-3, tube A). Next, cells from the culture containing ^{15}N DNA molecules were transferred to a medium containing ^{14}N exclusively. The DNA from bacteria resulting from one reproductive cycle were found to be of an intermediate weight, indicating that they were one strand of ^{15}N-DNA and one of ^{14}N-DNA (tube B). After a second generation both intermediate-weight molecules and light ones were present (tube C). The results indicated clearly that the DNA molecules replicated

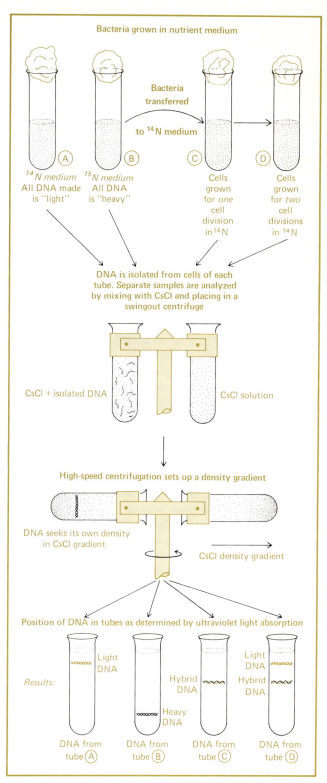

9-3 A diagram based on the Meselson-Stahl experiment that established semiconservative replication of bacterial DNA. DNA isolated from bacterial cells grown on ^{14}N or ^{15}N media was spun at high speeds in a cesium chloride (C_3Cl) solution until a density gradient was established. The densities of the DNA from bacteria in the different growth experiments are shown in tubes A, B, C, and D.

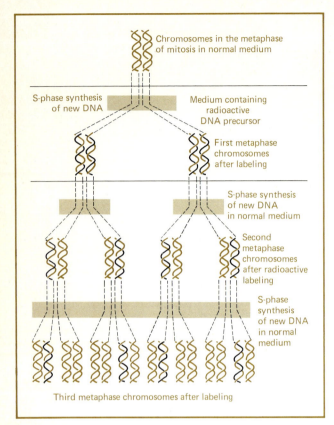

Chromosomes in the metaphase of mitosis in normal medium

S-phase synthesis of new DNA

Medium containing radioactive DNA precursor

First metaphase chromosomes after labeling

S-phase synthesis of new DNA in normal medium

Second metaphase chromosomes after radioactive labeling

S-phase synthesis of new DNA in normal medium

Third metaphase chromosomes after labeling

9-4 A diagram of chromosome replication, as shown by labeling experiments in which chromosomal DNA was replicated semiconservatively in a medium containing a radioactive DNA precursor and subsequently allowed to replicate two additional times in normal medium.

in a semiconservative manner in these procaryotic cells.

In eucaryotic cells, the semiconservative replication of chromosomal DNA was demonstrated by J. H. Taylor in 1957. He grew cells in a medium containing a radioactive precursor of DNA, thymine, until all their chromosomes were radioactive. When the cells were allowed to reproduce in a nonradioactive medium, only one in each pair of daughter chromosomes in the divisions that followed was radioactive (Figure 9-4).

9.5 It is not precisely clear what triggers the progression from stage to stage in the cell cycle.

Although experiments have determined exactly when and how chromosomal DNA replicates, we still do not know what initiates the process. Similarly, it is unclear what event or signal initiates the G periods, although it is known that when a cycle is very long, it usually has a long G_1 period and a moderately long G_2 period. Also, once a cell enters S, the period of DNA synthesis, it is known to be committed to division within the next 24 hours, and once its chromosomes begin to condense it will *usually* proceed with mitosis within a few hours.

The discovery of the mechanism triggering these phases, especially the crucial synthesis of DNA, may well hold the key to solving the cancer problem. Some of the most successful weapons against cancer are drugs that interfere with DNA synthesis. Unfortunately these drugs do not discriminate between cancer cells and the healthy cells of the body, which also need DNA synthesis to divide.

9.6 Mitosis involves the separation, transport, and precise distribution of chromosomes to daughter cells.

Now that we have considered the replication of chromosomal DNA, which is the necessary first step in reproduction, we will proceed to mitosis. The process of mitosis is divided, for convenience of discussion, into five phases. The details of mitosis differ somewhat for animals and plants. We shall describe it first as it occurs in animals. The first phase is **prophase,** during which the chromosomal threads begin to condense and the nucleus swells (Figure 9-5A). The chromosomal DNA has already replicated to form "double" chromosomes; that is, each chromosome is composed of two strands, or **chromatids.** By the end of prophase, each chromosome (still double) is constricted in at least one place. This constricted region is called the **kinetochore,** and it is here that a spindle fiber (called a **kinetochore fiber**) is usually attached in a later phase.

Also during prophase there is considerable activity outside the nucleus. In almost all animal cells there are two tiny cylindrical bodies called **centrioles** lying at right angles to each other just outside the nucleus. In most cells each centriole has replicated during interphase, but still appears single. When the two centrioles separate, each becomes surrounded by a growing star-shaped burst of microtubules known as an **aster.** As prophase begins, one aster remains where it is, and the other migrates to a position some 180 degrees around the nucleus. Their positions mark the future poles of the dividing cell.

When the asters are in place, the nuclear envelope ruptures and usually disappears from view. This marks the beginning of the **prometaphase.** During this phase the **mitotic spindle** (which may have begun to form in prophase) develops in full

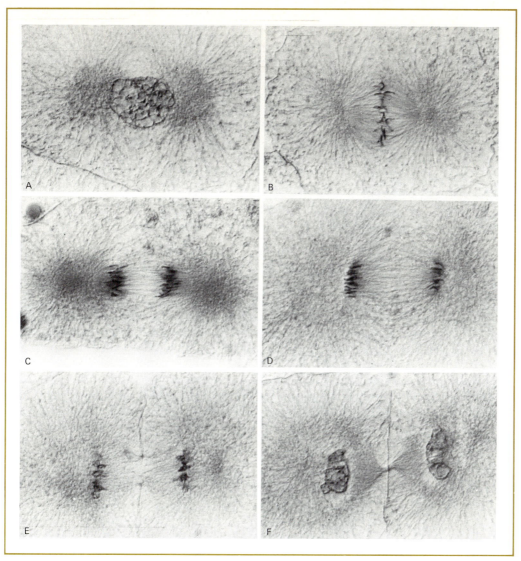

9-5 Stages of mitosis in sections of dividing whitefish blastula cells that have been fixed and stained: *A.* Late prophase, showing the two asters on opposite sides of the nucleus. *B.* Metaphase stage, in which the paired chromatids lie on the equatorial plate of the spindle. *C.* Anaphase, the stage during which the chromatids migrate to the poles of the spindle. *D.* The end of anaphase, at which time the chromatids have reached the poles. *E.* Early telophase, showing the advancing cleavage furrow. *F.* Late telophase, when the cleavage furrow has passed through the remnant of the spindle. (Photos by author.)

between the two asters. The spindle with asters at both ends is called the **mitotic apparatus.** The spindle itself (Figure 9-5B, C, and D) is a football-shaped construction of microtubules and other structures which extends from pole to pole. As it is forming, the chromosomes migrate to the **equatorial plate** of the spindle, the region midway between the poles (Figure 9-5B). Here they remain for some time. Near the end of **metaphase,** the two

chromatids in each sister chromosome abruptly separate. This slight separation occurs even if the spindle is poisoned or destroyed, indicating that a passive pulling apart of chromosomes is not responsible for it.

Next the cell enters **anaphase** (Figure 9-5C and D), which is characterized by poleward movement of the chromatids along the spindle axis. In this phase, the precisely equal distribution of daughter chromosomes to daughter cells is accomplished. When the chromosomes have reached their destination, the cell is in **telophase** (Figure 9-5E). The end of telophase is marked by the reconstitution of the nuclear envelope and the complete separation of the cytoplasm of the daughter cells, a separate process known as **cytokinesis** (Figure 9-5F).

The overall phases of mitosis in plant (Figure 9-6) and animal cells are the same, but some details are

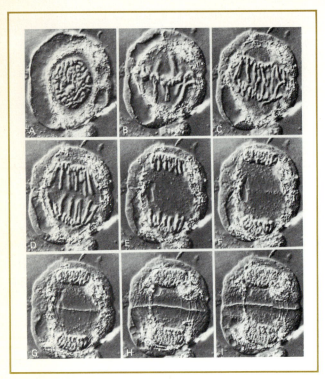

9-6 Stages of mitosis in a living African blood lily endosperm cell. *A.* Prophase: Note the spindle forming outside the nucleus. *B.* Metaphase. *C.* Early anaphase. *D.* Late anaphase. *E.* Early telophase: Note vesicles coming into the spindle remnant, how termed the *phragmoplast. F.* Vesicles accumulate in the middle of the phragmoplast. *G.* Vesicles begin to fuse to form the *cell plate,* which separates the daughter cells. *H* and *I.* The separation becomes complete, and the chromosomes become less distinct as the nucleus begins to resemble that of an interphase cell (compare Figure 9-2). (Courtesy of Robert Hard.)

different due in part to the structural differences between plant and animal cells. For example, plant cells rarely have centrioles but have some other microtubule-organizing centers instead. As a result animals may have, but plants generally lack, asters at the poles of the spindle. Perhaps for this reason, plant and animal spindles are somewhat differently shaped.

9.7 In plant and animal cells, cytokinesis accomplishes the division of the cytoplasm in different ways.

In most cells, cytokinesis follows mitosis without a pause, and, in fact, the division of the cytoplasm is usually initiated during anaphase.

In plant cells the remnant of the spindle that exists as anaphase nears completion is called the

phragmoplast (Figure 9-6E). This structure now assumes a new function, relevant to the formation of cell membranes that will close off the daughter cells. Plant cells have many small vesicles or membranous sacs, each containing the essential materials for cell wall construction. These are called **Golgi vesicles** because they are budded off from the Golgi apparatus (section 2.5). The phragmoplast, containing parallel microtubules, transports the Golgi vesicles by an unknown mechanism to a point midway between the two daughter nuclei. The vesicles line up and fuse to form the **cell plate**, which in turn becomes the **middle lamella** of the daughter cells. The phragmoplast is remarkably precise in its alignment of the Golgi vesicles.

Unlike plant cells, animal cells divide their cytoplasm by a constriction called the **furrow** that passes through the middle region of the spindle remnant (Figure 9-5). The furrow gradually deepens until the daughter cells are separated. What causes the furrow to deepen is a ring of microfilaments composed of actin, which contracts.

9.8 Cytokinesis does not always divide a cell equally.

The details and perhaps even the mechanisms of cytokinesis vary widely. *Amoeba proteus* stops its locomotion during mitosis and then resumes locomotion with two nuclei. Before long, however, the binucleate cell pulls itself in two parts by pseudopodia oriented in opposite directions. The two division products may differ in mass by as much as 10 percent.

In egg cells, cleavages are programmed to be either nearly equal or very unequal. In the eggs of mollusks and annelids, for example, unequal cleavage is an essential part of normal development. In all higher organisms utilizing sexual reproduction, special meiotic divisions (sections 11.4 and 11.5) occur that reduce the amount of genetic material in the **gametes** (eggs and sperm) to half that found in the somatic (body) cells. (If this did not occur each offspring would have double the number of chromosomes found in each of his parent's cells, an unmanageable situation). However, in the meiotic divisions leading to the formation of an egg, the cytoplasm is divided grossly unequally. These unequal divisions provide one cell with most of the cytoplasm and allow unnecessary genetic material to be discarded.

9.9 The mitotic spindle is one of the most beautiful and intriguing organelles in the cell, and the least understood.

The spindle appears only once in the division cycle of all eucaryotic cells. It transports chromosomes with precise equality to daughter cells. It is, however, the only organelle that has not been made to function at all outside the cell after isolation. How the spindle works is still a mystery. The spindle is composed mainly of microtubules and probably other, as-yet-undiscovered components that are important in its function. There are several theories about how the spindle works, but the evidence is both incomplete and contradictory.

SUGGESTED READING

Bloom, W. and D. W. Fawcett. *A Textbook of Histology.* 9th ed. Philadelphia: Saunders, 1968.

Brachet, J., ed. *The Living Cell.* San Francisco: Freeman, 1961.

Brachet, J. and A. E. Mirsky. *The Cell,* vols. 1–5. New York: Academic Press, 1961.

Clayton, R. K. *Light and Living Matter: A Guide to the Study of Photobiology.* 2 vols. McGraw-Hill, New York, 1970.

De Robertis, E. D. P., F. A. Saez, and E. M. De Robertis. *Cell Biology.* 6th ed. Philadelphia: Saunders, 1975.

Dickerson, R. E. and I. Geis. *The Structure and Action of Proteins.* New York: Harper & Row, 1969.

Dupraw, E. J. *Cell and Molecular Biology.* New York: Academic Press, 1968.

Dyson, R. D. *Cell Biology.* Boston: Allyn & Bacon, 1974.

Fawcett, D. W. *An Atlas of Fine Structure — The Cell.* Philadelphia: Saunders, 1966.

Galston, A. W. *The Green Plant.* Englewood Cliffs, N.J.: Prentice-Hall, 1968.

Jensen, W. A. *The Plant Cell.* Belmont, Calif.: Wadsworth, 1964.

Jensen, W. A. and R. B. Park. *Cell Ultrastructure.* Belmont, Calif.: Wadsworth, 1967.

Ledbetter, M. C. and K. R. Porter. *Introduction to Fine Structure of Plant Cells.* New York: Springer-Verlag, 1970.

Loewy, A. G. and P. Siekevitz. *Cell Structure and Function.* 2nd ed. New York: Holt, Rinehart and Winston, 1969.

Novikoff, A. B. and E. Holtzman. *Cells and Organelles.* New York: Holt, Rinehart and Winston, 1976.

Porter, K. R. and M. A. Bonneville. *Fine Structure of Cells and Tissues.* 3rd ed. Philadelphia: Lea and Febiger, 1968.

Racker, E. *Mechanisms of Bioenergetics.* New York: Academic Press, 1965.

Ris, H. et al. *Topics in the Study of Life.* New York: Harper & Row, 1971.

Rothstein, H. *General Physiology.* Waltham, Mass.: Xerox College Publishing Co., 1971.

Swanson, C. P. *The Cell,* 3rd ed. Englewood Cliffs, N.J.: Prentice-Hall, 1969.

White, A., P. Handler, and E. L. Smith. *Principles of Biochemistry.* 4th ed. New York: McGraw-Hill, 1968.

Wilson, E. B. *The Cell in Development and Heredity.* 3rd ed. 1925. Reprint. New York: Macmillan, 1953.

Wolfe, S. L. *Biology of the Cell.* Wadsworth, Belmont, Calif.: 1972.

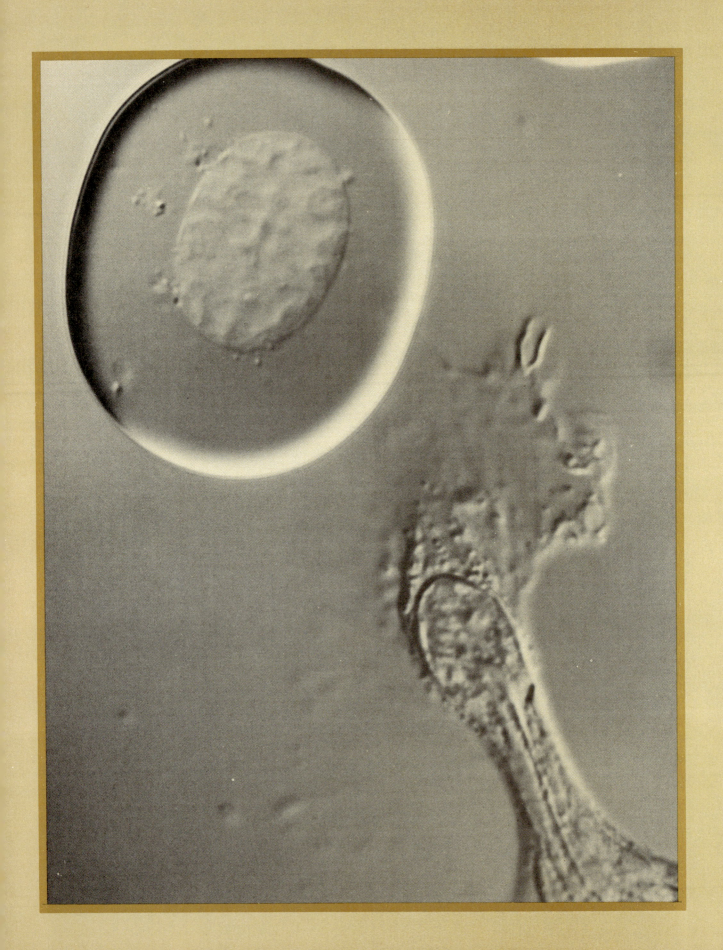

II DEVELOPMENT: FROM CELLS TO ORGANISMS

In Part I the foundation was laid for an understanding of how organisms develop. Some simple organisms are either single cells or acellular—not made up of cells. Chapter 9 has prepared the reader to understand in a general way how these organisms reproduce.

Many multicellular organisms reproduce sexually and in the process retrace part of their probable evolutionary history—the development of the multicellular condition; for the eggs of these organisms begin as single cells and become multicellular by repeated cell divisions while remaining attached. In addition, the cells of these developing organisms become different from one another (i.e., they differentiate) and become specialized in both structure and function, until the organism reaches adulthood.

Like the study of the cell, developmental biology is an important research frontier. In the past decade, descriptive studies have largely been replaced by analytical ones in which the causes of various developmental processes are explored. Such investigations inevitably lead to the molecular level, where the control of gene action determines how the tissues of the developing organism interact. The nature of these interactions cannot be understood without some background in molecular and cell biology (Chapters 3, 7, and 9) and a general descriptive knowledge of the changes that take place in developing organisms.

10

Asexual Reproduction

What are the principal methods of asexual reproduction?

In what ways is fission different in protists and in flatworms?

What are the advantages to an organism in producing a large number of spores?

How is budding different from fission?

How do sponges reproduce?

What are some examples of vegetative propagation in plants and animals?

What are the advantages and disadvantages of reproducing asexually?

Reproduction is the process by which life is maintained from generation to generation. In all cases its objective is to separate a viable bit of genetic material (DNA) from the parent organism in such a way that a new organism can develop according to its genetic instructions.

In **asexual reproduction** genetic material from a single parent is supplied to one or more daughter cells. This is accomplished without the union of special sex cells—eggs and sperms—and hence without the mixing of genetic material. Therefore, the generations of asexually reproduced cells and organisms tend to resemble one another closely.

There are several different methods of asexual reproduction, including fission, budding, the for-mation of spores, and several other forms of vegetative propagation.

10.1 Fission is the formation of two identical individuals by the splitting of a single individual.

Fission, the simplest method of asexual reproduction, is widely used by bacteria and also by protists (single-celled eucaryotic organisms). Depending on the species, protists may undergo **transverse binary fission,** which is a splitting into two cells across the short axis of the parent cell; **longitudinal binary fission,** which is splitting in two lengthwise (Figure 10-1); or **schizogony,** splitting into numerous daughter cells. Except in the case of procaryotes, fission is the result of the processes of mitosis and cytokinesis described in Chapter 9.

Fission is also used by some multicellular organisms to reproduce, and in these cases it is not the result of any single cell division. Free-living planarians (flatworms), for example, undergo fission by breaking transversely, and in some of these animals a second set of eyes and digestive structures appear in the posterior half of the worm prior to fission (Figure 10-2).

Fission may also occur as a part of sexual reproduction in protists and in some more complex organisms. In humans and many other animals, identical twins result when two of the fission products from a single fertilized egg become permanently separated. Each of these separated cells contains all the information necessary to make a new individual.

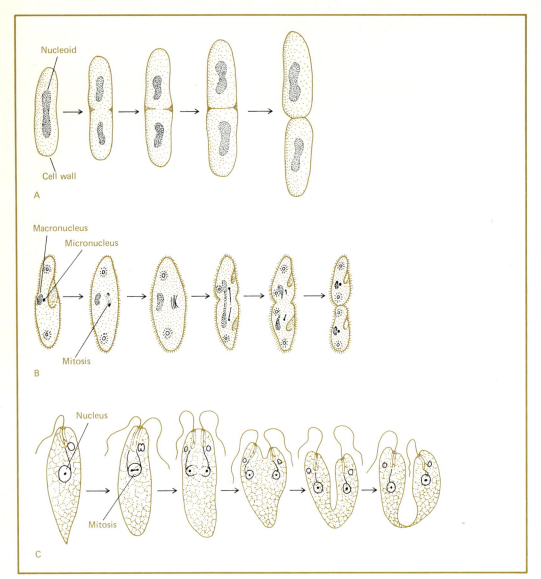

Nucleoid

Cell wall

A

Macronucleus

Micronucleus

Mitosis

B

Nucleus

Mitosis

C

10-1 Asexual reproduction by fission. *A.* Transverse fission in a typical bacterium. *B.* Transverse fission in *Paramecium:* The micronucleus divides mitotically and the macronucleus separates into two parts. *C.* Lontitudinal fission in *Euglena.*

10.2 Budding is the pinching off of a small portion of a parent organism to create a new individual.

Budding is another relatively simple method of asexual reproduction used by both unicellular and multicellular organisms. For example, in many yeasts, which are unicellular fungi, budding takes place during one phase of the life cycle (Figure 10-3). Before the budding process, the DNA of the nucleus replicates. Soon afterward, a small protu-berance begins to form at one end of the cell. The nucleus begins to divide and moves toward the bud until it lies in the plane of constriction. Gradually the cytoplasm shifts until the volume of the mother cell and that of the bud are approximately equal, and at the same time, the nucleus divides mitotically. Subsequently the two parts of the yeast cell separate, and each complete daughter cell is free to grow until it attains the size of the original mother cell.

Budding is a very effective form of reproduction. The generation time may be as short as 30 minutes.

Budding also occurs in multicellular organisms such as *Hydra*. This animal, named after the mythological serpent that grew two heads for every one cut off, first gives indication of budding by the ap-

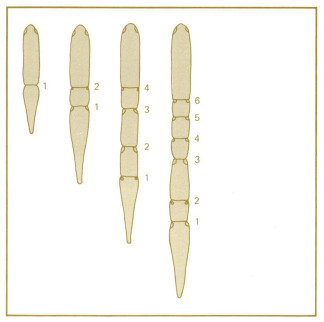

10-2 Fission in a flatworm, *Stenostomum*. The numbers show the sequence of successive fissions.

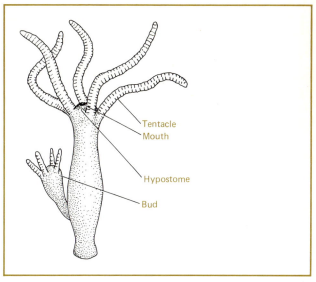

10-4 Budding in *Hydra*.

pearance of a small lump of dividing cells on the side of its saclike body. With continued cell proliferation, the bud elongates and sprouts tentacles at its free end (Figure 10-4). The bud develops all parts characteristic of a mature *Hydra* and while still attached may independently trap and consume smaller animals. Eventually it separates from the parent.

Plants also use budding as a means of asexual reproduction, and many have localized nodes or groups of cells capable of producing entire new plants. Striking examples of this are found in the genus *Kalenchoë*, also called bryophyllum or "air plant." The leaves are notched at the margins. At the base of each notch is a small group of specialized cells. As the leaf ages, or after it drops from the plant, these cells differentiate and form small plants, each with root, stem, and leaf (Figure 10-5).

10.3 The formation of spores—sporulation—is a kind of asexual reproduction used by bacteria, fungi, and a few protists.

A **spore** is a special kind of cell, usually with a resistant covering, that can remain dormant for long periods when conditions are unfavorable for its growth. When these conditions change, the spore germinates and develops into a new individual. In some organisms spores are formed in a specialized structure called a **sporangium**. When released they may be passively distributed by air or water currents. **Zoospores** (independently motile spores) have flagella to help propel them through water.

Figure 10-6 illustrates **sporulation** in the common bread mold, *Rhizopus*. This fungus, like many others, produces a fuzzy growth of **rhizoids** ("roots") and sporangiophores, topped by minute black beads, the sporangia, from which spores are released. These are dispersed by air currents and if

10-3 Stages in the budding in yeast.

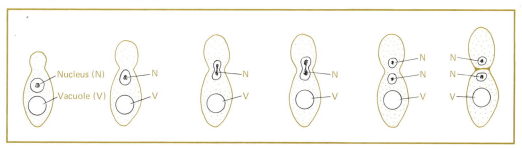

10-5 A *Kalenchoë* plant showing the growth of tiny plantlets from the leaves. Note that two have fallen and taken root in the soil. (Courtesy of Robert Spech.)

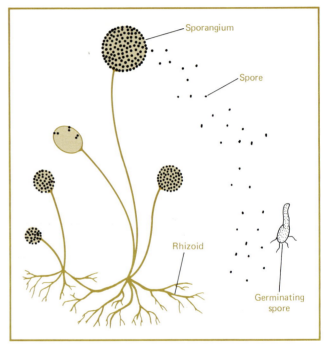

10-6 Sporulation in the common bread mold, *Rhizopus*.

10-7 Some examples of vegetative propagation in plants. *A.* Stolon, or runner, forming a new strawberry plant. *B.* A grass propagating by means of a rhizome. *C.* Layering: formation of a new plant from a branch that touches the ground. *D.* Propagation by an underground stem called a *tuber,* as in potatoes.

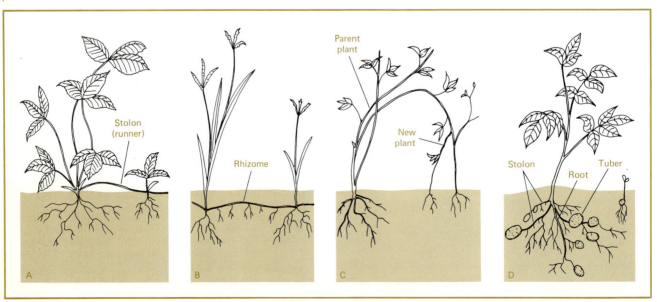

they encounter an appropriate material, develop more rhizoids.

10.4 Vegetative propagation is the production of new individuals from stems, branches, roots, and leaves.

The most familiar examples of asexual reproduction among plants fall into the category of **vegetative propagation** (Figure 10-7). There are several ways in which this can occur.

In strawberry plants, for example (Figure 10-7A), the plant propagates by means of a runner, or **stolon**, which is an extension of the stem. The runner grows over the soil surface, and at some distance from the parent plant, develops roots and leaves. This creates a new satellite plant, which will then survive even if the stolon is destroyed. Many grasses (Figure 10-7B) also propagate from stem outgrowths, but their stems, called **rhizomes**, grow horizontally underground. Another method of vegetative propagation using underground stems is the development of **tubers** (e.g., potatoes), which are enlarged portions of an underground stem or stolon. Each tuber has numerous "eyes" (or buds), from each of which a new plant can grow (Figure 10-7C). Upright stems or branches are used by still other plants to produce new plants by **layering** (Figure 10-7D). Branches or stems may bend, touch the ground, and then develop new root systems, and eventually new plants. Extensions from a root may be utilized in much the same manner. Silver poplars

and some other trees can propagate from their roots.

Some **sessile** (essentially stationary) animals also use vegetative propagation to reproduce. For example, some aquatic organisms, including jellyfish, corals, hydra, and other relatively simple multicellular animals, send out stolons along which small buds develop. Each bud grows into an animal similar to the parent, and the parent and offspring comprise a **colony** (section 25.7).

10.5 Some organisms rely upon accidents to initiate asexual reproduction.

There are plants and animals that increase their numbers by regenerating an entire individual from a fragment of a parent organism. Starfish are able to regenerate from any fragment that includes an arm and a part of the base. (Oystermen, who hate starfish for the damage they do to oysters, used to try to kill the starfish they inadvertently brought up in their dredges by chopping them up and dumping them overboard. When they learned they were actually *increasing* the starfish population they began boiling them instead.)

Asexual reproduction is used by many organisms and is a sure method of producing offspring that will be exactly like the parent genetically. In a sense this is a shortcoming, for although asexual reproduction is a successful and efficient method, it lacks flexibility. Generations that are identical genetically with preceding ones have little ability to adapt to a changing environment.

11

Sexual Reproduction

QUESTIONS TO KEEP IN MIND

What are the advantages, disadvantages, and risks of sexual reproduction for the individual? for the species?

What method of reproduction best serves an organism in a changing environment?

What is a mutation?

How do homologous chromosomes behave in meiosis?

What is synapsis, and what does it accomplish?

What is parthenogenesis? How does parthenogenesis serve bee societies?

11.1 The essence of sexual reproduction is its capacity for generating variability, some of which may have survival value.

The vast majority of living organisms, be they bacteria, plants, or animals, reproduce sexually, although some may reproduce asexually also. Why this should be so may not be immediately apparent, since asexual reproduction is so efficient.

Why should thousands upon thousands of species run the greater risks involved in forming male and female sex cells and trying to get them together under favorable circumstances? Yes, as a human you're aware that it's more fun that way, but sexual reproduction has evolved for survival value. The male praying mantis, stimulated to copulate by decapitation (performed by his mate), does not have the opportunity to savor the joys of sexual reproduction. In spite of the risks, sexual reproduction is the best way to insure the success of the species in a continuously changing environment. Why? Because sexual reproduction provides millions of unique combinations of genetic material from pairs of *unidentical* parents and thus allows for variety in future generations. Some of the variations produced may be just those needed for the species to survive under changing environmental conditions. An asexual organism cannot adapt in this way. To illustrate, if a damp area such as a marsh begins to dry up over a long period of time, the species native to the area will ultimately be destroyed unless drought-resistant survivors of the original species can, in time, persist and repopulate the area.

11.2 Organisms that reproduce sexually or asexually may change by mutation.

An inheritable change in the makeup of a DNA molecule, such as can be caused by irradiation, is called a **mutation**. Such changes are essentially irreversible, and all the cells or individuals arising from the mutated cell will carry the change. For asexually reproducing organisms a mutation results in a sudden change which, whether beneficial or detrimental to the organism, will be passed on to succeeding generations. If it is beneficial, fine; if detrimental, the mutant's progeny will usually die off. Sexually reproducing organisms, however, receive genetic material from two parents. Therefore mutations will be tempered by the "normal" partner's genetic material. Thus sexual reproduction allows for variety in the long run, yet counteracts drastic changes (mutations) over short periods of time.

11.3 Sexual reproduction involves the sorting and recombination of chromosomal DNA.

As we discussed in Chapter 9, genetic information is contained in threadlike strands called chromosomes. Many years ago it was noticed that the number of chromosomes in any cell is frequently even. Moreover, nearly all the cells in an organism have the same number of chromosomes; and this number characterizes the cells of all the organisms belonging to that species. Further, it was noted that chromosomes most often exist in pairs—two chromosomes of similar size and shape that contain similar genes. These are called homologs.

Examining the 46 human chromosomes, an experienced person can recognize and sort out each pair of homologs and can give them distinguishing numbers (Figure 18-5). It has been shown in several ways that each parent contributes one chromosome to each pair for the new individual. For purposes of convenience we say that the total even number of chromosomes in the cell is the diploid number. The haploid number is half that number, that is, one from each pair. Each parent contributes a haploid number of chromosomes to the offspring.

11.4 Chromosomes are transmitted from one generation to the next in the nuclei of specialized sex cells called gametes.

In simple organisms, the sexes often look alike. Their sex cells, called gametes, are similar also. Such gametes are called isogametes, and when two come together, isogamous fertilization is said to occur. The unicellular flagellated alga *Chlamydomonas* reproduces this way. Sexes are referred to as mating types rather than male and female.

In humans and other complex forms of life, there are notable differences between the sexes, and each produces a distinctive gamete. In animals, the female produces a large macrogamete, incapable of locomotion, called an ovum or egg. Males produce a small, highly motile microgamete, or sperm. In higher plants, the macrogamete is also called the egg, whereas the microgametes in pollen are the sperm nuclei.

Two gametes fuse during sexual reproduction, yet the number of chromosomes in a species remains constant generation after generation. Obviously, therefore, there must be some process by which the normal diploid number of chromosomes in each parent is reduced to haploid number in the gametes. This process is called meiosis and is part of gametogenesis, the formation of gametes.

In multicellular animals gametes are produced by the primary sex organs, called gonads. The female gonad is the ovary; the male gonad, the testis. It is usually in the gonads that the meiotic divisions, which reduce the number of chromosomes, take place. Here also occurs the differentiation that produces the special physical characteristics of egg or sperm. In the eggs of some species, meiotic divisions occur after ovulation, the release of eggs. Whereas an egg needs a large store of macromolecules for rapid development after fertilization, a sperm must develop a flagellum and accessory structures for motility (Figure 11-3).

11.5 Meiosis involves two successive special cell divisions which produce gametes, each with a haploid number of chromosomes.

Superficially, each of the successive cell divisions involved in meiosis resembles mitotic division. Each has the same phases of nuclear division as mitosis (prophase, prometaphase, metaphase, etc.), and in both the cytoplasm is divided (cytokinesis). There are two main differences between these two types of cell division processes: (1) the first meiotic division the pairs of homologous chromosomes come together and adhere laterally in a very precise way, a process known as synapsis (Figure 11-2). (2) The genetic material is replicated only once for the two meiotic divisions. During synapsis, an exchange of genetic material occurs between the homologous chromosomes. The exchange of genetic material is called crossing-over. Its results can be seen in the shapes of meiotic chromosomes (Figure 11-1). Crossing-over is of widespread occurrence and is an important factor contributing to genetic variation in sexual reproduction. Meiotic chromosomes possess a unique structure called the synaptinemal complex, that apparently brings synapsis about, although the force which attracts the homologs remains unknown. (3) In most organisms there is essentially no interphase or prophase before the second meiotic division.

Synapsis performs two important functions in sexual reproduction. First, it ensures that each sex cell formed during meiosis will receive one of each homolog. Second, synapsis sets the stage for a pre-

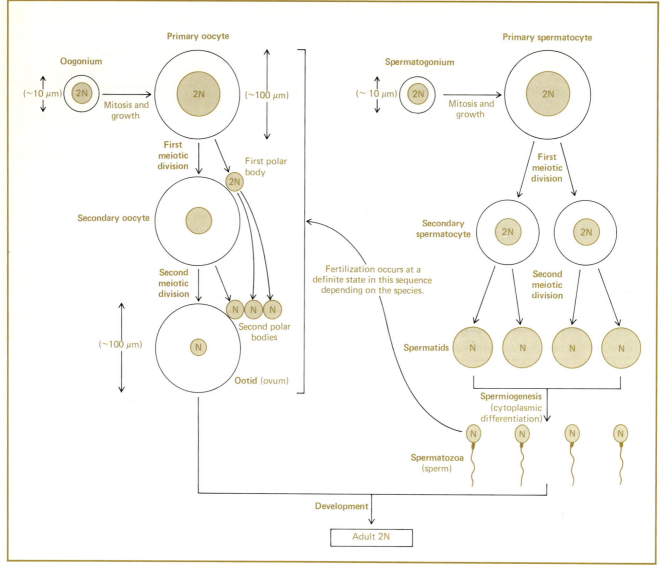

11-1 Crossing-over between maternal (*color*) and paternal chromatids during the prophase of the first meiotic division.

cise reduction in chromosomes (during the second meiotic division) by bringing together pairs of homologous chromosomes so that they behave as a single unit. Because each of the homologous chromosomes brought together has previously replicated and is made up of two chromatids, these pairs are called **chromatid tetrads.** (They are also called **chromosome bivalents**). *During synapsis the diploid number of replicated chromosomes becomes a haploid number of chromosome bivalents or chromatid tetrads.* During the second meiotic division, these bivalents will be halved, producing gametes with a haploid number of chromosomes.

Synapsis occurs in the prophase of the first meiotic divisions (meiosis I). The resulting tetrads mi-

grate to the equatorial plane, become attached to spindle fibers, and subsequently split—each into two **dyads** (a two-chromatid chromosome). Cytokinesis then forms two cells, each with a haploid number of dyads. In the second meiotic division (meiosis II) each of these cells divides without replication of the genetic material. In meiosis II each dyad splits to form two **monads;** thus four cells have been produced from the original one. Each carries a unique variation of the parent's genetic material because of crossing-over (and also because of independent assortment, discussed in section 15.5).

Actually it is not accurate to say that four sex cells are produced from one in all cases of meiosis

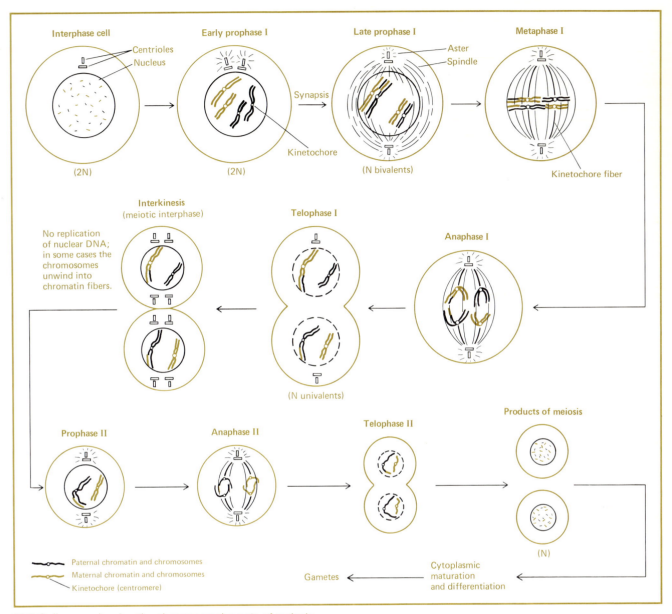

Interphase cell

Centrioles
Nucleus

(2N)

Early prophase I

(2N)

Synapsis

Late prophase I

Aster
Spindle

Kinetochore

(N bivalents)

Metaphase I

Kinetochore fiber

Interkinesis
(meiotic interphase)

No replication
of nuclear DNA;
in some cases the
chromosomes
unwind into
chromatin fibers.

Telophase I

(N univalents)

Anaphase I

Products of meiosis

Prophase II

Anaphase II

Telophase II

(N)

Paternal chromatin and chromosomes
Maternal chromatin and chromosomes
Kinetochore (centromere)

Gametes

Cytoplasmic
maturation
and differentiation

11-2 A diagram showing the chromosomal events of meiosis.
Chromosomes, which have replicated in the S phase of the
preceding interphase and are thus already double, pair with
their homologs to form chromatid tetrads. These are split
twice in two successive divisions with no further DNA replication. The result is a halving of the original diploid (2N) chromosome number. Note the crossing-over on one of the tetrads.

11-3 Meiosis plays an important part in the processes leading
to the formation of sperm in the testis and eggs (ova) in the
ovary of animals by reducing the number of chromosomes
from the diploid (2N) to the haploid (N) number. Primary and
secondary spermatocytes and spermatids remain connected
during the meiotic divisions. Similar processes occur in protists, plants, and fungi, although the details differ.

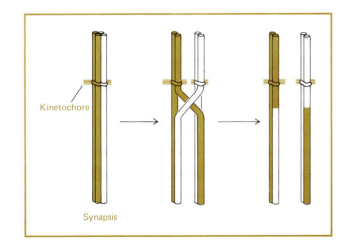

Kinetochore

Synapsis

in animals. In the formation of sperm it is true—one cell divides twice meiotically to produce four sperm; but in the formation of eggs (**oogenesis**) one cell yields only one egg and two or three tiny polar bodies, "dead-end cells" that play no further role in development. The object of oogenesis is not to produce four tiny eggs, but one big one with a large supply of the essential materials needed for development upon fertilization. The nutrients that might be divided among four cells are stored in one, the egg.

11.6 Fertilization is the union of a male and a female gamete or of two isogametes.

When fertilization occurs, the nuclei of two gametes, each with its haploid number of chromosomes, unite, and the normal diploid number is reestablished. It is obvious that fertilization represents still another method for mixing genetic material.

For some organisms—for example, marine invertebrates such as sea urchins, clams, and starfish—fertilization appears to be a wasteful process. Each adult uses a tremendous expenditure of energy to produce a large mass of eggs or sperm. Only a part of this is successfully joined into fertilized eggs, and subsequent predation on eggs, larvae, and juveniles is so great that a fraction of 1 percent of the original eggs may finally become adult organisms. Although this seems wasteful in terms of the individual, many species survive in this manner, a fact which argues for the system's ultimate success.

Many other animals, especially on land, have evolved methods of internal fertilization which avoid much of this waste.

11.7 Parthenogenesis is the development of unfertilized eggs.

Of the many organisms that undergo sexual reproduction, some produce eggs that develop without fertilization by a sperm, and this type of virgin birth is called **parthenogenesis**.

Colonies of bees contain both sexually produced and parthenogenetic animals. Both arise from eggs laid by the queen bee. The queen mates with a male drone only once and thereafter carries within herself a reservoir of sperm. The eggs she produces that are fertilized by the sperm develop into the diploid female workers (and possibly future queens), whereas the eggs that are deposited unfertilized develop into haploid drones whose role it is to provide sperm for the next queen.

Spontaneous parthenogenesis is characteristic of some higher animals as well. There are several varieties of lizards and fish in which males are unknown. Females can produce young despite prolonged isolation from other animals. Even in some strains of turkeys, eggs can develop parthenogenetically. The number that survive to maturity is small, and all that do are males, some of which have fathered offspring. Some eggs that are not naturally parthenogenic can be made to develop if stimulated chemically or physically—a discovery made in 1898 by J. Loeb.

12

Sexual Reproduction and Development in Plants

QUESTIONS TO KEEP IN MIND

How do the processes of mitosis and meiosis fit into the alternating sexual and asexual phases of plant reproduction?

In what ways do the sporophyte and gametophyte generations of *Ulva* (sea lettuce) differ?

Is a fern that you might pick in the forest in the sporophyte or gametophyte generation? How does it give rise to the other generation?

What male and female sex organs are present in a flower?

What structures constitute the male and female gametophyte generations in a flowering plant?

How do pollination and fertilization differ?

What are the principal parts of the embryo of a flowering plant?

How do the nutritional requirements of plant embryos change during development?

Of what advantage to a plant is the dormancy of its seeds? What factors may promote germination?

What are meristems? Where are meristematic tissues found in developing plants?

The advantages of sexual reproduction are as important to plants as they are to animals, but because higher plants are stationary they must find special ways to bring two gametes together to form a new individual. After fertilization, the development of an embryonic plant begins. Within a remarkably short time the single-celled fertilized egg develops into a fully formed plant.

12.1 Most plants alternate between sexual and asexual methods of reproduction.

Almost all types of plants reproduce in a cycle that includes both sexual and asexual phases. In general, the union of two gametes produces a diploid cell called a **zygote**, which develops into a diploid plant of the asexual **sporophyte** generation. The sporophyte plant is so named because it forms haploid **spores** by meiosis. Each spore produces a haploid plant of the sexual **gametophyte** generation, so named because it produces haploid gametes. The cycle begins again when two gametes fuse to form a zygote and reestablish the diploid number of chromosomes. The spores never fuse, but develop directly into plants by mitotic cell division.

A relatively simple example of the **alternation of generations** exists in the common alga *Ulva*, or sea lettuce. (Algae include some of the simplest forms of plants. They range in size from single microscopic cells to multicellular strands of kelp some 75 feet long.) *Ulva* grows in shallow portions of the sea as thin, nearly transparent green sheets of cells. Some of these plants are diploid sporophytes, while others are haploid gametophytes. At certain times of the year some of the cells of the diploid sporophyte divide and undergo meiosis within the old cell walls, forming numerous haploid spores that break out of the surrounding cell wall and move through the water by means of four flagella (Figure

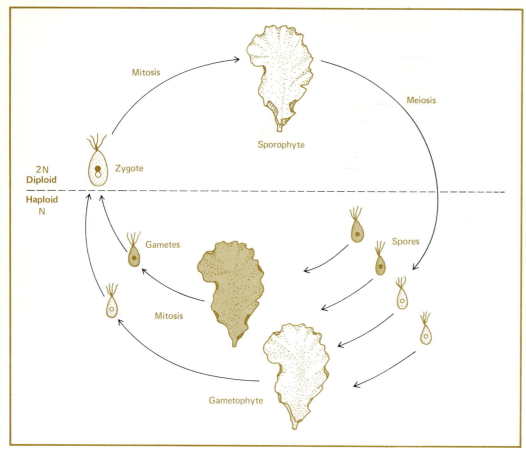

12-1 The life cycle of *Ulva*, sea lettuce. Color indicates mating types.

12-1). The motile spores all look alike and have no tendency to join together. If a spore settles down on a suitable surface, it produces an *Ulva* plant of the gametophyte generation, which is identical to the sporophyte from which it was originated except that it has a haploid number of chromosomes in its nuclei. Eventually, the haploid plants produce isogametes with only two flagella each. If isogametes from two different plants come into contact with each other, they form a zygote which can develop into a diploid sporophyte.

12.2 In land plants the alternation of generations is more complicated, in part because spores and gametes have more trouble moving through air than through water.

In nonflowering plants such as ferns, the plant you recognize is the diploid sporophyte. The spores develop after meiosis in the brown sporangia on the underside of the fern leaflets, and after the spores fall to the ground, they develop into small, flat, heart-shaped gametophytes (Figure 12-2). Gametes are produced in specialized structures on the undersurface of these miniature plants. One type of gamete, is the small, and motile **microgamete**, or sperm. The other is the large, nonmotile **macrogamete**, or egg. In ferns, both eggs and sperm are formed on the same gametophyte.

When the sperm are released (during a rainy period), they swim through water that surrounds the gametophyte. The eggs produce a specific substance that attracts them, and the sperm swim to the eggs and fuse with them. The zygotes that result grow and destroy the gametophyte. The new plants are the sporophytes.

12.3 In higher plants the reproductive organs are contained in the flower.

Plants living in terrestrial environments are frequently exposed to extremes of heat and cold and

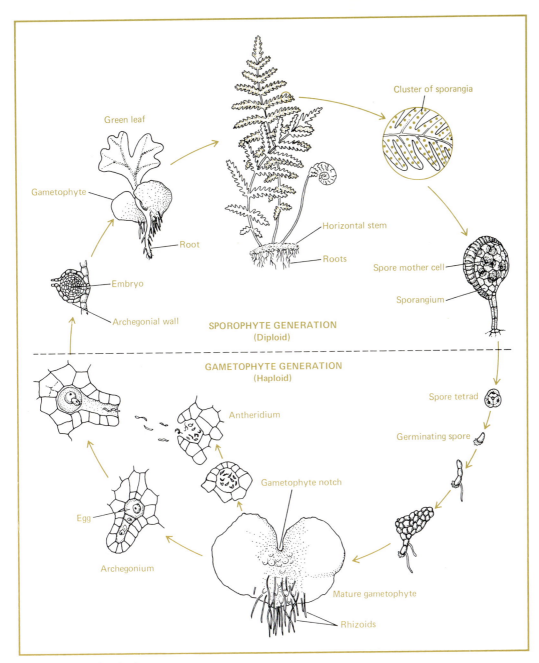

Cluster of sporangia

Green leaf

Gametophyte

Root

Horizontal stem

Roots

Spore mother cell

Sporangium

Embryo

Archegonial wall

SPOROPHYTE GENERATION
(Diploid)

- - - - - - - - - - - - - - - -

GAMETOPHYTE GENERATION
(Haploid)

Spore tetrad

Antheridium

Germinating spore

Gametophyte notch

Egg

Archegonium

Mature gametophyte

Rhizoids

12-2 The life cycle of a fern.

wetness and dryness, and although mature plants have means of adapting to these conditions, most zygotes could not survive if exposed to such extremes. In many plants, therefore, zygotes are contained in and protected by the tissues of the mature plant. Fewer gametes and zygotes are produced but each has a better chance of survival.

To furnish this protection special structures and mechanisms are required: flowers. A "perfect" flower contains all the elements required for sexual reproduction (Figure 12-3).

The accessory components of a flower are the **sepals**, which protect the bud during its formation, and the **petals**, which often give a flower its spectacular display. The essential components for reproduction include the male **stamens**, made up of **anthers** borne upon **filaments**, and the female **pistil**, with its **stigma**, **style**, and **ovary**. Microgametogenesis occurs in the anthers. It is within the ovary that macrogametogenesis occurs, and where fertilization and zygote formation take place.

In some species of plants, self-fertilization occurs

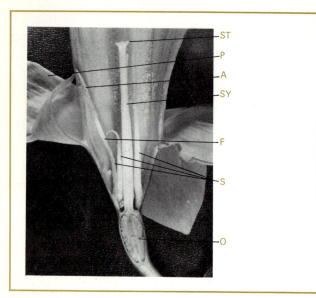

12-3 The flower of a daffodil cut open to show its parts: petals (*P*); stamen (*S*), consisting of an anther (*A*) on a filament (*F*); the pistil, consisting of the stigma (*ST*), style (*SY*), and ovary (*O*). (Photograph by the author.)

—that is, the microgametes from the anthers can fertilize the macrogametes in the ovary of the same flower. In others, "selfing" is prevented by physiological barriers such as the formation of the micro- and macrogametes at different times during the flowering season. In a third group of plants, male and female flowers are borne on separate plants, and selfing is impossible. Furthermore, some plants that reproduce only vegetatively produce sterile flowers.

Whether or not self-fertilization takes place, the gametophyte generation in flowering plants is not a separate plant but part of the flower. In fact, the male gametophyte is a **pollen grain,** and the female gametophyte is the **embryo sac.** Within the anther, four haploid pollen cells, or **microspores,** are produced by meiosis from each diploid **pollen mother cell.** This establishes the male gametophyte generation. The walls of the pollen cells thicken, and the nucleus divides mitotically to produce a mature pollen grain containing two nuclei. When the anther ripens it opens to expose the pollen grains, so that they can be carried to other flowers by the wind, insects, or other means.

The female gametophyte develops in the plant ovary, completely sealed off from the environment. In the ovary there are usually numerous growth centers that form structures called **ovules,** in which the female gametophyte tissue will form. At the center of the ovule, a **macrospore mother cell** pro-

duces four haploid macrospores by meiotic division. Only one of the four survives, and that one divides mitotically to form an eight-nucleated embryo sac (Figure 12-4B). Portions of the embryo sac cytoplasm segregate around six of the nuclei, forming cell-like units. One of these new cells will become the egg cell, which is formed toward one end of the embryo sac, near an opening in the ovule called the **micropyle.** The ovule cells which surround the embryo sac form the protective coats or **integuments.** Unlike the male pollen grains, the embryo sac does not move.

12.4 Pollination, the transfer of pollen grains from anther to stigma, precedes fertilization, which is the union of egg and sperm nuclei.

Shortly after a pollen grain lodges on the stigma of a flower, one of the two nuclei in the pollen divides to form two **sperm nuclei.** The pollen grain produces a tube that grows down the style towards the ovule. The two sperm nuclei move along just behind the advancing tip of the tube, a trip that ordinarily takes hours or days, but may take several months in some plants (Figure 12-4A).

The pollen tube passes through the micropyle of the ovule, and as it enters the embryo sac, its tip bursts and the two sperm nuclei move into the sac. One sperm nucleus joins with the **polar nuclei** of the ovule to form **endosperm,** a nutritive tissue that arises from the fusion of three nuclei and thus contains three or more times the haploid number of chromosomes. The other sperm nucleus enters the egg and fuses with the egg nucleus to form a diploid zygote.

12.5 Once fertilization occurs an embryonic plant begins to grow.

A striking characteristic of plant embryos is their capacity to develop without moving cells about. Multicellularity and **morphogenesis,** the establishment of form, are accomplished by cell division alone and, in the embryonic plant, form is created by differences in the rate of cell division in different regions and by differences in the planes of division which orient cells differently relative to the surface of the plant. (Once a plant is mature, its continued growth depends upon cell divisions that are usually concentrated in undifferentiated tissues called

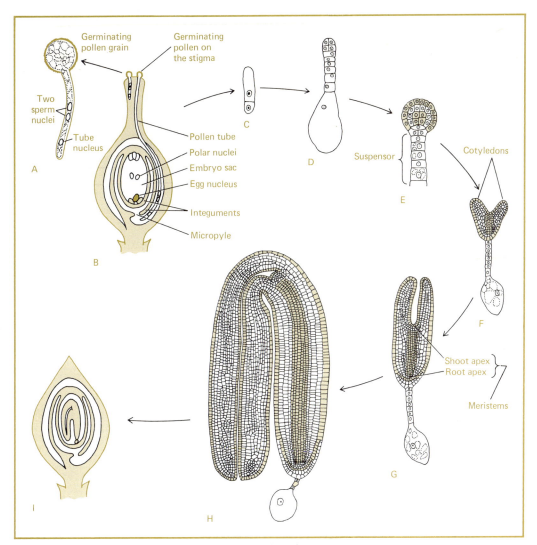

Germinating pollen grain

Germinating pollen on the stigma

Two sperm nuclei

Tube nucleus

A

Pollen tube

Polar nuclei

Embryo sac

Egg nucleus

Integuments

Micropyle

B

C

D

Suspensor

Cotyledons

E

F

Shoot apex

Root apex

Meristems

G

H

I

12-4 Fertilization and early development of a flowering plant. *A.* A germinating pollen grain. *B.* A pollen tube approaching the micropyle of the ovary of a flower, through which it must pass to fertilize the ovule. *C–G.* Stages in the early development of the plant embryo, showing the suspensor, cotyledons, and apical meristems. These stages all take place inside the ovule. *H.* The mature embryo, curved to fit into the ovule. *I.* Stage *H* shown inside the ovule.

meristems. These are, therefore, often considered persistent regions of localized embryonic development.)

Once fertilization has occurred, the endosperm nuclei divide rapidly to produce a large number of nuclei within the embryo sac. Later, cell walls appear between these nuclei. The endosperm cells thus formed contain most of the cytoplasm of the embryo sac. The endosperm is a nutrient tissue for the embryo. It is often liquid, as in the case of coconut milk.

After the endosperm is well established, the zygote begins to divide. It forms two structures— the embryo (Figure 12-4C–H) and a **suspensor** (Figure 12-4E), which pushes the embryo into the endosperm. The embryo develops rapidly, and the cells that will become the long axis of the developing plant are soon discernible. At one end of the axis is the embryonic shoot, or **epicotyl**, and at the other is the embryonic root, or **radicle.** The first embryo leaves or **cotyledons** (Figure 12-4F) will absorb the endosperm and transfer the stored material directly into the embryo. In the meantime, the integuments (coverings) of the ovule change in composition and texture to become the tough seed coat which helps protect the embryo in the seed until upon germination it is ready to produce a young plant or **seedling.**

12.6 As embryos develop their requirements become simpler.

At one time, only relatively mature embryos could be reared successfully in cultures, for the younger the embryo, the more complicated the medium it requires. Then solutions containing coconut milk were found to support development of early embryos because coconut milk is a form of liquid endosperm and contains plant hormones and organic nutrients embryos need. More recently these hormones have been identified and purified, and now very young plant embryos can be grown in purely artificial media.

These experiments indicate that the synthetic capability of a very young embryo is extremely limited. The embryo relies upon its environment for the complex as well as the simple substances it needs. As its development advances, so does its synthetic capacity. Ultimately the embryo satisfies its need for complex materials by manufacturing them from simpler ones.

12.7 Mature embryos often enter a dormant period.

An embryo within a seed is often capable of surviving conditions that kill the plant that produced it, and consequently many plants use seeds not only for reproduction but also for bridging unfavorable environmental conditions. Some seeds can lie dormant for many years and still germinate. When lotus seeds over 1000 years old were planted, they germinated and produced mature lotus plants.

Dormancy prevents seeds from sprouting at the wrong time. In our temperate climate many seeds formed in late summer lie inactive on the ground next to the plant that produced them. Conditions are still satisfactory for the mature plant, but the seed does not germinate until subjected to prolonged cold. Such seeds are thus protected from germinating in autumn or during a warm spell in January when the following could would kill them.

In the desert some plants are very short-lived, thriving only during a brief rainy season. Such plants drop seeds which will not germinate until a certain amount of rain has fallen during the following rainy period. The water in the soil is then sufficient to support the new plant until it matures and drops seeds for the next generation. The device in the seed that seems to measure rainfall is a water-soluble chemical that inhibits germination. Only after sufficient rain has fallen to dissolve and wash away the chemical inhibitor can germination occur.

12.8 Dormancy over, the seed germinates, and the embryonic plant develops a root.

The germination of seeds involves growth of the root, the shoot (or stem), and the leaves. Once begun, the development and expansion of these regions continues for the life of the plant and may be interrupted only by periods of seasonal dormancy. Each season, a plant adds new stem, new leaves, and new roots and, in woody plants, increases the stem diameter. It appears that this persistent growth and development often involves the same processes that take place in the embryo.

Roots develop first. The normal development of

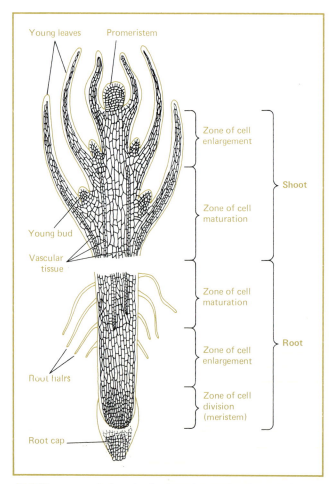

12-5 Diagrammatic longitudinal section of the shoot and root of a flowering plant.

a young plant usually requires that the root be established in the soil, where it can extract water and minerals. The cells at the root tip are the youngest, and those closer to the base of the root are oldest (Figure 12-5). The **root cap** at the end appears to be a living shield which spares the more organized, sensitive region behind it from damage by soil particles.

Behind the cap is the **zone of cell division**, the root meristem, where practically all new cells in a growing root are formed. Although this is the only zone of great mitotic activity, it is *not* the place in the root where the most conspicuous growth takes place. Immediately after division the cells in the root meristem are cube-shaped, and it is only after more divisions have displaced them to a position some distance back from the tip that the cells elongate, thus making the root lengthen. Cell elongation and the uptake of water provide the force that pushes a root through the soil. If marks are placed on growing roots, the **zone of elongation** becomes apparent.

Above the zone of elongation is the **zone of maturation**, where the cells differentiate into specialized types (section 13.2). This occurs by selective gene action, meaning that differentiating cells use only a part of their genetic information. After cells differentiate, they normally retain their specific characters for the life of the plant.

Even though the root of a mature plant is composed of differentiated cells, the meristem at the tip of the root (the apical meristem) retains the same embryonic character that existed in the germinating seed. Certain circular layers of cells within the mature root also retain meristematic activity. Such a circular layer is a **cambium.** There is usually a **cork cambium,** which forms and maintains the bark of the root, and a **vascular cambium,** which continually adds cells to the conducting tissue. Thus a root can continually grow in length and circumference.

12.9 Shoot development, like root development, occurs in an apical meristem.

The growth that occurs at the tip of the developing shoot produces new tissue for both stem and leaves (Figure 12-5). This region is recognized as the plant organization center. The basic organizational activities of shoot development take place in the dome-shaped region at the very tip, which is usually referred to as the **promeristem.** The types of cells within the promeristem differ in different plant groups.

As in the roots there are other regions of shoot growth, namely **cambia** (Figure 12-6). In the stem of a tree or woody shrub, there is the cork cambium, located just beneath the bark and producing bark. Deeper within the stem is a second meristematic

12-6 Diagram of a radial solid sector of a woody stem showing the meristematic layers (*in color*) and the conductive and supporting elements they give rise to during growth of the plant.

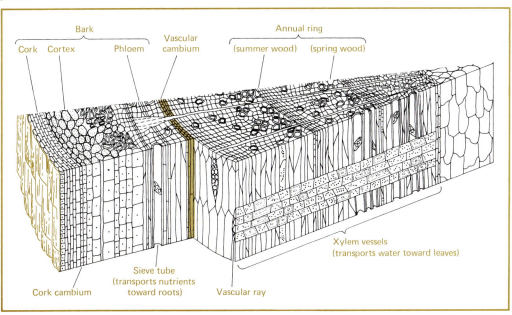

Bark
Cork   Cortex   Phloem   Vascular cambium
Annual ring
(summer wood)   (spring wood)
Xylem vessels (transports water toward leaves)
Cork cambium
Sieve tube (transports nutrients toward roots)
Vascular ray

cylinder, the vascular cambium. The division plane of its cells is usually parallel to the outer surface of the stem. The daughter cells that lie inside the ring of vascular cambium form the **xylem** tissues. These move water upward from the roots. Xylem tissues are added in concentric rings and constitute the main bulk of the tree. In temperate climates where temperature and rainfall fluctuate seasonally, the annual growth periods are distinct from one another, and the xylem appears organized into a series of concentric **annual rings** in a cross section of a stem. The daughter cells that lie outside the vascular cambium form **phloem** tissue. This moves fluids, primarily sap, downward from the leaves. The rate of cell division in the cambium depends upon temperature and available moisture.

13

Sexual Reproduction and Development in Animals

QUESTIONS TO KEEP IN MIND

What are the two major functions of the gonads?

What effect does photoperiod have on reproductive cycles?

How does the structure of a sperm differ from that of an egg?

What events occur in a sperm when it makes contact with an egg?

How does an egg respond to the attachment of the fertilizing sperm and others that arrive subsequently?

What effect does the quantity of yolk in an egg have on the way it becomes a blastula?

How does gastrulation differ in sea urchins, frogs, chicks, and mammals?

What are extraembryonic membranes?

How is the body plan of a vertebrate embryo established by the positions of the three primary germ layers after gastrulation?

What are the roles of the three primary germ layers in the formation of the major organs of the body?

Animals, like plants, have adapted themselves to almost every type of environment and have therefore evolved a vast variety of reproductive methods. A few examples will be given here, followed by a more detailed discussion of the development of animal embryos from the zygote to the mature individual than was given in Chapter 11.

13.1 Animal gametes (sex cells) develop in the gonads; gonads sometimes also produce hormones.

Even in earliest embryonic stages a great many animals set aside future sex, or **germ,** cells from the general, or **somatic,** cells of the body. Later, gonads, organs in which these germ cells develop, are formed. Female gonads are called ovaries, male gonads, testes. It is in these that the mature gametes —eggs and sperm—develop.

The degree of complexity and permanence of the gonads varies greatly among animals. When the gonad is permanent, it frequently serves as an **endocrine organ**—a site of hormone production—as well as a place in which gametes develop. In other animals the gonads may only function seasonally or in extreme cases (that is, in some tunicates, primitive chordates) may disappear completely between breeding seasons.

In most animals, the gonads persist for the life of the individual but show seasonal variations in their activity. (Primates—monkeys, apes, and man —are among the exceptions to this rule, as they are able to breed at any time of the year.) In many animals reproductive activity is correlated with the period of daylight, or **photoperiod.** Birds are among these: It has been found that the breeding cycle of captive birds can be controlled by manipulating the apparent day length with artificial lighting. Animals with this kind of regulation have become adapted to the seasons in such a way that their young are produced when the environment is least hostile to them. Thus, breeding in rapidly developing ani-

BIOEPICUREAN DELIGHTS

Sautéed Shad Roe

Shad are an ocean fish that come up rivers to spawn in late winter or early spring. At this time their ovaries are swollen and packed with eggs, called roe. Two pairs of ovaries will serve four. Slit and remove the ovarian membrane, and sprinkle the roe with salt and pepper; then dip in flour. Melt a stick of butter in a skillet of appropriate size and wait until the foam subsides, then sauté the ovaries 6 to 7 minutes on each side, taking care that the heat is not excessive. The roe should brown gradually and evenly. To make a sauce, add one teaspoon of Worchestershire sauce, the juice of half a lemon, chives, parsley, and dill to the butter in which the roe has been sautéed, and pour over the roe. Add slices of crisp bacon or capers for flavor.

mals such as birds and small mammals occurs in spring and early summer. The young are born soon after and are able to cope with the following winter. Some larger animals, such as deer, however, breed in the fall. Their offspring grow and develop for several months in the **uterus** or womb, are born in early spring, and mature enough during the summer so that by the following winter they have a reasonable chance for independent survival.

13.2 The ovary is the female gonad wherein the germ cells form ova or eggs.

When eggs are mature, they are different from all other cells of the female. They consist basically of a haploid nucleus, some "active" cytoplasm, and varying amounts of stored nutrient material which may be lumped under the term **yolk.** Yolk contains fats, proteins, and carbohydrates; it supplies the embryo with the large amount of energy and raw materials it requires for development.

The size of eggs varies tremendously. Some jellyfish eggs are only about 30 micrometers in diameter, while the volume contained in an ostrich egg is about a quart. (African bushmen use ostrich egg shells for water storage.) The manner in which the yolk is distributed (Figure 13-1) also varies. In sea urchin eggs, the yolk and active cytoplasm are almost homogeneous. In frog eggs, the nucleus and most of the active cytoplasm are in the **animal hemisphere,** which is the upper hemisphere; most of the heavier yolk is in the lower **vegetal hemisphere.** The pattern of yolk distribution has little or nothing to do with the appearance of the adult organism.

13.3 The male gonad, or testis, is the site of the production of sperm.

Sperm structure varies so greatly among animals that the shape and size of a sperm can be used as a positive method of species identification (Figure 13-2). Most sperm are motile. Their motility is usually achieved with a flagellum.

The sperm's haploid nuclear material is typically borne in an enlarged head. At the tip of the head in nearly all species is a small body, the **acrosome,** which contains enzymes that play an important role in egg-sperm union.

13.4 Fertilization is a complicated process in which egg and sperm interact with each other.

In some animals, it is the sperm that shows the first visible response to the proximity of the egg, by undergoing an **acrosomal reaction.** Microfilaments of some acrosomes are somehow rapidly reorganized to produce the **acrosome filament,** which impales the egg. The sperm of some species also produce enzymes which assist in penetrating the jelly and other extraneous layers of the egg.

The egg's response to penetration is the formation of a **fertilization cone** of outer egg cytoplasm (Figure 13-3) that appears to climb up the sperm

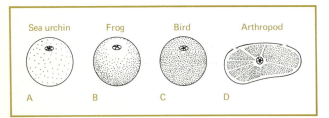

Sea urchin	Frog	Bird	Arthropod
A	B	C	D

13-1 Yolk distribution in different kinds of eggs. *A.* Isolecithal (yolk evenly distributed). *B.* and *C.* Telolecithal (yolk at one end). *D.* Centrolecithal (central yolk mass surrounded by a thin layer of clear cytoplasm). Approximate sizes: *A.* 0.1 mm; *B.* 2 mm; *C.* 20 mm; *D.* 0.5 mm.

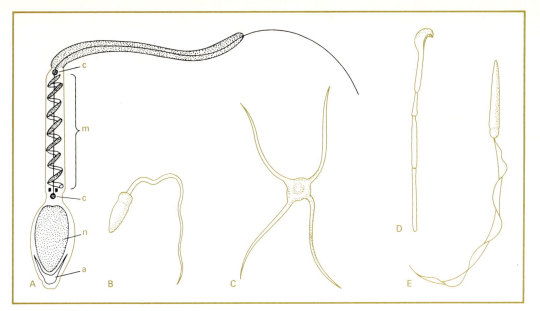

13-2 Some different types of sperm and the anatomy of a mammalian sperm. *A.* Mammalian sperm: acrosome (a), nucleus (N), proximal (near) and distal (far) centrioles (c) and midpiece (mp), mitochondrion (m), tail (t). *B.* Sea urchin. *C.* Crayfish. *D.* Guinea pig. *E.* Toad.

13-3 A diagram showing the reactions of sea urchin sperm and eggs at fertilization. The details vary even in closely related animals. *Top:* acrosome reaction in sperm. *Bottom:* the cortical reaction leading to the formation of the fertilization membrane.

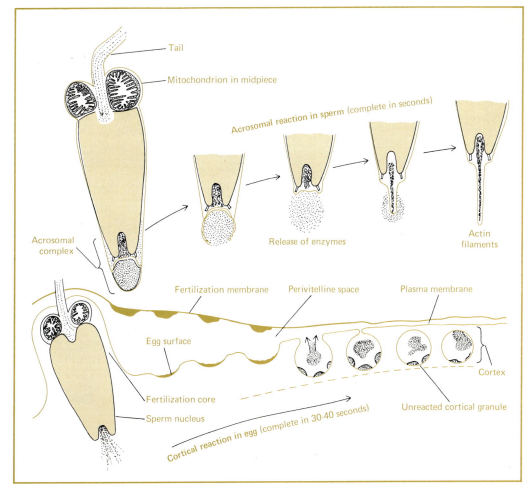

head and engulf it. The surface layer, or **cortex,** of the egg responds to the activating effect of the sperm by a series of dramatic structural and chemical changes called the **cortical reaction.**

Within minutes after sperm penetration a great variety of changes in the structure and action of the egg take place. The rate at which it consumes oxygen and fluids usually changes (increasing in some species, decreasing in others). In some species a **fertilization membrane** rises from the egg surface and forms a protective capsule. In most species the egg also loses responsiveness to additional sperm.

It is convenient to consider that the sperm in fertilization has two roles: First, the sperm activates development; second, the sperm contributes its haploid set of chromosomes to form a diploid individual. When sperm are injected directly into eggs, development is not activated. The intact sperm thrashes about in the cytoplasm for some time, but no development takes place. It appears that the early events at the egg surface—**activation** —are essential to development.

The many ways in which sperm and egg are brought together generally fall into two broad categories—**external fertilization,** where the gametes of aquatic animals are shed together; and in-**ternal fertilization,** where copulation introduces sperm into the body of the female.

13.5 After fertilization, the single-celled egg converts itself, by cleavage, into a population of cells called a blastula.

The basic process of cell division during the early stages of embryo development (**cleavage**) is essentially the same as it is in the adult. Cytokinesis follows mitosis. However, cleavage differs in that the daughter cells remain smaller than the parent cell, and the division rate is usually much faster (sometimes only 15 minutes between cleavages).

The pattern of cytokinesis varies among embryo species. The most important factor in determining the different patterns appears to be the way in which the yolk is distributed.

Sea urchin eggs have a modest amount of yolk that is fairly evenly distributed throughout the egg. The nucleus is near the center, and the first two divisions produce cells of approximately equal volume (Figure 13-4). These cells or **blastomeres** divide synchronously. After the fourth division, one group of 4 cells, the **micromeres,** is much smaller than the other 12 cells. Their size is not correlated with yolk content but rather with some property of the egg surface. The blastomeres all have some part of their surface exposed to the outside and form a loose, hollow ball called a **blastula.** The space in the center of the ball is filled with liquid and is called the **blastocoel.**

13-4 Cleavage of a sea urchin egg into 2, 4, 8, 16, 32, and 64 cells; and a ciliated blastula. The cleavage planes for the first three divisions are shown.

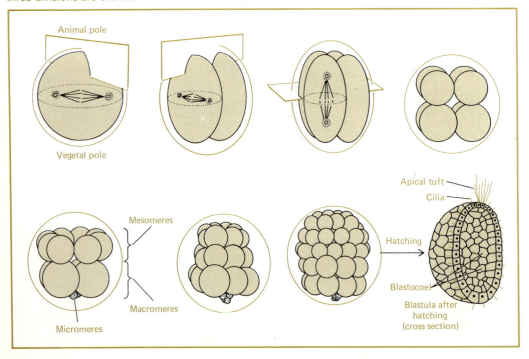

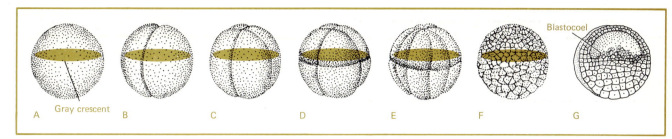

13-5 Cleavage of an amphibian egg. After fertilization, a gray crescent (*in color*) is formed opposite the site of sperm penetration when some pigment moves toward the animal pole. The gray crescent marks the future tail of the embryo.

Amphibians such as frogs and salamanders produce eggs with more yolk particles than do sea urchins. The yolk is more densely concentrated at the **vegetal pole,** whereas the nucleus lies closer to the **animal pole.** The first cleavage furrow slowly divides the egg along the animal-vegetal axis, the animal pole being at the top and the vegetal pole at the bottom in Figure 13-5. The second division begins before the first furrow has divided the egg into completely separated cells. The furrows of the third cleavage are at right angles to the plane of the first two and somewhat above the equator. Subsequent cleavages systematically divide the egg substance

13-6 Cleavage in the blastodisc of a bird's egg.

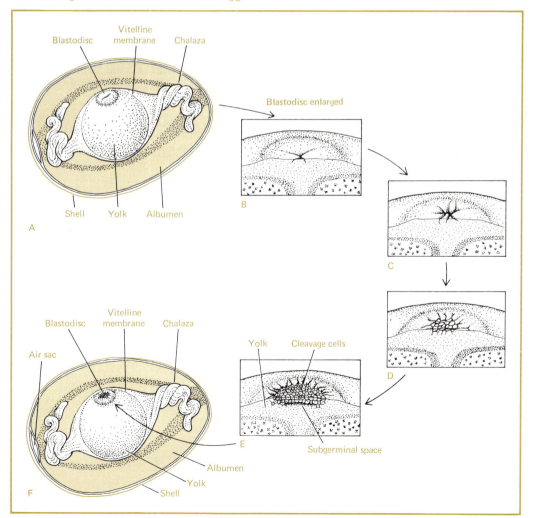

into smaller units, but the division rate is slower in the vegetal area. It has been suggested that the furrowing process in the vegetal area may be physically impeded by the large amount of yolk. Whatever the mechanism may be, the slower division rate of vegetal cells results in a gradation of cell size from animal to vegetal pole. Internally a fluid-filled blastocoel develops.

The cytoplasm of a bird's egg is a tiny, yolk-free **blastodisc,** about a millimeter in diameter. It is the tiny white speck that usually rotates to the top side of the yolk when an egg is in a frying pan. The blastodisc undergoes several divisions before the egg is laid. The early cleavage planes are at right angles to the yolk surface, and the embryo develops as though it were attempting to live in only two dimensions (Figure 13-6).

As cleavage continues and cells become smaller, the disc becomes several cells thick. Soon the lower cells of the disc form a loose but definite layer, the **hypoblast,** which is separated from the upper cell layer, the **epiblast,** by a fluid-filled space. This is the chick's equivalent of the blastocoel (Figure 13-7). During this period the thick, soupy yolk has sagged away from the blastodisc, leaving the fluid-filled **subgerminal space.**

The egg of a mammal resembles that of the sea urchin in that the small amount of yolk is evenly distributed and the entire egg divides into a group of cells of fairly uniform size. In humans and some other mammals, however, these cells are very soon partitioned into two groups. One is the source of cells for the embryo, the other for the **extraembryonic** membranes and the **placenta,** necessary for the survival of the embryo in the body of its mother.

In mammals, fertilization takes place near the mouth of the oviduct, the passage leading from each ovary to the uterus, and the embryo is developing as it moves down to the uterus. The solid ball of cells produced by cleavage acquires fluid-filled spaces which fuse to form a relatively large central cavity. The embryo mass at this time is called a **blastocyst.** The cells which will give rise to the entire embryo are concentrated in a knoblike thickening, the **inner cell mass,** in the blastocyst wall (Figure 13-8). The other cells of the blastocyst constitute the **trophoblast** region. These will give rise to the extraembryonic membranes and placenta.

After the blastocyst has implanted in the uterine wall, changes occur in the inner cell mass. A space, the **amniotic cavity,** appears between the inner cell mass and the blastocyst. Cells of the mass that are adjacent to the blastocyst cavity grow down around the inside of the trophoblast, making a new inner layer, the **yolk sac.** Part of the inner cell mass has become a disc which separates the amniotic cavity from the larger yolk sac. This disc is called the **embryonic shield.** Only this part will give rise to the embryo.

Thus far, development in the groups described has hardly been spectacular. Cell division has led

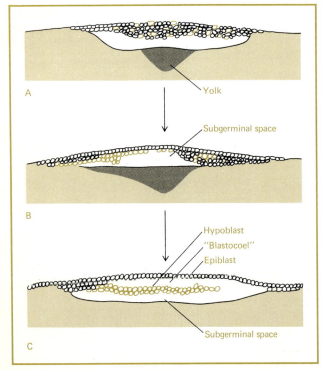

13-7 The formation of the first two embryonic cell layers in the bird egg.

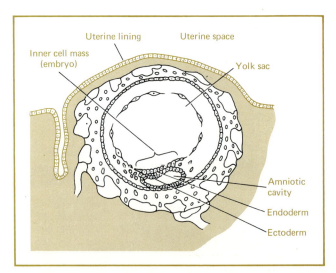

13-8 Diagram of a mammalian embryo about 12 days old that has implanted itself in the wall of the uterus.

to the formation of a blastula or similar structure, and this sets the stage for the remarkable processes that follow.

13.6 Gastrulation is the process whereby embryos establish a gut and subsequently reorganize themselves into three germ layers.

Most multicellular animals are constructed as a tube within a tube—a gut surrounded by a body. The gut tube may open to the outside at both ends or only at one end, but in either case an animal can take in chunks of food, hold them until digested, and expel the insoluble remains either through the opening by which they entered (as in *Hydra*) or, in more complex organisms, through the far end, where they leave through an anus.

Although the gut is not used as such until late in embryonic development, its early establishment is an important event that clearly reveals the future symmetry of the animal. The process whereby embryos reorganize themselves into several layers as they establish the gut is called **gastrulation,** and the embryo at that stage is called a **gastrula.**

The simplest way to make a sphere into a tube within a tube is to push one side inward as the whole structure elongates. In many animals such as sea urchins with large blastocoels, formation of

the embryonic gut, or **archenteron,** closely resembles this simple process (Figure 13-9). Cell division is practically absent during gastrulation. Part of the process is due to the independent activity of inward-moving cells. As the gut of a sea urchin embryo develops, long, thin pseudopodia, or **filopodia,** can be seen joining the tip of the archenteron to the wall at the animal pole. These can shorten and develop some of the force required for gastrulation.

The formation of the archenteron not only establishes the embryo's basic body plan, but also reorganizes the embryo into an inner and an outer layer. A third layer then develops between them from a loose grouping of cells that move in around the blastopore (Figure 13-9). The sheet of cells forming the embryonic gut wall is called the **endoderm** ("inner skin"); the layer of cells left on the outside after the gut is formed is the **ectoderm** ("outer skin"); and the group of cells which comes to lie between the ectoderm and endoderm is the **mesoderm** ("middle skin"). The three cell sheets are usually referred to as the embryonic **germ layers.**

These germ layers develop into similar structures in a wide variety of animals, although it is not true that in all animals the same structures always arise from the same germ layer. In general, however, the lining of the gut and of the glands and organs associated with it develop from the endo-

13-9 Sea urchin development from gastrulation to the pluteus larva.

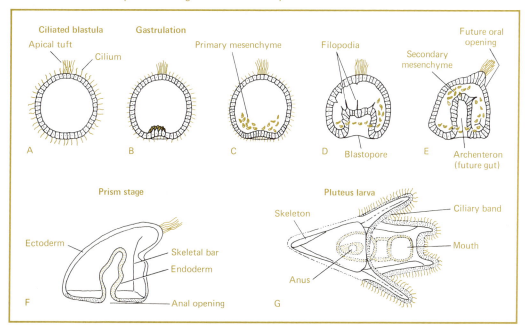

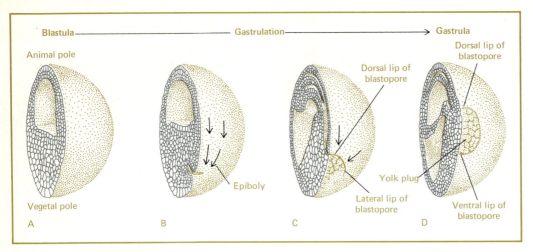

13-10 Gastrulation in an amphibian egg.

derm; the blood and circulatory system, and the muscles, skeleton, and other supporting tissue develop from the mesoderm; while the nervous system, skin, and similar outside structures are formed from the ectoderm.

Thus gastrulation in sea urchins has established a body plan typical of many animals. The opening of the gut, established as a part of gastrulation, is called the **blastopore.** In the sea urchin, but not all animals, it persists throughout life as the anus. The mouth forms as a new opening, where the anterior end of the archenteron attaches to the body wall (Figure 13-9). Sea urchin eggs develop within a few days into a **pluteus larva.** Further development into an adult sea urchin takes a much longer time and involves sweeping changes in body form, or **metamorphosis.**

The gastrula of frogs and salamanders in general resembles that of the sea urchin, but the details of gastrulation are different. In the amphibians a line of cells below the equator of the egg moves inward, creating a notch or fold on the surface. (Figure 13-10)

Gastrulation begins at a spot containing unpigmented cytoplasm, called the **gray crescent,** located just below the egg equator, usually opposite the point where the sperm entered. This place is also the future dorsal (back) part of the blastopore. The blastopore is posterior and its dorsal lip can be determined; the anterior, posterior, dorsal, ventral, right, and left regions of the embryo are also easily determined.

The line of inward-moving cells is arc-shaped, and as gastrulation becomes more extensive the arc elongates and eventually forms a circle. The cells which turn in at the blastopore are still connected to those which have already moved in and to those which are still outside. Thus gastrulation is accomplished by a sliding sheet of cells.

By the end of gastrulation, the cells of the amphibian embryo are profoundly rearranged. The outside of the completed gastrula, the ectoderm, comes from cells that previously covered about half of the blastula surface. Considerable stretching must take place. The cells which passed over the dorsal lip of the blastopore form the temporary roof of the archenteron and will become lined by endodermal cells which move up the archenteron walls. Most of the cells which move in over the lips of the blastopore later become mesoderm.

The blastocoel, meanwhile, has been pushed out of position and compressed. It gradually disappears. It serves in amphibia, as did the sea urchin, as a place into which cells could move during gastrulation.

Although the yolk mass changes its appearance superficially, the amphibian gastrula is basically like that of the sea urchin. But the mechanisms whereby these two kinds of eggs accomplish this drastic form change are obviously very different.

In birds, reptiles, and some mammals the behavior of the discoidal blastodisc during gastrulation resembles in a general way that of the blastula in sea urchins and amphibians, but the flattened nature of the blastodisc requires certain differences. Recall that the layer of cells lying above the "blastocoel" is called the epiblast, while that forming its floor is called the hypoblast (Figure 13-7). In gastrulation, the hypoblast does not move much; it becomes the endoderm. The epiblast gives rise mainly to the ectoderm and mesoderm.

At the future posterior end of the blastodisc, the

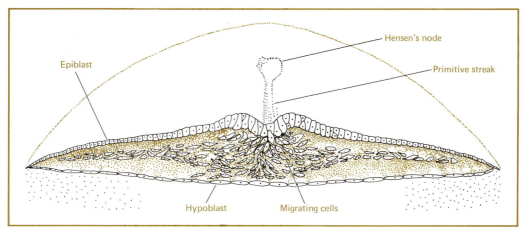

13-11 Gastrulation in a bird egg.

cell layers are somewhat thicker, and gastrulation involves the shifting of the right and left halves of the epiblast in that area toward each other. They meet in what will be the animal's midline (Figure 13-11). In the line where the two sheets of cells meet there forms a shallow, elongate groove terminating in a depression known as **Hensen's node** with a low ridge on each side. This groove-like structure is the **primitive streak.** It is the place where cells moving from each side meet and migrate downward. They then move out toward the edges of the blastodisc between the epiblast and the hypoblast.

Most of the cells moving downward at the primitive streak remain in a sheet between the epiblast and the hypoblast. This sheet is the mesoderm. A few of the internal cells join the hypoblast, which is then known as the endoderm. The cells which do not turn inward but remain on the upper surface constitute the ectoderm. The primitive streak is thus the equivalent of the blastopore, and the embryo forms anterior to the primitive streak.

In this description of germ layer development, similarities have been emphasized. There are, however, many animals that gastrulate by different means, although the end result is similar. Blastocoels, for instance, are not always present. In embryos where this is the case, small cells derived from the egg's animal pole slide down over the larger cells of the vegetal area. The gut may be formed by a population of cells which move inward or as a solid group and only later form an archenteron cavity with an opening to the outside. It appears that the form of the embryo after gastrulation, rather than the way it gastrulates, is fundamentally important.

13.7 After gastrulation, tissues differentiate and organs develop.

Immediately after gastrulation one can predict which cells will become brain, skin, and so forth, but such predictions are based upon the position of the cells rather than on any known special feature of form or chemical composition.

After gastrulation, these cells rapidly differentiate and develop into organs. This process is known as **organogenesis.** It will be described as it occurs in vertebrates. Keep in mind that although you can read about changes in organs only one at a time, many important events occur simultaneously in the embryo.

13.8 The central nervous system develops from the ectoderm.

A primary contribution of the ectoderm to the developing vertebrate embryo is the formation of the nervous system, notably the formation of the central nervous system with its brain, spinal cord, and peripheral nerves. In most vertebrates the hollow dorsal nerve cord is established when a part of the dorsal ectoderm forms an elongated plate which rolls up to form a tube. In frogs and salamanders this is a spectacular event, some aspects of which can be seen without a microscope (Figure 13-12).

On the future dorsal side, a flattened shield-shaped area marked off by low ridges appears shortly after the yolk plug is pulled into the blastopore. This is the **neural plate.** It will give rise to

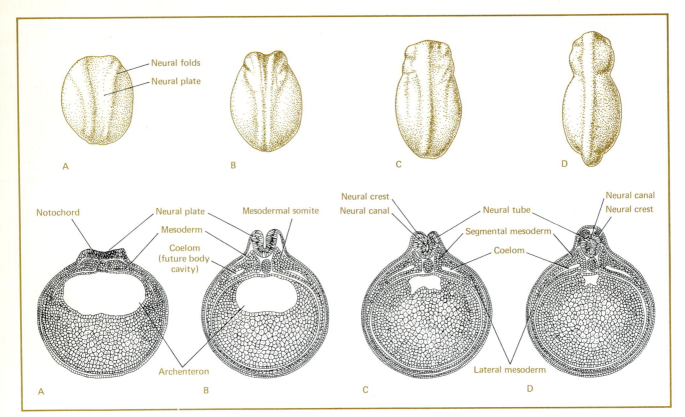

13-12 Stages in neurulation (formation of the neural tube) and in establishment of the notochord and mesoderm in an amphibian embryo. *Top:* Whole mounts; *bottom:* cross-sectional views.

the brain and spinal cord. As the ridges (**neural folds**) on either side shift toward the center, the middle of the neural plate drops to form a longitudinal trough at the midline. The ridges meet in the middle, converting the plate into a **neural tube** lined with cells that were formerly on the outside of the gastrula.

During neural tube formation, the cells of the plate undergo changes in shape. Soon after the tube closes, different parts expand at different rates, so that swellings, constrictions, and protuberances appear. The most striking protuberances are the paired **optic vesicles,** or future eyes, which appear at the anterior end of the brain.

13.9 The heart develops from patches of mesoderm.

Vertebrates have but one heart, which lies approximately in the ventral midline. The heart usually develops by fusion of a pair of tubular structures that have developed from patches of mesoderm. After gastrulation these structures lie just beneath

the anterior ridges of the neural plate. As these patches slowly move toward the midline, each develops into a thin-walled tube. These tubes fuse into a single tube when they meet in the anterior part of the embryo's thorax, or chest region, (Figure 13-13). Shortly thereafter, the heart begins to beat. While it beats, it curls back upon itself. Thereafter, partitions form, and the chambers are created.

13.10 The notochord, somites, and limbs develop from mesoderm.

The mesoderm forms a tubular supportive structure beneath the neural tube called the **notochord.** This is the structure peculiar to the phylum Chordata, for which the phylum is named. In higher vertebrates its supportive role is later taken over by the vertebral column.

On either side of the notochord are blocks of mesodermal tissue called **somites** (Figures 13-13 and 13-14), which soon develop into bone, cartilage, muscle and the dermis layer of the skin.

In the regions where the fore and hind limbs or arms and legs will develop, there appear mounds of mesodermal tissue, the **limb buds.** These grow

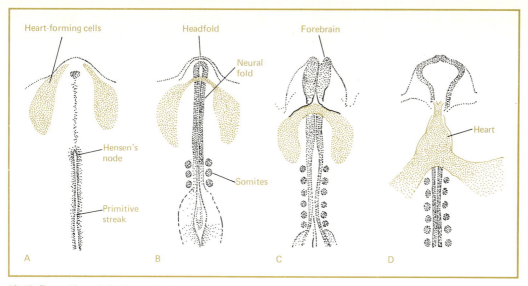

13-13 Formation of the heart in the chick embryo. The rudimentary parts of the heart are shown in color.

13-14 *Left:* A 48-hour chick embryo, showing the elements of the nervous and vascular systems and the paired somites. *Right:* Cross sections at the levels of the fourteenth and eighteenth somites.

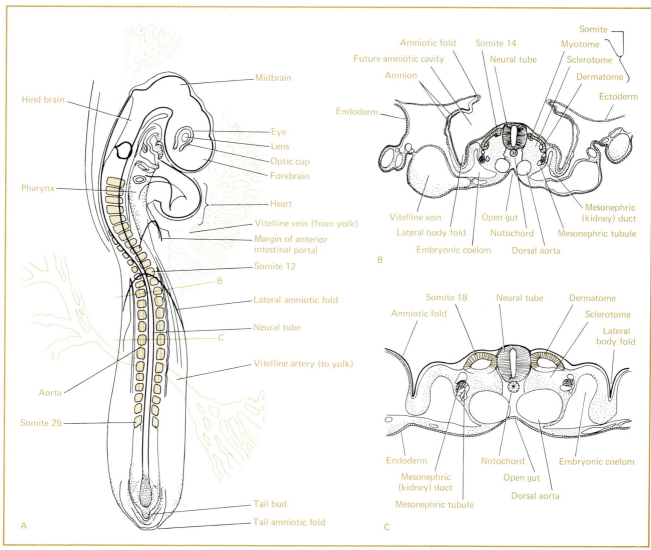

into shapes resembling cones or paddles; cartilages appear that are the forerunners of the long bones, smaller bones, and digits. The shaping of the limbs and digits is accomplished in part by localized **cell death** (Figure 13-15).

13.11 The gut forms from the endoderm.

In embryos that gastrulate in the amphibian pattern, the gut is in its tubular form shortly after gastrula-

tion. In birds and mammals, however, its formation is delayed. The gut does not remain a simple tube for long. In different regions it evaginates (turns outward) to form buds that will develop into the respiratory system (larynx, trachea, bronchi, and lungs) and the digestive glands (liver, gall bladder, and pancreas), as shown in Figure 13-16. In the anterior region of the guts of all vertebrate embryos, **gill arches** and **gill clefts** form. In fishes these structures remain throughout the life of the animal; in amphibia they function only until metamorphosis; and in mammals, including man, the gill arches represent a temporary phase in the development of various parts of the throat and middle ear.

13.12 The development of organs requires the cooperative interaction of embryonic germ layers and tissues.

It is important to emphasize that it is the *lining* of the developing gut that is derived from the original endoderm and forms buds that develop into the *lining* of the respiratory system and digestive glands. The muscular walls, connective tissues layers, and blood supply of the visceral (body

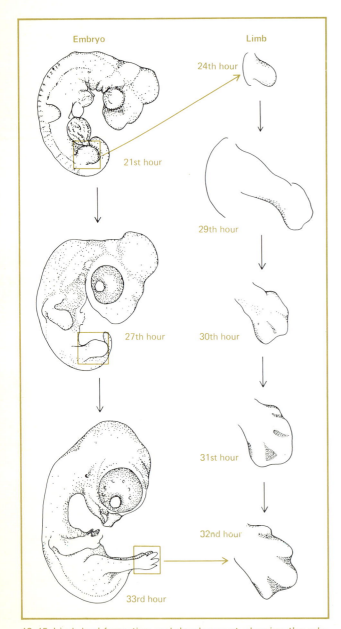

13-15 Limb bud formation and development, showing the role of local cell death (*in color*) in forming the digits.

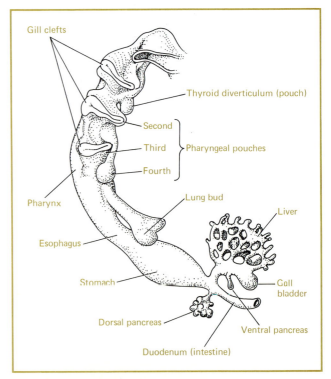

13-16 The various structures formed as buds from the embryonic gut of a mammal.

cavity) organs are derived from the mesoderm. Similarly, the bulk of the nervous system, which is of ectodermal origin, develops from the neural tube, which consists almost exclusively of differentiating nerve cells or **neuroblasts.** As development proceeds, the nervous system is supplied with blood and connective tissue derived from the mesoderm. The largest component of the nervous system, the brain, is an organ derived in part from the neural tube. Its vascular structures, however, originate in the mesoderm.

The development of virtually every organ in the body of a vertebrate results from interactions between mesoderm and endoderm or between mesoderm and ectoderm.

14

Principles and Processes of Development

What is meant by the ability of an egg (or one of its blastomeres) to "regulate"?

How is it known that a cell has differentiated? Is the process irreversible in both animals and plants?

What is differentiation? Do the nuclei of cells, in developing embryo, merely specialize in reading the genes of the species, or do they actually "forget how"?

If a sea urchin egg is bisected, does it matter to the larvae that will develop in what plane the egg is divided?

What is meant by the establishment of polarity?

What is embryonic induction? How do tissues of the developing embryo influence one another?

What role do hormones play in developmental processes?

In Chapters 12 and 13, development in representative plants and animals was described and illustrated without any attempt to analyze the principles and processes involved. This chapter will outline some of the more important generalizations that can be drawn from experimental analysis of the processes of development. The frontier of research in developmental biology is largely concerned with the analysis of cellular and molecular mechanisms underlying the processes that can be seen in embryonic development.

14.1 The genome of a developing organism is like a library from which groups of cells read different instructions at different times and therefore develop differently.

The DNA of a zygote contains more genetic information than is ever used by any one cell of a developing embryo, yet each cell contains a complete genetic complement, called its **genome.** Very early in development some cells of animal embryos retain the ability to **regulate,** that is to be able to read any part of the entire genetic library or genome and follow its directions for the synthesis of proteins required for development. Consequently, a sea urchin egg, if bisected in a certain way, will develop into two normal pluteus larvae and eventually into two complete sea urchins (Figure 14-1A). The cells of mammalian embryos also retain their ability to regulate until about the time of implantation into the wall of the uterus. Until this time two embryos can be fused, or any part of an embryo can form either placenta (an extra embryonic membrane) or embryo proper. Twinning can also be induced experimentally.

Usually, however, cells gradually lose the ability to regulate as the embryo develops: Cells in different parts of the embryo lose their ability to read some parts of the genome. These cells are said to **differentiate;** that is, they become different by expressing different genes. As will be shown in greater detail in Chapter 17, genes control the synthesis of enzymes and other proteins, the presence or absence of which makes one cell different from

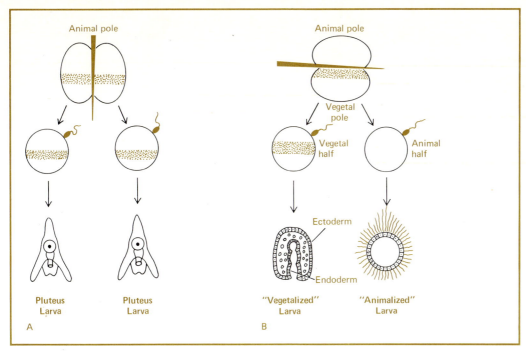

14-1 The experimental bisection of unfertilized sea urchin eggs which are subsequently fertilized leads to (A) normal development only if the pigmented band is bisected. Otherwise (B) the result is "animalized" and "vegetalized" larvae that develop abnormally.

another. Cells have elaborate mechanisms for controlling gene expression; genes are most often **repressed,** or "turned off." However, in any given cell of an embryo, genes are selectively **derepressed,** or "turned on" in a specific sequence, in order that required enzymes and structural proteins can be synthesized.

14.2 Experiments show that the genetic information in the nuclei of differentiated cells has not been lost.

In experiments in the laboratory of J. B. Gurdon, nuclei from intestinal cells of *Xenopus* (clawed toad) tadpoles were transplanted into eggs from which the nuclei had been removed. The transplanted nuclei were sometimes able to successfully guide the development of the eggs into normal tadpoles. These experiments show that genetic instructions are generally not lost in differentiated cells; they are simply not read.

In animal embryos differentiation seems to be an irreversible process. Cells that have already begun

to look different can be cultured for many generations without losing their differentiated properties.

The cells of many plants, on the other hand, even though visibly differentiated, retain the ability to regulate under the proper environmental conditions. It has been found that single cells teased from the root of a carrot can either produce unchanging, differentiated cells or regulate to form growing groups of cells capable of developing into a complete carrot plant. The key factor in determining the ability of carrot cells to regulate is their chemical environment. The same may well be true of animal cells, but it is probably the cytoplasmic chemistry that is important, rather than the cell's external environment.

14.3 Developing embryos pass certain "points of no return," at which time their polarity, axes of symmetry, and future parts are determined.

The eggs of two common marine invertebrates illustrate the extremes of early and late **determination.** The sea urchin egg is an example of late determination. If at the two-cell stage the egg's two blastomeres are separated, each will develop into a complete adult. The same is true if blastomeres are separated from the four-cell stage. On the other hand, the separated blastomeres of a clam egg in the same two- and four-cell stages will develop into incomplete clam larvae that cannot develop into adults. The fates of the two cells of the clam egg are said to have been determined even at this early stage.

The sea urchin egg is said to be a **regulatory egg**, because each early blastomere can regulate or utilize all the genetic information within it and form a complete organism; but this is not to say that determination of another sort has not occurred. It has, as can be shown by a simple experiment. If a sea urchin egg is cleaved artificially with a needle, the ability of each half to develop normally depends on the plane along which the egg was cut. The eggs of one species of Mediterranean sea urchin, *Paracentrotus lividus*, for example, have a pigmented band, which is bisected by the first and second naturally occurring cleavages (Figure 14-1). If one of these eggs is artifically bisected along one of the first two planes of cleavage, both parts (each containing a nucleus) will develop normally. On the other hand, if the egg is bisected *parallel* to the pigment band, either half can continue to divide, but each develops into an abnormal embryo *of a different type*. The "upper hemisphere" of the bisected egg develops into a ciliated blastula without any endoderm or gut; the "lower hemisphere" forms mostly endodermal cells.

Experiments of this kind (which work with fertilized *or* unfertilized eggs) show that the cytoplasm of the sea urchin egg is not homogeneous; it has **polarity** even before the cell divides. The rather imaginative names *animal* and *vegetal poles* have been given to the two poles. Although not clearly understood, it seems that the cytoplasms at the poles possess different chemical properties. By the addition of certain chemicals which imitate one or the other of the polar substances, a whole sea urchin egg can be made to develop into an "animalized" or "vegetalized" larva.

As a normal embryo develops, other points are reached at which a "decision" or determination is made—perhaps to make this side the head, the other the tail; or one side the stomach, the other the back. In more scientific terms the **axes of symmetry**, anterior-posterior and dorsoventral, are established. In addition it is determined which organs of the adult are going to be derived from which cells of the embryo. In most cases the points at which significant determinations are made have been identified by careful descriptive and experimental studies, but the underlying reasons why determination has occurred at a particular stage are not known.

14.4 Morphogenesis, the emergence of form in an embryo, is accomplished by cellular processes and interactions.

Developmental processes are more dynamic than they may appear to the unaided eye. The development of a clam or sea urchin zygote, when viewed by time-lapse cinematography appears to be a process in which forces of various kinds are almost continuously applied to constrict, invaginate, or otherside change the overall shape of the embryo. These changes in embryonic form are spoken of as **morphogenesis.**

Cleavage is a morphogenetic process that divides the egg into two cells and then into progressively larger numbers of cells. The blastula that results may consist of a single layer of cells or of many layers, but it is generally hollow. At some point during the cell divisions leading up to the blastula state, the individual cells in an animal embryo become capable of changing shape and amoeboid movement, which are also morphogenetic processes.

Whole layers of cells also undergo changes in shape and adhesion when the cells within them aggregate, enter or leave the layer, form a double layer with a cavity in between, fold, or form **invaginations** (inpocketings) or **evaginations** (outpocketings).

One of the earliest and most important examples of an invagination is that which occurs at gastrulation. The "flask cells" of the blastula that first invaginate at the dorsal lip of the blastopore of amphibian eggs will give rise to the lining of the primitive gut, or archenteron. This changes the embryo from one layer of cells to two—an outer layer of ectoderm and an inner layer of endoderm. In all but the most primitive animals, similar cellular events give rise to a third layer of mesoderm between the other two.

14.5 In vertebrates, including man, organ formation involves a highly specific sequence of tissue interactions on the part of the primary germ layers of the embryo.

Ectoderm, endoderm, and mesoderm—the three primary germ layers of the animal embryo—are established very early in development by gastrulation and its immediate consequences.

As the cells from each of these germ layers de-

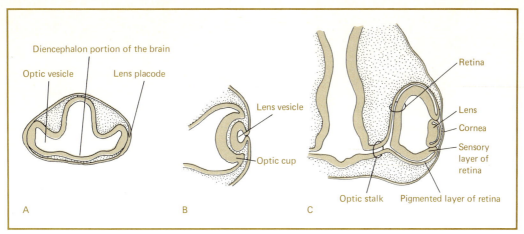

14-2 Stages in the formation of the vertebrate eye, as seen in cross sections through the head.

velop into organs they are affected not only by their own genes but also by interactions with neighboring cells that are reading a different portion of the genome. **Tissue interaction** and **embryonic induction** involve one group of cells "inducing" another to develop in a certain way. The development of the vertebrate eye offers the clearest example of embryonic induction and of another morphogenetic process, the folding and bending of cell layers. The eyes first become apparent as two symmetrical evaginations of the brain called **optic vesicles** extend toward, and come in contact with, the overlying region of head ectoderm. Once induction has occurred, invaginations take place in both the ectoderm and optic vesicle to form the **lens vesicle** and **optic cup** (Figure 14-2). These two structures then induce the overlying tissue to form the transparent lens (cornea) of the eye and induce the surrounding mesoderm to form the outer protective coatings of the eyeball. If the optic vesicles are prevented from coming in contact with the ectoderm, however, no lens or coating forms. The important developmental principle is: If one step involving cellular interaction in the development of the eyeball is omitted, the other steps do not occur.

14.6 Cellular interactions may occur at a distance. Hormones play important but different roles in the development of plants and of animals.

A **hormone** is a substance produced, often in minute quantities, by a cell (or group of cells) that has an effect on a receptive cell, called the **target cell** (or cells), often some distance away. Hormones, then, are chemical messengers. These substances are classified by their effects as either **developmental** or **regulatory** hormones, the former being involved in the development of plant or animal embryos, the latter in the regulation of the metabolic processes. Some hormones may be primarily developmental at one stage in an organism's life and regulatory at another. The effects of hormones in development seems to be the calling forth of the activity of specific genes.

Since plants do not have nervous systems to coordinate their activities, they rely heavily on hormones. The principal growth hormones in plants are **auxins,** named by F. W. Went, who showed in the 1930s that these formerly unidentified chemical substances cause cell elongation. He incorporated extracts of growing plants in small agar blocks and bound each of these to one side of an oat coleoptile (a special leaf that encloses the shoot). Auxin activity was detected when the oat seedlings bent away from the blocks, because auxins, diffusing into the cells on their side of the coleoptile, caused these cells to elongate. Later it was discovered that the most common auxin is indole acetic acid (IAA), although other substances exhibit auxin activity also.

Auxins also correlate growth processes in plants so that the lengths of roots and branches are in proper proportion to their thicknesses and so that the development of the whole branch system is in balance with the root system. In addition, auxins induce mitosis in lateral root primordia (tissue capable of generating new roots).

With another hormone, **ethylene,** they also regulate the process of **abscission,** the falling of leaves.

Ethylene is a volatile gas produced in minute quantities by many fruits, whose ripening it aids; but it may inhibit the sprouting of potatoes.

Another plant hormone is **cytokinin**, which stimulates cell division in meristems. In combination with auxin, cytokinin produces a wide variety of differential effects on plant growth, depending on the amounts of the two hormones supplied.

By far the most striking effects on plant growth and development are made by the **gibberellins,** substances first discovered in Japan in the course of research on a fungus, *Gibberella*, that causes a disease of rice seedlings. Giberellins are synthesized in actively growing regions of plants and transported to other areas. They are active in extremely minute amounts and often have profound effects. For example, certain dwarf mutants of corn grow to a normal height if given a few micrograms of gibberellin. Gibberellins play an important role in the very early development of some plants such as cereal grasses. The embryo releases gibberellin, which diffuses through the endosperm to the outer layer of the seed, where it induces the synthesis of enzymes which digest the nutrients stored in the endosperm. The developing embryo feeds on the products of this digestion.

Hormones are important also in the development of many animals. Even more mysterious than the complex changes that produce an eye or ear in a mammalian embryo are the complete changes of form that transfigure a tadpole into a frog, a caterpillar into a moth, or a lobster larva into an adult lobster. Such extensive changes, common in amphibians, insects, and many marine invertebrates, are called **metamorphosis** (radical change of form). The process is often regulated by hormones.

The metamorphosis of the *Cecropia* (silkworm) moth has been extensively studied. Three hormones regulate it. First a moth egg hatches into a small larva (caterpillar) which begins to feed and grow. A pair of glands behind the brain produces **juvenile hormone** (common in many insects), which in a sense acts to keep the larva in its juvenile form. Working against this are two others (i.e., the juvenile hormone's antagonists). Certain brain cells produce a hormone that stimulates the **prothoracic glands** to produce a second hormone mixture known as **ecdysone**. Ecdysone acts to induce molting (shedding of the exoskeleton) and metamorphosis.

Thus, in its early life a larva has a high level of

14-3 Hormonal control of metamorphosis in the silkworm, *Cecropia*. JH represents the juvenile hormone; E, the ecdysone, or prothoracic gland hormone.

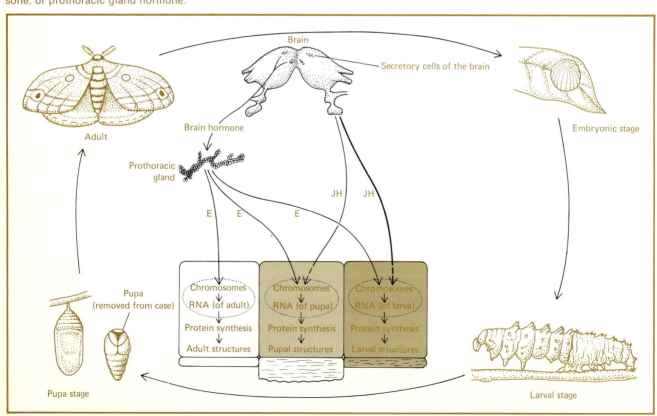

juvenile hormone and ecdysone's only effect is to allow it to molt in order to grow; but later, as the level of juvenile hormone drops, ecdysone's effects become more pronounced. The larva pupates (i.e., forms a hardened case) and eventually emerges from its cocoon as an adult moth (Figure 14-3).

The effects of ecdysone can be dramatically demonstrated in the developing embryo of a two-winged fly. In the salivary glands of the larvae of the fruit fly, *Drosophila*, and other dipterans (two-winged flies), there are giant cells containing giant chromosomes on which the position of various genes can be seen as bands (Chapter 16). During development, first one and then another of these bands develops a "puff" or swelling that has been shown to be associated with transcription of a gene onto messenger RNA (section 7.3). The injection of exceedingly tiny amounts of ecdysone into dipteran larvae has been shown to cause a specific sequence of chromosomal bands in salivary gland cells to undergo puffing, showing that a specific set of genes has been "turned on."

A second example of metamorphosis is the transformation of a tadpole into a frog. In this case the pituitary gland initiates the process by releasing a hormone which stimulates the thyroid gland to produce a thyroid hormone, *thyroxin*, which governs metamorphosis. In the tadpole this involves regressive changes—the resorption of gills, rasping teeth, and tail; and constructive changes—the development of legs and, in fact, all the adult characteristics. If a bit of thyroid gland is implanted in a tadpole, or purified thyroxin is injected, metamorphosis will begin immediately in the normal fashion.

SUGGESTED READING

Balinsky, B. I. *An Introduction to Embryology.* 3rd ed. Philadelphia: Saunders, 1970.

Ballard, W. *Comparative Anatomy and Embryology.* New York: Ronald Press, 1964.

Barth, L. J. *Development: Selected Topics.* Reading, Mass.: Addison-Wesley, 1964.

Bell, E., ed. *Molecular and Cellular Aspects of Development.* New York: Harper & Row, 1965.

Bellairs, R. *Developmental Processes in Higher Vertebrates.* Coral Gables, Fla.: University of Miami Press, 1971.

Berrill, N. J. *Developmental Biology.* New York: McGraw-Hill, 1971.

Bodemer, C. W. *Modern Embryology.* New York: Holt, Rinehart and Winston, 1968.

Ebert, J. D., and I. M. Sussex. *Interacting Systems in Development.* 2nd ed. New York: Holt, Rinehart and Winston, 1970.

Galston, A. W. and P. J. Davies. *Control Mechanisms in Plant Development.* Englewood Cliffs, N.J.: Prentice-Hall, 1970.

Gilcrist, F. G. *A Survey of Embryology.* New York: McGraw-Hill, 1968.

Grobstein, C. *The Strategy of Life.* San Francisco: Freeman, 1964.

Hamburgh, M. *Theories of Differentiation.* New York: Elsevier, 1971.

Kerr, N. S. *Principles of Development,* Dubuque, Iowa: Brown, 1967.

Kuhn, A. *Lectures on Developmental Physiology.* 2nd ed. Translated by Roger Milkman. New York: Springer-Verlag, 1971.

Nalbandov. A. V. *Reproductive Physiology.* 2nd ed. San Francisco: Freeman, 1964.

Patten, B. M. and B. M. Carlson. *Foundations of Embryology.* 3rd ed. New York: McGraw-Hill, 1974.

Rugh, R. *Vertebrate Embryology.* New York: Harcourt Brace Jovanovich, 1964.

Saunders, J. W. *Animal Morphogenesis.* New York: Macmillan, 1968.

Saunders, J. W. *Patterns and Principles of Animal Development.* New York: Macmillan, 1971.

Spratt, N. T. *Developmental Biology.* Belmont, Calif.: Wadsworth, 1971.

Thomas, J. B. *Introduction to Human Embryology.* Philadelphia: Lea and Febiger, 1968.

Torrey, T. W. *Morphogenesis of the Vertebrates.* New York: Wiley, 1962.

Waddington, C. H. *Principles of Development and Differentiation.* New York: Macmillan, 1966.

Watson, J. D. *Molecular Biology of the Gene.* New York: W. A. Benjamin, Inc., 1965.

Willier, B. H., and J. M. Oppenheimer, eds. *Foundations of Experimental Embryology.* Englewood Cliffs, N.J.: Prentice-Hall, 1974.

III GENETICS: HEREDITY, VARIATION, AND CONTROL OF DEVELOPMENT

Genetics is the study of heredity and variation. The breeding of plants and animals to bring out desirable traits in the offspring is probably as old as human civilization. However, genetics as a science began a little more than a century ago: The experiments of Gregor Mendel led to the discovery of the basic laws which made it possible to predict the outcome of breeding.

By the beginning of the twentieth century, studies of the mechanisms of cellular reproduction had advanced to the point where they could explain Mendel's laws—and some apparent exceptions to them—in terms of the structure and behavior of chromosomes. By the middle of this century, genetics was an advanced and sophisticated science that could account for almost all aspects of inheritance except its molecular basis.

The great breakthrough came in 1953 with the double-helical model for the structure of DNA. The implications of that model were so fundamental that it is responsible for new insights nearly a quarter of a century later.

Chromosomal genetics (cytogenetics) and molecular genetics are having a profound effect on our understanding of human heredity and its role in health and disease. Various disorders that have only recently been recognized as heritable can now be treated by special diets or chemotherapy, and many such disorders can be prevented by genetic counseling. Many molecular geneticists have predicted that some genetic disorders may eventually be corrected by a kind of "molecular surgery." As more is learned about the molecular mechanisms of inheritance, various kinds of corrective action for genetic defects (often called "genetic engineering") may be possible.

15

Mendelian Genetics

QUESTIONS TO KEEP IN MIND

What was Gregor Mendel's contribution to the science of genetics?

Why was the pea plant a fortunate choice of material for Mendel's work?

Why would Mendel's work have been important for Charles Darwin to know about?

How are breeding experiments on pea plants done?

What is meant by *multiple alleles*? By *multiple genes*?

The study of genetics deals with the inheritance of traits or, more precisely, with the transmission and functioning of the genes that control these traits. Every sexually reproducing organism, with the exception of identical twins, inherits a different constellation of genes; these are recombined and transmitted to the following generations in a complex but somewhat predictable pattern.

15.1 The fundamental rules governing heredity were first discovered by Gregor Mendel.

The next time you have peas for dinner, chew them tenderly and with respect. They helped provide mankind with knowledge of the basic laws governing the transmission of hereditary traits from parent to offspring. The discoverer of these laws was an Austrian monk named Gregor Mendel (1822–1884). Mendel was a substitute teacher for several years,

but twice failed examinations for appointment as a regular teacher. Ironically, he failed the natural history part of the examination, and his supervisor commented that he "lacked the necessary clarity of comprehension."

In 1856, Mendel began to crossbreed different varieties of the garden pea with the simple aim of determining whether there was a "generally applicable law governing the formation and development of hybrids." (Hybrids are those plants or animals descended from a male of one variety and a female of another.) Nine years later, he presented two lectures to members of a natural science society concerning his garden pea crosses, and his results were published. His work received no more recognition than would a new recipe for apple pie, yet here for the first time was a clear statement of the rules governing heredity.

One man who apparently never learned of Mendel's work but who was passionately interested in the mysterious workings of heredity was Charles Darwin (1809–1882). In his book *The Origin of Species* (1859), Darwin argued that there were *inherited* variations among organisms, and that all organisms produced more offspring than could possibly survive. This, he wrote, results in the "natural selection" of the more advantageous inherited variations, which are thus passed on to the next generation. Eventually, these adaptive variations would prevail in the population, and a new species would slowly evolve.

A major difficulty in Darwin's theory was that he could not adequately explain the basis for inherited variation that was so essential for natural selection. Although Mendel had the answer by 1865, Darwin, unaware of it, searched for a solution for the rest of his life.

Mendel himself died on January 6, 1884, possibly believing that the great discovery he had made applied only to garden peas and a few other plant groups.

In 1900 Mendel's insights were independently rediscovered by three scientists: Hugo DeVries in Holland, Carl Correns in Germany, and Erich von Tschermak-Seysenegg in Belgium. Their results and interpretations paralleled Mendel's. Searching through earlier scientific literature on the subject, they found Mendel's paper and gave proper recognition to his work.

Thus, after one of the most unbelievable oversights in the history of science, the fundamental principles of genetics were rediscovered.

15.2 Mendel studied the inheritance of seven traits in garden peas.

Mendel made an intelligent choice of an experimental organism for a study of heredity, for there are many true-breeding varieties of peas that transmit, unchanged, several clearly distinguishable traits from generation to generation. He could easily cross these varieties and obtain offspring that could themselves successfully reproduce.

It is useful to know how pea plants are crossed. The pea flower contains both the male and female reproductive organs of the plant. Since its petals completely enclose these, self-fertilization normally occurs (section 12.3), and accidental fertilization by pollen from another plant does not occur. When plants are to be crossed, however, the anthers are removed from a plant before they produce mature pollen. Then, when the pistil of this flower is mature, it is sprinkled with pollen from a plant of another variety which is to be the other parent. The ease and precision with which this can be done contributed to Mendel's success.

Mendel first made certain that each variety he wished to use was true-breeding (that is, would transmit, unchanged, its distinguishing traits from one generation to the next). He did this by raising each variety for two years and observing the constancy of traits. Then he chose varieties that among them included seven contrasting traits.

Mendel's experiments were carefully planned. He crossed two varieties that differed in a clearly definable trait, then raised the first generation of the cross (known as the **first filial** or F_1 **generation**), and carefully recorded the absence or presence of the

traits being studied in each of the offspring. Then he allowed the F_1 plants to **self-fertilize** and produce the **second filial generation** (F_2). Again, he classified all offspring with respect to the traits he was interested in. He continued to raise successive generations and analyze them in the same manner. "The object of the experiment," he wrote, "was to observe these variations in the case of each pair of differentiating characters and to deduce the law according to which they appear in the successive generations."

After studying one trait at a time, Mendel made crosses between plants that differed in several traits. Again, he raised F_1 and F_2 generations and followed the mode of inheritance of these characters.

15.3 Dominant traits appeared in the F_1 generation, and both dominant and recessive traits appeared in the F_2 generation in a ratio of 3:1.

When Mendel crossed plants differing in any single trait, he found that all the F_1 individuals closely resembled one of the parental plants. The plants had the same **phenotype** or appearance. The trait they exhibited was said to be **dominant,** whereas the parental trait that did not appear was called **recessive.** The F_1 plants self-fertilized, and in the second generation (F_2) *both* original parental traits reappeared. The dominant trait, however, was three times more common than the recessive one (Figure 15-1). (Note that in the study of genetics, the symbol "X" means "crossed with.")

Mendel's interpretation of these results was brilliant. He realized that he could explain what he found only if each parent plant had a pair of "factors" which together determined a specific trait. Mendel used the term *factors;* the modern term is **genes.**

If the letter R is used to represent a gene responsible for round seeds (a dominant trait), and r represents a gene for wrinkled seeds (recessive), the **genotype,** or composition, of the gene pair determining the specific trait is RR in the true-breeding round parental type, and rr in the wrinkled type. When an organism has two genes of the *same* kind for a trait, it is said to be a **homozygote** with respect to that gene pair. When Mendel crossed round and wrinkled parent plants he assumed, correctly, that each plant contributed one of the pair of factors (genes) to the offspring. Thus each F_1 individual receives an R gene from the round parent and an r gene from the wrinkled parent. The genotype of

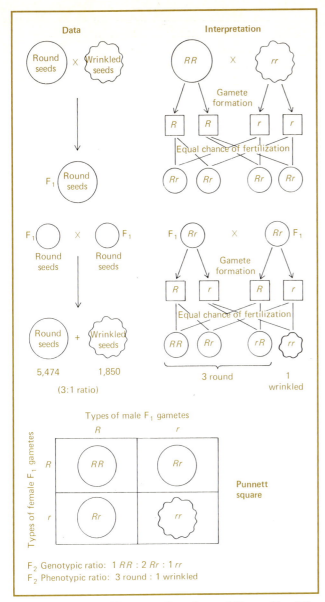

Data   **Interpretation**

F_2 Genotypic ratio: 1 RR : 2 Rr : 1 rr
F_2 Phenotypic ratio: 3 round : 1 wrinkled

15-1 The results and interpretation of one of Gregor Mendel's experiments. First, true-breeding lines of peas with round and wrinkled seeds were crossed to produce the F_1 generation. Then F_1 offspring self-fertilized to produce the F_2 generation. A Punnett square illustrating the F_2 combination is included.

every F_1 individual is Rr. Such individuals, with two *different* genes for a trait, are said to be **heterozygous** with respect to that gene pair.

Mendel did not know that his "hereditary factors" occurred on chromosomes. Since we know that they do and that a diploid parent has pairs of homologous chromosomes, it is now clear that each gene of a gene pair, such as Rr, is carried on one chromosome of a homologous pair. If such genes occupy the same position, or locus, on the homologs, they are called alleles or allelic pairs. An

allele is best defined as a special condition of a gene at one locus. Since a gene may exist in many different conditions, a population of organisms may have several alleles of a gene designated by the same letter (R, R_x, R_y, r, r_p, etc.) at a particular locus; these are called **multiple alleles.**

15.4 Mendel's law of segregation states that two different genes in the F_1 heterozygote segregate to different offspring according to simple laws of chance.

Mendel assumed that the Rr individuals of the F_1 generation produce two kinds of gametes containing genes R and r, in equal numbers. He further correctly assumed that when an F_2 is raised from self-fertilization of the F_1, these two kinds of gametes combine at random. This means that every kind of male gamete (R and r) has an equal chance to combine with every kind of female gamete (also R and r).

One way to see the kinds and ratios of offspring that theoretically result from such a random combination is to construct a **Punnett square** (Figure 15-1). The types of female gametes and their frequencies are written on the left side of a checkerboard square, and the types of male gametes and their frequencies are written on an adjacent side. The squares show the types of combinations possible and their frequencies. As you can see, one-fourth of the F_2 generation is RR, one-half is Rr, and one-fourth is rr, in a ratio of 1:2:1. This is called the **genotypic ratio.** The **phenotypic ratio,** on the other hand—in this case, the ratio of round peas to wrinkled peas—is 3:1, because roundness is dominant and both RR and Rr plants have round peas. (Both these ratios are approximate and a large number of counts of offspring are required to observe the closeness to simple arithmetical ratios.)

The law which serves as the basis for calculating genotypic combinations is the same as that used in determining what the frequency of the heads-heads, heads-tails, and tails-tails combinations will be when two coins are tossed. It is the product law of probability, which states that if two events are independent, the chance that they will occur together is the chance of one of the events occurring multiplied by the chance of the other event occurring.

Mendel did not state the relevant law of chance in this way, but his understanding of it enabled him to formulate his **law of segregation:** the two different genes in the F_1 heterozygote segregate, pure

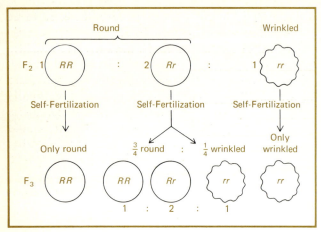

15-2 The F₃ generation of a cross between round and wrinkled peas.

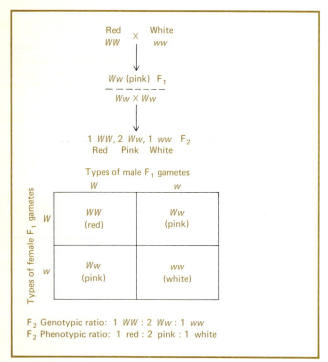

15-3 Results of a cross between red and white snapdragons, an example of incomplete dominance.

and uncontaminated, to different offspring according to simple laws of chance.

With his F₂ generation, Mendel showed that there is a predictable regularity in heredity, at least as far as the seven contrasting traits in the pea plants are concerned. But raising an F₂ generation did not satisfy Mendel's desire to be sure of his assumptions. He proceeded to raise an F₃ generation (Figure 15-2). Because his ideas on the heredity of these traits were correct, he was able to predict the makeup of the F₃ generation.

The seven traits that Mendel had been working with in peas were all cases of **complete dominance**; that is, the dominant trait completely masked the recessive one, and no peas were produced that were slightly wrinkled or half smooth. There are many instances, however, of **incomplete dominance**, both in plants and animals. For example, if a cross is made between snapdragons that have red flowers (*WW*) and ones that have white flowers (*ww*), the F₁ individuals (*Ww*) are all pink (Figure 15-3). At first glance, it might seem that the genes for red and white flowers had blended together to form pink F₁ individuals. However, when an F₂ generation is raised, red-, pink-, and white-flowered plants occur in a ratio of 1:2:1. This shows that the genes *W* and *w* did not blend together in the F₁ individuals; they remained unaffected by each other and then segregated into different gametes forming the F₂ generation. In a *Ww* individual, the *phenotypic* effect is a blend, although the genes remain distinct and unaffected by the mixture.

15.5 When two pairs of genes are involved in crosses, each pair may segregate independently of the other pair.

After Mendel had experimented with single pairs of contrasting traits he began to work in more complex crosses that involved two pairs of traits. For example, he crossed true-breeding pea plants with round, yellow peas (both dominant traits represented by *RRYY*) with plants having wrinkled, green peas (both recessive, or *rryy*). As with his earlier experiments he raised an F₁, F₂, and F₃ generation and classified the results. The results of this cross are illustrated in the Punnett square in Figure 15-4.

As Mendel predicted, gene pairs are inherited independently of each other, which is to say that when a heterozygote such as *RrYy* forms its gametes, dominant traits will not link up making only *RY* and *ry* gametes possible. On the contrary, the genes for color and shape are distributed independently, and all possible combinations exist in the gametes: *RY, Ry, rY,* and *ry*. In the F₂ generation the genes recombine according to the same simple rules that govern the inheritance of single traits. Mendel's description of this independent behavior of different gene pairs became known as his **second law**, that **of independent assortment.** It is applicable only when the pairs of genes are located on different chromosomes. When the pairs are **linked** (located on the same chromosome), they

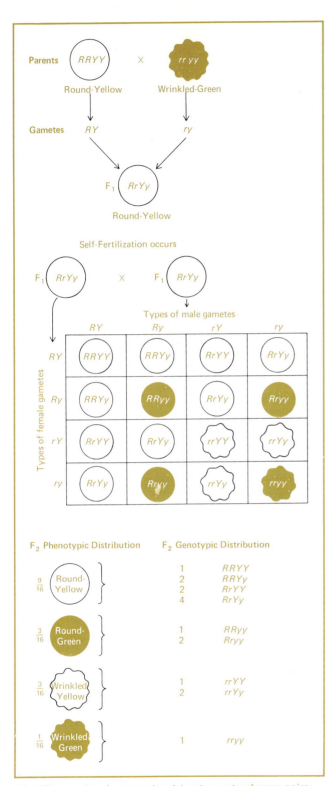

Parents RRYY X rr yy
Round-Yellow   Wrinkled-Green

Gametes RY   ry

F₁ RrYy
Round-Yellow

Self-Fertilization occurs

F₁ RrYy X F₁ RrYy

Types of male gametes

	RY	Ry	rY	ry
RY	RRYY	RRYy	RrYY	RrYy
Ry	RRYy	RRyy	RrYy	Rryy
rY	RrYY	RrYy	rrYY	rrYy
ry	RrYy	Rryy	rrYy	rryy

Types of female gametes

F₂ Phenotypic Distribution   F₂ Genotypic Distribution

9/16 Round-Yellow   1 RRYY
                         2 RRYy
                         2 RrYY
                         4 RrYy

3/16 Round-Green   1 RRyy
                         2 Rryy

3/16 Wrinkled-Yellow   1 rrYY
                         2 rrYy

1/16 Wrinkled-Green   1 rryy

15-4 The results of a cross involving two sets of gene pairs: round-wrinkled and yellow-green.

are not distributed independently, but tend to be inherited together (section 15.6).

A further test of Mendel's laws involves making either a **backcross** or a **testcross**. In the former, an F₁ individual is crossed with a parent organism, and in the latter, an F₁ individual is crossed specifically with the recessive parental type. In a testcross involving the round-wrinkled trait an F₁, *Rr*, would be crossed with an *rr*. About half the offspring would be round, *Rr*, and half wrinkled, *rr*. A testcross can determine whether the genotype of an individual with a dominant phenotype is homozygous or heterozygous.

15.6 Many traits are controlled by more than two genes and are said to be the consequence of multiple gene inheritance.

In the first decades of the twentieth century, when genetic experiments were just beginning, it seemed that many traits in plants and animals were of a kind that could not be explained by Mendel's laws. Instead of being tall *or* dwarf, the way a pea plant was, most organisms exhibited a finely graded range of such traits as height, color, and weight. Such traits are called **quantitative traits**. Their widespread occurrence posed a serious problem for geneticists until the idea was advanced that such traits are controlled by many genes, (or **multiple genes**), each having a small effect on the phenotype.

To illustrate the principle of quantitative inheritance, let us imagine that there are three genes, *A*, *B*, and *C*, determining the height of a plant: Each contributes two centimeters (cm) of height when present. The respective alleles, *a*, *b*, and *c* contribute no growth to the plant. In our model, we shall consider that the genes act cumulatively, each contributing an equal amount of height, and we shall ignore any environmental effects on gene expression. These assumptions greatly oversimplify the true situation, but are helpful for understanding the principle involved.

If a tall plant, *AABBCC*, were crossed with one 12 cm shorter, *aabbcc*, the F₁ individuals (*AaBbCc*) would all be of intermediate height. The F₂ plants, however, would show six different gradations, the whole range from tall to short.

If 20 genes were involved instead of 3, the number of gradations in height in the F₂ generation would increase. If enough genes were involved, there would be so many different heights represented that we would have what appeared to be a continuous distribution.

16

Cytogenetics: The Chromosomal Basis of Heredity

QUESTIONS TO KEEP IN MIND

What discoveries in cell biology paved the way for the appreciation of Mendel's laws?

How was the chromosomal basis of inheritance discovered?

What properties of the common fruit fly made it useful for genetic research?

How is sex inherited?

How was sex linkage discovered?

What is nondisjunction?

Does Mendel's second law apply when two genes are members of the same linkage group?

How are genes mapped on chromosomes?

Although Mendel described the mathematical rules governing the transmission of traits from parent to offspring, he knew nothing about what these hereditary factors were or where they were located in the organism. Nor did he seem to be aware of the important discoveries in the field of genetics that were being made concerning the components and functions of the cell.

After Mendel's death, while his work was still undiscovered, further discoveries were made. By 1900, the details of mitosis, meiosis, and fertilization were known, and it was suspected that the physical basis of heredity was in the nucleus. The biological world was ready to benefit from Mendel's laws.

16.1 Sutton suggested correctly that chromosomes constitute the physical basis of the Mendelian laws of heredity.

A major breakthrough came in 1902 and 1903 with the publication of two papers by W. S. Sutton. Sutton was a graduate student at Columbia University studying chromosomes and cell division in grasshopper testes. He noted a great similarity between the behavior of chromosomes during meiosis and the pattern of segregation of genes as postulated by Mendel. Sutton concluded that:

1. Chromosomes occur in pairs, just as Mendelian factors were believed to occur in pairs; presumably one chromosome of each pair comes from the mother and the other comes from the father.
2. Chromosomes segregate to different gametes at meiosis, just as Mendelian factors segregate.
3. Members of one chromosome pair segregate independently of other chromosome pairs, just as Mendel's factors show independent assortment.
4. Among all of the cell's organelles, only the chromosomes are accurately divided during cell division, and their individuality is retained from generation to generation, just as one would expect Mendelian factors to be.

In his 1902 paper, Sutton wrote: "I may finally call attention to the probability that the association of paternal and maternal chromosomes in pairs and their subsequent separation during the reducing division as indicated above may constitute the physical basis of the Mendelian law of heredity."

Sutton had the experimental observations and the insight to crystallize a major concept in genetics,

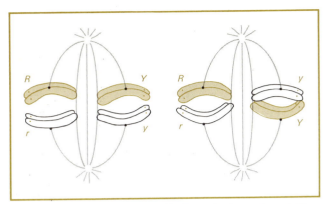

16-1 The chromosomal basis of independent assortment. The diagram shows two tetrads in meiosis I and illustrates the fact that the two tetrads align themselves independently.

namely that *genes are probably parts of chromosomes*. In the late nineteenth century, other biologists had suspected that the physical basis of heredity might be the chromosome, but Sutton turned the suspicion into a working hypothesis. Furthermore, he linked the fields of **cytology,** the study of cell structure, and genetics to produce the new discipline of **cytogenetics.** Subsequent investigations substantiated Sutton's belief that genes were parts of chromosomes, and added details to several of his assumptions.

As was discussed in section 11.5, meiosis involves the halving of the diploid number of chromosomes. In prophase I the homologous chromosomes synapse and form tetrads, and in metaphase I these line up at the cell's equator. The tetrads line up independent of one another. Thus if two pairs of genes, Rr and Yy, are carried on separate pairs of chromosomes, the way in which the Rr pair orients itself at the equator is independent of the way Yy orients itself (Figure 16-1). Thus the gametes YR and yr will be produced just as often as Yr and yR.

16.2 Early geneticists discovered that the common fruit fly offers unique advantages for the study of heredity.

A few years after Sutton did his classical work on grasshopper testes, other scientists discovered that the fruit fly *Drosophila melanogaster* is a fine experimental organism for the study of heredity. Here are some of the reasons:

1. Large numbers of flies can be easily cultured on a mixture of cereal, molasses, agar, and yeast.

2. The flies are simple to maintain and handle. For example, when a new generation with particular traits is desired, the flies are anesthetized with ether and examined under a dissection microscope. Males and virgin females are selected and put into a new jar, where they start a new generation of flies.

3. Hundreds of offspring can be obtained from a single female mated to a single male. Moreover, the generation time is less than two weeks at 25° C. The generation time can be lengthened by raising the flies at low temperatures, or speeded up by raising the temperature. Rapid breeding and many offspring are desirable for an experimental organism.

4. Many distinct traits are available for study in *Drosophila*. Since work began on the fruit fly early in the twentieth century, literally thousands of hereditary variations in *Drosophila* have been reported, and stocks of flies exist with an amazing variety of inherited abnormalities.

16.3 Other characteristics that recommend *D. melanogaster* to geneticists are the presence of only four pairs of chromosomes and the giant size of these in certain cells.

The chromosomes in the salivary glands of fruit fly larva are very large and are consequently easy to study. When stained, these chromosomes have a characteristic pattern of dark and light crossbands (Figure 16-2), which correspond to the location of genes.

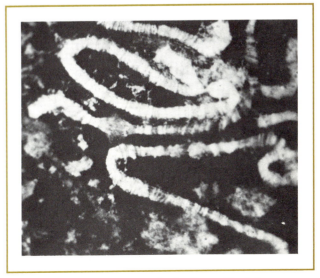

16-2 A fluorescence micrograph of salivary-gland chromosomes of *Drosophila*, showing the bands that mark the positions of various genes. (Courtesy of R. R. Cowden.)

These large chromosomes have acquired their size by virtue of another unusual characteristic: They replicate many times, but their products do not separate. The cells of the glands themselves grow, but do not divide. After a while, each of these cells has produced roughly 10^2 meters of DNA, in the form of "many-threaded," or **polytene**, chromosomes, that is, chromosomes that may have up to 512 strands of DNA lying together side by side.

Each of the four chromosomes can be identified by its banding pattern. Each band seems to correspond to the location of a particular gene, such as that for eye color or wing shape.

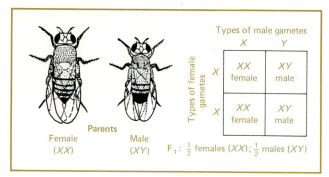

16-3 The inheritance of sex chromosomes in *Drosophila*.

16.4 T. H. Morgan used fruit flies to unravel the mystery of sex-linked genes.

For more than 20 years, Nobel Prize winner T. H. Morgan and his students worked at Columbia University on fruit flies. Many of the traits that he and his coworkers investigated in the "fly room" followed the simple patterns of Mendelian inheritance. Other traits, such as white eye color, seemed to be at variance with the Mendelian pattern (normally the eyes are red).

To understand how Morgan's work reconciled the apparently aberrant inheritance with Mendelian principles, it is necessary to understand how sex is determined. In many organisms sex is determined by the number and kind of chromosomes found in the fertilized egg. In the fruit fly, as in man, the sex chromosomes of the female are said to consist of two similar **X chromosomes**, whereas in the male there is one X chromosome and a different **Y chromosome**. Chromosomes other than the X and Y chromosomes are called **autosomes**. As a result of meiosis, every egg contains an X chromosome, but half the sperm carry an X chromosome and half carry a Y chromosome. An egg fertilized by an X-bearing sperm will be an XX zygote and will develop into a female, and an XY zygote will become a male (Figure 16-3). In some other organisms sex is determined by the male or the female having one extra chromosome.

With this introduction we can follow Morgan's investigation of what later became known as **sex-linked traits**. About 1909, he found a single white-eye male *Drosophila* in a culture of normal red-eyed flies. This unusual eye color was presumably the result of a mutation (a sudden heritable change in an organism). To determine the pattern of inheritance of white eye color, Morgan mated the white-eyed male with a red-eyed female. In the F_1

generation, all the flies had red eyes. When an F_2 generation was raised, a ratio of three red-eyed flies to one white was obtained. One unusual observation, however, indicated that white eye color was not due to a simple recessive gene. Instead of the white eye color in the F_2 generation being about equally distributed between male and female flies, all of the F_2 females were red-eyed, while half the males had red eyes and half had white eyes (Figure 16-4A).

Morgan's results became even more confusing when he performed a reciprocal cross, that is, when he mated a white-eyed female (obtained from backcrossing an F_1 female with the white-eyed male [Figure 16-4B]) with a red-eyed male (Figure 16-4C). Instead of an F_1 generation of red-eyed flies, all the males had white eyes and all the females had red eyes. Furthermore, the F_2 generation did not exhibit the typical 3:1 ratio; half the males and half the females were white-eyed.

Morgan knew that the sex of *Drosophila melanogaster* is determined by an X chromosome and a Y chromosome in the male. He could explain his results by assuming that the gene for white eye color was on the X chromosome and that the Y chromosome did not carry an eye-color allele. The pattern of inheritance of the gene for white eyes was the same as the pattern of inheritance of the X chromosome.

If you follow Figure 16-4C, for example, you can see that a white-eyed female, whose genotype is *ww*, produces eggs each of which has the recessive gene on an X chromosome. The red-eyed male, however, produces sperm half of which have a dominant gene for red eye color, *W*, on an X chromosome; and half of which have no gene for eye color, for they carry the Y chromosome, which lacks a gene of this type. When such eggs and sperm unite to form the F_1 generation, all the females must be red-eyed, because each gets an X chromo-

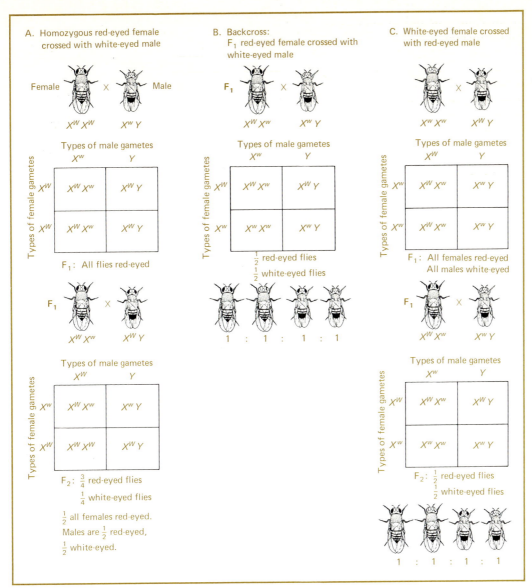

16-4 Crosses illustrating the inheritance of white eye color in *Drosophila*.

some from her father, and all his X chromosomes carry the dominant red-eyed gene. All the males must be white-eyed, because each gets his X chromosome from his mother and a Y chromosome with no eye-color gene from his father. These males will in turn produce gametes half with recessive genes on the X chromosome and half with "empty" Y chromosomes. Their sisters will produce eggs half with the dominant gene, W, and half with the recessive gene, w. When these eggs and sperm unite to produce the F_2 generation, four different combinations of genes are formed (ww and Ww in females, W and w in males). Half the members of either sex will be red-eyed, and the other half white-eyed.

Morgan performed many other crosses with red- and white-eyed fruit flies. They bore out his assumption that the gene for white eye color is located on the X chromosome. Subsequently, many sex-linked genes were discovered in *Drosophila* and in other organisms, and these followed the same pattern of inheritance Morgan had illustrated. It was also found that the Y chromosome is not totally devoid of genes. Genes on the Y chromosome are inherited in a logically predictable pattern: male-to-male transmission. The male offspring receive their X chromosomes from their mothers and Y chromosomes from their fathers. Female offspring receive an X chromosome from each parent. Genes on the X chromosome are said to be X-linked; genes on the Y chromosome are Y-linked.

16.5 Further anomalies in the eye color of fruit flies led to the discovery of nondisjunction, the failure of chromosomes to separate during gametogenesis.

American geneticist C. B. Bridges studied the genes for eye color in fruit flies and discovered still another quirk in the pattern of its inheritance. Occasionally when he crossed a white-eyed female with a red-eyed male, instead of getting all red-eyed females and all white-eyed males he observed some *white*-eyed females and some *red*-eyed males (Figure 16-5). After studying the offspring of these anomalous flies and examining their chromosomes microscopically, he determined that in these rare cases the X chromosomes of the female parent had failed to separate during meiosis. This is called **nondisjunction**. It leads in this case to the production of an egg with two X chromosomes instead of one, or an egg with no X chromosomes, depending upon whether the egg or the polar body receives the two X chromosomes.

When the egg with two nondisjoined X chromosomes is fertilized by an X-bearing sperm from the father, we get an individual with three X chromosomes. This is known as a "super-female," but it usually does not survive. If a two-X egg is fertilized by a sperm carrying the Y chromosome, we get an XXY individual (Figure 16-5). Such a fly is a female, and since it has two genes for white eyes and no corresponding allele on the Y chromosome, it has white eyes. Here is the exceptional white-eyed female. (In *Drosophila*, sex is determined not by the Y chromosome, but by the ratio of X chromosomes to autosomes, or nonsex chromosomes. When the normal number of autosomes is present, two X chromosomes result in a normal female; one X chromosome results in a male.)

When an egg with no X chromosome is fertilized by an X-bearing sperm, we get an XO individual, which is the exceptional red-eyed male (Figure 16-5). This male looks normal but is sterile, for although the Y chromosome in *Drosophila* is not important for sex determination, it is needed for male fertility. Finally, if an egg with no X chromsomes is fertilized by a Y-bearing sperm it does not develop.

16.6 Mendel's law of independent assortment does not always apply, the exception being when the genes considered are linked.

Genes on the same chromosome behave as though **linked** to one another. They are said to belong to the same **linkage group.** There are as many linkage groups in an organism as there are pairs of chromosomes. Thus *Drosophila melanogaster* has 4 linkage groups, whereas man has 23. (Do not confuse this form of linkage with sex linkage. The latter term refers specifically to genes that are on the sex chromosomes.) Genes located near one another in a linkage group tend to segregate as a unit.

It was soon found in *Drosophila melanogaster* that the organism's several hundred genes were divided among four units and inherited in this way. T. H. Morgan worked on this problem and explained the patterns of inheritance that would result from linkage groups. For example, if the genes for body color and for wing length are located on the same chromosome, that is, are in the same linkage group, then these genes will tend to be inherited together. They will not show independent assortment. Letting *Bb* represent body color (gray is dominant, black recessive) and *Vv* represent wing

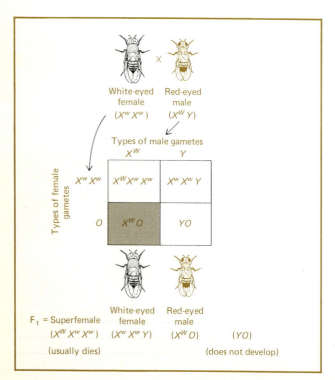

16-5 The inheritance of sex chromosomes and white eye color in *Drosophila* in the case where nondisjunction occurs in oogenesis.

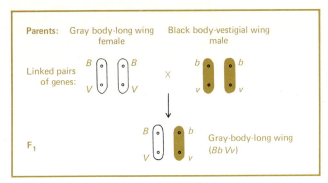

16-6 A cross between *Drosophila* strains homozygous for two linked pairs of genes.

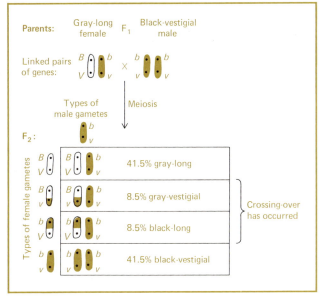

16-7 A backcross (testcross) to show linkage and crossing-over.

length (long wings are dominant, vestigal wings recessive), then Figure 16-6 illustrates what presumably happens in a cross if the genes are linked. The genotypes for the F_1 flies are written BV/bv to indicate that the B and V genes will on one chromosome and their respective alleles, b and v, will be on the corresponding homologous chromosome.

It would be logical to expect that half the gametes of an F_1 female would contain BV and half bv. This may not happen because of the process known as crossing-over (section 11.5). During the prophase of meiosis, when the chromatids of the homologous chromosomes intertwine, equal pieces may be exchanged between maternal and paternal chromatids. When such crossing-over occurs, a chromatid consists of a mixture of paternal and maternal genes. This type of genetic recombination is extremely im-

portant in explaining the great amount of variation that exists among sexually reproducing organisms.

Figure 16-7 shows that crossing-over has occurred in the production of 17 percent of the female's eggs. Thus four different gene combinations exist—BV, bv, Bv, and bV—but not in equal proportions. To see if the inheritance of linkage groups and the process of crossing-over has actually occurred as predicted, the F_1 female can be backcrossed to a $bbvv$ male, whose sperm will all carry bv. The results are shown in Figure 16-7: Just as the female gametes showed unequal proportions, so do the traits in the F_2 generation.

Further experiments showed that crossing-over does not occur in male *Drosophila*.

16.7 The frequency of crossing-over between two genes is directly proportional to the distance between them.

One of the most exciting discoveries made in genetics in this century was the realization that the frequency of crossing-over between any two linked genes is a function of their distance apart on the chromosome. This has enabled scientists to map the relative positions of genes.

Morgan and his colleagues had been testing linked genes, such as those controlling body color and wing length, and had found that crossing-over occurred with a consistent frequency, which varied depending upon which genes were being studied. Thus, when a gray, long-winged female BV/BV was crossed with a black male with vestigial wings bv/bv, the F_1 generation females BV/bv produced gametes that were not all BV or bv but included 8.5 percent Bv and 8.5 percent bV. This indicated that crossing-over had occurred between the B locus and the V locus 17 percent of the time. Other pairs of genes consistently crossed over 2 percent of the time, or perhaps 12 percent. (If genes are located far apart on a chromosome, more than one crossover may occur between them in the same cell during meiosis.)

Morgan hypothesized that the farther away two genes were on a chromosome, the more frequently they would cross over. Thus the percentage or proportion of crossovers between any two linked genes, their **crossover value,** was an index of the distance between them. For example, if B and V cross over 17 percent of the time, they may be said to lie on their chromosome 17 centimorgans away from each other. If B and C, a gene for another trait, cross over with a frequency of 20 percent, then they are separated by 20 centimorgans (Figure 16-8). To find out whether *C* lies on the same side of *B* as *V*, crosses are made involving the traits controlled by *C* and *V*. Their crossover distance may be either 3 or 37 depending on the sequence of these genes on the chromosome. From information of this kind, linkage maps were made showing the relative positions of hundreds of the known genes of *Drosophila melanogaster.*

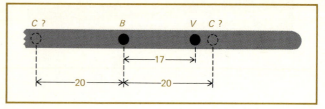

16-8 A generalized gene map, showing the two possible locations for gene *C* on a chromosome. Measurements are in centimorgans. The text explains how to determine which is the actual position.

17

Molecular Genetics

What evidence led to the general belief that the genetic material is DNA?

How do bacterial transformation and transduction differ? What do they accomplish in common?

How did the Watson-Crick model of DNA structure suggest a way in which mutations might be inherited?

What is meant by *chromosomal mutation* and *point mutation*?

What evidence suggested the "one gene–one enzyme" and "one cistron–one polypeptide" theories?

Why was *Neurospora* such favorable material for early studies of molecular genetics?

Why was it necessary to refine the term *gene* by introducing the terms *cistron, mutron,* and *recon*?

How are genes turned on and off?

How does the operon theory account for the control of inducible and repressible enzyme systems?

Up to this point we have been discussing genes in terms of the traits they transmit. This is a perfectly valid operational definition, but if we are to understand *how* genes control the transmission of flower color, wing length, hair color, and all the thousands of other characteristics expressed by living organisms, then it is necessary to study genes on the molecular level. What are genes made of? How do they actually operate within the cell?

17.1 The discovery that DNA is the hereditary material resulted from a series of experiments on bacterial transformation.

The fact that the genetic material is composed of DNA is the very basis of molecular genetics. This knowledge resulted from a whole series of discoveries made over three-quarters of a century, starting with the discovery of DNA in 1871. At first, scientists suspected that proteins, a seemingly more complex and therefore more variable component of the cell than DNA, would prove to be the genetic material. Even in the first decades of the twentieth century, when the patterns in which genes transmitted traits from generation to generation were carefully worked out, the chemical nature of the gene was unknown. Then, in 1928, the English scientist F. Griffith made an exciting discovery about bacterial heredity that eventually led others to the identification of DNA as the genetic material.

Griffith was working with two different strains of *Pneumococcus,* a type of bacterium that causes pneumonia. One strain, known as **type III (S),** is virulent (able to overcome the body's defenses) and causes pneumonia. It can easily be identified by its smooth appearance, which is due to the presence of a smooth capsule around each cell. The other strain, **type II (R),** is nonvirulent and does not cause pneumonia. It has a rough appearance, since each cell lacks the capsule found in the virulent type II form. S cells give rise only to other S cells, and R cells give rise to R cells, so their various features are inherited traits.

Mice injected with S cells developed pneumonia

and died, whereas mice injected with R cells did not die, nor did they die when injected with S cells that had been killed by heat. Griffith also injected mice with two types of pneumococcus at the same time, the R type and the heat-killed S type. Since the rough form did not cause pneumonia when injected alone, and since the smooth form did not cause pneumonia when heat-killed, these mice were not expected to get pneumonia. However these mice developed pneumonia and died. Moreover, Griffith was able to isolate *live* smooth pneumococci as well as the rough forms from the bodies of these mice. The most reasonable explanation seemed to be that some sort of interaction occurred between living rough type II and dead smooth type III, such that the hereditary properties of the dead S organisms were transferred to the living R organisms. Somehow, R cells had been transformed into S cells by some "transforming substance" from dead S cells. Griffith did not know the exact nature of this transforming substance.

Some time later, M. H. Dawson demonstrated the **transformation** of R cells into S cells in vitro, that is, in a test tube. Next, J. L. Alloway demonstrated that a crude extract of smooth cells was capable of causing this transformation of R cells to S cells in vitro. Thus the intact S cells were not necessary. Yet neither Dawson nor Alloway could tell what the transforming substance was.

In 1944, O. Avery, C. Macleod, and M. McCarty succeeded in isolating, purifying, and identifying the chemical substance responsible for transformation. It was DNA. Thus for the first time, scientists had identified a chemical substance that could carry hereditary information.

17.2 In addition to transformation, the somewhat similar process of transduction provided more evidence that DNA is the genetic material.

In transformation scientists introduce genetic material from one kind of bacterium into another kind of bacterium. **Transduction** is a similar phenomenon, except that it occurs naturally and a virus serves to transfer the genetic material between bacteria.

In order to understand how transduction can occur, it is necessary to understand what happens when a virus infects a bacterial cell. *Escherichia coli*, the bacterium that inhabits our intestine, can be attacked by certain bacterial viruses, or **bacteriophages**, known as **T phages**. The T4 virus consists of a head, and a tail portion surrounded by a sheath (Figure 17-1A and B). The head of the T4 phage is composed of a core of DNA surrounded by a protein coat. Like other viruses, bacteriophages are on the border between the living and the nonliving. They can only reproduce within living bacterial cells that they infect. When infection occurs, the virus takes over the metabolic machinery of the infected cell and causes it to produce new virus particles. About 13 minutes after infection, the first new virus has been manufactured. In another 10 minutes or so, approximately 100 virus particles are present in the cell. The cell soon undergoes **lysis** or breaks open, releasing its newly synthesized viruses, which may then infect other cells. The DNA of the attacking virus has the genetic information necessary to direct its own replication, using the enzymes of the bacterial cell as a "factory" (Figure 17-1D).

In 1952 N. D. Zinder and J. Lederberg reported that these bacteriophages could transfer a small amount of bacterial genetic material from one strain of bacterium to another. In other words bacteriophage could transfer specific hereditary traits (i.e., some bacterial genes) from one bacterium to another. The process is called **phage transduction**.

17.3 It was found to be the DNA of the virus, not its protein coat, that carried genetic instructions.

In the early 1950s it was still not absolutely clear whether it was the DNA or the protein coat of the bacteriophage that carried the genetic instructions. When the phage infects a bacterium, it attaches itself by its tail to the bacterial cell wall. The tail contracts so that it penetrates the bacterial wall and allows the viral DNA to enter the cell (Figure 17-1C). It seemed possible that the protein coat might also enter the infected cell and take control of the metabolic machinery.

In 1952, A. Hershey and M. Chase answered this question. They worked with the T2 phage which infects *E. coli*. The protein coat of this phage contains sulfur but almost no phosphorus; the DNA is rich in phosphorus but contains no sulfur. Hershey and Chase took advantage of this distinct chemical difference between the DNA and the protein coat of the T2 virus. First they grew bacteria in a medium containing a radioactive form of phosphorus, ^{32}P. The bacteria incorporated the ^{32}P into their structures. After several hours, T2 phages were added to the system. The phages immediately in-

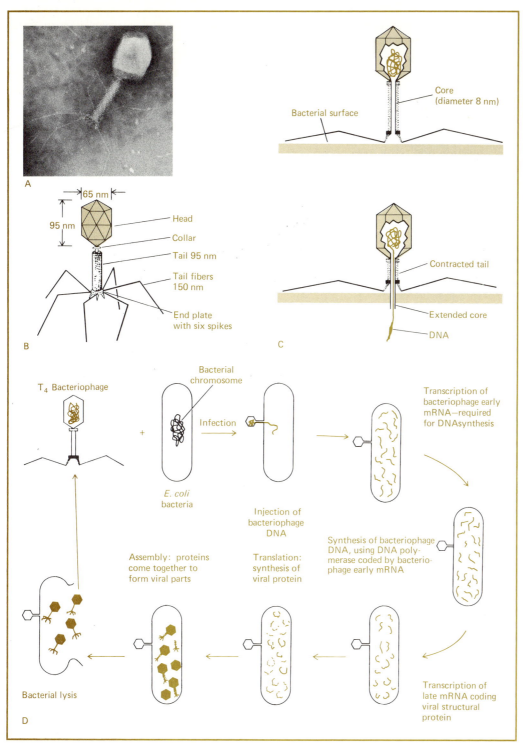

A

B
65 nm
95 nm
Head
Collar
Tail 95 nm
Tail fibers 150 nm
End plate with six spikes

Bacterial surface
Core (diameter 8 nm)

C
Contracted tail
Extended core
DNA

T_4 Bacteriophage

Bacterial chromosome

Infection

E. coli bacteria

Injection of bacteriophage DNA

Transcription of bacteriophage early mRNA—required for DNAsynthesis

Synthesis of bacteriophage DNA, using DNA polymerase coded by bacteriophage early mRNA

Translation: synthesis of viral protein

Assembly: proteins come together to form viral parts

Transcription of late mRNA coding viral structural protein

Bacterial lysis

D

17-1 A. Electron micrograph of a T4 bacteriophage. (Courtesy of R. Williams and H. Fisher.) B. An interpretation of its structure. C. The injection of bacteriophage DNA into E. coli. D. The reproductive cycle of a bacteriophage.

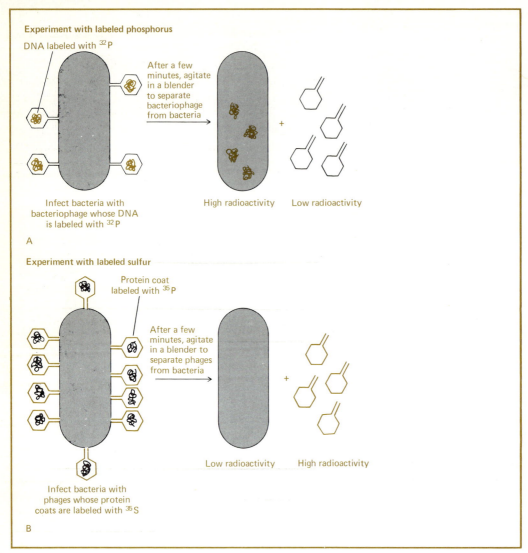

Experiment with labeled phosphorus

DNA labeled with ^{32}P

After a few minutes, agitate in a blender to separate bacteriophage from bacteria

Infect bacteria with bacteriophage whose DNA is labeled with ^{32}P

High radioactivity   Low radioactivity

A

Experiment with labeled sulfur

Protein coat labeled with ^{35}P

After a few minutes, agitate in a blender to separate phages from bacteria

Infect bacteria with phages whose protein coats are labeled with ^{35}S

Low radioactivity   High radioactivity

B

17-2 Diagram of the Hershey-Chase experiments explained in the test. Color denotes radioactivity.

fected the bacteria and began directing the synthesis of more viruses. As they reproduced, these viruses incorporated the radioactive ^{32}P of the host bacteria into their own DNA. The protein coats did not pick up ^{32}P and were nonradioactive. Thus the experimenters had complete T2 phages with radioactive DNA and nonradioactive protein coats. Similarly, by adding T2 phage to bacteria grown in a medium containing radioactive sulfur, ^{35}S, they obtained phage with radioactive protein coats and nonradioactive DNA.

The phages with radioactive DNA were then used to infect nonradioactive bacteria, but only for a few minutes. The phages which had not yet infected bacteria were removed. When the bacteria were examined, it was found that most of the radioactivity formerly associated with the phage DNA was now associated with the infected bacteria. This suggested that ^{32}P-labeled phage DNA entered the cell. Furthermore, multiplication of phage could proceed in the infected bacteria. When these cells burst, phage particles released were found to contain 30 percent or more of the original ^{32}P. This reinforced the conclusion that the infective material directing the synthesis of new viruses was indeed DNA (Figure 17-2). A complementary experiment using the phages with radioactive ^{35}S protein coats indicated that most of this protein remained outside the cell and was not involved in phage reproduction.

17.4 In some viruses, RNA is the genetic material.

Not all viruses have DNA. Those that cause influenza, poliomyelitis, and encephalitis (inflammation of the brain) in man are composed of an RNA core surrounded by a protein coat. For these, RNA is the genetic material.

The RNA viruses are not the only exception to the general rule that genetic information is transferred from DNA to RNA and from RNA to proteins. Recently H. Temin discovered that information can flow "in reverse" from RNA to DNA in the infection of certain animal cells with tumor viruses. In these viruses, RNA can serve as a template for the synthesis of DNA which is subsequently integrated into the chromosomes of the infected cell. DNA synthesis during the infective process is catalyzed by an enzyme called either **RNA-dependent DNA polymerase** or **reverse transcriptase**.

17.5 DNA exists as a double helix.

As you may remember from section 3.6 DNA is composed of three types of ingredients: phosphate, a five-carbon sugar (deoxyribose), and the nitrogenous bases adenine (A), guanine (G), thymine (T), and cytosine (C). For years it was not known how these components were actually arranged in the molecule.

In 1953 Watson and Crick, using the X-ray diffraction studies of Wilkins, interpreted the structure of DNA to be a double helix that was "right-handed with two chains running in opposite directions." With this discovery, new dimensions for the study of genetics were revealed, and the secrets of the gene could be investigated in terms of its molecular structure. The momentous work of Watson, Crick, and Wilkins earned them the Nobel Prize for Medicine and Physiology in 1962.

17.6 The Watson-Crick model of DNA provides ways of explaining mutations.

In a broad sense, the term *mutation* stands for any process which results in a change of the hereditary material. **Chromosome mutations** are changes in chromosome structure or number. For example, the addition of chromosome sets (**polyploidy**), loss or addition of genes or chromosomes, or changes in the arrangement of genes in a chromosome are chromosome mutations.

However, the term mutation is generally used to describe molecular changes *in the gene itself*. Such internal changes in the gene are heritable and are known as **point** or **gene mutations**. The various eye-color genes in *Drosophila* and blood group genes in man illustrate gene mutations. Ultimately, gene mutations must be traced to changes in the DNA itself.

One of the many gratifying features of the Watson-Crick model of DNA is that it provides ways of explaining gene mutations. An inherited change results from an alteration of the information content of the DNA molecules. A change in the sequence of DNA bases alters the DNA coded message. This "wrong" message will be passed on to RNA, which may cause the synthesis of a defective enzyme or other protein. Such a molecular deficiency can have a noticeable effect on the organism.

There are several ways in which the DNA coded message can be distorted. A mistake can be made during replication, and a "wrong" nucleotide base may be substituted for a usual one. Such a **base substitution** would change the meaning of a codon of which it was part so that it coded for either a wrong amino acid or no amino acid. In either case a faulty protein would be produced by this segment of DNA. Mistakes of this sort result in functional disorders like sickle cell anemia (section 18.3).

A second kind of distortion of DNA occurs with the addition or deletion of a single base. The genetic message is presumably translated three letters at a time. Thus, to give an example, the subtraction of the first letter G changes *GCA GCA GCA* to *CAG CAG CA . . .* , which is not at all the same thing.

Any inversion of the order of bases in DNA also produces distortion, since the sequence of bases in DNA dictates the sequence of amino acids in a protein. Suppose a piece of DNA containing several nucleotides became switched around 180 degrees in the molecule. This **inversion** in part of the DNA molecule would disrupt the usual base sequence and could result in a malfunctioning gene.

All of the above mechanisms for mutation are consistent with the Watson-Crick model of DNA. They all illustrate how changes in the sequence of bases in DNA alter the coded message. This distorted message is passed on to the next generation when the changed DNA replicates. Messenger RNA reflects this change, and an abnormal protein may result.

17.7 Early in the twentieth century a relationship was perceived among genes, enzymes, and hereditary biochemical deficiencies.

In 1908, an important clue about how a gene controls a trait came from the work of Sir Archibald E. Garrod, an English physician. He suggested that certain hereditary diseases in man characterized by chemical abnormalities were due to the absence of particular enzymes in the body. He thus hinted at some sort of connection among genes, enzymes, and metabolic diseases, the exact nature of which he did not know. In the 1940s, G. W. Beadle and E. L. Tatum more clearly defined this speculation with their experiments on the bread mold *Neurospora*.

Neurospora, although itself complex, does not have complicated requirements for growth. It will thrive in a test tube if provided with a simple mixture called a **minimal medium**. This contains only salts, sugar, and a single vitamin called biotin. *Neurospora* can survive and grow on this very simple medium only because it has the ability to manufacture all of the complex biochemicals from basic ingredients.

Beadle and Tatum produced inherited defects in *Neurospora* by bombarding the spores with X-rays or ultraviolet light. They isolated the mutant strains that could no longer grow on minimal media. They could grow in a **complete medium** which contained amino acids, vitamins, and so on, but somehow these strains had lost the ability to synthesize some of the materials necessary for their growth.

To find where the *Neurospora* assembly line had been sabotaged, Beadle and Tatum added different biochemicals singly to the minimal medium and tested these media for mutant growth. For example, Beadle and Tatum found a mutant that could not grow on minimal medium, but could grow on minimal medium to which the amino acid arginine was added. Presumably, irradiation had resulted in a hereditary defect that made the affected strain lose the ability to synthesize this amino acid, although it could still make the other vitamins, amino acids, and so forth. This arginineless condition was found to be inherited as a Mendelian gene. A host of other biochemically deficient mutants were identified in a similar manner.

Complex organic compounds such as amino acids do not suddenly appear in a cell. They are the result of a series of chemical reactions. The **biosynthetic pathways** that produce these compounds involve changes from one intermediate substance to another until the end product is reached. Each step in the synthesis of a particular compound is mediated by a specific enzyme.

In *Neurospora* the amino acid arginine is the end product of a series of biochemical steps:

$$\text{precursor substance} \xrightarrow{\quad \text{enzyme 1} \downarrow \quad} \text{ornithine} \xrightarrow{\quad \text{enzyme 2} \downarrow \quad}$$

$$\text{citrulline} \xrightarrow{\quad \text{enzyme 3} \downarrow \quad} \text{arginine}$$

As we can see, each step of this biosynthetic pathway involves a specific enzyme. Beadle and Tatum tested the *Neurospora* mutants that required arginine to grow in a minimal medium, plus either ornithine or citrulline. They found three different types of arginine mutants: One required the addition of arginine or citrulline or ornithine; a second required citrulline or arginine; and a third required arginine. Thus mutation in three different genes gave rise to *Neurospora* that required arginine for growth: The first mutant could not carry out reaction 1, but could carry out 2 and 3. The next mutant could not carry out reaction 2, but could carry out reaction 3. The last mutant could not carry out reaction 3. Since each of these reactions is controlled by an enzyme, Beadle and Tatum assumed that a reaction block was due to a deficient enzyme. Mutation in the three different genes had led to three different enzyme deficiencies.

17.8 Genes play a role via enzymes in the stepwise synthesis of organic compounds.

The results of Beadle and Tatum's research led them to suggest that the primary function of a gene is to control the development of a single enzyme, as illustrated below:

$$\text{precursor substance} \xrightarrow[\text{enzyme 1} \downarrow]{\text{gene 1} \downarrow} \text{ornithine} \xrightarrow[\text{enzyme 2} \downarrow]{\text{gene 2} \downarrow}$$

$$\text{citrulline} \xrightarrow[\text{enzyme 3} \downarrow]{\text{gene 3} \downarrow} \text{arginine}$$

If a mutation occurred in a gene, a defective enzyme might be produced. This could block a particular step in the biosynthetic pathway.

Beadle and Tatum's "one gene controls one enzyme" hypothesis was a substantial advance in our understanding of genetics. Their hypothesis was slightly modified to "One gene controls one polypeptide," since some enzymes are apparently constructed from more than one polypeptide. This work of Beadle and Tatum provided a biochemical explantion of gene action, and they were awarded the Nobel Prize in 1958 for their outstanding accomplishment.

17.9 A gene is a segment of a DNA molecule.

Working with bacterial viruses, Seymour Benzer developed a molecular definition of a gene. He viewed the gene as a segment of a DNA molecule. He coined the word **cistron** to mean a **unit of function** in the DNA molecule; in other words, a cistron is a sequence of DNA nucleotides that determines one polypeptide chain. How long is a cistron? It may be hundreds of nucleotides long. Since proteins may consist of more than one polypeptide chain, more than one cistron may be involved in the manufacture of a particular protein.

In a functional sense, we already know that a mutation can involve a change in a single nucleotide in a sequence of DNA nucleotides. Therefore, a **unit of mutation** is much smaller than a cistron. Benzer used the term **muton** to refer to the smallest part of a DNA molecule that, changed, could result in a mutation. Hence, a cistron would consist of many mutons, or sites of mutation.

Benzer used a third term, **recon,** to describe the smallest unit of DNA capable of recombination. Since recombination can occur between any two adjacent nucleotides, a recon is essentially the same size as a muton, namely one nucleotide.

In summary, Benzer proposed three subsidiary meanings for the term *gene*. We might think of a gene as a cistron, a large segment of a DNA molecule capable of directing the synthesis of one polypeptide chain; as a muton, or mutation site; or as a recon, a recombination site.

Benzer's view of the gene does not invalidate the "one gene–one polypeptide" model of gene action presented by Beadle and Tatum. A primary function of the gene is still considered to be its control of a specific protein. However, since we now know that more than one cistron may comprise what we consider the classical gene, we must refine this hypothesis to say, "One cistron, one polypeptide chain."

17.10 Genes turn on and off, a process which controls the transcription of messenger RNA, which in turn regulates the production of proteins.

All the cells of an organism contain the same kind and number of chromosomes and genes; yet one develops into a muscle cell, while another develops into a nerve cell. This problem of cell differentiation has puzzled scientists for many years. Today it is known that in cells containing the same sets of genes, different ones are active or inactive at different times. Presumably, there are mechanisms for regulating gene action: Genes are turned on and off. Consequently, different enzymes and proteins can be produced in different cells during development and during maturity depending upon their changing requirements.

Evidence concerning gene regulation has come from the study of certain enzyme systems in bacteria. It has been observed that certain enzymes seem to be present in a cell all the time and show little variation in amount. These enzymes are always being synthesized and are known as **constitutive** enzymes. For example, the enzyme necessary for the breakdown of glucose is present in the bacterium *Escherichia coli* regardless of whether glucose is present in the medium.

Other enzymes are synthesized only when the substrate they work on is present. Such enzymes are known as **inducible enzymes.** An example of an inducible enzyme in *E. coli* is beta galactosidase (β-galactosidase). This enzyme is responsible for the breakdown of the sugar lactose to glucose and galactose. When bacteria are grown in a medium without lactose, there is no sign of β-galactosidase being synthesized. However, if we transfer these cells to a lactose medium, β-galactosidase is synthesized in great quantities. When we remove the cells to a nonlactose medium, the synthesis of β-galactosidase stops. Thus, β-galactosidase is produced only when needed.

Based on studies of bacterial mutations, F. Jacob and J. Monod proposed a model for the regulation

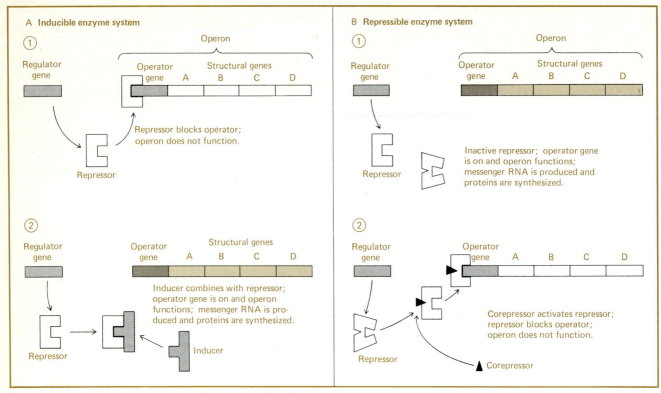

17-3 A diagram of the operon theory of gene control as it applies to inducible and repressible enzyme systems in bacteria. Color indicates functioning operon.

of gene activity. They identified three different types of genes that work together in controlling enzyme synthesis. These are known as regulator genes, operator genes, and structural genes. A **structural gene** is a sequence of DNA nucleotides that specifies an enzyme. A linear sequence of several structural genes may sit side by side on a chromosome. The enzymes specified by such a linear sequence of structural genes may be involved in a single metabolic pathway. An **operator gene** is located immediately adjacent to the series of structural genes. The operator gene controls the turning on or off of the sequence of structural genes adjacent to it. When the structural genes are turned on, they synthesize mRNA, which dictates the formation of the proteins (enzymes) needed to catalyze the reactors of the biochemical pathway. When these structural genes are turned off, the whole biochemical pathway is turned off. An operator gene and the adjacent structural genes that they control are known collectively as an **operon** (Figure 17-3).

How is the operator gene controlled? According to the Jacob-Monod model, a **regulator gene,** which may be located close to or far from an operon, exerts control over the operator. The regulator gene synthesizes mRNA which in turn directs the production of a **repressor protein.** This binds onto the operator gene with numerous hydrogen bonds and thus causes it to turn off the structural genes.

There are two ways in which the model can work, depending on the particular system being studied. In both situations, the natural state of the operon is considered to be turned on and it stays on unless a repressor complex is formed to block the operator and prevents its structural genes from functioning.

In one type of system, involving an inducible enzyme such as β-galactosidase, the repressor protein is fully active as soon as it is synthesized (Figure 17-3A). It immediately bonds onto the operator gene, insuring that the structural genes stay turned off. Then, when an **inducer** molecule is present, (e.g., lactose in the case of the lactose operon), this molecule or one formed from it will combine with the repressor and inactivate it. When the repressor is changed in this manner, it can no longer bond onto the operator, and the latter is free to turn on. The structural genes transcribe mRNA, a process which results in enzyme synthesis. When the inducer is used up, the repressor proteins are again in the right configuration to bind onto the operator gene and turn the system off.

In another type of system, involving a **repressible enzyme** (Figure 17-3B), the regulator gene codes for a **repressor** substance that must combine with another molecule, known as a **corepressor,** before it can be activated to repress an operon. The corepressor may be an end product of the functioning operon. In this way, an excess of this operon product would act to stop its own further production.

Recently, W. Gilbert and others have succeeded in isolating repressor proteins that combine with any inducers present, or with the operator gene in the absence of inducers. For this and other reasons the Jacob-Monod operon theory is generally accepted as an explanation of the control of gene action in bacteria. The extent to which this theory may apply to the control of gene action in eucaryotic cells is still uncertain. Because of the importance of the control of gene action in embryonic development, this question is under very active investigation at the present time.

18

══════════════════════════════════════

Human Heredity

QUESTIONS TO KEEP IN MIND

Why are comparative studies of identical and fraternal twins useful?

Why does pedigree analysis lend itself better to genetic studies on humans than the methods applied to fruit flies or bread mold?

Do Mendel's laws apply as well to humans as to pea plants?

What human diseases are inherited?

In what manner are color blindness and hemophilia inherited?

How are abnormalities in the distribution of chromosomes brought about? What disorders result?

How frequent are genetic disorders in humans? What measures can be taken for their prevention and/or treatment?

Studying patterns of heredity in fruit flies, peas, and other such organisms is relatively simple compared to studying heredity in humans, for at least two reasons. First, we can arrange matings among these organisms at will; and second, we can obtain many progeny in a short period of time. Obviously, this approach cannot be used with humans. Aside from the obvious moral and social difficulties, humans have small families and a long generation time. In consequence, human heredity is studied by the pedigree method and to a lesser extent by the examination of identical twins.

18.1 Studies of identical twins help distinguish between traits that are inherited and those that are environmentally determined.

Since **identical twins** result from the splitting of an early embryo, the twins have identical genes. Thus, any differences we find between identical twins may generally be attributed to the effects of environment. Through studies of twins, we can obtain an estimate of the extent to which a trait is inherited and the extent to which it is environmentally determined. To do this, scientists observe how often the trait appears in identical twins as compared with its occurrence in fraternal, or nonidentical twins. **Fraternal twins** are derived from two distinct eggs that happened to be released from the ovary at about the same time and were fertilized by different sperm. Except for the fact that fraternal twins share the same uterine environment, they are no more alike than ordinary brothers and sisters.

Suppose we wanted to estimate the extent to which an eye-color trait is dependent upon the genotype. We would then observe as many identical twins as possible and find out in how many instances both twins are alike in eye color and how often they are different. Such a study has been made, and in 99.6 percent of the cases studied, when one identical twin had a certain eye color, the other had the same eye color. This was true in 28 percent of the cases with fraternal twins. These results indicate a strong hereditary component in eye color.

18.2 A useful technique in studying human heredity is pedigree analysis.

A pedigree is simply a family history. Yours would include all your relatives, present and past. If we are interested in whether or not certain traits or diseases are inherited and whether these show particular patterns of inheritance, we can look at a family pedigree and see who had the trait or disease. Thus, instead of arranging specific matings, as might be done with *Drosophila,* the results of matings are examined in retrospect.

In the pedigree chart shown in Figure 18-1, males are represented by squares, females by circles. Individuals having the trait under consideration are shaded; individuals without the trait are left uncolored. A marriage is indicated by a horizontal line joining the husband and wife. Children of this marriage are on the next horizontal line, joined to the parents by a vertical line.

Each generation is designated by a Roman numeral. Each individual of that generation is also given an Arabic number, ascending from left to right. Thus any given individual can be identified by a Roman numeral and an Arabic number: I-1, I-2, for example.

In the same pedigree, a male, I-1, has the particular trait we are interested in. He married a woman, I-2, who does not have the trait. They have four children, shown in generation II, identified as II-2, II-3, II-4, and II-5. There are two males and two females. Notice that of the four children of the original parents, two have the trait and two do not. Male II-2, who does not have the trait, marries a female, II-1, of another family. They have five children, none of whom have the trait. These children are shown as III-1 to III-5. Offspring III-2 and III-3 are identical twins. This is indicated by their con-

nection to the same vertical stem extending below the horizontal line. (Fraternal twins from two different eggs fertilized at the same time would be shown by oblique lines that met at the horizontal line, without the vertical stem.) Notice that female II-5, who has the trait, married a male from outside the family, II-6, and had two daughters, III-6 and III-7. One of the daughters, III-6, has the trait.

A pedigree is analyzed according to the laws of Mendelian genetics discussed in Chapter 15. If the trait is due to a dominant gene (*T*), it is likely to appear in the offspring whenever one or both parents have the trait. If the gene is dominant, then male I-1 is either *Tt* or *TT*, whereas female I-2 has both recessive alleles (*tt*). Suppose male I-1 is homozygous *TT*. Then all the first generation children should be *Tt*, which means they should show the trait. The sample pedigree shows that this is not the case. However, suppose the male parent is heterozygous *Tt*. Then about half the children would be *Tt* and show the trait, although there are never sufficient children in a family to express a reliable ratio. If the trait on the pedigree is a dominant one, it should show up in each generation, and not "skip" a generation. We might also predict that, barring new mutation, parents without the trait should give rise to children without the trait. At least one of the parents would have to show the trait in order for it to appear in the offspring.

A trait caused by a recessive gene may skip a generation and will usually appear less frequently in the pedigree. Whereas a dominant gene will show up in both heterozygous and homozygous condition, a recessive gene will only show up in homozygous condition. When marriages occur between close relatives, a recessive gene has a better chance of appearing in homozygous condition.

Using the pedigree method, examples have been found in man of all types of inheritance patterns—simple Mendelian inheritance, sex-linked inheritance, multiple factor inheritance, and so on. Much of the research using the pedigree method has concerned hereditary diseases and disorders.

18.3 Some traits, controlled by a single pair of genes, are inherited according to Mendel's laws.

Simple examples of heritable traits include polydactyly (extra digits), a dominant trait, and one form of albinism (no pigment in the hair, skin, and eyes), which is recessive. Of much greater import are genes that cause diseases such as **Huntington's**

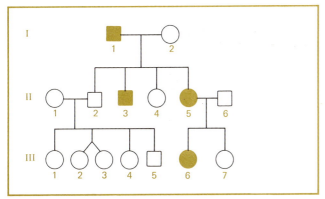

18-1 A sample pedigree (see text for explanation).

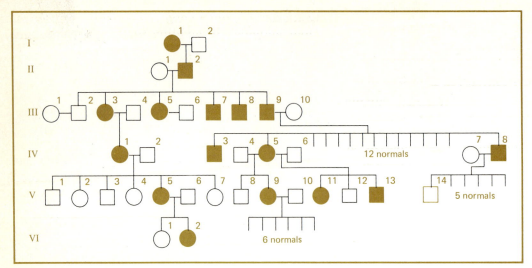

18-2 An abbreviated family pedigree showing the inheritance of Huntington's chorea. Couples A and B had too many off-spring to draw on the pedigree. Couple A had 9 children and 53 grandchildren, all lacking the disease. Couple B had 11 children, 3 of whom were affected. Individuals V-1, -2 and -3 produced a total of 14 children, all normal.

chorea, which is caused by a dominant gene and is inherited in the same relative pattern that holds true for seed shape or flower color in garden peas.

Huntington's chorea is a disease of the central nervous system that results in twitching movements of the limbs and body, progressive mental and physical deterioration, and finally death. Although the symptoms may occur in young and old individuals, the disease most commonly strikes when individuals are in their thirties and forties, often *after* they have had children.

The pedigree of a family in which Huntington's chorea appears is shown in Figure 18-2. The trait is due to a dominant gene, but because victims of this disease can develop symptoms at almost any time in life, the pedigree is difficult to interpret. It may be that certain individuals in the pedigree were characterized by the investigator before they reached the age when their symptoms appeared, or it may be that persons with the gene died from unrelated causes before developing the disease. This may explain the observation that a marriage between V-9 and V-10 produced 6 children, all apparently normal. However, such an outcome may occur by chance. The observation that III-9 and III-10 produced 12 normal children and only 3 with Huntington's chorea, when a 1:1 ratio would be expected, may also be explained by the variable age factor or by chance.

Other diseases that occur late in life may have

an inherited basis, but difficulties similar to those discussed for Huntington's chorea are encountered in charting a pedigree for such diseases. Even without this problem there are many complicating factors in pedigree analysis.

Sickle cell anemia, a fatal disease, is controlled by a single gene and appears in homozygous recessive individuals. In the blood of such persons the red blood cells become distorted into a sickle shape. A heterozygous person (a carrier) shows none of the symptoms and may unknowingly marry another carrier. The couple has one chance in four of producing a baby who will die during childhood of the disease, and one chance in four of producing a baby who will not be a carrier for the disease. Carriers can now be identified, because their cells can be made to sickle under laboratory conditions.

The cause of sickle cell anemia has been traced to the substitution of one amino acid (valine) for another (glutamic acid) in the beta chain (one of the polypeptide chains) of the hemoglobin molecules of persons with the disease. The mutation responsible for this trait can be merely the substitution of one nitrogen base (adenine) for another (thymine) in one codon of the DNA in the cistron specifying the amino acid sequence of the hemoglobin beta chain.

One would expect the frequency of such a gene to decrease in a population because of the failure of homozygotes to successfully reproduce. However, in some parts of Africa, the frequency of this gene approaches 40 percent of the population. The reason for this is that heterozygotes are apparently more resistant to malaria than homozygous normal individuals.

18.4 Some human traits are controlled by multiple alleles. Blood type is an example.

At the beginning of the century, K. Landsteiner experimented with mixing the red blood cells and the blood serum of different individuals. (Serum is blood plasma from which clotting agents have been removed.) In some instances, **agglutination,** or clumping together of red blood cells occurred; in other instances, no clumping occurred. On the basis of these reactions, mankind can be divided into four major blood groups: **A, B, AB,** and **O.** The agglutination reaction is due to the presence of certain substances called **antigens** on the surface of the red blood cells. These antigens determine the blood group of the individual. A person with type A antigen has type A blood; a person with B antigen has type B blood; the presence of both A and B antigens means type AB blood; the absence of both A and B antigens characterizes type O blood. A person with type O blood is called a **universal donor.**

The blood plasma (fluid portion) of an individual either contains or lacks **antibodies,** which complement the blood cell antigens. For example, the plasma of blood-type A persons would contain anti-B antibodies, which would react with B antigen and therefore agglutinate AB or B red blood cells. Similarly, a person with B blood has antibodies in his plasma which would cause clumping of AB or A cells, which contain the A antigen.

The inheritance of the ABO blood groups is determined by three allelic genes which are usually designated by I^A, I^B, and i. The existence of more than two forms of a gene is an example of **multiple alleles.** The gene I^A is responsible for the production of antigen A; the gene I^B directs the production of antigen B; gene i is incapable of producing either antigen A or B. For simplicity, we shall call these alleles A, B, and O instead of I^A, I^B, and i, respectively. Only two of these alleles can be present in any normal individual, one allele from the mother and one allele from the father. Thus, we find individuals whose genotypes are AA, AB, AO, BB, BO, and OO. Both gene A and gene B are dominant over gene O. However, when genes A and B are present in an individual, both genes express themselves fully, and both antigens A and B are made. The genes are said to be **codominant.** Thus a person who is phenotypically blood type A can be genotypically either AA or AO; a type B person is either BB or BO; a type AB person is genotypically AB; a type O individual has the genotype OO. The situation is somewhat more complicated than indicated.

For example, the A antigen is actually divisible into four antigens. In addition there is an H antigen, which is found rarely.

18.5 The Rh factor in man is controlled either by multiple alleles or by three closely linked genes.

Many blood group systems other than ABO have been discovered in man. One of the most interesting is the **Rh** (Rhesus) system. In 1940, Landsteiner and A. S. Wiener injected blood from Rhesus monkeys into rabbits. Antibodies were formed in the rabbit. These antibodies could agglutinate the red blood cells of the Rhesus monkey and the blood cells of about 85 percent of the Caucasians tested; 15 percent of the humans sampled showed no reaction. Presumably, 85 percent of the population had a certain antigen which stimulated the reaction. It was called an Rh antigen; people having it are designated Rh positive (Rh$^+$); those who do not have the Rh antigen are called Rh negative (Rh$^-$).

The genetic story of the control of the Rh factor is complex. Many different Rh antigens have been discovered since 1940, and one group of geneticists proposes that inheritance of these antigens involves multiple alleles. Another group maintains that three closely linked genes control the inheritance. In any event, only one of the many antigens has a major medical importance. The presence of this antigen is determined by a dominant gene D; its recessive allele is d. Individuals who have the genotype DD or Dd will be Rh positive; individuals who are homozygous dd lack the Rh D antigen and are Rh negative. These genes are inherited according to simple Mendelian rules.

The medical significance of Rh is that an Rh-negative mother married to an Rh-positive husband used to run a risk in having more than one or two Rh positive children. Depending on her husband's Rh genotype, such a woman will either assuredly have Rh-positive children or will run a 50-50 chance of conceiving one. If she does, some of the Rh positive blood from the fetus (which usually remains separate) may "leak" into the mother's bloodstream at birth, when the placenta is ruptured. The Rh antigen is recognized as a foreign substance by the mother's body, and she starts producing antibodies against it. The first Rh-positive child is usually born before any ill effects occur. However, if a second or third Rh-positive child is born, the high Rh antibody concentration already in the mother's blood may diffuse across the placenta into the fetus's blood and destroy its red blood cells. The child may be stillborn or may be born with a severe anemia originally known as *erythroblastosis fetalis*, but now called **hemolytic disease of the newborn.**

To save the newborn baby, a complete exchange of blood may be performed; it is even possible to perform the transfusion while the child is still in the uterus. Once born, the child is no longer exposed to an inflow of the mother's Rh antibodies. The ones already present gradually disappear and so do the symptoms of anemia.

It is now possible to prevent the occurrence of hemolytic disease of the newborn by injecting Rh antibodies into an Rh-negative mother shortly after birth of an Rh-positive baby. Since most of the leakage of Rh-positive cells into the mother's bloodstream occurs near birth, the mother forms Rh antibodies after the birth of the first child. The injection of Rh antibodies into the mother just after birth (before the mother produces her own Rh antibodies) destroys any Rh positive cells that may have leaked into the mother's bloodstream. Thus, the mother's body is prevented from forming her own permanent Rh antibodies. This injection can be given after the delivery of each Rh-positive child, and thus protection can be provided.

Difficulties concerning Rh occur only when an Rh-positive father and Rh-negative mother produce an Rh-positive fetus. When the fetus is Rh negative and the mother is Rh positive, there may be some leakage between circulatory systems, but the fetus is unable to produce antibodies until several months after birth, so none are formed to endanger the mother.

18.6 Sex-linked genes control human traits such as red-green color blindness and diseases such as hemophilia.

Among the best known of the more than 100 examples of sex linkage in humans is inheritance of red-green color blindness. Before 1800, it was known that the inability to distinguish between red and green color was inherited. However, it was not until Morgan explained sex linkage in the fruit fly that the mechanism of inheritance of red-green color blindness could be understood. Actually, there are several kinds of red-green color blindness, and the genetic situation is complex. For simplicity, we can consider that red-green color blind-

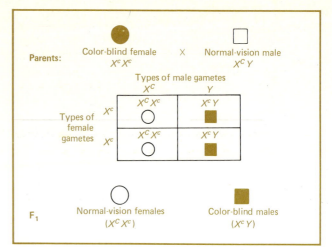

18-3 Results of a cross between a color-blind female and a normal human male.

ness is due to a single recessive sex-linked gene (c). Assuming this, we should be able to predict the presence of the trait in the offspring of particular marriages. If a color-blind woman ($X^c X^c$) marries a man with normal vision ($X^C Y$), her sons will be color-blind, but her daughters will have normal vision (Figure 18-3).

Another rare sex-linked gene in man results in a deficiency in an enzyme known as **glucose-6-phosphate dehydrogenase.** Persons who have this gene are perfectly normal until they are subjected to certain conditions. Afflicted persons suffer **hemolytic anemia** when they eat a type of bean known as the fava bean. Also, such persons react abnormally to an antimalarial drug and to high concentrations of naphthalene and sulfanilamide. Here we see that what genes an individual inherits may determine how he responds to environmental conditions.

Hemophilia is also due to a recessive sex-linked gene. Individuals having this disease cannot carry out the series of chemical reactions leading to blood clotting because they lack the **antihemophilic factor** (AHF). Consequently, they bleed profusely even from a minor cut and do not often survive to maturity unless given repeated injections of AHF.

Queen Victoria was a carrier of the hemophilia gene. Since none of her ancestors showed the trait, it is suspected that a mutation of the recessive hemophilia gene occurred in one of the gametes from Queen Victoria's parents that gave rise to Queen Victoria. Analysis of the pedigree of descendants from Queen Victoria clearly demonstrates that the disease is due to a sex-linked recessive gene (Figure 18-4). For example, Queen

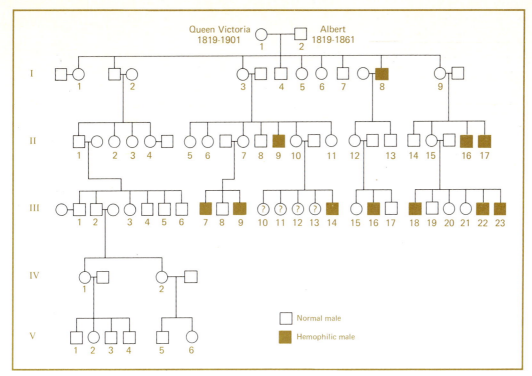

18-4 Pedigree of the descendants of Queen Victoria, showing the incidence of hemophilia.

Victoria married Prince Albert and bore nine children, five daughters and four sons. One of the sons, Leopold, Duke of Albany, received the fatal sex-linked gene from his mother and died of hemophilia at age 31. Two of Queen Victoria's daughters, Alice of Hesse and Beatrice, presumably received the hemophilia gene from their mother and a normal allele from their father, and were carriers. This is evidenced by the occurrence of hemophilia among several of their male descendants.

As might be expected, hemophiliacs are almost exclusively males. In order to produce a hemophiliac female, a hemophiliac male must marry either a hemophiliac or carrier female. Since male hemophiliacs usually died young, the likelihood of one surviving to parenthood is slight. Then, even if the male survives, the likelihood of his meeting and mating with a carrier female is very small, unless marriages occurs among related individuals. Queen Victoria's son Leopold survived to age 31 and had one son and a daughter. Since the daughter, Alice of Athlone, received her father's X chromosome and a normal X from her mother, she was a carrier of the gene for hemophilia. Alice passed her X chromosome (and hemophilia gene) to her son Rupert, who died from hemophilia.

18.7 Nondisjunction of chromosomes results in an extra chromosome and may cause serious problems.

In 1866, a British physician named John Langdon Down described the symptoms of an unusual medical condition. It was characterized by folds at the inner corners of the eyes; mental retardation; underdeveloped sex organs; certain facial features; and other defects. Because of the folds in the eyes, Down named the condition **mongolism.** Since these folds are totally unrelated to the eye folds in Orientals, the condition is more appropriately called **Down's syndrome.** A syndrome is a unique combination of symptoms. The cause of this syndrome remained unknown for many years.

In the mid-1950s, J. H. Tjio and A. Levan developed a new technique for preparing human chromosomes for observation: They obtained stained metaphase chromosomes which were spread out and could be counted easily. They discovered that 46 chromosomes were the normal number of chromosomes in a human cell—44 autosomes and 2 sex chromosomes. Using enlarged photos, the chromosomes in a single cell could be cut out and arranged in homologous pairs, according to size of the chromosome and position of the kinetochore. Such a graphic representation of the chromosomes of a cell is called a **karyotype** (Figure 18-5). Using this new technique to study skin cells from patients with Down's syndrome it was found that afflicted individuals had 47 chromosomes in their cells instead of 46. There were 3 number-21 chromosomes instead of the normal pair. The extra complement of genes was responsible for the characteristic syndrome of effects (Figure 18-5).

This abnormal number of chromosomes can result from nondisjunction in the formation of eggs or sperms. Because of nondisjunction (which was discussed for *Drosophila* in section 16.5) the two chromosome 21's pass into the same gamete. At fertilization, one chromosome 21 is donated by the normal gamete and two by the abnormal gamete. Thus, chromosome 21 is represented three times in each cell arising from divisions of the original fertilized egg.

Down's syndrome can also be caused by a chromosomal abnormality known as a **translocation.** For example, most of chromosome 21 can become attached to a large segment of chromosome 15, resulting in a 15/21 chromosome. A gamete resulting from meiosis can thus end up with a usual chromosome 21 and another chromosome 21 which has become attached to chromosome 15. Fertilization introduces a third chromosome 21, and Down's syndrome occurs. The translocation form of Down's syndrome tends to recur in families. However, most cases of Down's syndrome are believed to be caused by nondisjunction.

The incidence of afflicted offspring increases with the increasing age of the mother. In fact, it has been estimated that half of all cases of Down's syndrome result from pregnancies of women over the age of 35. Many medical geneticists now recommend examination of the karyotype of fetuses carried by older mothers. Fetal cells are obtained by **amniocentesis** — insertion of a hypodermic needle through the mother's abdominal wall and uterus in order to obtain a sample of the fluid from the amnionic cavity around the embryo. The amniotic fluid always contains living cells that have sloughed from the skin and oral epithelium of the embryo. These cells can be cultured and karyotyped. In case of a positive diagnosis for Down's syndrome or some other genetic abnormality, the mother may elect to have the fetus aborted in the hope of later starting a normal pregnancy.

18-5 Karyotype of a patient with Down's syndrome having 47 instead of the normal 46 chromosomes. The extra chromosome is a third number 21.

18.8 Nondisjunction of sex chromosomes during meiosis causes other disorders.

As in *Drosophila,* it is possible for humans to exhibit nondisjunction of sex chromosomes during meiosis. If this occurs as a part of egg formation, an egg with two X chromosomes results. If this egg can be fertilized by a Y-bearing sperm, an XXY individual results; if fertilized by an X-bearing sperm, an XXX individual results. Alternatively, this nondisjunction can lead to an egg without any X chromosomes. When fertilized, such an egg will produce an XO or a YO individual. Thus, by nondisjunction of sex chromosomes during meiosis, XXY, XXX, XO, or YO combinations are possible.

Persons with an XXY chromosomal complement show symptoms collectively known as **Klinefelter's syndrome.** Unlike the situation in *Drosophila,* the Y chromosome in humans plays a positive role in determining maleness. Whereas an XXY fruit fly is a fertile female, an XXY human is male. However, the XXY male has underdeveloped testes and is sterile, body hair is sparse and breasts may be enlarged; also, many XXY individuals are mentally defective.

An individual with three X chromosomes has the **triplo-X syndrome.** Triplo-X fruit flies are weak and inviable. A triplo-X human may be mentally retarded and sterile. However, most triple-X humans are fertile and may have normal intelligence.

A person with only one X chromosome and no Y has **Turner's syndrome** (XO). Such persons are females but have rudimentary ovaries and are sterile. They show a peculiar webbing of the neck, are short in stature, and may have impaired intelligence, although usually not. Interestingly, whereas XO humans are females, XO fruit flies are sterile males.

As in *Drosophila,* the YO combination in humans is yet to be observed, presumably because the presence of the Y chromosome alone is not sufficient for development of a viable individual.

Another interesting sex chromosome abnormality in humans is the XYY male, produced if nondisjunction of Y chromosomes occurs during the second meiotic division in sperm formation. It was once reported that 4 percent of the inmates at a maximum-security prison in Scotland were XYY. Although this and other studies had suggested an association between the XYY genotype and violent behavior, subsequent research has failed to confirm any relationship between XYY and criminality. Many XYY males are, however, generally found to be more than six feet tall, mentally dull, and prone to acne. Since the Y chromosome determines maleness in humans, it is suspected that an extra dose of a Y chromosome may result in an abnormally high male sex hormone level, a condition which may be responsible for excessive aggressiveness. There is considerable controversy, however, about the relation between an XYY chromosomal constitution and aggressive behavior, and there are probably many XYY individuals in the population at large who conform to normal behavior patterns. There are also vast numbers of tall, mentally dull individuals with a tendency for acne who have a normal karyotype!

18.9 Sex chromosome abnormalities can be identified by observing the presence or absence of one or more Barr bodies.

Cheek cells taken from the lining of the mouth and stained offer a simple way of identifying an abnormal number of X chromosomes. The interphase nuclei of most cells from females prepared this way show the presence of a darkly stained body called the **sex chromatin body** or the **Barr body.** This body is located just inside the nuclear membrane. The nuclei of male cells do not have a Barr body. There is a definite relationship between the number of X chromosomes present in a cell and the number of Barr bodies: The number of Barr bodies is one less. Thus the cells of a triplo-X female (XXX) contain two Barr bodies, cells of an XXY individual have one Barr body, and so on.

In the early 1960s, Mary Lyon and others proposed an interesting hypothesis to explain this relationship. The Lyon hypothesis maintains that only one of the two X chromosomes in somatic cells of a normal female is active. The other X chromosome becomes condensed and inactive during embryonic development and this is the Barr body. Hence, their cells will have one X chromosome and one Barr body. In females, either the paternal or the maternal X chromosome may become inactive in different cells. Which one becomes inactive seems to be a matter of chance.

Studies of cells of individuals with abnormal numbers of sex chromosomes supported the Lyon hypothesis, for it seemed that only one X chromosome was active in such cells, regardless of the number of X chromosomes present.

18.10 Defective genes resulting in defective enzymes produce a variety of metabolic disorders.

Many hereditary metabolic diseases are known to involve specific metabolic blocks in the biochemical pathways of metabolism. These can be traced to an enzyme deficiency or lack due to gene mutations. For example, phenylalanine is an **essential amino acid,** meaning that it cannot be synthesized by the body, but must be supplied in the diet. There are several possible normal fates for phenylalanine in the body. Most of it is converted to an amino acid known as tyrosine. Through a series of intermediate steps, tyrosine may be converted into the skin pigment, melanin; tyrosine also serves as a precursor for the thyroid hormone, thyroxine, and for the adrenal hormone, epinephrine. It may also follow a pathway toward formation of an important metabolite, the organic acid known as acetoacetic acid (Figure 18-6).

Normally, the enzyme which catalyzes the conversion of phenylalanine to tyrosine (*phenylalanine hydroxylase*) is produced by the liver. Individuals

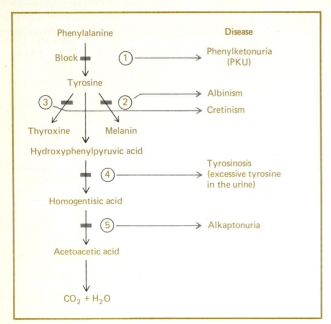

18-6 Biochemical disorders due to inherited blocks in the metabolism of phenylalanine and tyrosine in man.

homozygous for a certain recessive gene lack the ability to synthesize this enzyme. The pathway to tyrosine is thus blocked. Phenylalanine then follows the little-used alternate pathway to phenylpyruvic acid. Phenylpyruvic acid and unused phenylalanine accumulate in the blood. These substances have toxic effects on the central nervous system and result in a disease known as phenylketonuria (PKU). This disease is characterized by severe mental deficiency and is especially damaging to the nervous system during the first six months after birth. The incidence of this disease has been roughly estimated as one in about every 10,000 births in the United States.

Luckily, simple tests have been developed to detect the presence of this disease in newborn babies. In many states, this test is compulsory for newborn babies. When PKU is detected, treatment involves a rigid, distasteful diet, which has a very low amount of phenylalanine. If treatment is begun early, severe mental impairment may be avoided.

Albinism is the result of homozygosity for a certain recessive gene that leads to deficiency in an enzyme involved in the conversion of tyrosine to melanin. Consequently, albinos completely lack this dark pigment, a lack that is easily observable in their skin and hair and in the pinkish appearance of their eyes.

Another metabolic block in the biochemical pathway normally followed in phenylalanine metabolism causes alkaptonuria. Homozygosity for a reces-

sive gene results in absence of the enzyme needed in the conversion of homogentisic acid to maleylacetic acid. Consequently, homogentisic acid accumulates and eventually appears in the urine, causing the urine to turn black on exposure to air. Other symptoms of this condition are the blackening of cartilage and a tendency toward arthritis in later life.

An inherited metabolic block may also occur in the pathway leading from tyrosine to thyroxine. Such a genetic defect may result in a deficiency of thyroxine and consequent deformity and mental impairment. This disease is known as cretinism. Damaging results may be prevented if the disease is detected early enough and thyroid extracts are administered.

18.11 Genetic diseases in man are being attacked through basic research, genetic counseling, and prenatal diagnosis; genetic engineering is a future possibility.

The science of genetics, which could be said to have begun at the turn of the century with the rediscovery of the work of Gregor Mendel, has now progressed to the point where it has enormous potential application to medicine. There are now about 3 million live births per year in the United States alone. About 1 percent of these infants are afflicted with single gene disorders and about an equal number with chromosomal defects. An additional 10 percent suffer from polygenic ("many gene") disorders, many of which do not cause disease until later life. Examples of polygenic disorders are diabetes mellitus ("sugar diabetes") and rheumatoid arthritis.

Within the past half-century the number of disorders that have been recognized as being of genetic origin has steadily increased. In a recent survey, V. McKusick cited 498 known (and 588 suspected) autosomal dominant disorders, 419 known (and 448 suspected) autosomal recessive disorders, and 92 known (and 65 suspected) X-linked defects. Of these 1009 known and 1101 suspected genetic diseases, about 100 have been investigated sufficiently that their biochemical bases are understood. Medical geneticists are endeavoring to characterize an increasing number of these disorders so that the persons afflicted with them can be appropriately treated. In recent years, these diseases have been studied by culturing cells from individuals with genetic defects.

Many genetic disorders can now be prevented in

the next generation by counseling couples known or suspected to have genetic defects in their pedigree. Individuals who are carriers of certain recessive genetic defects can in many cases be identified by simple laboratory procedures and can be counseled regarding the risks of producing handicapped children.

Couples with genetic risks but wishing to have children have several options. They may adopt a family or elect to initiate a pregnancy by artificial insemination with semen from a genetically screened male. Some couples may prefer to accept the genetic risk and rely on prenatal diagnosis to determine whether a pregnancy should be aborted or allowed to proceed to term. As the techniques of detecting genetic carriers of disease become perfected, this may be the preferred procedure. Increasing attention is being directed toward the prevention of genetic disorders, because at present about 20 percent of all hospital beds are occupied by genetically handicapped individuals.

The techniques for modifying the genes of both procaryotic (bacterial) and eucaryotic (including human) cells are now available, and the potential for projects of both medical and economic benefit are enormous. Not only could a number of genetic diseases be corrected, but new blends of microbes could be produced for economic or ecological benefit. For example, bacteria have been produced that can clean up oil spills by metabolizing various products of petroleum. Unfortunately, however, the techniques of **genetic engineering** also have the potential for the accidental production of lethal **pathogens** (disease-producing organisms) that could be difficult to control once released. Consequently, scientists working on genetic engineering research have pressed successfully for the establishment of rigid **containment** procedures that reduce the risk of accidental production of pathogens to an acceptable minimum. This is but another example of the need for society to monitor the uses to which the results of scientific research are put.

SUGGESTED READING

Drake, J. W. *The Molecular Basis of Mutation.* San Francisco: Holden-Day, 1970.

Hayes, W. *The Genetics of Bacteria and Their Viruses.* Oxford, England: Basil, Blackwell, and Mott, 1964.

Levine, R. P. *Genetics.* 2nd ed. New York: Holt, Rinehart and Winston, 1968.

Mendel, G. "Experiments in Plant Hybridization." An English translation of Mendel's original article, reprinted in *Classic Papers in Genetics*, edited by J. A. Peters. Englewood Cliffs, N.J.: Prentice-Hall, 1959.

Sager, R. and F. J. Ryan. *Cell Heredity.* New York: Wiley, 1961.

Sinnott, E. W., L. C. Dunn, and T. Dobzhansky. *Principles of Genetics.* New York: McGraw-Hill, 1950.

Srb, A. M., R. D. Owen, and R. S. Edgar. *General Genetics.* San Francisco: W. H. Freeman, 1965.

Stahl, F. W. *The Mechanics of Inheritance.* 2nd ed. Englewood Cliffs, N.J.: Prentice-Hall, 1969.

Stent, G. S., ed. *Papers on Bacterial Viruses.* Boston: Little, Brown, 1960.

Stent, G. S. *Molecular Genetics.* San Franciso: Freeman, 1971.

Stern, C. *Principles of Human Genetics.* 2nd ed. San Francisco: Freeman, 1960.

Sturtevant, A. H. *A History of Genetics.* New York: Harper & Row, 1965.

Watson, J. D. *The Double Helix.* New York: Atheneum, 1968.

Watson, J. D. *The Molecular Biology of the Gene.* 3rd ed. Menlo Park, Cal.: W. A. Benjamin, 1976.

Yanofsky, C. "Gene Structure and Protein Structure." *Scientific American*, May 1967. Offprint 1074.

Zubay, G. L., ed. *Papers in Biochemical Genetics.* New York: Holt, Rinehart and Winston, 1968.

IV ORGANISMS: BIOLOGICAL DIVERSITY

The sheer numbers and the enormous diversity of living things on the earth are such that no human could possibly recognize and "know" all of them. Even specialists in the study of one group of animals, plants, or microorganisms frequently encounter organisms that they do not recognize; quite often these are "new" organisms that were previously unknown or at least unreported. A system of classification makes it possible to deal with all of this diversity. The human mind rests easier when the number of types of organisms is known to be finite. For this reason, the authors have taken pains to present a necessarily simplified picture of the diversity of living things in the framework of a generally accepted classification scheme, in which the relatedness of organisms is stressed.

The most fundamental aspect of organismal biology is the relationship of structure, function, and behavior. The differences among microorganisms, plants, fungi, and animals are such that this relationship is best treated in somewhat different ways. For example, Chapter 23 deals with certain common aspects of the functions of higher plants. On the other hand, it is most useful to discuss the fundamental aspects of the physiology and behavior of animals by means of representative examples, with emphasis on how each representative animal meets the problems posed by its environment. Because of the number of representatives that must be considered in order to gain any realistic impression of the diversity of organisms, the depth with which function can be treated is unavoidably superficial. Each of these representatives could be the subject of an entire volume.

In Part V on Man, greater space is given to exploring these relationships between form, function, and behavior in a single organism.

19

Biosystematics: An Interpretation of the Origins and Relationships of Living Things

QUESTIONS TO KEEP IN MIND

Why is it necessary to name organisms and classify them?

Who was the person that initiated the use of binomial nomenclature?

What are the advantages and disadvantages of the five-kingdom classification system in relation to other systems?

What characterizes a moneran?

What is the probable evolutionary history of the protists?

What kinds of evidence do taxonomists trust in classifying organisms?

No single person has seen more than a small fraction of the millions of species of organisms alive in the world today. And these are themselves only a fraction of the total that have existed throughout the eons of time, to judge from fossil remains. How can the biologist interested in the diversity of life begin to cope with the problem of classifying these organisms and understanding the relationships among them? On what basis—structure, function, ancestry, or molecular makeup—can they be sorted into groups? Perhaps no classification plan will ever be wholly satisfactory. The best at any given time will most nearly express the relationships which have existed and exist now among organisms. Those who attempt to improve on the classification of organisms are called either **taxonomists** (a general term for all classifiers) or more specifically **biosystematists**.

19.1 Any classification system constitutes a hypothesis to be tested by future observations.

Giving names to organisms and assigning them to groups based on similarities and differences is not a real science unless the system that emerges can be tested. Most systematists agree that a successful classification of organisms should show evolutionary relationships by grouping organisms according to their common ancestors. Consequently, the classification system as a whole should reflect our best guess as to how various forms of life originated on our planet. The ways of testing whether a classification system meets this criterion will be discussed in sections 19.7 and 19.8.

Aristotle of ancient Greece was probably the first serious proponent of any system of taxonomy. He attempted to classify all animals into groups of opposites—for example, winged and wingless. However, bats, butterflies, and bluebirds obviously should not belong in the same taxonomic grouping. Aristotle faced another difficulty. Among the ancients there were engaging story tellers of vivid imagination and they were pleased to describe for Aristotle the curious beasts of which they had heard. Consequently, among Aristotle's surprisingly accurate descriptions of real animals were descriptions of griffins, dragons, and similar creatures of fancy.

19.2 The first successful classification scheme was introduced by Linnaeus in the eighteenth century.

The Swedish naturalist Carl von Linné (1707–1778) originated the system of **binomial nomenclature** by which an organism is identified by its **genus** and **species.** Because Latin was the language of educated Europeans at the time, the genus (plural *genera*) and species (singular and plural) were always given in Latin. In fact, the texts of most scientific papers of that time were written in Latin, and the name Linné is better known in its latinized form, *Linnaeus.* The rules of nomenclature, which Linnaeus helped to establish, were aimed at making certain that only one name would be applied to one kind of plant or animal. This was necessary so that scientists in different countries could understand one another and avoid duplication.

In the time since Linnaeus, biosystematics has become a highly sophisticated branch of theoretical biology. Systematists often differ in their systems of classification. A system that seems sufficient today may well have to be discarded tomorrow, when new evidence is found to confirm or deny an unsuspected relationship among organisms.

19.3 Modern taxonomic schemes are hierarchical.

To say that modern taxonomic schemes are hierarchical in structure means that they set up a few broad categories, each of which is progressively divided into more numerous categories. At the "top" there are three to five **kingdoms** (depending on the classification being used) and at the "bottom," tens of millions of species.

A species consists of the members of a population of organisms that actually do, or are at least capable of, interbreeding, or if asexual, that are alike in structure and function as far as can be determined. Species that are similar are grouped into the same genus, similar genera into a **family,** similar families into an **order,** similar orders into a **class,** similar classes into a **phylum,** and similar phyla into a kingdom. Additional subheadings (suborder, subfamily, subspecies or race, etc.) are used in some cases.

This system of hierarchical groupings is employed whether plants, animals, or microorganisms are being considered. Table 5 shows the classification of seven common organisms: man, chimpanzee, cat, mouse, sea pork, amoeba, baker's yeast, and corn. The first five of these organisms were chosen to illustrate how an increasing gap in relatedness is reflected by organisms' membership in different taxonomic categories at increasingly higher levels.

19.4 The number of kingdoms reflects current views regarding the origin and evolution of life on earth.

It was obvious to the ancients that plants and animals were the two major divisions of living things, and for centuries classifications divided all organisms into one of these two kingdoms. When protozoa (one-celled animals) were first seen in the seventeenth century, they were classed as tiny animals, because they moved and movement was considered to be an animal characteristic. By the

TABLE 5 CLASSIFICATION OF EIGHT SPECIES

	Man	Chimpanzee	Cat	Mouse	Sea pork	Amoeba	Bakers' yeast	Corn
Kingdom	Animalia	Animalia	Animalia	Animalia	Animalia	Protista	Fungi	Planta
Phylum	Chordata	Chordata	Chordata	Chordata	Chordata	Sarcodina	Eumycota	Tracheophyta
Class	Vertebrata	Vertebrata	Vertebrata	Vertebrata	Urochordata	Rhizopoda	Ascomycetes	Angiospermae
Order	Primates	Primates	Carnivora	Rodentia	Aplousobranchia	Amoebida	Endomycetales	Graminales
Family	Hominidae	Pongidae	Felidae	Muridae	Synoicidae	Amoebidae	Saccharomycetaceae	Graminiae
Genus	*Homo*	*Pan*	*Felis*	*Mus*	*Amaroucium*	*Amoeba*	*Saccharomyces*	*Zea*
Species	*sapiens*	*troglidites*	*domestica*	*musculis*	*stellatum*	*proteus*	*cerevisiae*	*mays*

1830s, microscopes had been improved so that scientists could see some of the structures inside protozoa. Since some of these internal structures seemed more typical of plants than animals, a heated controversy arose as to whether various of these protozoa were in fact animals or plants, single-celled or more complex. A Frenchman, F. Dujardin, argued that protozoa were single cells, each with a nucleus and cytoplasm, while a German scientist, C. G. Ehrenburg, insisted they were tiny multi-cellular organisms with complex tissues and organs inside. At the same time, it was seen that many unicellular creatures were green like plants, although some of these same ones moved like animals. It was clear to some scientists at the time that the "unicellular way of life" should be the basis for a third kingdom, Protista.

As mentioned above, whether organisms moved or not had much to do with early ideas about how they should be classified. Nearly all plants had cell walls and were sessile (not free to move about). Largely for this reason, fungi were classified for a long time as plants. So were bacteria, in part because so many were nonmotile and in part because some were able to photosynthesize and had cell walls.

In the second half of the nineteenth century Ernst Haeckel, a German naturalist, suggested that because they had no true nuclei, bacteria and blue-green algae were probably the most elementary forms of life and should be placed in a special division of the protists, namely, Monera. The idea that life evolved from extremely simple organisms resembling a modern form enjoyed tremendous popularity among some scientists and briefly launched a bizarre search for the "primordial ooze from which all life ascended."

In 1868 the famous British naturalist T. H. Huxley reexamined some samples of deep-sea muds which had been collected years before; he found lying above them a layer of mucuslike jelly. When examined microscopically, the jelly was seen to move slightly, apparently engulfing bits of mud (Figure 19-1). Huxley named this elemental sheet of protoplasm *Bathybius haeckelii*, after Haeckel, who believed it to be the root of the evolutionary tree. Seven years later this "protoplasmic" jelly was discovered actually to be a chemical precipitate caused by the prolonged contact of sea water with alcohol. With the *Bathybius* theory dispelled, the name *Monera* sank into relative obscurity, where it remained until quite recently.

Even in the first years of the present century when a great many new facts that might aid classi-

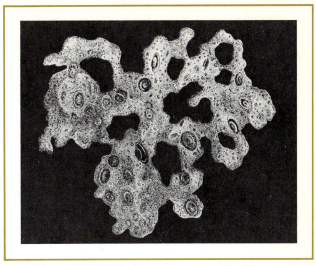

19-1 A famous nonexistent organism, *Bathybius haekelii*. (See text for explanation.)

fication were being discovered, the two- or three-kingdom model (plants, animals, and sometimes the protists) was rarely challenged. Since the 1950s, however, quite a variety of new schemes have been suggested.

The simplified taxonomic scheme presented in this book is the one considered by the authors to be most nearly in accord with the currently accepted views of the origin and evolution of life on earth. In 1969, R. H. Whittaker proposed a comprehensive taxonomic scheme based on *five kingdoms*: **Monera, Protista, Fungi, Planta,** and **Animalia.** These five kingdoms and the major groups of organisms they comprise are shown in Figure 19-2 as a kind of "evolutionary tree." The roots are the chemical events leading to **biogenesis,** the origin of life on earth approximately 3 billion years ago. At the top are the most advanced forms of life within each kingdom. Arrows show current views regarding the evolution of living things from the simplest unicellular organisms to such dominant forms as humans and insects, which are believed by some to occupy exalted positions at the top of the evolutionary tree.

19.5 The name *Monera* now designates the kingdom comprising all procaryotic organisms. These are relatively simple in structure and function, but amazingly diverse chemically.

The kingdom Monera is comprised of the bacteria and the blue-green algae. All monera are procary-

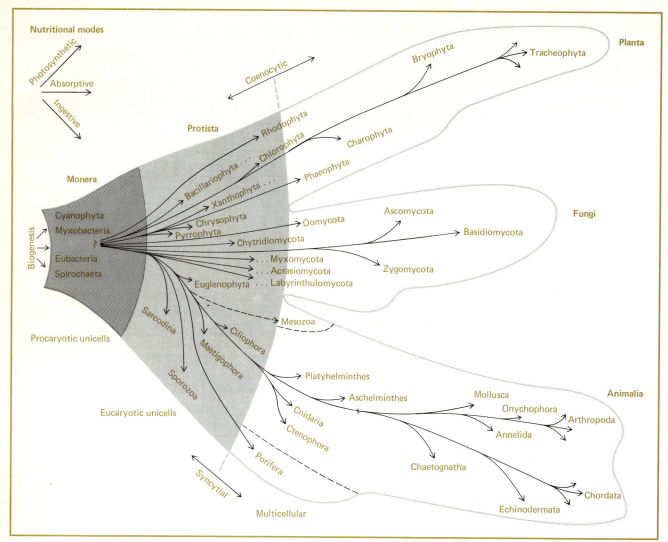

Nutritional modes

Photosynthetic

Absorptive

Ingestive

Biogenesis

Monera

Cyanophyta
Myxobacteria
?
Eubacteria
Spirochaeta

Procaryotic unicells

Protista

Bacillariophyta . . .
Xanthophyta . . .
Chrysophyta
Pyrrophyta
Chytridiomycota

Rhodophyta
Chlorophyta
Phaeophyta
Charophyta

Oomycota

. . . Myxomycota
. . . Acrasiomycota
Euglenophyta . . . Labyrinthulomycota

Coenocytic

Mesozoa

Sarcodina
Mastigophora
Ciliophora
Sporozoa

Eucaryotic unicells

Syncytial

Porifera
Ctenophora
Cnidaria
Platyhelminthes
Aschelminthes

Multicellular

Chaetognatha

Planta

Bryophyta
Tracheophyta

Fungi

Ascomycota
Basidiomycota
Zygomycota

Animalia

Mollusca
Onychophora
Arthropoda
Annelida
Chordata
Echinodermata

19-2 The five-kingdom scheme of classification adopted in this book showing the presumed evolutionary positions of the major phyla.

otes, and there are no procaryotes outside the kingdom Monera (section 2.5). Procaryotic cells by definition have a single genetic unit (or "chromosome") consisting of a single DNA strand. Procaryotic DNA is not complexed with the basic **histone proteins** always found in close association with eucaryotic chromosomal DNA. Procaryotic cells have no nuclear envelope, no mitochondria or plastids, endoplasmic reticulum, Golgi bodies, or vacuoles. They never form a mitotic spindle for division, and their flagella differ from those found in eucaryotes.

What the monerans do possess is far more impressive than the list of features they lack. The bacteria alone show a most impressive breadth of diversity, comparable to that in any group of or-

ganisms in the other kingdoms. The diversity is not so much structural as it is physiological and chemical.

Bacteria are primarily **degraders**; that is, the majority obtain their energy by breaking down organic molecules in their environment. Presumably the earliest bacterial life appeared spontaneously in the watery pools, rich in organic and inorganic nutrients, that are believed to have existed on the primitive earth. Early bacteria lived by degrading these nutrient compounds and later evolved related forms with enhanced synthetic abilities that could live on simpler energy sources, such as the oxidation of hydrogen and iron. Photosynthesis evolved in both the bacteria and blue-green algae, made possible by the development of photosynthetic pigments. Some pigments have been found as residue in rocks as old as the first traces of life on earth.

The chemical diversity that makes the monerans so fascinating to microbiologists and so important to all of us is better understood with a background in organic chemistry and biochemistry. Consequently this book does not devote much space to them. Nevertheless, many of the fundamental processes in biochemistry and genetics that are discussed in Chapters 6, 7, and 17 were elucidated in large part by studies of bacteria and bacterial viruses. More importantly perhaps, bacteria play enormously important roles in our lives that we rarely appreciate. A vast number of bacteria cause diseases, from dental caries (cavities) to streptococcal tonsilitis; but an even greater number are directly involved in the production of cheese, vinegar, yogurt, and butter, and other food products. Bacteria are indirectly involved in a far, far greater number of human activities. For example, meat comes largely from animals that feed on grass and cereals. These plants must have nutrients from the soil which can be procured only by the process of **bacterial decomposition.** Without bacterial decomposition, the world would be sterile and unfit for human habitation.

19.6 The protists are the descendants of the first eucaryotes, and comprise groups with surprising structural and chemical diversity.

There are two ways of looking at protists: either as unicellular organisms or as acellular organisms. The first view implies that they function as single cells, just as single cells from higher organisms can sometimes function if isolated under appropriate conditions. The second view considers them as organisms that did not find it adaptively expedient to become compartmentalized into smaller cellular units.

According to current theories, the eucaryotes probably first evolved as symbiotic associations of procaryotic cells, some of which were originally **endosymbionts,** that is, organisms that live inside others for their mutual benefit. Then, according to these theories, they evolved into organelles. This idea seems plausible because many protists alive today contain a variety of endosymbionts instead of or in addition to the organelles characteristic of eucaryotes.

There are both "animal-like" and "plant-like" protists. The most important groups (phyla) of "animal-like" protists include Mastigophora, Sarcodina, Sporozoa, and Ciliophora. The phylum Sarcodina includes free-living and parasitic amoebae, heliozoans, foraminifers, and radiolarians. Ciliophora is represented by ciliates, such as *Paramecium, Stentor,* and *Euplotes,* and by suctorians. Sporozoans are mostly parasitic, and many cause diseases in humans and domestic animals. The plantlike protists are an equally large and important group including the phyla Chrysophyta, or golden algae; Bacillariophyta, or diatoms; Xanthophyta, or yellow algae; the Pyrrophyta (dinoflagellates); and Chlorophyta (green algae).

The protists are regarded as the forerunners of the first fungi, plants, and animals. Their diversity presumably allowed the forces of evolution to guide their fates in many directions. Even though present-day forms obviously differ from early protists, they do give us some idea of the probable origins of the multicellular kingdoms: Planta, Fungi, and Animalia.

19.7 The evolution of multicellular forms of life proceeded in three directions, each dependent upon a different form of nutrition.

Higher green plants are nearly all characterized by their ability to synthesize food from sunlight: Their nutritional mode is said to be photosynthetic or **photoautrophic.** This mode of nutrition, plus multicellularity (or at least the possession of many nuclei), has served to characterize the kingdom Planta in many taxonomic schemes.

Members of the kingdom Fungi are characterized by their absorptive mode of nutrition, passed down to them by their protistan and bacterial ancestors: They absorb organic molecules. Some fungi are now believed to have evolved from more than one group of colorless flagellates with a similar nutritional mode, while others may have evolved from red or green algae. Fungi characteristically invade their food supply, promote its decomposition, and then absorb the organic nutrients released.

Fungi were for a long time classified as plants, but there are several fundamental reasons, in addition to nutrition, why they are now recognized as a separate kingdom. For example, fungi have a unique form of sexual reproduction, in which two nuclei coexist for some time as a so-called **heterokaryon** without nuclear fusion.

Most fungi are not multicellular in the same sense as are higher plants and animals. Often they are **coenocytic** (a term synonymous with **syncytial**), meaning that their many nuclei have at least some freedom to move throughout the organism. However, coenocytic or syncytial organisms occur also in the protists, lower plants, and lower invertebrates: There are many examples of related multi-cellular and syncytial organisms. Multicellularity, therefore, is not a very good dividing line between kingdoms; yet it may be useful if its limitations are recognized.

Members of the kingdom Animalia are mostly ingesters and consumers; that is, instead of manufacturing food or absorbing it, they eat, or ingest, it. Consequently, animals have evolved increasingly efficient nervous systems and mechanisms of movement so that they may search for and capture food. Circulatory, respiratory, and excretory systems of increasing complexity and capacity evolved to accompany these developments.

19.8 Biosystematics draws upon many kinds of evidence to establish similarities and differences among organisms.

Early taxonomists relied heavily on the structure of organisms to determine taxonomic relationships. Leaf shape, number of appendages, presence or absence of certain organs or organelles were often factors used to decide the taxonomic position of an organism. In the fungi, life cycle and manner of reproduction have proven to be important.

As physiology and biochemistry developed, additional factors were considered. The type of photosynthetic pathway and accessory photosynthetic pigments are one such chemical-physiological characteristic used to determine plant relationships. Similarly, respiratory pigments or excretory systems found in animals can provide clues to taxonomic relationships.

Within the past decade it has become possible to compare the structure and composition of the protein and DNA components of supposedly related organisms. The more their macromolecules resemble one another, the closer these organisms are related. The techniques of DNA comparison (DNA hybridization) are now being refined so that they may be able to provide stronger evidence in some instances of suspected relationships and add new knowledge in areas heretofore inaccessible by traditional approaches.

20

Unicellular Organisms: Monera and Protista

QUESTIONS TO KEEP IN MIND

What features characterize the major divisions of the kingdom Monera?

What factors in the earth's early environment may have favored diversity among procaryotes?

Are there factors other than the environment that may have influenced the evolution of protists?

What are the major divisions of protists?

The five-kingdom scheme of classification is based on the belief that all multicellular organisms—plants, fungi, and animals—evolved long ago from the eucaryotic unicellular organisms that existed then and would now probably be classified in the kingdom Protista. These protists had themselves previously evolved from the early procaryotic unicellular organisms, presumably similar to the modern-day kingdom Monera. To understand how these complex evolutionary steps might have occurred in the distant past, it is necessary to review some of the general properties of the procaryotes before discussing some of the features they evolved that might have contributed to the rise of the protists.

20.1 The procaryotes of the kingdom Monera are divisible into four phyla: Eubacteria ("true" bacteria); Myxobacteria; Cyanophyta (blue-green algae); and, Spirochaeta (spirochetes).

The phylum Eubacteria contains an enormous number of species and doubtless many more remain to be characterized. They exist in the shape of rods, spheres (**cocci**), or spirals and have stiff cell walls. Some are nonmotile, while others have from one to many **bacterial flagella**. Bacterial flagella are naked filaments composed of the protein **flagellin**; they are helical and are caused to rotate by a "motor" associated with the bacterial membrane. In contrast, the flagella of eucaryotic cells consist of an axoneme with its "9 + 2" arrangement of microtubules (section 8.10) protruding from the cell and surrounded by an extension of the plasma membrane.

Eubacteria have evolved an amazing number of ways to extract energy from chemicals in the environment. There is scarcely an organic compound known that cannot be decomposed and utilized by some species of bacterium. Among the eubacteria are some aerobic (oxygen-using) forms that have respiratory pigments incorporated into their membranes—as do the mitochondria in eucaryotic cells (sections 6.3 and 6.4). Some eubacteria are photosynthetic.

The phylum Myxobacteria represents another large group of organisms, closely related to the blue-green algae. They have relatively thin cell walls, and some species exhibit an interesting kind of "gliding" movement that often involve changes in cell shape. It is not known whether this kind of movement bears any relationship to amoeboid movement or cytoplasmic streaming, both of which are found throughout the protista.

Representatives of the phylum Cyanophyta (blue-green algae) are similar to myxobacteria in many respects, but they possess chlorophyll and other pigments which give them their character-

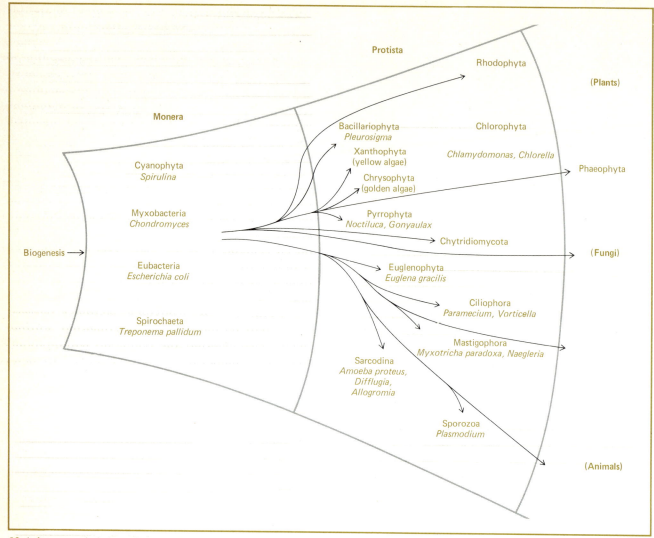

20-1 An expanded view of two of the five kingdoms, showing the positions of major phyla and representative organisms discussed in the text.

istic colors. In many blue-green algae the photosynthetic pigments are localized in lamellar membranes. Some of the blue-green algae can fix atmospheric nitrogen—that is, synthesize nitrogen containing materials from nitrogen gas in the air. Because of this ability they help to fertilize rice crops in the Orient.

Many filamentous blue-green algae move by gliding. One of the most intriguing forms of movement is seen in *Spirulina*, a filamentous form shaped like a long screw. Depending on which way the screw turns, the chain of cells moves forward or backward. These filaments grow as the cells of which they are composed divide. As more and more cells are added, the filaments intertwine and twist around one another and eventually break into

shorter lengths. Continued growth, breakage, and random locomotion all contribute to their dispersal.

The members of the phylum Spirochaeta (spirochetes) are organisms of helicoid shape coiled around an **axial filament.** They often possess threadlike or fibrillar structures. Many are harmless symbionts, such as those commonly found in the human mouth and in the termite gut, respectively. One infamous spirochete, however, is anything but harmless. *Treponema pallidum* causes syphilis, a venereal disease which has spread great suffering throughout the Western world. Once on the verge of being controlled by antibiotics, it has made a tragic comeback. More permissive sexual attitudes and an unfortunate reluctance of persons with the disease to identify their sexual contacts so that effective public health measures can be taken have contributed to its resurgence.

20.2 Evolution among the procaryotes is believed to have been brought about by natural selection acting on populations having wide genetic variations.

Genetic studies on bacteria have shown that mutations occur quite frequently in large populations—once in 10^6 to 10^9 divisions. Considering the short cell cycle of many bacteria (from 20 minutes to a few hours), mutations may occur once every several hours or days. Physicians treating bacterial infections with penicillin have been distressed to learn how rapidly bacteria can evolve penicillin-resistant strains by mutation. In the early days of penicillin therapy, infections were often treated for too short a time, and the leukocytes of the blood did not have time to destroy all of the infecting bacteria. (Penicillin does not kill bacteria, it merely halts their reproduction, allowing the host leukocytes to phagocytize them.) When therapy was termined too soon, the infection flared up again and some of the surviving bacteria were found to be penicillin-resistant. Physicians had to substitute other antibiotics and use prolonged therapy.

Presumably conditions prevailing on earth at an earlier time were even more conducive to the occurrence of mutations. At that time, oxygen was in short supply in the atmosphere, and the earth lacked its present protective envelope of ozone, which is formed from oxygen and absorbs rays of ultraviolet light. Ultraviolet light is readily absorbed by nucleic acids and causes changes in their base composition, resulting in mutations. The early evolution of bacteria may have been greatly speeded by the occurrence of such radiation-induced mutations, and some of these early bacteria probably produced cells with some attributes that we now consider eucaryotic.

20.3 The evolution of eucaryotic Protista from procaryotic Monera is believed to have been brought about by two processes, natural selection and hereditary symbiosis.

There is a large gap both in structure and in organization between the most complex procaryotes and the simplest eucaryotes. In the past decade theories have been advanced that permit us to understand in a general way how the protists might have evolved.

Many scientists believe that some of the charac-

teristics of eucaryotic cells developed by the process of **hereditary symbiosis.** Many forms of symbiosis (meaning "life together") are not hereditary, in which case two organisms that live together for their mutual benefit reproduce independently. Their descendants reassociate with one another at some later phase of their life cycle. In hereditary symbiosis, however, the two species of organisms have developed a more complete association with each other and stay together generation after generation. *Paramecium bursaria* (Figure 20-2), for example, is a protist that has cells of the green alga *Chlorella* living in its cytoplasm. By experimental means, *P. bursaria* can be deprived of its *Chlorella* symbionts, and the latter can also be isolated and raised separately. If the paramecia are presented with free-living *Chlorella* that have been collected from a pond, they will ingest the algae and then digest them. If, however, they are fed back symbiotic *Chlorella* that were removed from *P. bursaria*, these will be ingested but not digested. Furthermore, once ingested, the symbionts will reproduce only to a number of individuals genetically deter-

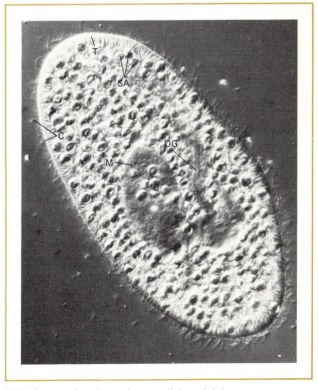

20-2 *Paramecium bursaria*, a protist containing many symbiotic *chlorella* algae (*SA*). Parts of *P. bursaria* visible include the cilia (*C*), oral groove (*OG*), trichocysts (*T*), macronucleus (*M*). The organism is slightly compressed to show its inner structures. (Photomicrograph by author.)

mined by the host. These results suggest either that the protist recognizes its symbionts or that the symbionts have acquired some method to avoid being digested. In either case, when *Paramecium bursaria* divides, its daughter cells inherit symbiotic *Chlorella* and along with them a mechanism for controlling their number.

Many protists have bacteria living symbiotically on their surface or in their cytoplasm. An example is *Myxotricha paradoxa*, a flagellate that lives in the gut of an Australian termite. When this protist was examined with an electron microscope, what had been believed to be its eucaryotic ("9 + 2") flagella beating in synchrony were found to be bacterial flagella belonging to spirochetes nestled in pits in the cell surface. Curiously, these spirochete flagella serve to propel their host. However, the synchrony of their beating is not due to any kind of nervous coordination but to hydrodynamic interactions between neighboring flagella. *Myxotricha* actually has two kinds of spirochete symbionts, and at least one kind of bacterial symbiont as well.

To explain these curious observations, and also the evolution of the earliest protists, it has been suggested that the very first eucaryotic cells, and therefore the first protists, were capable of ingesting various kinds of procaryotes, some of which resisted digestion and became endosymbionts. Because the simplest cells we know that can ingest other cells are amoeboid, it is suspected that this first eucaryotic protist might have been a very simple anaerobic amoeba. Such organisms exist today, and many contain bacteria and spirochetes living in them symbiotically. An example is one species of giant amoeba, *Pelomyxa palustris*, which lives in anaerobic mud at the bottom of many shallow ponds. It contains cytoplasmic symbiotic bacteria.

20.4 The mitochondria in protists may have evolved from aerobic bacteria acquired as endosymbionts.

In virtually all cells the ground cytoplasm contains all the enzymes necessary to catalyze the reactions of anaerobic glycolysis (sections 6.1 and 6.2). This is one of the simplest forms of energy metabolism and probably was the one used by early amoebae and other protists that lived in anaerobic environments (as some still do). It is now widely believed that certain amoebae developed the ability to profit by oxygen-rich environments and to metabolize their food more efficiently by acquiring aerobic bacteria as endosymbionts. These bacteria, similar

in many respects to ones alive today, may have evolved into mitochondria, structures that have features in common with bacteria. Mitochondria contain a single strand of DNA (similar to a bacterial "chromosome") on which are located genes which control the production of some of the enzymes needed for cellular respiration (aerobic glycolysis) and most of the structural proteins required for new daughter mitochrondria.

20.5 Centrioles, flagella, and cilia may have evolved from spirochete endosymbionts.

The **centriole** is a structure found in many protists and in nearly all animal cells. It exists in several forms and may serve either as a **basal body** from which cilia and flagella grow (section 8.10) or as a microtubular organizing center involved in the formation of mitotic spindles and asters (section 9.6). Although protists typically have centrioles or their derivatives, most cells of higher plants do not.

In a process resembling the evolution of mitochondria, primitive amoeboid eucaryotes may possibly have acquired a hereditary symbiont which evolved into the centriole. In this case the symbiont was probably an organism similar to some present-day spirochetes. Such symbionts would then have evolved in several ways inside their hosts to become a centriolelike organelle of ancestral eucaryotes. In modern flagellates alone, there are many different forms of centrioles and centriolar derivatives. In many kinds of protistan cells, centrioles and basal bodies are interconvertible and can replicate within the cell.

The evidence against the hypothesis that centrioles and their derivatives originated as endosymbionts is the fact that DNA cannot be detected in centrioles. This could mean either that it is (and perhaps always was) absent, that its quantity is too small to be detected with present methods, or that the DNA somehow became part of the host genome.

Whatever the origin of centrioles, the possession of one or more eucaryotic flagella derived from centrioles offered flagellated protists the advantages of motility. Superficially the symbiotic relationship could have resembled that of man and horse: transportation in return for food and shelter. From the diversity of flagellated organisms and the persistence of flagellated stages in the development of higher organisms, we can surmise that the flagellum was of enormous **selective advantage** in the course of evolution. That is, organisms with flagella were able to disperse into favorable environments and were therefore selected by natural forces to survive more frequently than their nonflagellated relatives.

20.6 The process of mitosis must have evolved early in the history of the protists and was essential for their further evolution.

Mitosis occurs in the reproduction of all eucaryotic cells, both in protists and in cells of multicellular organisms. However, among the protists there is much greater diversity in the details of mitosis than is found in any of the three kingdoms of multicellular organisms. Because of this it is believed that mitosis probably evolved very early, and its evolution took a very long time. Perhaps a period of a billion years was necessary for the distinctly different mechanisms of animal and plant mitosis to evolve. As we shall see later, the divergent details of mitosis and cytokinesis provide clues as to how the land plants may have evolved (section 21.9).

Mitosis is a nuclear phenomenon and the nucleus of eucaryotes differs from the nucleoid of procaryotes in several important ways. Procaryotes have a single unassociated strand of DNA. Eucaryotes, in contrast, may contain up to several hundred chromosomes in which the DNA is associated with histones and other proteins. Also, eucaryotic chromosomes are all contained within a nuclear envelope but the procaryotic nucleoid is not. It is not known how the eucaryotic nucleus and the process of mitosis arose, but one result of its presence is clear: protists could undergo extensive evolution once they had acquired a mechanism for the precise apportionment of more than one replicated chromosome to daughter cells. The mitotic spindle made this precise apportionment possible.

Meiosis and sexual reproduction greatly enriched the variability of populations by causing recombination of genetic material. Meiosis in some form exists in representatives of all of the major groups of protists, and probably contributed significantly to the evolution of diversity within the kingdom.

20.7 Today the kingdom Protista contains more than 35,000 species of enormous diversity in structure and function.

The protists are commonly divided into nine or more phyla (Table 6). One of the largest of these is the phylum Mastigophora, the zooflagellates.

TABLE 6 UNICELLULAR ORGANISMS

Kingdom Monera (unicellular, procaryotic organisms)	Phylum Eubacteria ("true" bacteria) Phylum Myxobacteria (myxobacteria) Phylum Cyanophyta (blue-green algae) Phylum Spirocheta (spirochetes)

Kingdom Protista (unicellular, eucaryotic organisms)	Phylum Sarcodina (pseudopodial organisms)
	  Class Rhizopoda (naked amoebae)
	  Class Testacea (shelled amoebae)
	  Class Foraminifera (shelled reticulopodial organisms)
	  Class Radiolaria (reticulopodial organisms with a skeleton)
	  Class Actinopoda (heliozoans, or "sun-animalcules")
	Phylum Ciliophora (organisms moving by cilia)
	  Class Ciliata (ciliates)
	  Class Suctoria (suctorians—have ciliated juvenile stage)
	Phylum Mastigophora (zooflagellates)
	Phylum Sporozoa (sporozoans—parasitic, nonmotile or gliding)
	Phylum Euglenophyta (euglenoids, photoheterotrophic flagellates)
	Phylum Pyrrophyta (dinoflagellates)
	*Phylum Chrysophyta (golden algae)
	*Phylum Xanthophyta (yellow algae)
	*Phylum Bacillariophyta (diatoms)
	*Phylum Chlorophyta (green algae)

*Groups that contain multicellular forms.

Another large and important group is the phylum Sarcodina, which includes amoebae, planktonic organisms such as foraminifers and radiolarians, heliozoans, and many others. There is also the phylum Sporozoa, the phylum Ciliophora, and the phylum Pyrrophyta (dinoflagellates). Other phyla seem to bridge the transition zone between protists and fungi (e.g., the phylum Myxomycota, the slime molds) or between the protists and the plants (phylum Chrysophyta includes the yellow algae; phylum Bacillariophyta, the diatoms; and Chlorophyta, the green algae).

Examples will be given of particular organisms characteristic of several of these phyla, in addition to general comments on each phylum as a whole.

20.8 *Amoeba proteus* is a modern representative of the phylum Sarcodina.

Protists that move by means of pseudopods alone are classified in the phylum Sarcodina. In addition to naked amoebae there is a wide variety of shelled amoebae.

20-3 *Amoeba proteus* as observed in a differential interference contrast microscope. Structures visible include pseudopods (*Ps*), nucleus (*N*), contractile vacuole (*CV*), uroid (*U*), and many different kinds of cytoplasmic inclusions. (Photomicrograph by author.)

Although amoeboid movement presumably evolved early in the history of protists, a few sarcodines probably represent relics from that time. Many modern-day amoebae such as *Amoeba proteus* (Figure 20-3) are incredibly complex, both in structure and in their life cycles.

Amoeba proteus is found in unpolluted lakes and ponds. It and many other species of free-living amoebae can be collected by brushing aquatic plants gently in a dish of pond water. *A. proteus* is so large that it can be seen as a white speck on the bottom of a dish with the naked eye. Its surface is covered with a remarkably impermeable membrane, the **plasmalemma**, the outside of which has a very thin, fuzzy mucoprotein coat, the **glycocalyx.** The cytoplasm contains thin filaments of F-actin and thick filaments that are aggregates of myosin (section 8.3). These are used in cytoplasmic contraction, pseudopod formation, and probably phagocytosis (cell eating) and pinocytosis (cell drinking) as well. The cytoplasm of *A. proteus* contains mitochondria, Golgi bodies, vesicles, refractile bodies, and crystals of the nitrogen excretion product, **triuret.** Bacterial endosymbionts are also usually present. The nucleus is bean-shaped and surrounded by an unusual "honeycombed" nuclear envelope. Just inside the nuclear envelope are numerous nucleoli.

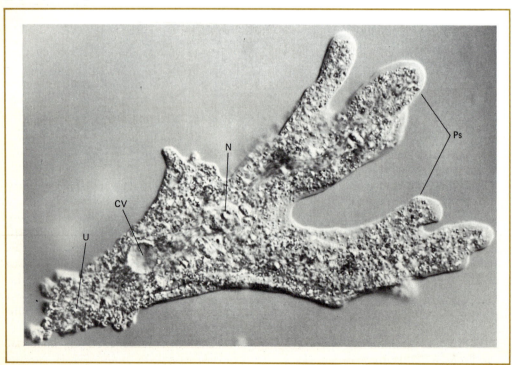

A. proteus feeds primarily on ciliates and flagellates that it entraps in **food-cup pseudopods.** In some cases these food cups are formed in advance and then closed when prey happen to wander in. In other cases the amoeba extends a pseudopod in the direction of a prey and forms a food cup of just the right size and shape to fit over or around it. The ingested organisms are surrounded by a section of the plasmalemma that becomes the membrane of a food vacuole. Lysosomes fuse with the membrane of the food vacuole and provide enzymes that digest the macromolecules in the prey. Any undigested material is **egested** (cast out) by fusion of the food vacuole membrane with the plasmalemma.

Amoebae are so small that gas exchange can be accomplished effectively by diffusion. Nitrogenous wastes from the breakdown of proteins and their amino acids may either diffuse out of the cell (as ammonia or urea) or may pass out of the cell via the **contractile vacuole.** This structure, found most frequently in the tail of the amoeba, gradually fills with water and dissolved cytoplasmic substances over a half-minute or so, then contracts to empty its contents through a temporary opening in the plasmalemma.

The breakdown products of nucleic acid accumulate in amoeba cytoplasm as crystals of triuret (formed from nitrogen-base breakdown) and as spherical refractile bodies (containing polyphosphates). Amoebae apparently seldom dispose of this unusual metabolic "solid waste"; they simply "outgrow" it.

A. proteus moves by means of pseudopods, temporary projections which may be extended or retracted. The extension of pseudopods is caused by the contraction of the cytoplasm that flows into them as it reaches the tip of the pseudopod (section 8.7).

Behavior of amoebae is not limited to the extension and retraction of pseudopods; they respond in a consistent manner to a number of environmental stimuli. For example, intense white or blue light shone on the tail region causes the amoeba to move rapidly away from the light. Irradiation of pseudopod tips causes retraction. Pseudopod tips also respond to mechanical stimuli (touching) by stopping; however, the posterior region of the cell, or uroid, is quite insensitive to touch. Chemical stimuli can cause amoebae to stop, speed up, or to drink in some of the fluid from their environment by pinocytosis. Amoebae respond to prey and to substances diffusing from prey by forming food cups.

The mechanisms responsible for amoeba behavior are incompletely understood. Stimuli may act either on the membrane or on the contractile mechanism in the cytoplasm. It now seems quite certain that the regulation of the calcium concentration in amoeba cytoplasm controls cytoplasmic contractility.

Amoebae reproduce by mitotic division of the nucleus followed by fission of the entire cell. The amoeba ceases movement and assumes a rounded and "rough" appearance during mitosis but resumes apparently normal shape and amoeboid movement after nuclear division. While temporarily binucleate, the cell literally pulls itself apart into two uninucleate amoebae by directing pseudopods in opposite directions.

20.9 Other sarcodines differ from amoebae in pseudopod structure, life cycle, and mode of defense.

The closest relatives to the naked amoebae are those that secrete a hard shell or else form a shell or **test** of solid particles collected from the environment. These testaceans or shelled amoebae (Figure 20-4A) are very common in fresh water and are most easily detected in water samples by their characteristic shells. The amoebae that inhabit the shells are very similar to some species of naked amoebae. More distantly related to the above groups are the foraminifers, radiolarians, and heliozoans (Figure 20-4B, C, D). All three groups have thin, filamentous pseudopods, but their movement and manner of support differ.

20.10 Photosynthetic and heterotrophic flagellates of the phylum Mastigophora have developed a great variety of feeding habits which allow them to live in many places.

The flagellates have clearly been a successful group; both photosynthetic ("plantlike") phytoflagellates and heterotrophic ("animallike") zooflagellates exist in great variety. Some phytoflagellates can metabolize in the manner of animal cells when it is dark, and have mutant forms without chloroplasts. Thus the distinctions between the phytoflagellates and zooflagellates may have been overemphasized. Indeed, the two types are quite possibly closely related.

Of the zooflagellates, some feed on bacteria, while others absorb organic molecules of various

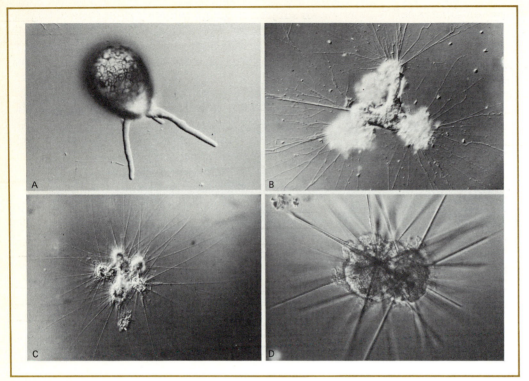

20-4 Some representative sarcodine protists. *A. Difflugia,* a testacean. *B. Allogromia,* a foraminiferan lacking a solid test. *C. Radiophrys,* a colonial heliozoan, and *D. Acanthometron,* a radiolarian. *A* and *C* are from fresh water; *B* and *D* are marine. (*A*, *B*, and *D* are photomicrographs by author; *C* is by C. W. Watters.)

kinds, decomposing them for a supply of energy. Many zooflagellates have taken up residence in the intestines of higher animals, either as symbionts or as parasites. One of the most remarkable assemblages of structurally diverse zooflagellates is found in the intestines of various species of termites. There they live symbiotically, digesting wood for their own use and providing nourishment for the termite, while in return the termite provides shelter, transportation, and a steady supply of wood. In one termite more than a hundred different species of zooflagellates may be found.

20.11 *Euglena* is a convenient example of a phytoflagellate of the phylum Euglenophyta.

Phytoflagellates probably arose originally from the ingestion of blue-green algae, which subsequently evolved into chloroplasts. A somewhat atypical but common representative example of a phytoflagellate is *Euglena gracilis* (Figure 20-5).

Euglenoids commonly inhabit fresh water rich

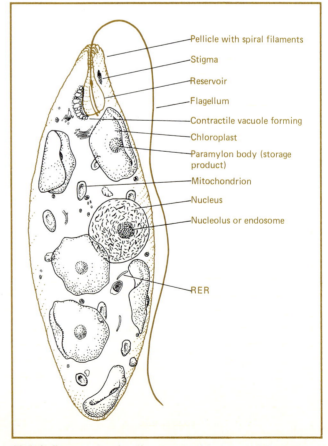

Pellicle with spiral filaments
Stigma
Reservoir
Flagellum
Contractile vacuole forming
Chloroplast
Paramylon body (storage product)
Mitochondrion
Nucleus
Nucleolus or endosome
RER

20-5 A diagram showing the structural features of *Euglena gracilis.*

in organic nutrients. They can move toward light. A euglenoid is an elongated cell with a well-formed nucleus, containing a nucleolus, two flagella (only one of which emerges), mitochondria, and a contractile vacuole. Each cell also contains several chloroplasts and can photosynthesize in sunlight. In the dark euglenoids live on dead organic matter.

Because euglenoids are small, diffusion suffices to bring about gas exchange with the environment. In the daytime, they utilize both carbon dioxide and oxygen. At night, they only respire. Like most fresh-water protists, euglenoids have contractile vacuoles that pump out of the cytoplasm the excess water that has entered the cell by diffusion.

The longer flagellum of euglenoids is used for locomotion. In addition the cortical region of the cell beneath the **pellicle** (outer coating) contains a spirally wound band of contractile material which causes peristaltic movements called "euglenoid movement." These movements occur when the cell is subjected to high-intensity illumination. *E. gracilis* has a red spot near the **reservoir,** called the **stigma.** This is a shading device for the photoreceptor which causes most of the Euglena's behavior.

20.12 The organisms of the phylum Ciliophora are probably descended from primitive flagellate ancestors, but they contain different specialized structures.

Members of the phylum Ciliophora are believed to have descended from primitive flagellate ancestors. Both groups possess structures derived from centrioles that have been modified in many special ways.

Cilia (Latin for "eyelashes") and flagella (Latin for "whips") were named long before it was discovered that the underlying similarities in structure

20-6 Some representative free-living ciliates, showing the diversity in form.

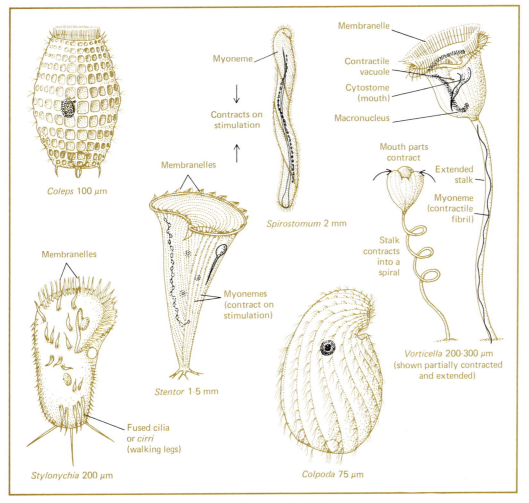

Coleps 100 μm

Membranelles

Membranelles

Stentor 1-5 mm

Fused cilia or *cirri* (walking legs)

Stylonychia 200 μm

Myoneme

↓

Contracts on stimulation

↑

Myonemes (contract on stimulation)

Spirostomum 2 mm

Colpoda 75 μm

Membranelle

Contractile vacuole

Cytostome (mouth)

Macronucleus

Mouth parts contract

Extended stalk

Myoneme (contractile fibril)

Stalk contracts into a spiral

Vorticella 200-300 μm (shown partially contracted and extended)

and chemistry were probably more significant than minor differences in movement (section 8.10). Despite the similarities of these two structures, however, ciliates and flagellates certainly do differ in a number of ways, suggesting that they diverged in evolution a very long time ago.

The members of the class Ciliata, or ciliates, form an enormously successful and diversified group that has adapted to such diverse habitats as the open oceans, drainage ditches, and animal intestines. Ciliates are classified according to the way their cilia are organized on their surfaces. Examples of the major groups are shown in Figure 20-6. The ciliates contain two kinds of nuclei—micronuclei and macronuclei.

Many ciliates reproduce asexually by **fission** (section 10.1) and engage in a kind of sexual process called **conjugation.** This involves the exchange of micronuclei between two organisms. Conjugation does result in genetic recombination and hence increased genetic diversity. Many ciliates possess permanent differentiated cytoplasmic structures as exemplified by those of the paramecium described in the next section.

20.13 *Paramecium* is the most studied ciliate, and a great deal is known about its structure, chemistry, genetics, and behavior.

One of the reasons *Paramecium* has been studied so extensively is that at least one of several species is found in almost every body of fresh water.

Paramecium multimicronucleatum (Figure 20-7), is a slipper-shaped animal with a permanent oral groove. Rows of cilia cover the entire body, including the oral groove. The most conspicuous internal structures are two contractile vacuoles, many micronuclei, and a single macronucleus. The cortical cytoplasm contains the **basal bodies** of the cilia, undischarged **trichocysts,** and various fibrillar structures. Trichocysts are organelles just beneath the cell surface that discharge threadlike structures that ward off predators or seize prey. The internal cytoplasm contains mitochondria, Golgi bodies, and other organelles common to most eucaryotic cells.

A paramecium feeds by creating currents with its cilia that bring food particles down the **oral groove,** where they enter the **cytostome** (cell mouth) and pass into a **food vacuole** surrounded by a membrane. Rotational cytoplasmic streaming, or **cyclosis,** carries the food vacuoles around and around the cell as digestion proceeds. Finally, the undigested portion of the food remaining in the vacuole is

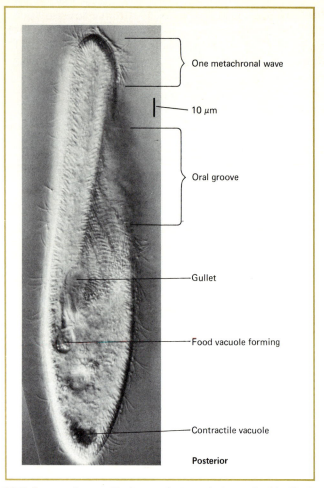

One metachronal wave

10 μm

Oral groove

Gullet

Food vacuole forming

Contractile vacuole

Posterior

20-7 *Paramecium multimicronucleatum* photographed by flash illumination to stop the metochronal pattern of ciliary beating. (Courtesy of H. Machemer.)

egested. Nitrogenous wastes may either leave the cell by diffusion or pass out with water when the contractile vacuole empties. The contractile vacuole is needed by freshwater protists to remove excess water diffusing into the cell (section 4.6). Paramecia are still small enough that gas exchange (cellular respiration) can occur by diffusion.

Paramecia rotate on their long axes as they swim through the water. Individual cilia beat like tiny oars in an **effective stroke** followed by a **recovery stroke.** The cilia beat in **metachronal waves,** meaning that at a given time, cilia along the body are in slightly different phases of their beat (Figure 20-7). The appearance is similar to that of a wheat field blown by gusts of wind.

Paramecia tend to move in nearly straight lines until they encounter an obstacle. If, for example, they bump into something, they reverse the direction of their effective stroke and paddle backward;

they then execute a turn of about 25 to 30 degrees and start forward again. This is called the **avoiding reaction.** Recently, R. Eckert and Y. Naitoh have discovered that paramecia accomplish the avoiding reaction by a change in the permeability of the plasma membrane that allows external calcium to enter. While calcium ions are present in the cortex, the animal swims backward; when they have been pumped out, the animal resumes its normal course.

Paramecia usually reproduce by regular cell division (fission) (Figure 10-1), but they may also exchange genetic material by conjugation. To accomplish this primitive sexual union, two organisms fuse temporarily in the region of the oral groove, then exchange micronuclei through the temporary bridge.

20-8 Life cycle of the malarial parasite *Plasmodium.* Sporozoites (*in color*) enter humans with the mosquito's saliva and move to the liver and lymph system. Here they develop into merozoites (*black*) which either infect more liver cells or move into blood cells. In the blood they reproduce and release more merozoites, along with a toxic substance that causes the chills and fever characteristic of malaria. This cycle may continue at 24-hour intervals. Some merozoites undergo meiosis, and two types of gametocytes are formed. When both types are ingested by a biting mosquito they fuse in the mosquito's stomach. The zygote lodges in the mosquito's intestinal wall and forms an oocyst. In the oocyst many sporozoites form by mitosis. These move into the salivary gland when the oocyst bursts.

20.14 The phylum Sporozoa consists of symbiotic and parasitic forms with complex life cycles; some cause serious human and animal diseases.

The phylum Sporozoa is so named because in one stage of their complex life cycles they exist as

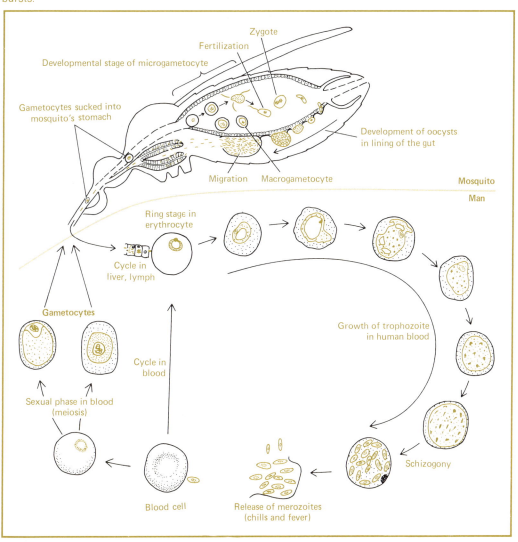

spores containing infective **sporozoites.** Members of this phylum are of considerable significance because of the diseases they cause in man and other animals. The life cycles of sporozoans may include both asexual and sexual reproduction and may be divided between two hosts.

Plasmodium is a sporozoan parasite that causes the disease malaria. Although malaria (originally meaning "bad air") is now restricted to tropical zones, half a century ago it was common well into the temperate zones of Europe and the United States. The disease is spread by the *Anopheles* mosquito, which is host to the parasite for half of its life cycle. The other half of its life cycle involves humans. The details of *Plasmodium's* life cycle are depicted in Figure 20-8.

Control of the disease requires control of the *Anopheles* mosquito. In the past this has been done with the insecticide DDT, but this agent is used much less because of its toxicity and its long-lasting effects on the environment. Efforts are underway to control *Anopheles* and other dangerous insects with hormones administered to male insects. When these males are released following treatment and copulate with females, they transmit enough insect hormone to the females (and to their subsequent sexual partners) to make successful reproduction impossible.

20.15 The "plant protists," also called the "protophyta" and "lower algae," pose a major problem for taxonomists: Are they protists, plants, or fungi?

In adopting the five-kingdom approach to taxonomy (section 20.4) the arbitrary decision has been made that unicellular eucaryotic organisms all belong in the same kingdom, Protista. This separation of unicellular Protista from multicellular Planta, Fungi, and Animalia is not always clear in nature, and this consequently poses problems for the systematist. No matter what boundaries between kingdoms, phyla, classes, and other subdivisions are drawn, they are artificial, man-made conveniences, and not all organisms will fit nicely into the established categories. The plant protists offer a fine example. These organisms can be divided in half a dozen ways among the kingdoms Planta, Protista, and Fungi. In this text the phyla Chrysophyta, Bacillariophyta, Xanthophyta, Pyrrophyta, and unicellular Chlorophyta (green algae) will be included among the protists. Discussion of the mul-

ticellular members of Chlorophyta, along with the brown and red algae, will be saved for the section on plants, and slime molds (phylum Myxomycota) will be discussed with the fungi in Chapter 21.

20.16 The phylum Pyrrophyta includes the dinoflagellates, which are very abundant in most phytoplankton collections.

Dinoflagellates, together with diatoms and other plant protists, make up the **phytoplankton,** or "grass of the sea." Most dinoflagellates have two flagella of equal length, one of which is wound around the midregion in a shallow groove. They secrete a cellulose covering, are often of irregular but species-specific shape, and many have armored plates (Figure 20-9). Their chloroplasts contain chlorophyll and other pigments that give them a reddish-brown hue. They contain a red spot similar to the stigma of euglenoid flagellates.

The better-known dinoflagellates include *Noctiluca* and *Gonyaulax*, the organisms responsible for luminescence in many parts of the sea. Members of the genus *Gonyaulax* are sometimes responsible for mussel and clam poisoning and for **red**

20-9 A group of marine dinoflagellates. (Courtesy of S. Honjo)

tides. In a red tide the number of *Gonyaulax* organisms increases 5 or 6 millionfold, coloring the sea red. Toxins secreted by the dinoflagellates kill fish, which decompose and rob the water of so much oxygen that *more* fish are killed. The decay provides nutrients that in turn sustain the red tide.

20.17 The phyla Chrysophyta (golden algae), Xanthophyta (yellow algae), and Bacillariophyta (diatoms) are important primary producers in aquatic environments.

The phyla Chrysophyta, Xanthophyta, and Bacillariophyta include tens of thousands of species, most of which are unicellular; but some are colonial or filamentous and multicellular. Their chloroplasts are yellow, brown, or goden brown and contain three different types of chlorophyll and several accessory pigments. They store food in the form of oil or leucosin, not starch, and have thick walls. The unique cell walls of diatoms are carbohydrate impregnated with silica (glass). When they divide mitotically each daughter cell takes one part of a cell wall and synthesizes the other. The shells of diatoms are symmetrical, highly ornate, and beautiful, especially when they contain a living cell (Figure 20-10). Shell morphology is the chief means of distinguishing species. Diatoms are extremely abundant and are responsible for much of the photosynthesis that occurs in the oceans. When diatoms die, they settle to the bottom, and in the course of millenia, their glassy skeletons have accumulated to great depths on lake bottoms and ocean floors. Over the years, some of these regions have risen with changes in the earth's crust, and huge deposits of diatomaceous earth have been revealed. These are now mined for use in many commercial processes, ranging from paper making and filtration to the production of foods.

20.18 The unicellular green alga *Chlamydomonas* is a phytoflagellate and green alga; some of its multicellular relatives demonstrate key steps in the evolution of plants.

The discussion of the phylum Chlorophyta, the green algae, will be divided between this chapter and the next, because by making **multicellularity** (or the possession of many nuclei) the basis of demarcation between protists and higher phyla, the green algae overlap two kingdoms.

Chlamydomonas is one of the simplest and most studied unicellular representatives of the Chlorophyta. It shares with all members of that phylum three characteristics: cell walls containing cellulose, a chloroplast containing chlorophyll, and stored starch. *Chlamydomonas* (Figure 20-11) is about 25 microns long and is found widely in fresh water. It has a rigid cell wall perforated by two flagella of equal length. It contains a single cup-shaped chloroplast which partly encloses the nucelus. Usually haploid, it reproduces asexually by mitosis. Occasionally daughter cells function as isogametes and fuse (initially by flagellar contact) to form a diploid zygote with four flagella. The zygote then enters a dormant stage for overwintering. Under favorable environmental conditions meiosis occurs and restores the normal haploid condition. This simple sexual cycle is believed to be representative of the earliest form of sexual reproduction in plants.

20.19 Some algae pose difficult systematic problems: How should colonial and filamentous algae be classified? And where do the slime molds fit?

Among the green algae in particular are many forms of closely related unicellular colonial and filamentous algae. In some, all of the cells are structurally and functionally alike, but in others, one or more cells become differentiated to form reproductive cells or holdfasts. At this point, the collection of cells is really no longer a colony but represents a very simple multicellular organism. This is the controversial interface between protists and plants, and in dividing the two, boundaries have to be drawn that are artificial.

To help establish the most "natural" categories, taxonomists try to reconstruct the probable pathways of evolution. Plant taxonomists have long been preoccupied with the origin of the land plants and have argued about the probable "evolutionary route" that led to them. While evolution probably proceeded by an almost infinite number of small heritable changes, only some of which were favored by natural selection, very few if any "living fossils" still remain to illustrate the pathways of evolution. Either organisms have kept on evolving or have become extinct. In other words, one major problem

20-10 A collection of marine diatoms showing considerable variation in shape and shell morphology. *A. Paralia sulcata, B. Navicula sp., C. Actinoptychus sinarilus, D. Thalassiosira, E. Planktoniella, F. Paralia Sp., and G. Coscinodiscus radiatus,* all photographed with a scanning electron microscope. (Courtesy of S. Honjo.)

facing taxonomists is that all they have to work with are the leaves of the "evolutionary tree," not the trunk or the branches which would show how the leaves are related. Nevertheless, it is believed at present that land plants evolved from ancestors of the green algae of the phylum Chlorophyta. Furthermore, out of the many groups of algae in this phylum, a few groups can be singled out as lying somewhere close to the evolutionary pathway that probably led to the bryophytes and higher land plants. Consequently, the unicellular members of the phylum Chlorophyta are considered here as protists ancestral to higher plants.

The other major problem in classifying protists is the interface between them—the red and green algae on the one hand and the fungi on the other. It now seems likely that slime molds are a transitional group. Both the phylum Myxomycota, the "true" (acellular) slime molds, and the phylum Acrasiomycota, the cellular slime molds, exist for part of their life cycle either as soil amoebae or as flagellates. This suggests they are protists. However, slime mold amoebae can (but do not necessarily have to) fuse into a multinucleate cytoplasm mass, or **plasmodium,** which grows and migrates in search of bacteria. In their plasmodial stage slime molds resemble other fungi. They keep on developing and help to decompose fallen logs when the air is moist after a rain. When the forest dries up and the sun appears, the plasmodium begins to dry up and forms tiny **fruiting bodies** containing spores, definitely a funguslike characteristic. When the spores are released, they blow away and are dispersed over a wide area. There they lie until moisture and the presence of nutrients favor another turn of the life cycle.

Clearly the slime molds are intermediate and can be considered as protists, as fungi, or as both. The origin of the higher fungi is uncertain. There are indications that possibly different groups of fungi originated from green algae, red algae, or zooflagellates.

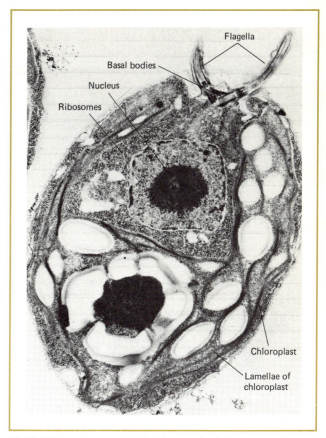

20-11 *Chlamydomonas reinhardtii,* a unicellular green alga. Structures visible in this transmission electron micrograph include the two flagella, nucleus, and chloroplasts. (Courtesy of D. Ringo.)

21

Fungi and Higher Algae

QUESTIONS TO KEEP IN MIND

What characteristics do fungi possess, and why are they classified in a kingdom separate from plants?

What is the nutritional life style of fungi? Are there exceptions?

On what differing bases are the algae and fungi classified?

What different connotations does the word *plant* have?

What are "algae," and of what importance are they?

What kinds of cellular differentiation have arisen in the algae?

What kinds of symbiotic relationships are found between algae on the one hand and either fungi or animals on the other?

The organisms considered so far arose as unicellular organisms and found it to their advantage to remain so. Fungi and higher algae, on the other hand, exemplify early experimental diversifications of form toward increased size and multicellularity. Many of these experiments were successful, and in consequence we find living organisms that have adapted to a great variety of environmental conditions. Less successful experiments are sometimes recorded in the fossil record, although more often than not conditions were not favorable for the preservation of ancient fungal and algal forms. The fungi and higher algae represent multicellular forms that have evolved heterotrophic and photosynthetic nutritional modes, respectively.

21.1 The fungi, both friend and foe of man, are sufficiently different from green plants to merit a separate kingdom.

The fungi, once classified as plants, are now recognized to differ from plants as much as they do from animals or bacteria. Many of the approximately 100,000 species of fungi impinge upon the human experience—delicious mushrooms, truffles, and puffballs are grown and eaten; but bread mold, dry rot, and athlete's foot are cursed and combated.

Generally speaking, fungi are mostly multinucleate, with eucaryotic nuclei dispersed in a mass of interwoven threadlike filaments (**hyphae**) called a **mycelium.** Fungi lack the plastids and photosynthetic pigments that characterize plants; most have cell walls containing **chitin,** the same polymer that imparts hardness to beetles. Fungi are mainly absorptive **heterotrophs,** although the amoeboid stages of slime molds ingest food in the manner of protists. Fungal mycelia typically invade their food supply and bring about its decomposition enzymatically. As a group, they rival the bacteria in their chemical versatility.

Most fungi are not truly cellular as are most plants and animals. Instead they are **syncytial,** meaning that they are multinucleate, but their cytoplasm is not compartmentalized. Even where there are **septa** (dividing walls) present, these tend to be incomplete, so that cytoplasm can often stream throughout the entire mycelium. Perhaps partly for this reason, growth of fungi is extremely rapid; proteins synthesized throughout a hypha stream to the growing tip. Cellular differentiation hardly occurs except in reproductive structures, the asexual **sporangia** and the sexual **gametangia.**

Fungi are of enormous economic and ecological importance. Soil in a meadow, hayfield, or vegetable garden may contain tons of fungi per acre. There they are involved in the decomposition and recycling of major nutritional elements (Chapter 48). Fungi also bring about most of the decomposition of trees that fall and animals that die in the forest.

The production of many food products and beverages utilizes the prodigious enzymatic capabilities of fungi. Yeasts are responsible for the fermentation of beer, wine, and all other alcoholic beverages, and also produce the carbon dioxide that causes bread dough to rise. Other fungi are involved in the curing of Roquefort and Camembert cheeses. The antibiotic penicillin was originally isolated from a mold.

Not all fungi are useful to man. In the 1840s, a million people starved in Ireland because of the potato famine caused by the **late blight fungus.** Lumber is another vulnerable target for a variety of fungi, and losses in the United States alone amount to several million miles of board each year. Lumber is by no means safe when incorporated into houses and other structures, and much of the costly damage ascribed to termites is due primarily to dry rot or other fungal growths which make the wood easier for termites to attack.

Annual crop damage attributable to fungal diseases is estimated to cause a loss of billions of dollars. Even the trees in our yards are not safe from the ravages of fungal diseases—Dutch elm disease, now sweeping across the northeastern United States, is only one example.

21.2 The fungi are classified according to their life cycles.

The kingdom Fungi naturally divides into three principal subkingdoms (Table 7): the Gymnomycota ("naked fungi"), the Dimastigomycota (with two flagella), and the Eumycota ("true" fungi). It is suspected that these groups may have evolved from more than one kind of protistan or algal ancestor (Figure 21-1); there are resemblances between some fungi and some algae suggesting a close relationship.

TABLE 7 THE MAJOR GROUPS OF FUNGI AND PLANTS

Kingdom Fungi	Subkingdom Gymnomycota (naked fungi)	
	*Phylum Myxomycota	acellular slime molds
	*Phylum Acrasiomycota	cellular slime molds
	*Phylum Labyrinthulomycota	cell-net slime molds
	Subkingdom Dimastigomycota (with biflagellate stage)	
	Phylum Oomycota	water molds
	Subkingdom Eumycota ("true" fungi)	
	Phylum Chytridiomycota	chytrids
	Phylum Zygomycota	conjugation fungi
	Phylum Ascomycota	sac fungi
	Phylum Basidiomycota	club fungi
Kingdom Planta	**Phylum Rhodophyta	red algae
	Phylum Phaeophyta	brown algae
	**Phylum Chlorophyta	green algae
	Phylum Charophyta	stoneworts
	Phylum Bryophyta	mosses, liverworts
	Phylum Tracheophyta	vascular plants
	Subphylum Psilopsida	psilopsids (extinct?)
	Subphylum Lycopsida	lycopods
	Subphylum Sphenopsida	horsetails
	Subphylum Pteropsida	ferns
	Subphylum Spermopsida	seed plants
	Class Gymnospermae	conifers
	Class Angiospermae	flowing plants

*Have unicellular stages in their life cycles.
**Have unicellular representatives.

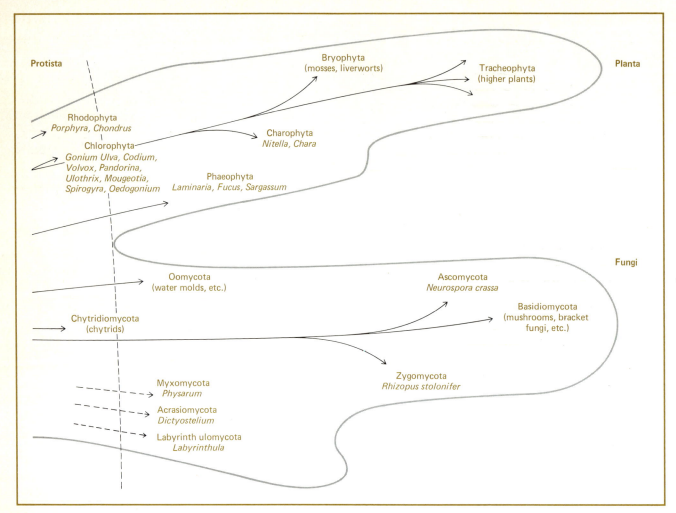

21-1 The major phyla and representative organisms of the
kingdoms Planta and Fungi.

The subkingdom Gymnomycota contains three
phyla: the acellular (or "true") slime molds (phylum Myxomycota), the cellular slime molds (phylum
Acrasiomycota), and the cell-net slime molds (phylum Labyrinthulomycota). These three phyla all
have life cycles involving separate (and potentially
independent) cells which aggregate into a differentiated cellular or syncytial mass called a plasmodium and sporulate. The life cycles of the cellular and acellular slime molds are shown in Figure
21-2.

The subkingdom Dimastigomycota contains the
phylum Oomycota, common representatives of
which are the water molds. The phylum is named
Oomycota because members form a large, nonmotile female gamete. This phylum also contains
some terrestrial forms, such as the one that caused
the Irish potato famine.

The subkingdom Eumycota contains over 50,000

known species, divided among four phyla. Members of the phylum Chytridiomycota, known as
chytrids, are microscopic fungi containing only a
few nuclei; they parasitize plant cells or decompose dead ones. Most chytrids reproduce vegetatively (asexually) by mitosis.

The phylum Zygomycota is comprised of terrestrial fungi that have evolved airborne spores that can
remain alive for months. These fungal spores are
produced in enormous numbers and can give rise
to a new plant whenever conditions are favorable.
The most familiar representative of the phylum
Zygomycota is *Rhizopus*, the black bread mold
(Figure 21-3) which can also decompose fruit and
other foods. *Rhizopus* is typical of the phylum in
remaining haploid during most of its life cycle.
However, sexual reproduction does take place by
conjugation between hyphae of two opposite
mating strains. A hormone is produced locally,

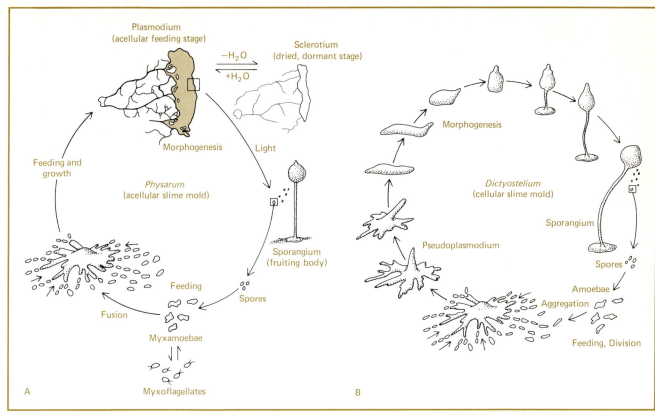

21-2 The life cycles of (A) *Physarum*, an acellular slime (phylum Myxomycota) and (B) *Dictyostelium*, a cellular slime mold (phylum Acrasiomycota).

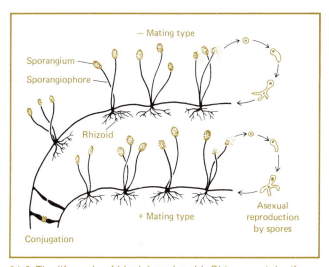

21-3 The life cycle of black bread mold, *Rhizopus stolonifer* (phylum Zygomycota).

causing hyphal outgrowths of the two mating strains to fuse. Nuclear fusion ensues, forming a **zygote.** Subsequent meiosis results in the production of **haploid spores,** each of which can form a new mycelium.

The **sac fungi** (phylum Ascomycota) comprise some 30,000 species, including yeasts, truffles, Dutch elm disease, and many other plant pathogens. Perhaps the most famous of the ascomycota is the orange-red bread mold, *Neurospora,* which Beadle and Tatum used to study the relationship between genes and enzymes (section 17.8). The life cycle of *Neurospora* (Figure 21-4) exhibits both sexual and asexual reproduction. Sexual reproduction forms a zygote called the **ascus,** from which the phylum gets its name. Asexual reproduction forms haploid spores called **conidia.**

The **club fungi** (phylum Basidiomycota) comprise some 25,000 species, including the familiar edible mushrooms and the highly poisonous toadstools, the puffballs, the bracket fungi that grow on trees, as well as the wheat rusts and the smut fungi that attack various other cereal grains (Figures 21-5 and 21-6).

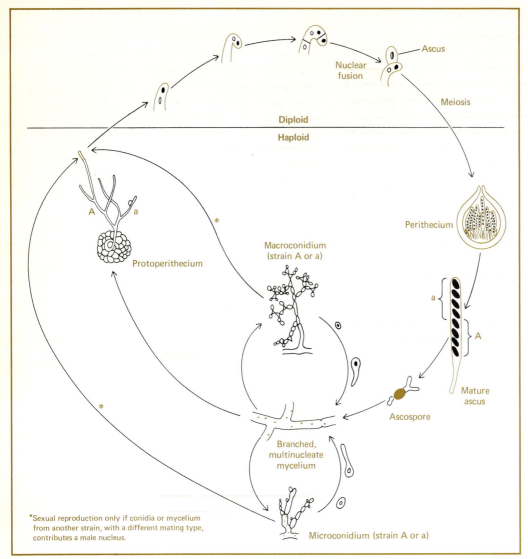

Diploid

Haploid

Ascus

Nuclear fusion

Meiosis

Perithecium

A   a

Protoperithecium

Macroconidium (strain A or a)

a

A

Mature ascus

Ascospore

Branched, multinucleate mycelium

*Sexual reproduction only if conidia or mycelium from another strain, with a different mating type, contributes a male nucleus.

Microconidium (strain A or a)

21-4 The life cycle of *Neurospora crassa* (phylum Ascomycota).

A typical mushroom results from the sexual union of two mycelia of opposite mating types that have fused beneath the soil (Figure 21-5). The fused hyphae grow upward through the soil to form the stalk (**stipe**) and fleshy cap (**pileus**) of the mushroom. The hyphae have **septa** that prevent the migration of nuclei, so these cells have two nuclei, one from each mating type. Eventually the hyphae terminate on the underside of the pileus in club-shaped cells, or **basidia,** in which nuclear fusion takes place to form zygotes. Each zygote undergoes meiosis, giving rise to haploid spores in astronomical numbers.

Not all fungi fit into the categories discussed so far. This is in part because the life cycles of these organisms are either incomplete or unknown. Consequently, there is a group of fungi known as the

Fungi Imperfecta, which might be described as "fungi waiting to be classified when more is known about them." This group includes *Penicillium* and the organisms responsible for athlete's foot and ringworm, many plant diseases, and some kinds of common molds and mildews, such as those that attack clothing in damp closets.

21.3 The words *plant* and *alga* have a variety of meanings; they are useful as scientific terms only if restricted in definition.

The word *plant* is popularly used to denote "green organisms"; this usage is certainly incorrect, as plants occur in many other colors. A generation ago,

21-5 Some representatives of the phylum Basidiomycota common after a rain in a northeastern forest: (A) *Russula* sp. (B, C) *Polyporus versicolor* on a birch trunk and a rotting stump. (D) A stipitate gill fungus growing in an old scar in the bark of a tree. (E) Coral fungus. (F, G, H) Three stages in the development of the poisonous mushroom *Amanita muscaria*, or fly agaric; photographs of three individual specimens. (Photographs courtesy of R. Speck.)

many biologists were content to classify all photosynthetic organisms, including bacteria and blue-green algae, as plants, and often the fungi (which, as you know, do not photosynthesize) were thrown in. Recently there has been a tendency to restrict the term *plant* to multicellular photosynthesizers. In following the five-kingdom taxonomy of R. H. Whittaker, we have followed his definition of the kingdom Planta as containing *multicellular eucaryotic organisms of photosynthetic nutritional mode*. Among the other characteristics of this group are

photosynthetic pigments contained within plastids; lack of motility (although there are exceptions); and in higher forms especially, a *structural differentiation of organs* for photosynthesis (leaves), anchorage (roots), and support (stems); and the differentiation of tissues specialized for photosynthesis, fluid transport (i.e., vascular tissue), and covering.

This restricted definition of "plants" has removed two groups that were once included: the blue-green algae (procaryotes of the kingdom Monera; see Chapters 2 and 20) and the plantlike flagellates of the kingdom Protista (Chapter 20). These two groups are often referred to together as the *lower algae*, while the term *higher algae* refers to the green, red, and brown algae, the majority of which are multicellular.

The term *protist* is now being redefined by some

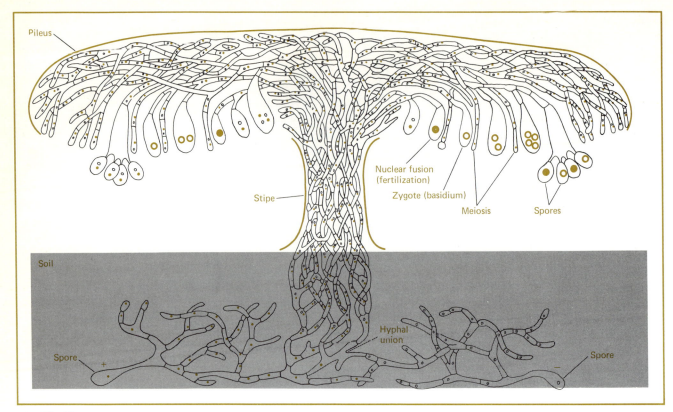

21-6 The life cycle of a typical mushroom (phylum Basidiomycota).

phycologists (algal specialists) to include not only the unicells but the higher algae as well. The rationale for classifying the higher algae among the protists is that these groups were evolutionary "blind alleys." Neither multinuclearity nor multicellularity without tissue and organ differentiation led to a way of life essentially different from the protists. According to this scheme, the term *plant* is limited to land plants (see Figure 19-3).

21.4 The criteria for classifying the higher algae are primarily based on the biochemical differences among various types of photosynthetic pigments, storage products, flagella, and cell walls.

Table 8 lists the major groups of photosynthetic organisms of the kingdoms Monera, Protista, and Planta and shows the major biochemical differences among them on which the classification is based. Although chlorophyll a is common to all these photosynthetic organisms (except for the photo-synthetic bacteria), chlorophylls b, c, and d (differing slightly in chemical structure) are restricted to one, two, or three phyla. The chlorophylls and the accessory pigments determine the distinctive colors of the green, red, and brown algae. The material of which the cell wall is made and the form in which metabolically derived energy is stored are indicators of the synthetic capabilities of these groups.

21.5 The red algae (phylum Rhodophyta) are seaweeds that occur at greater depths in the ocean than any other plants.

Red algae derive their color from the accessory pigments **phycoerythrin** and **phycocyanin,** which absorb in the blue-green region of the light spectrum. The sunlight that penetrates into the deeper layers of the sea has most of its energy in this region, so red algae absorb this energy and transfer it to chlorophyll for photosynthesis. Red algae collected in deep waters are always red; some collected in shallow water are more nearly green, but are often

TABLE 8 MAJOR GROUPS OF PHOTOSYNTHETIC ORGANISMS

Kingdom	Phylum (common name or examples)	Photosynthetic pigments	Storage products	Cell wall composition	Number of eucaryotic flagella
Monera	Photosynthetic Eubacteria	Bacteriochloro-phyll	Poly B-hydroxy, butyric acid	Mureins	0
	Cyanophyta (Oscillatoria)	Chlorophyll a, carotenoids, phycocyanin, phycoerythrin	Cyanophycean starch, Amylopectins	Gram negative mureins, mucocomplex substances	0
Protista	Pyrrophyta (Dinoflagellates)	Chlorophylls a and c, carotenoids	Starch, fats, oil	Cellulose, pectin	2
	Bacillariophyta (Diatoms)	Chlorophylls a, c, and e, carotenoids	Oils, leucosin, chrysolaminarin	Semicellulose, pectin, silica, $CaCO_3$	0
	Euglenophyta (Euglenoids)	Chlorophylls a and b, carotenoids	Paramylon	No true cell wall (periplast or pellicle)	2
Planta	Chlorophyta (*Chlamydomonas* and other unicellular, colonial, and multicellular forms)	Chlorophylls a and b, carotenoids	Starch within plastids, oils	Cellulose, pectin ($CaCo_3$*)	1, 2, 4, 7
	Rhodophyta (red algae)	Chlorophylls a and d, carotenoids, r-phycocyanin, r-phycoerythrin	Floridian starch	Cellulose, pectin, mucilages (agar), ($CaCo_3$*)	0
	Phaeophyta (brown algae)	Chlorophylls a and c, carotenoids	Laminarin mannitol	Cellulose, pectin, alginic acids, $CaCo_3$	2
	Charophyta (*Chara, Nitella*)	Chlorophylls a and b, carotenoids	Starch	Cellulose encrusted with lime	2
	Bryophyta (mosses, liver-worts)	Chlorophylls a and b, carotenoids	Starch	Cellulose, pectin, lignin	2
	Tracheophyta (vascular plants)	Chlorophylls a and b, carotenoids	Starch	Cellulose, pectin, lignin	mostly 0 (some 2)

"Algae" — "Higher algae" — Land Plants

*In some forms.

black. The same species may be red, green, or black, depending on where it grew. This is an example of **chromatic adaptation,** the regulation of the amount of an accessory pigment depending on the energy content of the light hitting the alga. Because of chromatic adaptation, red algae are able to live deep in the ocean (down to about 200 meters) where no other plants can make a living by photosynthesis.

Red algae are found in unicellular and multicellular forms. Their distinguishing features are chloroplasts of simpler design than other plant groups, nonmotile gametes, and no flagella at any stage of their life cycle. These characteristics suggest that red algae may have evolved independently of other plant groups. Some red algae extract calcium from sea water and deposit it in or around their cell walls. Red algae growing among corals on a reef do this to such an extent that they aid in the building of a reef. In the South Pacific, many reefs, such as the Onotoa reef in the Gilbert Islands, have been made exclusively by red algae.

Several of the red algae are of commercial im-

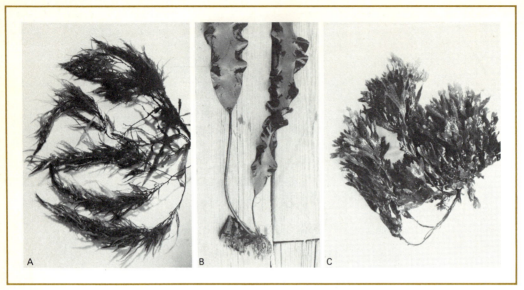

21-7 Three marine brown algae (phylum Phaeophyta): (A) *Sargassum*, (B) *Laminaria* (kelp), (C) *Fucus.* (Photographs by author.)

portance. *Porphyra* is farmed extensively in the warm coastal waters off Japan. It is dried and sold in flat stacks as *nori,* a delicacy eaten with boiled rice, or as the green wrapping around *sushi,* rice cakes containing raw fish, vegetables, or egg. *Chondrus crispus,* or Irish moss, is a source of **carrageenin,** used as an emulsion stabilizer and in making ice cream and puddings. Agar for bacteriological media is extracted from *Gelidium* and other red algae.

21.6 The brown algae (phylum Phaeophyta) are shallow-water seaweeds; some are economically important.

Brown algae are all marine multicellular seaweeds, except for a few rare freshwater forms. Their brown color is due largely to xanthophyll pigments. In some marine environments, brown algae are quite plentiful, as in the Sargasso Sea area of the Atlantic Ocean where *Sargassum* weed abounds. In certain cold-water intertidal zones off the coasts of Maine, Newfoundland, Puget Sound, Norway, Brittany, and elsewhere, **kelps** such as *Laminaria* (Figure 21-7) grow to considerable lengths and are harvested for animal feed or fertilizer. Alginic acid, a carbohydrate used to impart smoothness to ice cream and other desserts, is also prepared from kelps. In many areas of the northern seacoasts, the rockweed *Fucus* is found in abundance attached to rocks along the shore. Countless marine invertebrates and small fish take refuge among its fronds.

The large kelps, such as *Laminaria,* have a simple life cycle with alternation of generations (Figure 21-8). The kelp plants are members of the sporophyte generation, and at certain times of the year sporangia develop on the long, flat blade. Here meiosis occurs to produce haploid, motile zoospores capable of developing into a small haploid gametophyte plant of one sex or the other. Male and female gametes fuse to form zygotes, which remain attached to the gametophyte plant during early stages of development.

The life cycle of *Fucus* is quite atypical for an alga, and more closely resembles that of an animal. The rockweed itself is the diploid sporophyte generation, as with kelps. However, the haploid gametophyte generation is so short that only the gametes themselves are haploid.

21.7 The ubiquitous multicellular green algae (phylum Chlorophyta) exhibit many transitional stages from unicellular to colonial, multicellular, and coenocytic forms.

Green algae are the most ubiquitous of the algal forms. This phylum contains many unicellular forms of green algae which were classified with the protists. Some common multicellular marine representatives (Figure 21-9) are the green seaweed

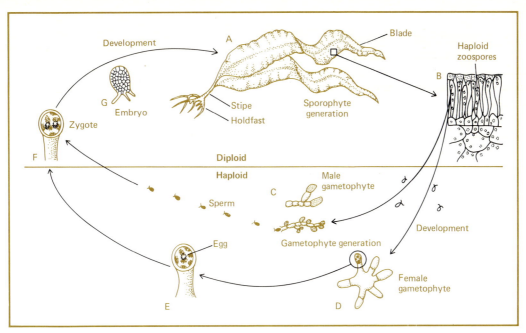

21-8 The life cycle of *Laminaria*, a kelp.

21-9 Two marine green algae (phylum Chlorophyta): *(A) Codium; (B) Ulva* (sea lettuce). (Photographs by R. D. Allen.)

Codium and sea lettuce, *Ulva* (the life cycle of which was discussed in section 12.1).

In pond water collections it is easy to distinguish three basic types of multicellular green algae. There are spherical colonies of similar cells, such as *Gonium, Pandorina, Eudorina,* and *Volvox,* in which the flagella of individual cells propel the entire colony (Figure 21-10). *Volvox,* the most complex of these organisms, has evolved some degree of cellular differentiation and specialization, and some cells are set aside for reproductive function. Another type of colonial form found in pond water is nonmotile. *Pediastrum* and *Hydrodictyon* (Figure 21-11) are such multicellular colonies. They exhibit a characteristic colony morphology but no cell differentiation. A third type of green algal colony is that exemplified by filamentous forms found in green scums on pond surfaces and attached to rocks in streams. Three examples of filamentous green algae are *Ulothrix, Oedogonium* and *Spirogyra* (Figure 21-12). These species show cellular and sexual differentiation.

Not all green algae are unicellular or multicellular; some are multinucleate and coenocytic like the fungi. An example is the marine alga *Caulerpa,* in which the cytoplasm is free to stream throughout the plant. *Acetabularia* is an alga that may be roughly an inch or two in length; at one stage of its life cycle it has a single nucleus in its rhizoid that controls an enormous volume of cytoplasm; during

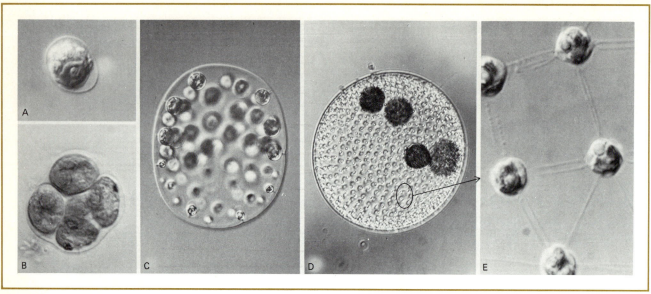

21-10 *Gonium*, a unicellular green alga (*A*) and three colonial forms, the cells of which are similar to *Gonium*: (*B*) *Pandorina*; (*C*) *Eudorina*, and (*D*) *Volvox*. (*E*) Interconnected cells of *Volvox*. (Photomicrographs by the author.)

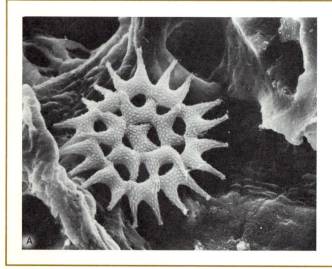
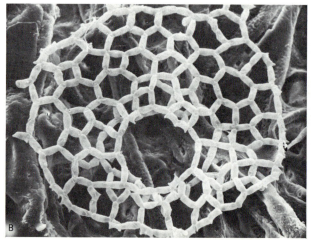

21-11 (*A*) *Pediastrum* and (*B*) *Hydrodictyon*, two colonial green algae. (Scanning electron micrographs by Harvey Marchant.)

its reproductive stage it becomes multinucleate and temporarily coenocytic before it forms germinating cysts.

Green algae are found on land as well as in freshwater and marine environments. In temperate and northern forests, algae tend to grow on the north side of tree trunks. After a heavy rain, a thin film of green algae may be found on porch railings, house paint, logs, and on the surface of soil and stones.

21.8 Many green algae have entered into stable symbiotic relationships with organisms of other kingdoms.

Many protists (e.g., *Paramecium busaria*, shown in Figure 20-2, and *Difflugia pyriformis*), animals (*Hydra viridis*, a coelenterate (section 24.4), and the sea slug, *Placobranchus ianthobapsus*) are

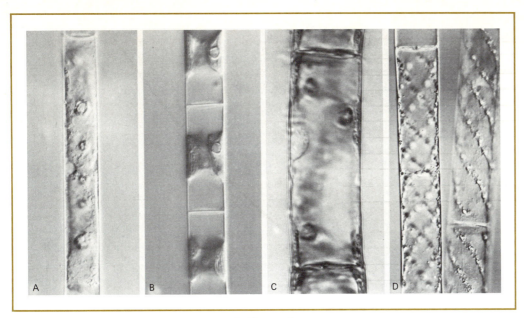

21-12 Examples of filamentous green algae: (A) *Mougeotia*, (B) *Ulothrix*, (C) *Oedogonium*, and (D) *Spirogyra*. (Photomicrographs by N. S. Allen.)

21-13 Two or more species of lichens on the bark of an oak tree. (Photograph by author.)

green because they possess algal endosymbionts living in their cells. Green algae have also associated with fungi to form **lichens** (Figure 21-13). Although lichens can survive some of the most extreme environmental conditions (such as at the summit of Mt. Washington, New Hampshire, where the temperature descends to −80° F and wind velocities of over 230 mph have been recorded), they are strangely sensitive to sulfur dioxide and other urban pollutants. Consequently, most large urban centers have a lichen-free zone around them which serves as an indicator of pollution level.

Lichens are usually combinations of one species of green algae with a fungal species of the phylum Ascomycota. In many cases the two associated organisms can be separated, grown separately, and then put back together. There are many thousands of organismal combinations, each of which is referred to as a species, for the reason that it reproduces itself consistently.

21.9 Stoneworts (phylum Charophyta) evolved cellular specializations that may have played a role in the evolution of land plants.

The stoneworts, such as *Chara*, *Nitella*, and *Tolypella*, are multicellular green algae that grow at the bottom of shallow fresh-water lakes and ponds. Systematists differ on whether to include them in the phylum Chlorophyta or place them in the separate phylum Charophyta.

A *Nitella* plant (Figure 21-14) is anchored to the bottom of the pond by a **rhizoid** (root) consisting of from one to several elongated cells lacking chloroplasts. The rest of the plant may be from several inches to several feet tall, with branches at evenly spaced **nodes.** The **internodal cells** are quite long (sometimes more than 10 cm), and extend from one node to the next. At the nodes themselves are some small **nodal cells,** and extending into the branches are **leaf cells.** All the cells in a stonewort are multinucleate.

Stoneworts are of considerable interest to scientists interested in plant evolution and the origin of the land plants because the details of mitosis and cytokinesis in stonewort cells resemble those in higher plants more than they resemble details of division in Chlorophyta. Thus one idea of the evolution of land plants suggests that some primitive representative of Charophyta may have been an intermediate form between algae and land plants.

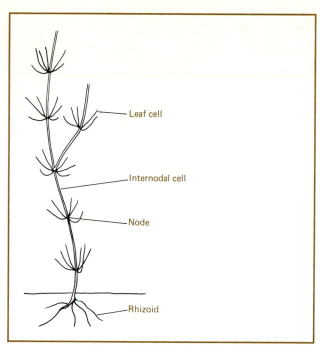

21-14 Diagram of the stonewart *Nitella.*

22

Higher Plants and Their Tissues

QUESTIONS TO KEEP IN MIND

What features of higher plants contributed to their success on land?

Why are mosses and liverworts destined to be smaller than most land plants?

What changes in life cycle accompanied the evolution of the vascular plants?

Why do seeds and pollen favor the survival of flowering plants?

In what ways are plants important to man?

What special relationships exist between flowering plants and insects, birds, and mammals? How did these relationships come about?

What are the major organs and tissues of higher plants? What are their functions?

Why are angiosperms considered more advanced evolutionarily than gymnosperms?

Some features that probably contributed to the success of the earliest land plants were (1) *multicellularity* and size, (2) *cellular differentiation,* (3) a *short life cycle with sexual reproduction* to maximize genetic variability for natural selection, (4) *structures and mechanisms to prevent dessication* while allowing gas exchange for photosynthesis and respiration, (5) *multicellular sex organs* for the protection of gametes, and (6) *internal fertilization* for the protection of embryos.

The first land plants to appear in the fossil record grew about 400 million years ago. Within 100 million years of their first appearance, many regions of the earth were covered by dense forests populated by representatives of plant groups that are now either extinct or represented by a very few surviving genera. The story of the evolution of land plants is a fascinating one, although many exciting questions are as yet unanswered.

The two phyla of higher plants, the Bryophyta (mosses and liverworts) and the Tracheophyta (ferns and seed plants), are believed to have evolved from a line of green algae (Chlorophyta) possibly through an early representative of the Charophyta. The biochemical evidence in support of this idea can be seen in Table 8. It is reasonable to suppose that the earliest land plants grew in shallow fresh water and that survival on land was initially an adaptation to drought.

22.1 One of the major divisions of the higher plants, Bryophyta, contains no vascular tissue. Consequently its representatives are limited in size.

The mosses, liverworts, and hornworts of the phylum Bryophyta possess neither roots nor true leaves, although they are anchored by a threadlike (but nonabsorptive) rhizoid and have flattened stems that are spoken of loosely as "leaves." The mosses absorb water and minerals through their stems and require a watery environment for fertilization. Motile sperm are produced in the male sex organs, or **antheridia**, of the moss plant, which is the gametophyte generation (Figure 22-1). After a rain, the sperm swim to the ova in the female sex organs, or **archegonia**. Fertilization results in a zygote which grows within the archegonium into the sporophyte generation. The latter develops a stalk with a sporangium on top containing spores, which are dispersed by wind. Each spore can form a **protonema**, a juvenile filamentous plant on which buds form and produce a gametophyte plant.

Although there are many thousands of species of mosses, perhaps the most prevalent in terms of biomass is *Sphagnum*, which grows in bogs from temperate climates well into the arctic. The layered annual growths of *Sphagnum* decompose into peat, an important fuel, especially in northern countries. Peat bogs yield important historical and scientific information, because of their preservative quality. They preserve pollen that has blown into them, and even human remains: in some bogs human bodies and artifacts have been found in such a perfect state of preservation that the stomach contents could be gleaned for information about human diets of several thousand years ago.

Liverworts are less conspicuous than mosses, and are usually found on rocks, tree trunks, or logs. They tend to be flat and "leafy" in appearance (Figure 22-2).

22-1 *A.* Photograph of a common moss, *Polytrichum* (phylum Bryophyta). *B.* Life cycle of a moss. (Photo by author.)

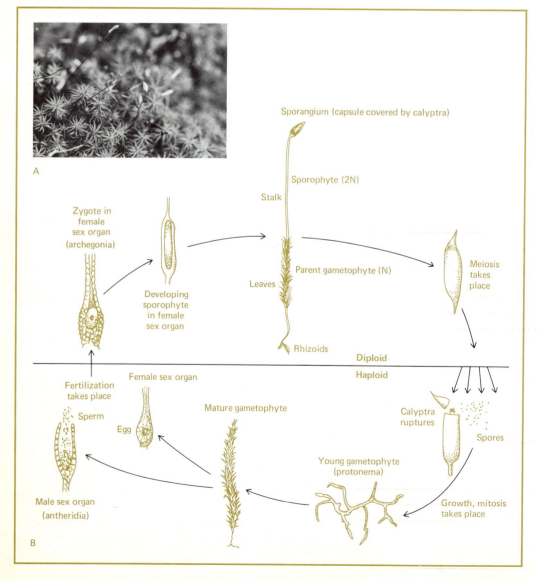

22-2 *Marchantia,* a common liverwort (phylum Bryophyta). (Courtesy of R. Speck.)

22.2 Vascular plants are the second major division of higher plants; their vascular tissue gave them a decisive advantage over the bryophytes in the competition for sunlight and space.

Unlike mosses and liverworts, which grow in dense mats along the ground, vascular plants (phylum Tracheophyta) mostly grow upward toward the sun. Although vascular plants range in size from the tiny water meal, *Wolffia,* about the size of a pinhead, to the giant sequoia tree over 100 meters high, they are alike in several ways. In general, each has (1) extensive systems of roots for both anchorage and absorption of nutrients from the soil, (2) broad leaves containing vascular tissue as organs of photosynthesis, and (3) strong woody stems capable of supporting their often considerable weight. It is clearly the vascular tissues of these plants that has freed them from restraints on size.

Vascular plants are, and have been for over 400 million years, the dominant form of plant life on land. Their success on land has been due in part to vascularization and in part to a fundamental modification of the generalized plant life cycle in which there is an alternation between asexual sporophyte and sexual gametophyte generations (sections 12.1 and 12.2). The tracheophytes represent a sudden departure from the bryophytes in that the gametophyte portion of the life cycle is greatly reduced. In mosses, the sporophyte generation is a separate plant, but it grows out of the female sex organ to release its spores and die. In ferns, the sporophyte is the dominant generation,

but the gametophyte persists as a separate, but much reduced, plant. This trend toward the reduction of the gametophyte generation established by the mosses and later by the ferns, is carried to its extreme in the seed plants, where the gametophyte exists as a parasite enclosed in the integument tissues of the sporophyte. The sporophyte produces two kinds of haploid spores: **megaspores** which give rise to ova and **microspores** which develop into pollen. The pollen and ova represent the plants of the male and female gametophyte generations respectively. Fertilization restores the diploid number of chromosomes in the plant embryo, which develops into the next sporophyte generation (section 12.4).

This basic alteration in life cycle undoubtedly contributed much to the success of seed plants over bryophytes, because pollen grains, which can withstand extremes of cold and dryness, are a more reliable means of dispersing gametes than motile sperm. The development of seeds also contributed much to the evolutionary success of the vascular plants, by ensuring germination only at times that were most favorable for development (section 12.8).

22.3 Five major groups of vascular plants have flourished on land in succession: One group is probably extinct and three others have survivors; while the seed plants, especially the flowering plants, thrive and continue to evolve.

The fossil record indicates that the first group of plants to invade the land were the **psilopsids** (subphylum Psilopsida). These were simple plants with no roots or leaves, but with branched stems that were green above the ground and capable of photosynthesis. The xylem of the stem was sufficiently strong to support growth to heights between 10 and 90 cm.

Until recently it was generally believed that two relics of this subphylum had survived: *Psilotum* (the whisk fern), which grows in the southeastern United States (Figure 22-3), and *Tmesipteris,* which is found in the South Pacific Islands. Recent evidence has suggested that these organisms may be primitive ferns rather than psilopsids. If so, the psilopsids would appear to be extinct.

The **lycopods** (subphylum Lycopsida) first appear in the fossil record between 350 and 400 million years ago. Within the space of about 100 million years, they had evolved into trees up to 30 m tall that grew in dense forests. Although these lycopod

22-3 A photograph of the whisk fern, *Psilotum nudum*. (Courtesy of R. Speck.)

22-5 Scouring rushes, *Equisetum*, near a redwood group in California. (Photograph by author.)

22-4 Ground pine, *Lycopodium*, showing terminal sporangia. (Courtesy of M. Stewart.)

trees had leaves and roots, they lacked some of the advantages possessed by the seed trees that evolved later. Consequently, they lost in the competition for sunlight and living space and became extinct. The most familiar surviving lycopod is the club moss (or ground pine) *Lycopodium*, often used as Christmas greenery (Figure 22-4).

The **horsetails** (subphylum Sphenopsida) left their mark on the fossil record at about the same time as the lycopods, and several representatives of these also evolved into large trees that were among the dominant forms in the forests of the Carboniferous age, the time at which coal deposits were being formed. Today, only a single genus survives, *Equisetum* (Figure 22-5). This plant is rarely eaten by animals, because its epidermal cells contain indigestible silica deposits; because of these, early American colonists called them "scouring rushes" and used them to clean pots and pans. *Equisetum* has roots; a jointed stem that is hollow, with peripherally arranged vascular tissue; and scalelike leaves radiating in whorls from the joints. The life cycle has distinct sporophyte and smaller gametophyte plants; both states are photosynthetic. The gametophyte produces motile sperm.

Ferns (subphylum Pteropsida) were also plenti-

BIOEPICUREAN DELIGHTS

Fougères au Beurre: Fern Fronds Sautéed in Butter

The uncurling fronds of the ostrich fern, *Matteuccia,* are a common sight during early spring in northern temperate forests. They make a delicious dish when rinsed in cold water for a few minutes, boiled for 8 to 10 minutes, drained, then sautéed briefly in butter with a few slices of orange peel.

22-6 A tree fern, *Cyathea,* and a ground fern, *Glycenia,* in a Puerto Rican rain forest. (Courtesy of M. Stewart.)

ful and prosperous during the Carboniferous age (Figure 22-6), and although they have been receding in importance, some 11,000 species are known, most of them tropical. They range in height from a few centimers to great towering tree ferns. The leafy fern plants are all sporophytes, with the gametophyte separate and much smaller (section 12.2).

The ferns represented a major advance over the sphenopsids and lycopods. They exposed more leaf area to sunlight and hence were more photosynthetically efficient for their weight. This adaptation doubtless gave them an advantage over other primitive competitors. Ferns were the first land plants to develop vascularized leaves, and such leaves must have been more efficient than the leaves of the more primitive sphenopsids. Fern stems are not very strong; consequently they are most frequently found in sheltered environments, such as gorges, ravines, and small valleys.

22.4 Seed plants, without question the most successful land plants, first evolved as gymnosperms.

The seed plants (subphylum Spermopsida) evolved two overwhelming advantages over earlier land plants, including ferns. The production of *pollen* freed most of them from dependence on a watery environment for sperm motility. The production of *seeds* allowed the species to survive hostile conditions (excessive heat, cold, dryness, or darkness) that the individual plant could not. Both pollen and seeds offered opportunities for new and efficient ways to facilitate sexual reproduction and species dispersal.

The class **Gymnospermae** ("naked seeds") were the first of the seed plants to evolve some 350 million years ago. Gradually this group came to dominate the forests, causing the extinction or near-extinction of all but a few lycopids, sphenopsids, and ferns. Much later these gymnosperms were themselves displaced by **angiosperms,** or flowering trees, in the more temperate forests.

Three groups of modern gymnosperms have familiar representatives. The **conifers** ("cone bearers") include the pines, firs, hemlock, spruce, cypress, redwoods, and junipers. In junipers, the cones have undergone a modification into "berries." Other conifers produce cones of two types: pollen cones and seed cones (Figure 22-7). The male and female gametophytes are the microspores and megaspores on the scales of the respective cones.

The conifers have been most successful in environments unfavorable to their hardwood angiosperm counterparts. Conifer forests extend from the arctic tundra south for many hundreds of miles be-

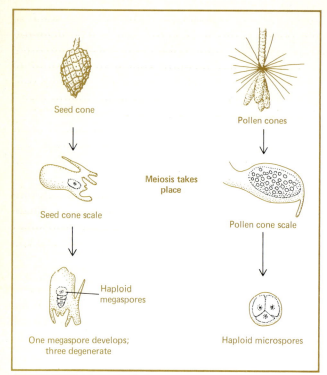

22-7 The two types of cones on a pine tree or other conifer—a seed cone and a pollen cone—and the development of the female and male gametophytes within them.

22-9 *Cycas*, a cycad growing in the Zombia Botanical Garden, Nyasaland. (Courtesy of M. Stewart.)

fore any hardwoods (angiosperms) are seen. In mountainous areas, the hardwoods at lower elevations give way at higher elevations to conifers, which extend to the timberline. Conifer forests are also seen bordering deserts, where the ground is too arid to support hardwoods.

Some conifers are among the largest and oldest plants alive. Sequoia in California have grown over the past 3,000 to 4,000 years to heights of over 100 meters, with trunks 10 meters in diameter. One such tree contains enough lumber to build a small housing project, a fact which has tempted lumbermen to exploit these limited resources. The lumber obtained from the sequoia, the Douglas fir, spruces, pines, and balsams accounts for more than three-quarters of the total supply of lumber in the United States.

22-8 A maidenhair tree, *Ginkgo biloba,* and one of its leaves. (Courtesy of T. H. Everett.)

Two other groups of gymnosperms are worth mentioning: the maidenhair tree, *Ginkgo biloba* (Figure 22-8), which might not have survived to the present had it not been cultivated as a shade tree, and the **cycads,** tropical "palmlike" trees with stems and broad leaves (Figure 22-9). (Real palms are angiosperms.) Both the ginkgos and cycads produce naked ova which are fertilized by motile sperm. In this respect these plants are unique among the gymnosperms.

22.5 The flowering plants (class Angiospermae) represent the very summit of evolution for the plant kingdom.

The flowering plants, or angiosperms ("covered seeds"), are relative newcomers on this planet, yet in the "short" space of about 130 million years, they have become the dominant form of land vegetation. Many of the first angiosperms were apparently trees, although it is possible that earlier, smaller representatives were not preserved in the fossil record. Many of the early angiosperm trees predated their herbacious relatives: Angiosperms were so successful that they diverged in evolution, giving rise to many familiar nonwoody plants, such as grasses, herbs, shrubs, and a host of flowering plants that took root in field, forest, swamp, desert, and on mountains and beaches. In all, roughly a quarter of a million angiosperms have been described and named, and many botanists suspect that an almost equivalent number are yet to be described.

22.6 Unlike the much older gymnosperms, the angiosperms were evolving at the same time as the insects, birds, and mammals; consequently they developed some very special interactions and relationships.

While early gymnosperms depended exclusively on wind and air currents for pollen and seed dispersal (we know this because land animals arrived on the scene less than 300 million years ago), the early angiosperms developed at a time when symbiotic relationships with insects, birds, and mammals could shape their evolution. Flowering plants evolved features that attracted the attention of insects, birds, and mammals. The production of nectar (which bees convert to honey) attracted insects which carried pollen from plant to plant, thus in-suring sexual reproduction. The fact that some of the pollen grains were consumed as food mattered little as long as some pollen grains were delivered.

T. Eisner has recently shown that pictures of flowers taken using ultraviolet light and a film sensitive to it reveal recognition patterns (Figure 22-10) which serve as signals to some insect that "food may be found here." The angiosperm-animal interaction may only involve one species of each, and if the plant and animal are separated, the plant can no longer reproduce.

Not all angiosperms depend on insects for pollination. Some depend on wind, a fact that is not happily accepted by the millions of people who suffer from hay fever. Although simple tests by an allergist can more precisely identify the source of the allergenic pollen, the time of year at which the malady strikes is a fairly good indication of the type of pollen responsible. Spring hay fever is usually due to tree pollen, summer sniffles to grasses, and early fall problems to ragweeds and sages (Figure 22-11).

22-10 Marsh marigold (*Caltha palustris*) (*A*) in visible light; (*B*) in ultraviolet light. The dark ultraviolet-absorbent basal regions of the petals are "nectar guides" that direct the pollinator to its "reward." (Courtesy of T. Eisner.)

22-11 Ragweed, *Ambrosia trifida,* the cause of widespread suffering to many afflicted with autumn hay fever. (Courtesy of R. Speck.)

Much more common than pollen dispersal is seed dispersal by wind among flowering plants; familiar examples are the dandelion and the maple.

Birds and mammals also play an important role in seed dispersal. What better way to spread seeds around than to have them spat out by monkeys (tropical fruits), carried about on bear fur (cockleburrs), stored underground but not all eaten by squirrels (cones, nuts, etc.), or passed through the gut and deposited in the fertile feces of many animals and birds.

One unique aspect of the reproduction of flowering plants (sections 12.4–12.9) needs to be stressed, because it was of considerable importance to the evolution of the group. This is **double fertilization,** the production of two sperm nuclei, one of which fertilizes the ovum, while the other fertilizes the endosperm, to provide nutrient medium for the embryo.

The various advantages enjoyed by angiosperms has led to the evolution of a remarkable diversity in form, function, and habitat (Figure 22-12). Many flowering plants are aquatic, like *Elodea* and *Lemna* (duckweed), and a few are marine, like eelgrass *(Zostera).* Dodder is a parasitic, nonphotosynthetic plant that winds itself around other plants to obtain its nutrients. Indian pipe is a flowering nonphoto-

synthetic plant that obtains nutrients from fungi. Sundews and Venus's-flytraps have assumed a carnivorous animal-like existence, living on captured insects.

22.7 The angiosperms are divisible into two orders, the monocots and dicots.

The earliest leaves to form from a seed plant embryo are called **cotyledons** (section 12.6). Approximately a third of all angiosperms have one cotyledon and are known as monocotyledonous plants (order Monocotyledonae), or simply as **monocots.** The group includes such common plants as grasses, sedges, rushes, irises and lilies, tulips, amaryllis, orchids, cattails, corn, onions, skunk cabbage, and palms.

The more numerous dicotyledonous plants (order Dicotyledonae), or **dicots,** have two cotyledons and include peas, carrots, magnolias, dandelions, squashes, sunflowers, and virtually all shrubs and trees. These two groups also differ in tissue arrangement.

22.8 Plants are indispensable to man and are used for food, shelter, and as providers of oxygen; they synthesize many products for our benefit—and a few that are deadly poisonous.

Students of prehistory tell us that man evolved as an herbivore and adopted a carnivorous or omnivorous diet only after the discovery of fire. Even today the greater part of the human diet consists of plants and plant products; and of course the animals that provide us with meat consume plants.

The plants that humans eat are mostly angiosperms. We eat *leaves* of lettuce, spinach, and cabbage, *stems* of celery and rhubarb, *roots* of carrots and radishes, *tubers* of potatoes, *flowers* of cauliflower and capers, *sepals* of artichoke, *ovaries* of tomato, *embryos* of corn and lima beans, *shoots* of alfalfa, *endosperm* of coconut (the milk), and *seeds* of the sunflower and pumpkin.

Virtually all food plants have evolved rapidly during the brief period of man's existence, and many changes in food plants have been brought about by careful breeding of plants selected for some desirable characteristic. Plant breeding was practiced successfully long before the science of genetics began. The process continues, and fruits, grains, and vegetables are being made tastier (some-

22-12 An assemblage of angiosperms showing some of their diversity: *A. Zostera marina,* eelgrass, one of the few marine representatives. *B. Elodea,* waterweed, often used in aquaria. *C. Iris,* a monocot. *D.* The sunflower, *Helianthus. E.* Pitcher plant, *Sarracenia. F. Cuscuta,* or dodder, a parasitic flowering plant. *G.* The sundew, *Drosera. H.* Venus's-flytrap, *Dionaea muscipula* (native to the Carolinas only); note the fly that was captured while the photograph was being taken.

times), more succulent, and more resistant to diseases.

Most of our beverages too are derived from plants. Coffee and cocoa come from beans, and tea from tea leaves that have been allowed to ferment.

Many medicines are derived from plants. Quinine (for the treatment of malaria and muscle spasms) comes from the bark of a tree; opium, from which sedatives and heroin are derived, comes

---

BIOEPICUREAN DELIGHTS

Some Edible Raw Plant Tissues

A bowl of crisp raw vegetables with a tasty dip makes a splendid appetizer. The secret of crispness is turgor pressure (section 4.5); soak cleaned vegetables in fresh water for about half an hour before serving. An attractive assortment might include young carrots or carrot sticks, celery stalks, cherry tomatoes, slices of turnip and cucumber, spinach and watercress leaves, and pieces of cauliflower, broccoli, and cabbage. Simple dips can be prepared from mixtures of sour cream with either clam or onion soup, and various herbs, especially dill.

---

from poppies; and digitalis, a heart stimulant and tonic, comes from foxglove. Many of the plant products used as drugs belong to a class known as the **alkaloids.** Tobacco smoke contains a very powerful and poisonous alkaloid, **nicotine,** which causes profound effects on the vascular system and the central nervous system. The widespread use of tobacco in smoking is causing increasing alarm among public health experts, because of evidence that between 50,000 and 100,000 Americans die each year from its insidious effects. If this toll in life could have been predicted before tobacco gained popularity, there is little doubt that its use would have been prohibited by law. Today, present levels of addiction in the population would make prohibition impossible both to legislate and to enforce.

A plant product that is replacing tobacco to some extent among younger Americans and Europeans is marijuana, which comes from the common weed *Cannabis sativa.* This drug has mood- and behavior-modifying capability and has sufficiently pleasurable subjective effect that many people develop a psychological if not a physiological dependence on it. Recent evidence suggests that permanent psychological damage and loss of sexual potency are possible effects if marijuana is used over a prolonged period. These preliminary indications and the disastrous experience with mortality from tobacco consumption suggest that the wisest course would be to restrain marijuana consumption by legal means until more is known about its long-term effects.

22.9 The success of higher plants can be understood only by knowing the structure of their cells and tissues.

All of the plants described in this chapter are built on the same general plan; that is, their cells are organized into tissues, and these into organs, each of which performs a particular function for the plant as a whole.

Roots, stems, and leaves are the principal organs for the maintenance of a plant. The sexual organs have been discussed in Chapter 12. Before examining plant organs and their functions, let's review the characteristics of a "typical," unspecialized plant cell. As you may remember, a plant cell is composed of cytoplasm surrounded by the plasma membrane and a rigid, nonliving cell wall. Everything inside the cell wall is known as the **protoplast.** A vacuole occupies the central portion of a mature cell and is separated from the cytoplasm by a **vacuolar membrane,** or **tonoplast** (Figure 2-9). Thus a plant cell may be thought of in principle (but not in shape) as a football—the leather covering resisting pressure from the inside is the cell wall, the bladder is the living cytoplasm, and the air in the ball represents the vacuole.

22.10 The cell wall is partly responsible for a plant's rigidity. It is composed of microfibrils embedded in a matrix of pectin and other substances.

The rigidity of a plant comes from a combination of factors—the composition of the cell walls and the turgor pressure of the cells themselves (section 4.5).

The first wall that forms about a growing cell is the **primary wall;** being relatively thin and elastic, it can grow with the cell to some extent. A major component of the primary wall is cellulose, a polymer of glucose. Cellulose is strung out into microfibrils which are "woven" together and embedded in **pectin,** another polymer of sugar units. This "fabric" of microfibrils and pectin substances forms a tough, resilient wall.

When plant cells mature, they usually form a new and stronger **secondary wall** inside the primary wall. The secondary wall is more resistant to stretching, and its greater rigidity is due to the

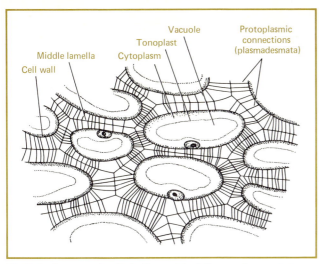

22-13 A diagram of cells in the nutrient tissue of a seed, showing the relationships of structures described in the text.

presence of substances such as waxes, lignin (a chemical substance), and cellulose fibers. The hardness of woody tissue is due to its high lignin content. Unlike cellulose, lignin is not "woven" into the wall, but moves into the spaces between the cellulose fibers.

Cell walls may be rigid, but they are by no means impermeable. Perforations in them enable adjacent cells to maintain bridges of protoplasm which keep them in communication. These protoplasmic connections are called **plasmodesmata** (Figure 22-13), and they extend through thin areas in the wall known as "pits."

Between the walls of neighboring cells is a layer of flexible "intercellular cement," the **middle lamella,** which is composed primarily of pectin. The middle lamella forms before the primary or secondary cell walls, for it is the structure that forms the first partition between dividing daughter cells.

22.11 Vacuoles act as storage areas and waste deposits; they also help regulate the movement of water from cell to cell.

Vacuoles, which may occupy as much as 90 percent of the volume of a cell, are used to store sugars, starches, and proteins that may be needed at some future time by the cell. Waste products may also be deposited in vacuoles, where they sometimes form crystals or granules of inert material.

Perhaps the most important function of the vacuole is its role in regulating water movement from cell to cell.

22.12 Protective tissue, fundamental tissue, and conductive tissue all arise from meristematic tissue.

As was discussed in section 12.5, there are certain regions in a plant where growth goes on indeterminately (without specific limits) whenever conditions are favorable. These are the **meristems,** and the cells of which they are composed are unspecialized, thin-walled, and contain many small vacuoles. These cells subsequently differentiate into either **protective tissue,** such as the surface layer, or **epidermis,** of the leaves and roots, and the cork of the stem; **fundamental tissue,** including the spongy **parenchyma,** in which photosynthesis occurs; and **conductive tissue,** such as **xylem** and **phloem,** which are concerned with transport.

22.13 The tissues in the stem are concerned with support, protection, growth, and the transport of food and water.

In addition to its most obvious task of supporting the leaves so that they receive adequate sunlight, the stem functions as a transport system carrying water and food to all parts of the plant.

Two kinds of conductive or **vascular** tissues are mainly responsible for this function. Xylem transports water and dissolved minerals upwards from roots to stems and leaves, and phloem transports organic materials, including the food manufactured by the photosynthesizing leaves, to the rest of the plant. Both these tissues develop from the apical meristems or from another meristematic tissue, the **cambium,** when present. In each growing season the cells which reside on the inner surface of cambium, towards the water-transport area of the tree, differentiate to form thick-walled water-conducting cells. These xylem cells form the "wood" of the tree. Xylem cells die as soon as they reach maturity, but even when dead, water and minerals can be passively transported through them.

Some long, slender xylem cells are called **tracheids** (Figure 22-14). These cells are lined up into long chains which overlap one another. Each tracheid is about one millimeter or less in length and is punctured with pits that allow water to move through the sides of the cells into surrounding tracheids, and finally to other parts of the plant. In this way, water is passed from cell to cell, from the roots up through the stem to the leaves.

A second type of xylem cell is called a **vessel**

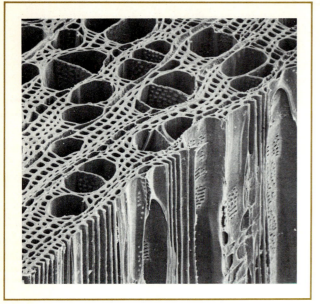

22-14 A scanning electron micrograph of the transverse and radial surfaces of the wood of a New Zealand beech, *Notho-fagus fusca* showing the slender tracheids and much larger vessels of the xylem. (Courtesy of R. R. Exley, B. G. Butter-field, and B. A. Meylan and the *Journal of Microscopy*.)

element; these resemble microscopic cylinders. They stack one on top of another, to form pipes called **xylem vessels.** Water passes through vessel elements without opposition, since the end walls of the mature cells are lost. The xylem vessels thus form straight pipelines that extend from roots to leaves. The secondary walls of xylem vessels are not as thick as those of tracheids and offer less support to the plant. Wood often contains both tracheids and vessel elements.

22.14 Whereas xylem conducts water and minerals, phloem conducts nutrients.

Phloem cells develop from the outer surface of the cambium. They do not have heavily thickened walls, nor do their end walls break down. They resemble groups of elongated cylinders stacked on top of one another and connected through pores in their end walls. These walls are called **sieve plates.** Although the nuclei of phloem cells disintegrate, their cytoplasm remains alive, and they actively transport food molecules. The cytoplasm of the cells above and below each other runs together through the openings in the sieve plates, thus forming a **sieve tube,** a continuous pathway along which food is transported (Figure 12-6).

Parenchyma cells (from the Greek "to pour in beside") are relatively unspecialized or "basic" cells found throughout the plant. Parenchyma cells of a special type, **companion cells,** are associated with sieve tube member cells. Unlike the sieve tube members, companion cells have functional nuclei, which are thought to govern the activities of nucleus-lacking sieve tube members. When a companion cell dies, the sieve member related to it also dies.

In the temperate climates, the xylem of woody plants (but not of annuals) is marked by a series of **annual rings.** Each year during the growing season the cambium produces new xylem and new phloem. In the spring, when water is plentiful, the xylem cells are large, but in late summer they are smaller. The spring and summer wood together constitute the annual ring; by counting these rings it is possible to estimate the age of a tree.

22.15 The bark serves to protect the stem.

The first bark of a woody stem is composed mainly of a meristematic layer called the **cork cambium,** and of the **cork** outside that (Figure 12.6). The walls of the cells of this outermost layer are impregnated with **suberin,** a waterproofing substance. This prevents the inner layers of stem from drying out, but also prevents the cork cells from taking in water. Consequently they die and slough off. New cork cells are produced from the cork cambium.

The organization of stem tissues in concentric cylinders is typical of the dicotyledons. The monocotyledons have the same tissues, but they are organized in a different manner: Bundles of conductive tissues are scattered throughout the **pith,** a central mass of loose parenchyma cells (Figure 22-15).

22.16 Tissues of the root bring in water and dissolved minerals and anchor the plant in the soil.

Roots are organized in much the same way as stems, except that they have no pith and their vascular system is arranged somewhat differently. At the very tip of the root is a protective **root cap,** which shields the **apical meristem** region of cell division. The new cells produced in the meristem elongate and eventually differentiate into xylem, phloem, epidermis, and other tissues. In this **region of maturation,** as it is sometimes called, the cells

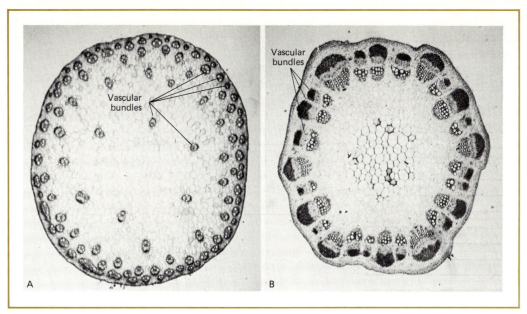

22-15 Cross section through the stem of (A) *Zea mays* (corn, a monocot); (B) *Helianthus* (sunflower, a dicot). (Photograph by author.)

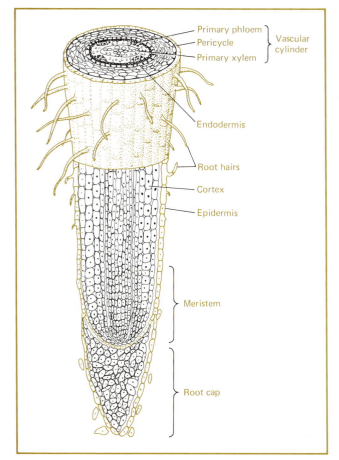

22-16 The structure of a root, shown both in transverse and longitudinal section.

around the periphery of the root develop long extensions of their cell walls, known as **root hairs** (Figure 22-16). These greatly increase the surface area of the root through which it can take in water and mineral salts. Inside the root epidermis is the **cortex,** a multicellular layer whose innermost layer of cells is the **endodermis,** which surrounds the vascular tissue. On the radial walls of the endodermal cells are **Casparian strips,** waxy thickenings that prevent water and minerals from flowing into or out of the plant without passing through the living endodermal tissue.

The root may also be a major storage site for starch and other materials.

22.17 The leaf, organized like a porous sandwich, has a filling of parenchyma cells between two layers of epidermis.

Leaves are formed from meristematic tissue which differentiates into specialized cells as the leaves grow and mature. Essentially, a leaf is composed of an upper and lower layer of protective cells, with photosynthesizing parenchyma cells and branches of the vascular system sandwiched between.

In leaves, the parenchyma cells have a large number of chloroplasts (Figure 22-17). The walls of parenchyma cells are thin and elastic, and secondary walls do not develop. Consequently, they remain flexible enough to permit the leaves to bend. Through the several layers of parenchyma cells

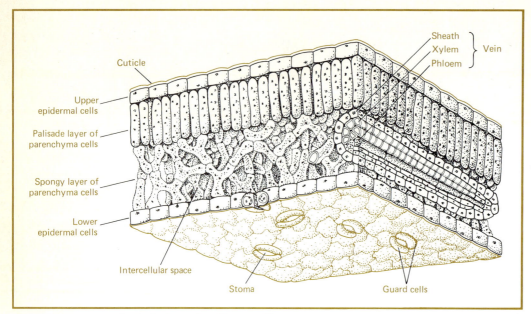

22-17 The structure of a leaf, shown in a diagram of a cutaway block.

run vascular bundles in the form of veins. These contain both xylem and phloem and are in direct contact with the main portion of the plant. The water coming through the xylem cells is particularly important for the leaves, for it enables them to exchange gases. In return, leaves send some of the sugars they manufacture (section 5.10) through the phloem to the rest of the plant.

On either side of the parenchyma cells is a layer of epidermal tissue composed of flattened cells that fit tightly together. The only spaces occurring between them are the pores, or **stomata** (singular, **stoma**), through which air enters and exits. The CO_2 in air is removed by the photosynthesizing leaves, and O_2 and water vapor are given off during the day (section 23.5).

Plants have special structures on their leaves that regulate the intake of CO_2 and the loss of O_2 and H_2O vapor. Pairs of special epidermal **guard cells** control the opening of **stomata** (apertures leading into the interior of the leaves where most of the plant's photosynthesis takes place).

The stomata open when the guard cells around them are turgid, and close when the guard cells are flaccid due to the loss of water. Since it is not advantageous for plants to photosynthesize when water is in short supply, the closing of stomata is an important protective mechanism against wilting and death. On the other hand, when water is plentiful, daylight causes the guard cells to become turgid because K^+ ions are pumped into them with

ATP produced by photophosphorylation during photosynthesis (section 5.9).

In one square centimeter of leaf surface from an apple leaf there are approximately 1000 stomata. Even so, the total area of all the pores when fully opened is small compared to the total area of the leaf.

The other cells of the epidermis are waterproof, due to a waxy covering of cutin on their outer surfaces. When the stomata are closed, water cannot leave, and the plant resists drying out.

22.18 Two types of fundamental tissue found throughout the plant are collenchyma and sclerenchyma tissues.

Collenchyma tissue (from the Greek word for "glue") is composed of elongated cells which have thick, sticky cell walls. **Collenchyma** cells are the first supportive cells formed by the plant. They can continue to elongate somewhat after their cell wall is laid down, and are thus useful for the support of leaves, roots, and stems. Another tissue, **sclerenchyma** (from the Greek word for "hard"), has thick, rigid walls and is used to support mature plants. When the cells reach maturity and stop growing, they often die and become hollow, but still support the plant.

23

The Physiology of Higher Plants

QUESTIONS TO KEEP IN MIND

What properties of water are important in understanding the physiology of plants?

How does water move into a plant? Are mechanisms other than osmosis required?

How do leaves use water?

For what functions do plants require minerals, and what minerals are required?

How do plants use nitrogen? Where do they get it?

In Chapter 22, it was shown that the greatest diversity and evolutionary success among plants was achieved in the subphylum Spermopsida, or seed plants, and especially in the flowering plants (angiosperms). Faced with this diversity, the biologist naturally asks how it came about. Plant physiology is one science, however, that asks not about plants in the past, but those of the present. How, for example, does water climb to the top of a towering redwood tree? To understand this and other important questions about plant physiology, it is necessary to learn a little about the physics and chemistry of water, an unusual substance from which plants are made, and of which they are largely composed.

23.1 The physical properties of water can be understood in terms of its molecular structure.

Water, unlike other common compounds, exists as a gas, liquid, and solid within the normal range of temperatures found on earth. Moreover, compared to compounds that are chemically similar, it is apparent to chemists that both the freezing and boiling points of water are abnormally high. These unexpected properties of water are due to its unusual molecular structure.

In section 4.1 it was explained that water molecules are polar by virtue of the fact that they carry a slight negative charge on the O atoms and a weak positive charge on their H atoms. Consequently the hydrogens are attracted electrostatically to the oxygens of other molecules, and weak hydrogen bonds form between them. If the temperature is low enough, the water molecules remain bonded to each other in an open latticework pattern (Figure 4-2) known as ice. Because the naturally occurring pattern is so open, ice is less dense than water in its liquid form. If the opposite were true, ice would sink, and form only on the bottoms of lakes. It is usually true for other compounds that the frozen form is denser than the liquid.

At temperatures above 100° C the hydrogen bonds among water molecules are constantly forming and breaking, but below 100° C, enough bonds exist at any one time to hold the water molecules together as a liquid. It is because of these hydrogen bonds that it takes a great deal of thermal energy to break water molecules apart and change water from either a solid to a liquid or a liquid to a gas. For example, 540 calories are required to convert just one gram of liquid water to water vapor. All this energy is spent breaking water molecules loose from one another: The temperature of the water is not changed. The 540 calories of heat is called the **latent heat of vaporization.** Warmed by the sun, the cells in leaves would "cook" were it not for the removal of heat by vaporizing water. A perspiring human body is cooled in the same manner.

Water is said to have a high **specific heat**, a statement which means that water can absorb a great amount of heat energy without itself experiencing a large change in temperature.

Still another consequence of hydrogen bonding and the tenacity with which water molecules hold together is water's high cohesiveness and surface tension. The rise of water from the root of a tree to a leaf 200 feet or more away, would be impossible if it were not for this cohesiveness of water.

23.2 Water moves in and out of plant cells by osmosis.

Having briefly examined the physical properties of water, we can now see how it behaves within plants. In general, a plant requires water so that its leaves or photosynthesizing portions can operate efficiently. As fast as water is lost from leaf cells it must be replenished, or the leaf will wilt and die. The water that leaves require comes in through the roots, but just how is not entirely understood. The main process involved is diffusion in response to an osmotic gradient.

In section 4.5, osmosis was defined as a special case of the diffusion of water through a semipermeable membrane in response to a difference in water concentration across the membrane. In terrestrial plants, osmosis occurs across the cell membranes of root hairs and other cells in the root. The osmotic pressure in these cells and in the cells that will pass the water from root hair to xylem to leaves, is much higher than that of the soil water, because of the high concentration of dissolved salts and sugars in the cells. Consequently, water tends to enter these cells, lowering their osmotic concentration. Before the water entering a root can reach the xylem in which it is transported, it must pass either through or between cells until it reaches the **endodermis**, a layer of cells through the membrane of which the water must pass to reach the xylem. The endodermal cells have the kind of semipermeable membrane required for osmosis to occur. Because there is a gradient in water concentration from the soil to the root hairs to the endodermal cells to the xylem, water does actually flow along this osmotic gradient under what is known as **root pressure**. Root pressure can be demonstrated by merely decapitating a tree and measuring the height to which the fluid exuded will rise in a slender tube attached to the cut base.

23.3 Neither root pressure nor suction nor activities in the stem are responsible for water transport. How then can water reach the leaves atop a 350-foot redwood?

Although root pressure can be demonstrated in a number of plants, the pressures are seldom large enough to cause a column of fluid to rise more than a few feet. Despite this fact, trees deliver tens of gallons of water to the tops of their branches on a hot summer day.

Suction does not account for water transport either. The hypothesis that plants have some system for pumping water from the roots by suction may seem like a possibility, at first glance. However, we know that when a pump sucks on a pipe (such as a xylem vessel), it is actually allowing atmospheric pressure to push water up the pipe. Atmospheric pressure will lift a column of fluid only about 32 feet; thus suction could not explain how water gets to the top of a tree 350 feet tall.

For a while the idea that plants might somehow actively pump water along their stems was popular, and some turn-of-the-century botanists claimed that a kind of pulsing rhythm or "vital movement" was involved. This was disproved by a German botanist who grandly cut down a 70-foot oak and somehow stood it in a huge tub of picric acid that he knew would kill the plant's cells as it moved up the tree. Three days later he replaced the acid with water; as he had predicted, the water rose to the top of the tree as easily as before. Clearly, active pumping, which would require metabolic activity of live stem cells, was not essential for the ascent of water.

23.4 The best explanation we have for water transport is the tension-cohesion theory.

The tension-cohesion theory of water transport, first outlined three-quarters of a century ago, is based on two properties of water that have already been discussed: its **cohesiveness**, or the degree to which its polarized molecules tend to attract one another; and its **adhesiveness** to hydrophilic substances. Taken together, these properties enable a thin, unbroken strand of water (such as exists in the tracheids of a pine tree or the more efficient vessel elements of a maple) to resist either breaking or pulling away from the sides of the xylem vessels. The water in these unbroken columns *moves* because the leaves are using the water at the top of the column (in their cells) for photosynthesis and losing some of the water through the stomata by evaporation. This water loss does not all represent waste, however, as it regulates the temperature of the leaves in hot weather.

Although plant physiologists quibble about the quantitative details of the tension-cohesion theory, its essential correctness can be demonstrated by a simple and rather dramatic experiment. Find a tall oak tree, preferably a young one, and insert a pin or other sharp instrument through the bark and into the xylem. As soon as the pin is withdrawn, you will be able to hear a prolonged "hiss," as air is drawn into the ruptured xylem vessels, breaking the column of fluid.

When leaves lose water by evaporation, a gradient of pressure is produced along this water column from the soil through the xylem to the leaves. When the loss of water is quite rapid, as it is on a hot summer day, pressure inside the xylem falls below that of atmospheric pressure and becomes a **tension.** When the tension in xylem sap exceeds the osmotic pressure in root cells, water in the roots is drawn up under tension. Along with the water are transported minerals needed by the leaves for their metabolic activities. It is important to emphasize that tension under these conditions is not equivalent to suction; suction will not suffice to explain water transport in plants, as has been pointed out above.

23.5 Leaves carrying on photosynthesis use water in several ways. Photosynthesis controls the rates at which water is lost and gases are exchanged through the stomata.

In the "light reactions" of photosynthesis, water is split by light to liberate gaseous oxygen and form materials which serve to reduce carbon dioxide in the "dark reactions." Water is thus an essential "raw material" for photosynthesis. Water also wets the surfaces of photosynthesizing cells inside the leaves so that the other raw material for photosynthesis, carbon dioxide, can be efficiently absorbed.

The total loss of water by evaporation from the surfaces of photosynthesizing cells in plant leaves is called **transpiration,** and its rate is governed by the activity of stomata. Stomata open in the light and close in the dark. Photosynthesis controls these movements by providing energy to transport potassium ions into stomatal cells. This transport in turn controls the osmotic potentials required to open the stomata.

Transpiration accounts for roughly 95 percent of the water absorbed by plants, the remaining 5 percent being used for growth and photolysis of water (Chapter 5). During its growing season, a single tomato plant may lose some 34 gallons of water through transpiration. A palm tree in a hot, dry climate where evaporation proceeds rapidly, may use 100 gallons *per day* for transpiration, photosynthesis, and temperature regulation. Although the water lost in transpiration is not directly used by the plant, it serves an important function of transporting dissolved minerals and other materials to the tissues that need them.

23.6 Plants require about 16 essential minerals to grow and to complete their life cycles.

Using water cultures, scientists in the last century determined that nitrogen, phosphorus, potassium, calcium, magnesium, sulfur, and iron are essential for plant growth. These elements are needed in large amounts by plants and are therefore termed **macronutrients.** Nine more elements, the **micronutrients** or trace elements, were isolated by water-culture experiments during the present century. These are needed in smaller quantities, and their functions in the plant are less clearly defined. They are iron, boron, manganese, zinc, copper, molybdenum, cobalt, chlorine, and sodium.

Among all these, nitrogen certainly plays one of the most important roles. A fair amount is known about how it functions in plants. Although free nitrogen (N_2) makes up 80 percent of the earth's atmosphere, higher plants have no mechanism for utilizing this exceptionally inert element. It must be obtained from the soil, where it exists as inorganic nitrate. Nitrate, unlike some essential nutrients, does not come from rocks which are weathered and broken down mechanically into the soil. Instead, its immediate source is plant and animal wastes, in which it exists in organic forms. These are freed from combination with carbon and hydrogen by soil bacteria and are thus converted to inorganic nitrate (Figure 23-1), and in this form nitrogen is absorbed by the roots of plants.

Much the same cycle occurs in aquatic environments, where plants take up inorganic nitrate and use it for amino acids, proteins, nucleic acids, and so on. Some of the plants are consumed by animals. Both plants and animals return the nitrogen in an organic form, as wastes and as their own bodies when they die. As on land, organic nitrogen is freed from combination with carbon and hydrogen by

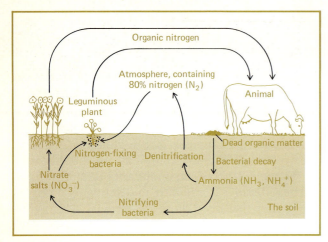

23-1 The nitrogen cycle on land.

bacteria capable of converting organic nitrogen compounds into inorganic nitrogen compounds.

A plant then takes in the nitrate and reconverts it to organic forms, generally amino acids. Amino acid production may occur in roots but is largely performed in the leaves after the nitrate has been transported there.

23.7 By transamination the few amino acids produced in the roots are converted into all the amino acids needed by the plant.

The amino acids, especially glutamic acid, produced in the roots travel up the plant with the water flow. There a process known as **transamination** occurs, in which the amino group (NH_2) of one amino acid is transferred to other compounds, thereby producing new amino acids. For example:

$$HOOC-CH_2-CH_2-\overset{\overset{\displaystyle NH_2}{|}}{\underset{\underset{\displaystyle H}{|}}{C}}-COOH \quad \text{glutamic acid}$$

+

$$HOOC-\overset{\overset{\displaystyle O}{\|}}{C}-CH_3 \quad \text{pyruvic acid}$$

↓

$$HOOC-\overset{\overset{\displaystyle O}{\|}}{C}-CH_2-CH_2-COOH \quad \alpha\text{-ketoglutaric acid}$$

+

$$HOOC-\overset{\overset{\displaystyle NH_2}{|}}{\underset{\underset{\displaystyle H}{|}}{C}}-CH_3 \quad \text{alanine}$$

In this way all 20 of the separate amino acids needed as protein building blocks are synthesized. Plants also synthesize the vitamins (cofactors) and metabolites they need. They are thus much more self-sufficient than animals. The latter cannot produce their full complement of amino acids by transamination and consequently require many of them in their diet.

23.8 All the nitrogen used by plants and animals originally comes from the atmosphere and is converted into usable forms by nitrogen-fixing bacteria or blue green algae.

Nitrogen fixation is the process by which certain blue-green algae and certain bacteria, notably *Rhizobium*, convert atmospheric nitrogen into usable nitrogen compounds such as ammonia.

Rhizobium and other nitrogen-fixing bacteria invade the root tissue of legumes such as clover, alfalfa, peas, and soybeans and cause the formation of irregular white knobs called **nodules** (Figure 23-2). The bacteria live within these nodules in symbiosis with the legume, taking dissolved sugars from the root and giving up nitrogen compounds in return. If the roots of the legumes are left in the soil after the growing season is over, they provide a rich source of nitrogen for the next crop sown in that field. Consequently, legumes are often used in crop rotation.

Most plants do not form symbiotic relationships with nitrogen-fixing bacteria and must obtain nitrates from the breakdown of other plant and animal matter going on in the soil around them. Under natural conditions a plant eventually returns all the nitrate it uses to the soil, but in farming, plants are continually removed from the land, thus depriving it of its natural complement of nitrates.

23.9 Nitrates and other minerals become available to plants by a process known as cation exchange.

The minerals a plant needs for growth occur in the soil in several forms. They may be in solution in the soil water, or be **adsorbed** on clay particles, or occur as constituents of organic and inorganic compounds. Minerals are most available to plants in the soil solution and on clay particles. In both cases the minerals exist as ions—positively charged **cations** such as Ca^{2+} (calcium ions), Fe^{3+} (iron ions), and K^+ (potassium ions), and/or negatively charged **anions** such as nitrates, sulfates, and phosphates.

The cations mentioned, and H$^+$ ions as well, tend to cluster around the negatively charged clay particles (Figure 23-3). The cations held by the clays are exchangeable—for example, H$^+$ may displace Na$^+$ (sodium), or Na$^+$ may displace H$^+$. The cations a plant absorbs from the surface of clay particles must be replaced by other cations, usually H$^+$. This process is called **cation exchange.**

23.10 Dissolved mineral salts enter the roots by diffusion and by some process of active transport.

Simple diffusion of minerals and water into the roots cannot account for all the mineral uptake that occurs in plants. Roots in water continue to absorb mineral ions even after the mineral concentration in the root tissue is greater than that in the water. If diffusion alone were operating, the ion flow into the roots would slow and eventually stop once the ion concentration in the root equaled that of the external soil water.

Energy from ATP is generally required to actively accumulate ions against a concentration gradient (section 4.6). If active transport is involved in mineral uptake, it must take place in the living root cells rather than in the dead xylem cells. It seems probable that the endodermis surrounding the root's vascular cylinder is the tissue that provides metabolic control.

The body of the root can be conceived as containing two portions: a nonliving portion called the **apoplast** and a living portion, the **symplast.** Water diffuses into the plant from the soil. It may pass first only through the apoplast—cell walls and the spaces between cells—until it reaches the endodermis; or it may reach the endodermis through living cells. Then water and dissolved minerals must pass across the endodermal layer, which means these materials must pass through the living endodermal cells, which are part of the symplast (Figure 23-4). These cells have waxy thickenings (Casperian strips) on their radial walls which block further intercellular transport. All water and minerals must therefore pass through a living cell membrane before entering the xylem. This gives the plant a chance to select the materials it takes in.

The theory is that the symplast, by active metabolic work, accumulates ions and moves them across the outer membrane of cells and then through

23-2 Nodules containing *Rhizobium* on the roots of a legume, the silk tree *Albizzia*. (Courtesy of R. Speck.)

23-3 A diagram showing the relationship between root hairs and negatively charged clay particles that attract cations which are plant nutrients.

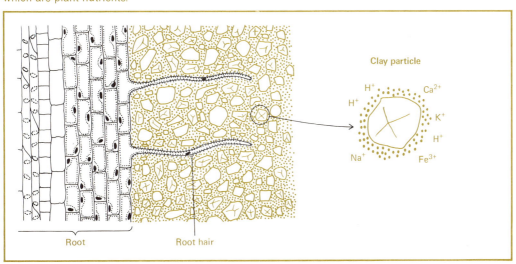

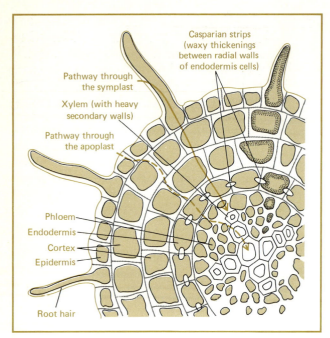

23-4 Diagram of a root showing the symplast (*color*) and apoplast, and the routes that water and ions may take to enter the root. The spaces occupied by cell walls have been expanded to illustrate these pathways.

23-5 An aphid on a basswood leaf with its proboscis (*arrow*) inserted into a phloem cell such as those shown in the diagram of a leaf (Figure 22-17). Note the droplet exuding from the abdomen. (Photograph by author.)

boscis, the pure sap which exudes from the severed proboscis can be collected.

Analysis of the sap shows it to contain about 10 to 25 percent sugar, a much higher concentration than occurs elsewhere in the plant. The movement of these sugars in the phloem is very fast—too fast, some investigators believe, to be accounted for by diffusion alone.

the cells. Once inside the endodermis, the dissolved substances have easy access to xylem vessels, in which they are carried passively to the stem and leaves.

23.11 Dissolved nutrients are carried throughout the plant with the water that flows through the phloem tissue.

The movement of sugars, amino acids, hormones, and even virus particles occurs primarily in the phloem tissues, which are present under the bark of trees and in the cortex of other plants. It is now believed that dissolved materials are carried in bulk along with the water that flows rapidly through the phloem. This is called **bulk flow**. The flow results from unequal osmotic conditions in different cells. For example, in a cell which is actively utilizing photosynthetic products, the osmotic conditions are different from those in a cell where photosynthesis is in process.

To get a pure sample of phloem content (sap) for analysis was very difficult until, in 1953, aphids were used for retrieving it. Aphids feed by inserting their tubelike proboscis into a single phloem cell (Figure 23-5). By cutting an insect free of its pro-

23.12 Plants and their cells exhibit many kinds of movements.

Although plants do not exhibit motility in the sense that most protists and many animals do, neither are they completely motionless. Cells isolated from a variety of higher plants exhibit cytoplasmic streaming (section 8.7) and the chloroplasts may move in response to changes in the intensity and quality of light, a process known as photodinesis (section 5.5). It is suspected, but not yet proven, that these types of movement may be related to those in animals and protists.

The bodies of higher plants move very slowly, that is, *grow* in the direction of light. Their roots grow downward into the soil rather than upward. These movements are referred to as **tropisms**, specifically **phototropism** (movement in response to blue light) and **geotropism** (movement in response to gravity). Plants also exhibit movements of a more rapid nature in response to light. Sunflowers in an open field face the light all day; leaves move up and down to capture the maximum amount of light. Some flowers open in the morning and close at night. Most of these movements are in response to changes in the turgor in cells caused by changes in photosynthetic rate and available water.

Some of these movements occur independently of light and are examples of **endogenous rhythms** or **biological clocks.**

23.13 Many plants exhibit photoperiodism; their flowering is controlled by the length of day and affected by the quality of the light.

While many plants flower regardless of the length of the light portion of their daily cycle and are called **day neutral plants,** others flower only if given either a long day or a short day. The light received affects the leaves, as has been shown in experiments in which leaves were grafted from one plant to another to change the latter's flowering response. Many photoperiodic effects have been traced to the action of light at the red and far-red portions of the spectrum on a protein pigment, **phytochrome,** found in many plant cells. This protein can exist in two forms, which differ in their physiological ac-

tivity and absorption. Plant physiologists have been able to duplicate the effects of long and short days by controlling the state of phytochrome in the cells of a plant using light of the proper wavelengths. Many aspects of the physiology of higher plants and of algae are controlled by mechanisms involving phytochrome. The study of these mechanisms represents one of the active frontiers of research in plant physiology.

This chapter may have helped to explain how a living higher plant functions. There are still many things about plants that are not well understood. For example, there have been repeated claims that plants possess supernatural or "psychic" powers, that they respond differently to "good" and "evil" people. Most scientists regard such claims and reports with healthy skepticism until they have been confirmed under rigorous scientific conditions by people who have been trained in the methods of scientific objectivity. To our knowledge, there are no scientifically accepted psychic phenomena that can now be attributed to plants.

24

Animals and Their Behavior: Acoelomates

How do the basic body plans of a sponge, hydra, and flatworm show evolutionary progression?

How is a colonial cnidarian such as the Portugese man-of-war organized? What advantages do colonial living offer?

What is the life cycle of the sea gull fluke?

Animals are multicellular, eucaryotic organisms, most of which ingest their food. Unlike plants and fungi, which are sessile, most animals either move about or possess mechanisms for pumping a fluid environment through their bodies. The ingestive mode of nutrition forces animals to consume other organisms in order to obtain a supply of energy and materials. Animals may be herbivores, eating plant tissues; carnivores, consuming meat; fungivores, feeding on fungi; omnivores, subsisting on all sorts of organisms; or scavengers, picking away on dead organisms. Each of these modes of nutrition has resulted in the evolution of special repertoires of behavior. However, parasitic animals that have adopted an **absorptive** mode of nutrition characteristically exhibit restricted and unusual behavior patterns.

As the need for locomotion has been met in thousands of different ways, so has the need for a digestive system capable of taking in tissues of some other organism, digesting them, and excreting the wastes. Much of the anatomy, physiology, and behavior of animals can be seen as adaptations for their role as consumers.

Many protists are consumers too, but in this book they are not considered as part of the animal kingdom. In part this is an arbitrary decision, for there are protists that are more complex in both structure and behavior than some of the very simplest animals (sponges, for example). With only these few exceptions to the contrary, however, the overwhelming majority of animals represent a significant departure from the protistan way of life; namely, they utilize a division of labor among cells.

Among the animals, the major evolutionary trends include (1) the development of **multicellularity** and increasing degrees of **cellular specialization** (i.e., the formation of tissues, organs, and organ systems); (2) the development of **bilateral symmetry** (roughly equivalent right and left halves); (3) a **tubular gut** with both mouth and anus; (4) a **lined body cavity (coelom)**; (5) **segmentation** and **appendages**; (6) **skeletons**; and (7) complex **nervous systems** providing the ability to react, learn, and finally to remember and reason.

Animals can be broadly classified according to their body plan before further sorting them into taxonomic categories. There are **acoelomates** (animals without a true body cavity), **pseudocoelomates** (those with a false body cavity), and **coelomates** (Figure 24-1).

As with protists, plants, and fungi, the classification of animals reflects our ideas of how they evolved. At present there are two major concepts of animal evolution. One theory holds that multicellular animals evolved from colonial flagellates which contained many cells and showed some cellular coordination. Other scientists suggest instead

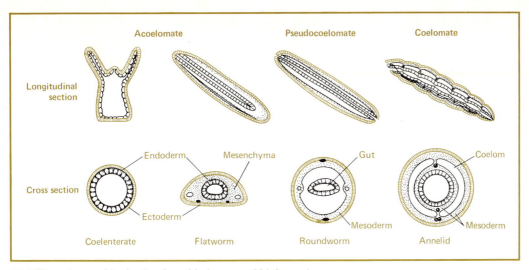

24-1 Three types of body plan found in lower and higher animals.

24-2 Simplified phylogeny of the kingdom Animalia, showing the relationships of the phyla and examples discussed in this section of the book.

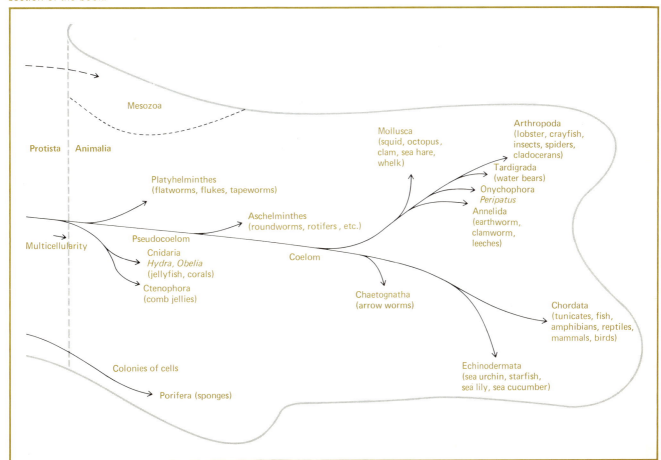

that multicellular animals evolved from multinu-cleate ciliates which formed membrane boundaries between the individual nuclei, yielding an animal which would resemble in many respects some of the simple flatworm groups now living.

Although both theories seem equally plausible, more research and information is needed before a definite choice can be made between the two. Figure 24-2 shows a simplified version of the generally accepted view of the evolutionary history of the animal kingdom. However, only the "major" phyla and their representatives discussed in this book are included.

In the remainder of this chapter and in the succeeding ones, some of the largest phyla of living animals will be discussed in detail. Many animals from numerous smaller phyla will not be mentioned but most bear some resemblances to organisms found in the major phyla, both in structure and function.

24.1 The sponges, phylum Porifera, are such primitive animals that they are sometimes referred to as parazoans, that is, "almost animals."

The great majority of the roughly 10,000 species of sponges known to exist are marine; a few live in fresh water. Sponges (Figure 24-3) are sessile organisms and except for a free-swimming larval stage, spend most of their lives attached to rocks and shells or anchored in mud on the floor of the sea. They feed by creating a current that passes through their body and from which they filter small organisms. Sponges consist of three layers of cells. On the outside of the body is a layer of thin **epithelial cells** (lining cells), perforated at regular intervals by **canals** through which water is continually drawn. Along the inside of the central cavity of the sponge are numerous **collar cells.** A flagellum protrudes through the collar of each of these cells, and the beating of many of these flagella causes water to be drawn in through the sides of the sponge and expelled through a large opening, the **osculum.** Food particles suspended in the water are filtered by the collar and enter the cytoplasm of the cell body by phagocytosis.

Between the epithelium and the collar cells lies a gel-like layer, the **mesenchyme.** In it are **amoebocytes,** wandering cells responsible for a primitive kind of communication; **scleroblasts,** which form hard spicules from either calcium carbonate or

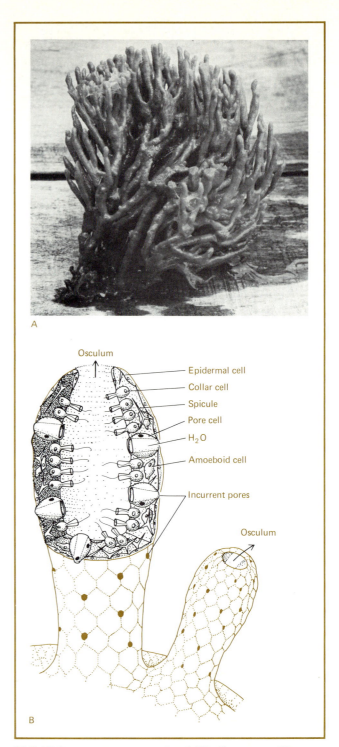

24-3 (A) A common orange-colored Atlantic sponge, *Microciona prolifera,* and (B) a diagram of the parts of the sponge discussed in the text.

silica, and/or **spongocytes**, which form tough but flexible protein fibers in some sponges; and other cells involved with asexual and sexual reproduction.

Sponges have a disconnected "skeleton" composed of either spicules or fibers (formed by the scleroblasts and/or spongocytes) imbedded in their walls. This allows them to stand upright and perhaps discourages other animals from eating them. In the case of so-called "bath" sponges collected commercially in warm ocean waters, the proteinaceous portion of the skeletal material is strong enough to be used repeatedly for cleaning purposes.

The various parts of a sponge are so loosely organized that it is possible to push a sponge through a fine sieve without killing it. After several hours the sponge cells reunite and form a new sponge. The wandering amoebocytes and the collar cells seem to be of particular importance in early stages of reaggregation. The former send out long **filopodia** (thin pseudopods) from their cell bodies, and some of these filopodia contact other cells in the area. If a filopodium sticks to another cell which it touches, it retracts, pulling the second cell toward the first. These two cells then send out other filopodia which "catch" other cells. The filopodia are quite selective in the cells they will stick to. Cells of two species in the same dish will sort out according to their origin.

24.2 The phylum Cnidaria includes hydras, jellyfish, and corals.

The cnidarians (also called coelenterates) are a phylum of some 2,000 species, whose members, like the sponges, live mainly in marine environments. Unlike the sponges, cnidarians have advanced along the evolutionary scale to the extent of having real organs and tissues. They possess epithelial cells with muscle fibers (or contractile fibers), **nematocysts** or stinging cells, nerve nets, a true mouth, and a digestive cavity with a single opening. Some have rather complex sensory structures.

The phylum contains three major classes—**Hydrozoa** (hydras and hydroids), **Scyphozoa** (jellyfish), and **Anthozoa** (corals and sea anemones).

Among the hydrozoa certainly the best-known animal is the freshwater hydra (genus *Hydra*) (Figure 24-4). An ordinary hydra is a thin, tubular animal that stands about half an inch tall and has a ring of five to eight tentacles at its mouth or uppermost end. At the other end is a base which it uses to adhere to any solid surface.

Essentially, a hydra has two layers of cells, one on the outside and one on the inside of its body. The outer cells are predominantly involved in protection and food capture, and the inside ones in digestion.

The largest cells in the outer (epidermal) layer

24-4 *Hydra*, seen in (A) longitudinal section and (B) cross section. (C) Details of cell structure in the ectoderm and endoderm.

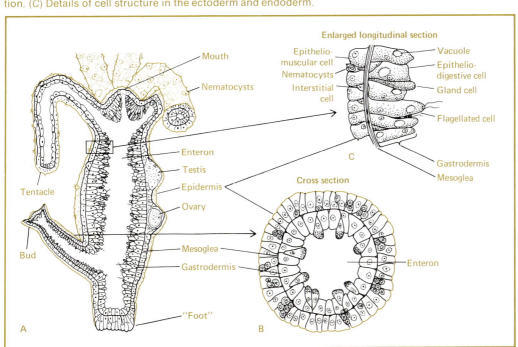

are **epithelio-muscular** cells, which provide much of the bulk of the animal and are active in movement. **Stinging cells** and **nerve cells** are also part of the epidermal layer. The other important epidermal cell type is the **interstitial cell,** which seems to retain certain embryonic characteristics in the adult hydra and consequently can give rise to several other epidermal cell types, such as eggs, sperm, nerve cells, and possibly mucous cells.

Between the outer and inner layers of cells is a thin layer of jellylike protein called **mesoglea.** The inner cell layer is composed mainly of **gastrodermal-digestive cells;** also present are **gland cells** and **mucous cells,** but they are confined mostly to the oral cone around the mouth and nearby regions of the body column.

Hydra has no anus and uses its mouth opening for both intake of food and discharge of indigestible materials. Violent contractions of the body tube periodically eject waste material out of the mouth to a distance where it will not foul the water in the immediate vicinity.

The regenerative powers of hydra are remarkable. The smallest piece of a hydra you can cut with a scalpel can regenerate into a complete animal (with the exception of pieces from the tentacles, oral cone, and base). If a single cross-sectional cut through the middle of the body column is made, both halves will regenerate their missing portions within a few days.

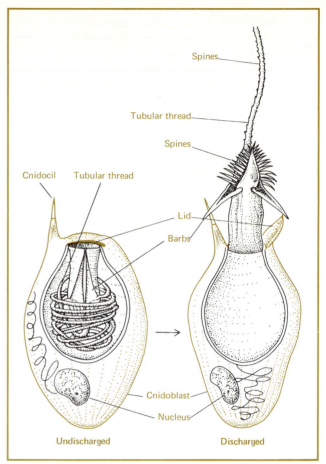

24-5 One type of nematocyst from *Hydra,* before and after discharge.

24.3 Hydra, in fact all cnidarians, have stinging organelles called nematocysts.

Cnidarians are the only major group of animals that possess **nematocysts,** eversible organelles that they use both to paralyze their prey and to hold onto it. Hydra have four types of nematocysts, which form originally in the body column and migrate to the tentacles. Each nematocyst is contained within cell which produced it, the **cnidoblast** (Figure 24-5). The largest nematocysts, called **penetrants** or **stenoteles,** are roughly oval in shape. When stimulated, they release a long, hollow, threadlike structure. Penetrants are used both in offense and defense by hydra and contain a paralyzing toxin that is injected as the nematocyst thread stabs into the predator or prey. The Portuguese man-of-war, a large relative of *Hydra,* can reputedly kill a two- or three-foot fish in a matter of seconds and can raise excrutiatingly painful welts on the bodies of swimmers. A smaller

cnidarian in the ocean around Australia possesses a particularly lethal toxin in its nematocysts, and this species has killed at least 50 humans in recent years.

There are three other types of *Hydra* nematocysts in addition to penetrants. One fires a long, straight thread which secretes a sticky substance at its tip. These are probably used for attachment. Cnidarians have altogether some 15 to 20 morphologically distinct kinds of nematocysts, but each species has only a few of these. Functionally, they mostly fit into the piercing-poisonous category or into the sticking-holding group.

24.4 Hydra is a carnivore.

There is some evidence that once a penetrant nematocyst has pierced the body of a prey animal, a chemical is released (either from the prey or from the nematocyst) which triggers a feeding reaction on the part of the hydra. The animal opens its mouth wide, and the tentacles move in toward the mouth and deposit the prey at the mouth opening. The oral cone under the mouth begins to move up and slide over the prey, aided greatly by large amounts of mucus secreted by mucous cells in that region. The prey is injested quickly: An adult hydra can eat 10 to 15 brine shrimp larvae in a few minutes.

24.5 The hydrozoan nervous system is often described as the most elementary nervous system known.

The series of actions involved in feeding show some of the fine coordination that hydra possess. They also contract their entire bodies rhythmically and sometimes move about by somersaulting. Other cnidarians, the scyphozoans (jellyfish) and the anthozoans (see anemones and corals), have more elaborate nervous systems and behavioral repertoires than do hydrozoans, the class to which *Hydra* belongs.

Hydra has a **nerve net,** a loosely connected chain of nerve cells, each of which has many short processes that are connected to other nerve cells. Impulses travel in only one direction along this network. *Hydra* also has ectodermal sensory cells that receive and transmit impulses along the nerve net. The strength of the stimulus determines how many cells in the net respond. There is no coordinating center such as a brain in this simple nervous system, so the hydra's behavioral repertoire is small.

24.6 *Hydra* reproduce by budding, an asexual method used by some animals in several phyla.

In *Hydra,* a small outgrowth begins on the parent animal about two-thirds of the way down the body column and grows to form a small new *Hydra* just like itself. All species of animals which reproduce this way are also capable of regeneration. Much has been made of this correlation, and in many ways it does appear that cells behave similarly during budding and regeneration. However, the triggers for the two processes are evidently different: In one process, an entire organism is formed, while in the other, only the missing parts are replaced.

With only a few exceptions, such as *Hydra,* all members of the class Hydrozoa are colonial, and in most cases all the members of a colony arise from a single embryo which settled, became an adult animal, and produced many new members by asexual budding.

24.7 *Obelia* forms colonies in which two distinct types of individuals live cooperatively.

Obelia, a colonial hydrozoan, begins as a hydralike polyp possessing a circlet of tentacles around its mouth. The polyp sends out rootlike structures, **hydrorhizae,** that secure the polyp to a stone or other suitable base. The polyp then grows and reproduces new members until a small, treelike colony is formed, with polyps at the end of each branch (Figure 24-6). Additional branches grow out, usually from the lower portions of the colony, that do not develop a mouth and tentacles. These, the reproducing members of the colony, are called **blastostyles.** They are dependent on the polyps for their nourishment. Each blastostyle produces a number of buds that gradually develop into small **medusae,** the alternate free-living jellyfish form of the animal that carries out sexual reproduction. (Such medusoid alternating forms are common among the hydrozoans and scyphozoans. See section 24.8.)

Other colonial hydrozoans may have four, five, or even more separate forms of individuals living in the colony, some for feeding and digesting, some for reproducing, others for protection, and still others for flotation. All of these individuals arise from a single fertilized egg and have identical genes. As in cellular differentiation, although we may know *how* genes are turned on and off (section 17.10), we do not know what coordinates the mechanism, thereby producing just enough of each different tissue in just the right place.

24.8 Most members of the classes Hydrozoa and Scyphozoa have alternating forms, or "generations."

A more widespread form of polymorphism than that exhibited by colonial cnidarians (coelenterates) is the alternation of polyp (hydralike) and medusa (jellyfishlike) generations that is common among hydrozoans and scyphozoans. The polyp is the sessile or fixed stage and its body, resembling a hydra, has the mouth and tentacles facing upward. In species that alternate forms in this manner the polyp gives rise to many medusae by asexual budding, which often occurs in spring as the waters warm.

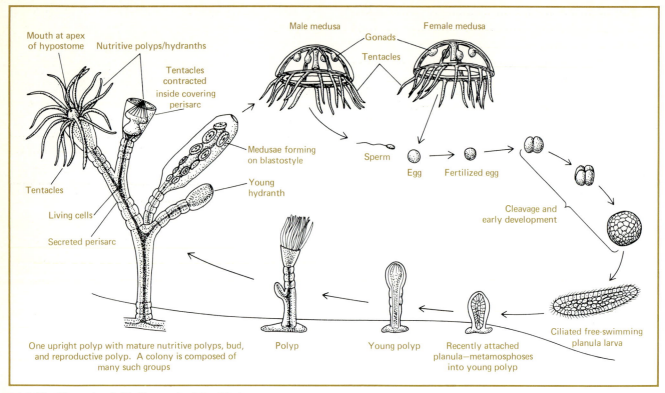

24-6 The life cycle of *Obelia*, a colonial cnidarian.

24-7 Life cycle of a scyphozoan jellyfish.

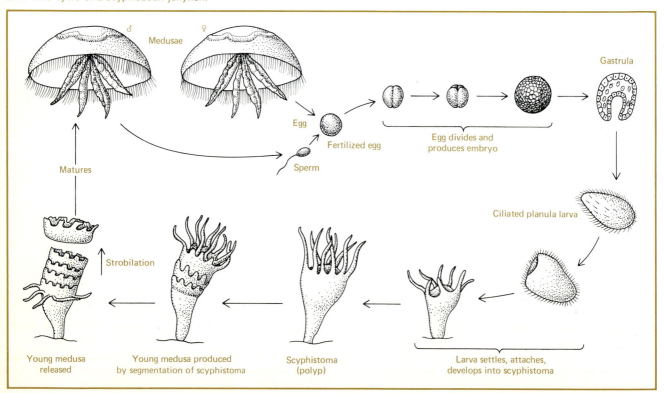

The medusae are umbrella-shaped, with mouth and tentacles facing downward, and are the free-swimming, sexually reproducing forms. Some release gametes and fertilized eggs into the water that develop into ciliated, free-swimming **planula larvae.** These settle to the bottom, attach themselves, and develop into polyps, which are better able to withstand cold winters than are medusae. A typical life cycle is shown in Figure 24-7. The complete life histories of many cnidarians are unknown, and many polyps have never been matched with their correct medusae. In many cases, the same animal has been given different generic and specific names for the two stages.

24.9 The jellyfish *Aurelia* is representative of the cnidarian class Scyphozoa.

All large jellyfish are scyphozoan medusae. *Large* refers to animals ranging from 7 feet across and weighing more than 100 pounds to ones a few inches across. Most species of scyphozoans are marine.

Aurelia (Figure 24-8) is among the smaller of the large jellyfish. Like other cnidarians, *Aurelia* has three layers of cells—the **epidermis; gastrodermis;** and jellylike **mesoglea,** or **mesodermis** (middle layer). As in sponges the mesoglea may contain some wandering amoeboid cells.

In jellyfish the mesoglea makes up the bulk of the animal. It is a proteinaceous jelly high in water content. In this case, *jelly* should not connote a soft, almost liquid material but a rather stiff, elastic substance. The jelly gives the animals much buoyancy while they are swimming or floating, and its elastic properties aid in swimming.

The musculature is so arranged that the bell of the jellyfish is compressed during the swimming stroke; the gelatinous-elastic skeleton tends to return to its original form when the muscles relax, creating a nonmuscular return stroke. The rigidity of the jelly allows an otherwise soft animal to grow to large sizes. In addition, the jelly is unpalatable to potential predators. Figure 24-8 shows the arrangement of tissues in the medusa stage of *Aurelia.* There is a polyp stage too, but it is small and less conspicuous than the medusa.

24.10 The anthozoans—corals and sea anemones—are considered the most advanced of the cnidarians.

Corals and sea anemones are marine animals that exist singly or in colonies all over the world. Unlike jellyfish, anthozoans exist only as polyps. These reproduce both sexually and asexually.

Sea anemones, a large and varied group, live singly and do not secrete a tough skeleton. Like *Hydra,* the sea anemone is mainly sessile, but it can creep over the sea floor. Anemones are structurally more complex than other cnidarians (Figure 24-9). Tentacles, often with stinging cells, surround the slitlike mouth, the only opening that leads to the gullet and digestive cavity, or **enteron** (simple

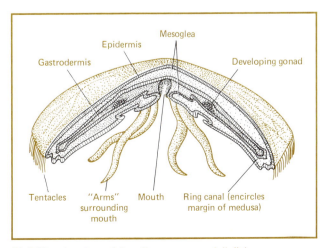

24-8 The structure of *Aurelia,* a common jellyfish.

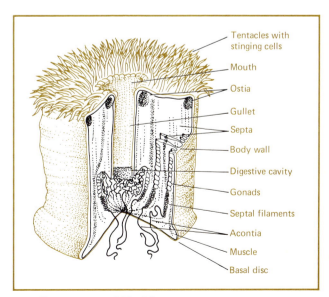

24-9 The structure of *Metridium,* a common sea anemone found in tidal pools or attached to wharf pilings.

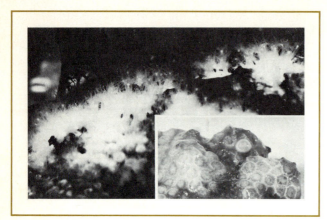

24-10 *Astrangia danae*, a coral common on the northeast Atlantic coast. (Insert shows the skeleton when the soft parts of the coral have withdrawn). (Photograph by author.)

gut). The anemone's blind-sac body cavity is divided by septa (walls) which increase its strength.

Among the most beautiful and best known of the anthozoans are the "true corals," members of the order Scleratinia (Figure 24-10). These include the reef-building corals of the tropics and subtropics, whose flowerlike shapes and brilliant coloring led early naturalists to classify them as flowers or flower-animals well into the eighteenth century.

Reef-building corals secrete a limy (calcareous) skeleton that is laid down by ectoderm cells, first as a plate that attaches the animal to the reef, then as a cup with radially arranged partitions that project inward. When the polyp dies its skeleton persists; and millions of these skeletons make up the great reefs and atolls.

24-11 Some representatives of the three classes of flatworms (phylum Platyhelminthes).

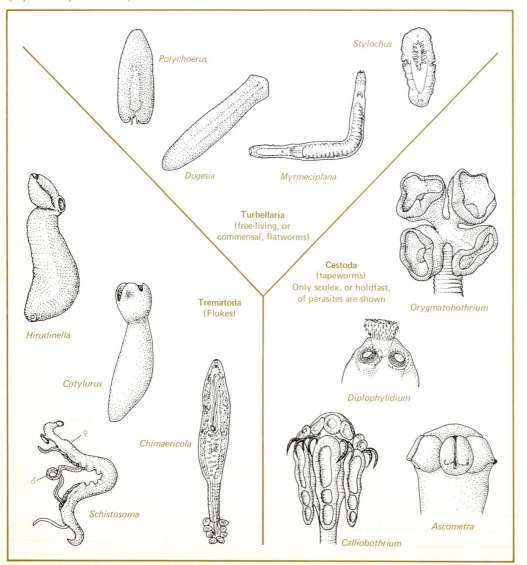

24.11 The phylum Ctenophora (comb jellies) is closely related to the cnidarians.

The comb jellies are common in coastal waters during the summer and are similar in many ways to the medusa stage of scyphozoans. Most scyphozoan medusae possess **radial symmetry,** which is to say that each animal has a top and bottom, but no front or back, right or left. If the animal is cut in half through the center at any point, the two halves will be mirror images of one another. In contrast, most higher animals are **bilaterally symmetrical** and have a distinct front and back and head and tail. Ctenophores are **biradially symmetrical,** meaning that while superficially radially symmetrical, they may have bilateral features, such as paired tentacles.

Comb jellies are named for their eight vertical rows of fused cilia, called **comb plates.** The beating of these comb-plate cilia moves the animal through the water.

24.12 The phylum Platyhelminthes (flatworms) is composed of elongated, bilaterally symmetrical animals.

There are approximately 13,000 species of flatworms in the phylum Platyhelminthes. They are divided into three major classes. The Turbellaria include free-living marine and freshwater flatworms, the best known of which are the planarians. The Trematoda (flukes) and Cestoda (tapeworms) are both parasitic forms. These worms (Figure 24-11) are important not only because they are probably the most primitive examples of bilaterally symmetrical animals, but more immediately because they cause some of the most widespread and debilitating diseases found in man and animals.

All flatworms are thin, a fact which allows for the diffusion of respiratory gases and other metabolic materials throughout their bodies. They have not evolved special respiratory or circulatory systems.

Like cnidarians, flatworms are composed of three layers. In the worms, however, the middle layer, or **mesoderm** (also called **parenchyma**), is more than a layer of jelly containing cells; it consists of **mesenchyme cells** that fill all spaces not occupied by organs. These are small, rather unspecialized, and loosely packed, so that materials diffuse readily among them. Mesenchyme tissue give support to the soft body of planaria and provide resistance to muscle contraction. They thus function in much the same way as the mesoglea of jellyfish. This layer also provides support for important organs such as the intestine (if present), the gonads, excretory system, and well-defined muscles (Figure 24-12). The complexity of these animals as compared to most coelenterates bespeaks the advantages of a third tissue layer.

24.13 Turbellarians (flatworms) have a simple central nervous system, an excretory system, muscles, and other systems, all indications of a division of labor among its cells.

Turbellarians have a well-defined head with rudimentary, light-sensitive "eyes" and chemoreceptive "auricles" (so called because they looked like ears). Cell bodies from these and other sensory structures are arranged into **ganglia** (collections of nerve cell bodies). The presence of a definite head end, which always encounters a new environment first and which is well equipped to sense what the environment is like, is called **cephalization** (from the Greek root *cephal,* meaning "head"). The culmination of this head-forming tendency is seen in vertebrates, especially in mammals, with their large, complex brains and sensory structures.

The best-known turbellarians are the planarians. Planarians that live in fresh water have developed an excretory system to eliminate excess water that enters the animal by osmosis. **Flame cells** (Figure 24-12) bearing beating cilia within them produce a water current that forces excess water into collecting tubules. These lead to minute pores on the surface of the animal where the water is discharged. Since freshwater planarians have no waterproof coating around their bodies, the entry of water and loss of ions is a problem. There is some indication that ions are selectively extracted by active transport from the water before it is driven out of the animal by flame cells. Marine planarians do not generally possess this type of osmoregulatory excretory system.

Planarians have three sets of muscles, which allow them various movements. Undulating waves of contractions, first by circular muscles and then by longitudinal ones, are responsible for rapid turning and swimming motions. The diagonal muscles apparently cause the animal to stiffen and allow it to raise its head or move other parts of its body.

In a planarian, the digestive organs consist of a **three-branched gut;** a muscular **pharynx,** which can be everted through a mouth opening on the ventral side to suck food up into the gut; and a **pharyngeal pouch,** in which the pharynx lies when not in use.

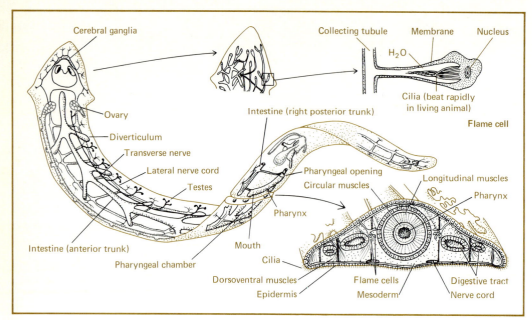

24-12 The anatomy of a planarian.

Other free-living flatworms may have a straight-tubed gut, no gut, or a many-branched gut; the degree of gut branching is the basis of classification in this class. None have an anal opening at the distal end of the gut.

24.14 Most planarians are hermaphrodites: they contain both male and female reproductive organs.

Hermaphroditism is the rule in planarians although self-fertilization is not. Mating results in mutual exchange of sperm between two partners. Eggs are fertilized internally, a yolk layer is deposited, and several eggs are wrapped up in a tough capsule or **cocoon.** Some eggs have a thin covering and hatch in the summer. Other eggs from the same animal have a thick capsule and hatch after a period of winter dormancy.

Planarians also undergo what is probably the simplest form of asexual reproduction known in a bilaterally symmetrical animal. The animal pulls apart at about the middle or two-thirds of the way to the tail and the anterior end separates from the rear part of its body. This portion regenerates a head, pharynx, and other necessary structures, and the parent head regenerates a new tail. Some species have never been observed to reproduce sexually and are thought to propagate entirely by such **transverse fission.** The regeneration process in planaria is slower than in cnidarians.

24.15 The other two classes of flatworms, the flukes and tapeworms, are evolutionary offspring of the free-living forms.

The worms in these two classes are entirely parasitic. Of the two groups, the flukes appear to be the more closely related to planarianlike animals. The internal structure is almost identical to that of planarians, but flukes have no eyes or auricles. They do have suckers, which most planarians lack.

The sea gull fluke, *Cryptocotyle,* is a typical parasite with three hosts. Immature stages of this animal live first in snails, then in fish (Figure 24-13). Almost all flukes follow a similar pattern, but the number of hosts involved in a given life cycle may vary. The parasites are usually named after the animal in which the sexually mature fluke lives.

The adult sea gull fluke is well suited for life in a sea gull's intestine—it has an outer epidermis which prevents it from being digested by the bird, and two suckers provide it anchorage. The mouth opens through one of these suckers, and a muscular pharynx sucks host fluids and cells into a branched gut. After copulation, flukes produce many encapsulated eggs each day, which pass out of the bird in its feces. The eggs mature in the capsule, and if they are eaten by the second host, a marine snail, they emerge and pass through a series of larval stages. Ultimately thousands of larvae with a tail and two suckers (cercaria) crawl out of the snail and swim off into the water. A small percentage en-

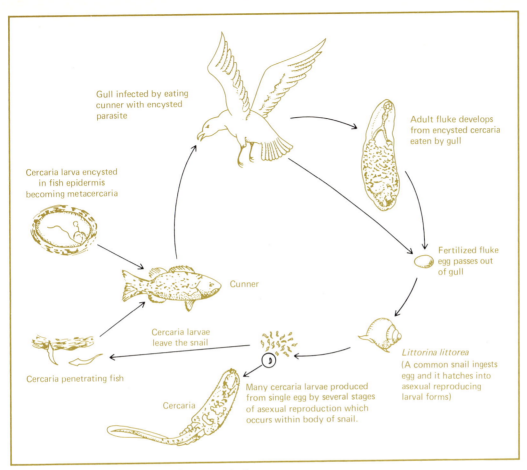

24-13 Life cycle of the gull fluke, *Cryptocotyle lingua.*

counter a fish, grab on with their suckers, burrow under the scales, and encyst. The fish responds by adding a tough envelope. Once a gull eats the fish, they mature in the bird's gut and renew the cycle. Only by producing great numbers of offspring from a single adult, both by sexual and asexual methods, can survival of the species be assured.

24.16 Some flukes, especially those belonging to the genus *Schistosoma,* are harmful to man.

In our society, *Cryptocotyle* apparently does no one real harm. Unfortunately, this is not always the case with flukes. There are many flukes that require mammals, including humans, as hosts or that live in animals which humans value as food or pets.

Every vertebrate group is parasitized by flukes, and just about every spot in a vertebrate is likely to harbor its very own species. It has been noted that there are 91 different places in the vertebrate body where flukes can live. Some flukes do con-

siderable harm to heir hosts; others affect them very little.

One of the most harmful groups of human parasites are the blood flukes, or **schistosomes.** These live in the bloodstream and can do enough damage to cause death, mainly by blocking and causing infections in the liver, kidneys, and other organs of the body which are well supplied with small capillaries. Usually the worms' spined eggs and cercaria larvae cause just enough local hemorrhaging to keep the host sick and weak but still alive.

The life cycle of schistosomes is fairly simple. The larvae develop in freshwater snails that abound in irrigation ditches, rice paddies, or other bodies of water. They emerge from the snail and either penetrate the skin of a person working or bathing in the water or are ingested. They migrate to large blood vessels, develop into adults, and reproduce sexually. Fertilized eggs are passed out with feces, or urine, which in many parts of the world are deposited in or near water holes, irrigation ditches, and the like. In water, the eggs develop into ciliated

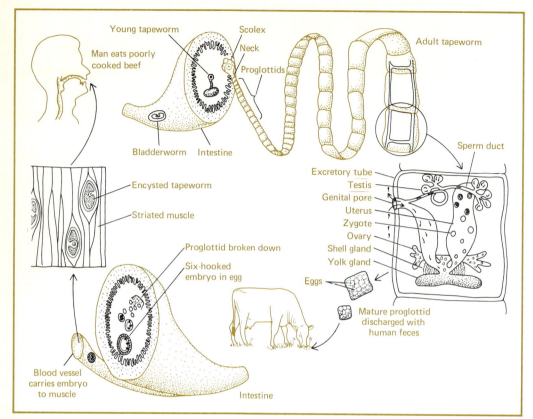

24-14 Life cycle of the beef tapeworm and structure of the scolex and proglottids.

larvae that swim around until they find a snail to enter, whereupon the cycle begins again.

In localized regions of central Africa and in many tropical Asian and South American countries schistosomiasis (blood-fluke infection) is estimated to affect between 75 and 100 percent of the population, depending on local methods of sanitation and farming. Dams such as the Aswan Dam in Egypt have created additional breeding places for snails and a lamentable increase in the schistosomiasis (section 50.3).

24.17 Tapeworms, class Cestoda, are all parasites of vertebrates.

A tapeworm has been described as "a hooked club with sacs of gonads stretched out behind," and in a way, this is an apt description (Figure 24-14). The anterior end of the animal consists of a club- or lightbulb-shaped holdfast or **scolex**, usually with a row of hooks around the distal end and suckers along the sides. The rest of the animal is divided into segments called **proglottids** that are essentially devoted to gonads and sexual reproduction.

The tapeworm has adapted so completely to parasitic life (mostly in intestines) that it no longer has a digestive system or even a mouth. It absorbs the food its host has digested directly through its body walls. The tapeworm also has virtually no sensory structures and the nervous system is reduced: it hangs on to the intestinal wall with its suckers and hooks and reproduces with the rest of its body.

The long, ribbonlike body of a tapeworm (ones up to 70 feet long have been found in man) is composed of hundreds of proglottids, each of which contains male and female reproductive organs. As each proglottid matures and produces thousands of fertilized eggs, it drops off the tail end of the worm and is passed out of the host with the feces. New proglottids are produced asexually behind the scolex. There is no known limit to the life span of an adult worm, but presumably it only dies if it loses its grip and is passed out of the host or if the host dies. It is relatively easy to rid humans of their tapeworms by oral administrations of chemicals and in countries practicing good hygiene, few persons in the population house these parasites.

The secondary hosts of many tapeworms are insects such as beetles and mites. In man, the most common source of tapeworm is undercooked infected beef. Fish and other animals also serve as intermediate hosts.

25

Pseudocoelomate Animals

QUESTIONS TO KEEP IN MIND

What accounts for the great success of roundworms?

What are the human diseases caused by roundworms, and how are they contracted?

One of the major trends in animal evolution is the development of a body cavity lined on all sides by mesoderm. We have seen that sponges, cnidarians, and flatworms have progressively developed a middle layer of true cellular mesoderm, but none has a body cavity lying between the gut and body wall. A flatworm, for example, is a simple tube. It has a hollow gut surrounded completely by tissues; whereas an earthworm (a true coelomate) is a tube within a tube: It has an inner tube, the gut, encased in muscle and surrounded by an outer tube, the body wall. The **coelom** is a mesodermally lined cavity between the two tubes (Figure 24-1).

There is an intermediate category between acoelomate and coelomate: the pseudocoelomate, animals having a false body cavity. Essentially, pseudocoelomate animals have a body cavity that is incompletely lined with mesoderm. Usually there is mesoderm along the outer boundary of the cavity but none along the inner, gut surface.

Most of the pseudocoelomates belong to the phylum Aschelminthes, which includes the enormously successful nematode worms and a smaller number of rotifers, hairworms, gastrotrichs, and related animals. All possess a false coelom which begins to provide a cushioned region for the organs that protrude into it. It can also function as a hydraulic system, in which water acts against the muscle layer and aids in movement. The aschelminthes also have a complete gut with both mouth and anus. Many are covered with a thick protective cuticle which they shed and replace as they grow, much as lobsters and other arthropods molt.

25.1 Nematodes, often called roundworms, are the most inconspicuous of the successful invertebrates.

No matter what criterion you use for success—number of species, number of individuals, diversity of habitats, or effect on other living organisms—the nematods, or roundworms, exceed minimal standards. Only the arthropods exceed them in number of species, but probably not in total number of individuals. There are estimated to be about 500,000 species of nematodes, some 15,000 of which have been described and named. Furthermore, roundworms are more numerous and found in more locations than any other group of multicellular animals. The number of worms found in any location is staggering. There can be as many as 90,000 in a single rotting apple, 1 million in a shovelful of rich garden dirt, and over 300 million of one species in an acre of sugar beets. And they are everywhere.

If all of the materials comprising the earth and all things on it were made transparent except for the roundworms, there would still remain an outline or shadow of everything that now exists. In other words, the earth's crust and waters and almost all living and nonliving objects are all infested with roundworms, yet most humans are hardly aware of their presence. Figure 25-1 shows a few kinds of nematodes.

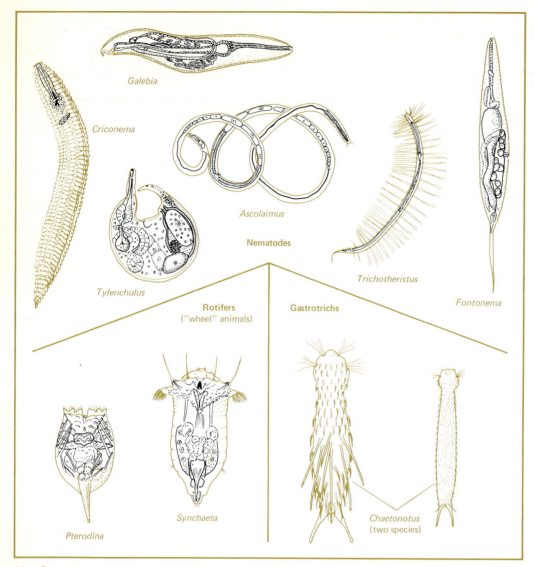

Galebia

Criconema

Ascolaimus

Nematodes

Tylenchulus

Trichotheristus

Fontonema

Rotifers
("wheel" animals)

Gastrotrichs

Pterodina

Synchaeta

Chaetonotus
(two species)

25-1 Some representative aschelminthes: nematode worms, rotifers, and a gastrotrich.

25.2 Vinegar eels are typical free-living nematodes, except that they live in an acid environment.

Vinegar eels *(Turbatrix aceti)* are normally cultured in regular store-bought vinegar, preferably nonpasteurized. Vinegar is quite acid and will kill most other animals placed in it even for a short time. The ability to thrive in a generally alien environment is typical of roundworms and is responsible for the success of this phylum. These organisms seem to have a fantastic resistance to the environment and can therefore live under diverse and taxing conditions. Many roundworms live in water too hot for a person to touch, and others can

survive in situations so adverse that bacteria are the only other living organisms found there.

Vinegar eels, like all roundworms are thin and elongate, lack a respiratory system, are pseudocoelomates, and are covered with a tough cuticle. The cuticle, which helps isolate them so effectively from environmental stresses, is composed of layers of complex protein. For all of its helpful aspects the cuticle has two major drawbacks. It is somewhat inelastic and impairs movement; and since it is quite rigid, it must be shed (molted) for growth.

The body cavity of the roundworm is filled with fluid that serves both to hold the walls and the internal organs in place, and to transmit any pressures exerted in one part of the animal to all other parts.

Many roundworms are **dioecious** (from the Greek for "two houses"), meaning that the two sexes are housed in separate individuals. Sperm are produced by males, eggs by females. Figure 25-2 shows longitudinal and cross-sectional views of a female nematode. In vinegar eels, the eggs develop inside the female after copulation. Small adultlike young hatch from eggs in the uterus and are born fully formed. Egg development requires about eight days, and the young are sexually mature in about a month. In some other nematodes, the eggs are deposited before the young hatch.

Vinegar eels are unusual among animals in being able to use the acetic acid in vinegar as food. Other nematodes can use materials for food which only bacterial metabolic systems can handle. Roundworms also excrete some rather unusual chemical compounds, otherwise known only from bacteria. This does not mean bacteria are closely related to roundworms, but just shows how two groups of organisms can use the same unusual paths to meet their food requirements and get rid of their wastes.

25.3 Some roundworms are parasitic; about 50 of these species live in man.

There are roundworm parasites in every species of vertebrate known and in many plants and most invertebrates as well. Among these capable of living in man are the hookworm, trichina worm, and *Wucheria bancrofti,* the nematode responsible for elephantiasis.

Hookworm is a very troublesome human parasite, especially in areas where human waste is used for fertilizer. Hookworm eggs hatch on the ground after passing from an infected person in fecal material. They undergo two molts after hatching and develop into juvenile forms which bore through human skin. The young parasites are then carried in the bloodstream to the liver, lungs, trachea, and throat of the host and are transported to the intestine, where the adults make their home. In their mouths, hookworms have sharp hardened structures with which they lacerate the intestinal wall of the host. The subsequent loss of blood often leads to a type of anemia. At any given time over 600 million humans in the world are infected. One of the best ways to avoid getting this disease is to wear shoes.

Some nematode parasites have intermediate hosts, *Wucheria bancrofti* being an example. Scores of these animals live in the lymph nodes and capillaries of the human body (section 36.6) and often produce tangled masses which block the flow of lymph. The blockage results in fantastic and hideous swellings of the body, usually in the extremities, scrotum, or breasts, which bloat to elephant-sized proportions. This condition is called **elephantiasis.**

The transfer of elephantiasis from person to person involves a secondary host and a unique biological rhythm. Swarms of minute young worms are produced by the female parasite within the lymph vessels of the host. They appear in the blood near

25-2 Longitudinal and cross sections of a typical nematode worm.

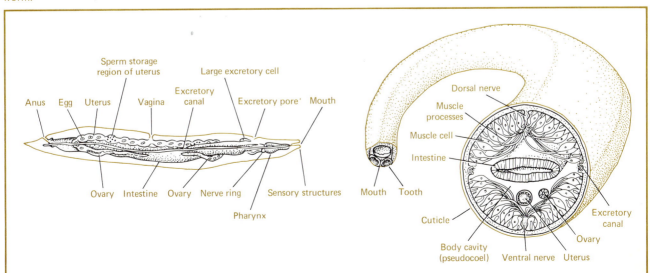

the surface of the human body every night from 10:00 P.M. to 2:00 A.M., but not at other times. It is precisely during these hours that mosquitoes that feed on human blood are active. This rhythm, therefore, enables these worms to reach their secondary host—the mosquito—efficiently. After changing in shape and size in the mosquito they are presented by injection to the next unfortunate human.

There are probably 30 million people in the world suffering from **trichinosis,** a nematode disease caused by eating incompletely cooked meat. Pork is usually the villain, although humans may get this disease from eating uncooked muscle taken from any meat-eating animal.

The adult male and female trichina worms live in the intestine of the human or other animal hosts, and apparently do no harm. After the worms copulate, the male worm dies and the female burrows deep into the intestinal wall. The females produce over a thousand **viviparous** ("born alive") young during the next three months; these are carried by the bloodstream of the host to all parts of the body. The juvenile worms begin to invade muscular tissues, especially skeletal muscles of the tongue, throat, diaphragm, and ribs. There they encyst. It is at this penetration-encystment stage that sufficient infection can cause muscle cramps, massive hemorrhaging, and even death of the host, which is often man. The encysted worms remain alive and infectious in even a dead host for a considerable time. If the infected meat is eaten by carnivores or fed as scraps to pigs and rats, the cysts release adult worms in the host intestine, and the cycle begins again. The encysted worms are killed by a thorough cooking of meat.

26

Annelid Worms

QUESTIONS TO KEEP IN MIND

Why are annelids considered among the "higher animals?

What special feature of annelids have contributed most to their success?

In what ways do earthworms serve man? How are they adapted to escape predation by birds?

In what unique way do certain marine annelids reproduce?

Members of the phylum Annelida ("little ring") are all segmented, soft-bodied worms. There are three classes (Figure 26-1): the **polychaetes** are marine worms, the **oligochaetes** are earthworms and related worms, and the **hirudineans** are leeches. In all, some 9,000 species have been discovered.

The annelids are true coelomates and have a body cavity lined with mesoderm (Figure 24-1). This lining, the **peritoneum**, suspends the internal organs within the fluid-filled coelom (body cavity), an arrangement that protects them, keeps them separate, and allows space for the development of the complex organ systems. The annelids and all other coelomates are considered higher animals.

The most striking characteristic of the phylum is, as its name suggests, the segmentation of these worms into small doughnut or ring-shaped segments. This segmentation is both internal and external: Most of the annelids' organs—muscles, nerve ganglia, excretory organs, blood vessels, and in some cases gonads—are segmented. Only the digestive system is unsegmented. It is not entirely clear what advantages segmentation gave to the very

earliest annelids, but it probably had something to do with locomotion. Annelid locomotion is much more rapid than that of flatworms or nematodes.

26.1 The common North American earthworm, *Lumbricus terrestris,* is a fairly typical example of an oligochaete annelid.

The exterior of an earthworm is round and relatively smooth. If it is examined carefully, however, small indentations that mark the limits of each segment can be observed. Under a dissecting scope it can be seen that each segment has four pairs of stiff bristles, or **setae,** protruding from the body wall. There is also a short, heavy belt, or **clitellum,** around the animal (Figure 26-2).

Internally, the compartmentalization of the worm is even more striking. There are about 150 small chambers or segments, cut off from one another by walls of connective tissue. Within each segment there are a pair of excretory tubules called **nephridia;** a set of nerve ganglia; muscles to control the setae; and branches of circulatory vessels. Segmented longitudinal and circular muscles are also prominent.

The major blood vessels and the intestine extend through each segment as a continuous tube. At the anterior end of the worm are several specialized regions of the intestine: the pharynx, esophagus, crop, and gizzard (Figure 26-2). The crop stores food, and the gizzard grinds it up into smaller pieces. Worms pass soil through their digestive tract and extract almost all the organic material. Five pairs of hearts located in the anterior region connect the dorsal blood vessels to the ventral blood vessels.

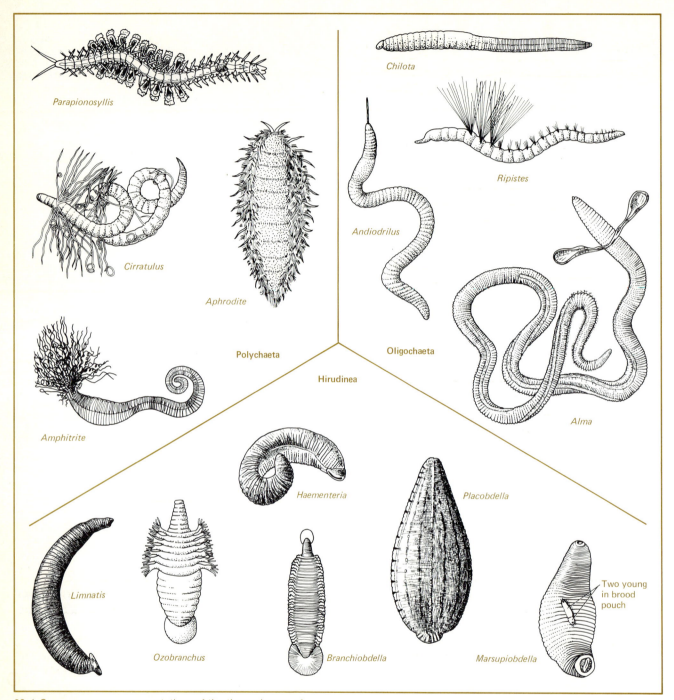

Parapionosyllis

Cirratulus

Aphrodite

Amphitrite

Polychaeta

Chilota

Ripistes

Andiodrilus

Oligochaeta

Alma

Hirudinea

Haementeria

Placobdella

Limnatis

Ozobranchus

Branchiobdella

Marsupiobdella

Two young
in brood
pouch

26-1 Some common representatives of the three classes of
annelid worms.

The actual number of earthworms living in the soil is hard to determine, but under optimal conditions, a cubic foot of soil can house a hundred or more. During the summer they live at a depth of up to 2 feet and in winter may be found 6 feet deep. Worldwide, *L. terrestris* is limited in its distribution. However, various species of terrestrial and freshwater oligochaetes live all over the world.

In the earthworm, the coelom is filled with fluid which contains cells of two types. One type is amoeboid and functions to ingest small foreign particles, parasites, and other potentially dangerous materials which reach the body fluids. The other major type of cell is rounded, usually yellowish, and filled with fat droplets. Upon irritation the earthworm squeezes out some coelomic fluid through

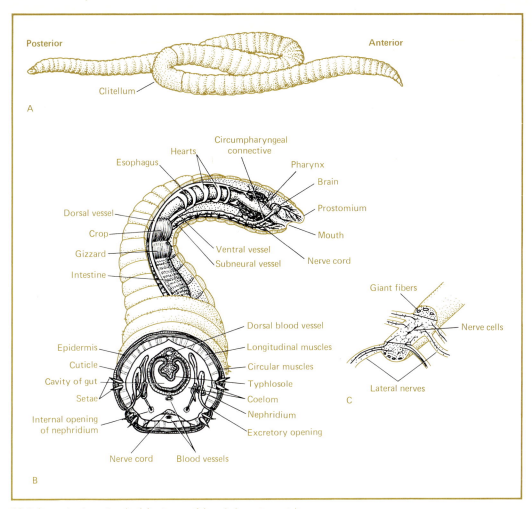

26-2 Important anatomical features of *Lumbricus terrestris* (earthworm): (*A*) whole animal; (*B*) cross and longitudinal sections; (*C*) nerve cord.

small dorsal pores. The fat droplet cells give it an unpleasant odor and make it distasteful to predators.

Like many organisms the earthworm exchanges respiratory gases through its body wall. Furthermore, as in all other organisms, these gases must cross the outer membranes in solution. If it is dry and the soil lacks moisture, the animal tends to lose water. Without a safeguard, the animal could eventually die. Earthworms secrete mucus through the epidermis to prevent desiccation.

26.2 Annelids have well-coordinated circular and longitudinal muscles, which they use for crawling and swimming.

All of the members of the annelid phylum are well equipped to move around. Beneath a thin cuticle and epidermal layer is first a rather thick band of circular muscles and then an even thicker layer of longitudinal muscle cells (Figure 26-2). These are innervated in such a way that only one layer of muscle is stimulated at a time.

The coelomic fluid provides a hydrostatic, or fluid, skeleton against which the actions of the muscles can be exerted. If the external circular layer contracts, the diameter of the worm will decrease, and the segment involved will elongate: The increase in coelomic fluid pressure during contraction of the circular muscle layer causes the longitudinal layer to stretch (the presence of the coelomic fluid also prevents total collapse of the animal during this contraction). Conversely, when a nerve impulse induces longitudinal muscle fibers to contract, a given segment becomes shorter and thicker and the circular muscles are stretched. Such a system of oppositely arranged muscles is called an **antagonistic system**.

Annelids do not contract all their circular muscles or all their longitudinal muscles at once. Rather, waves of muscle contractions pass down the worm. A photograph at any given moment of annelid movement would show several segments in a shortened, thick condition alternating with thinner units of elongation. At the next moment the thick and thin units would appear to move backward along the worm. The setae serves to anchor a part of the body to the ground for a short time, so it can progress forward.

26.3 The earthworm and other oligochaetes use internal fertilization and produce relatively few, well-protected eggs.

Elaborate reproductive structures and good protection for the developing young substitute for a large number of eggs in *Lumbricus*. The adults are **hermaphroditic** (every organism is both male and female) and reciprocally cross-fertilize at night by copulation. After exchanging sperm the worms separate. Later, the clitellum of each animal secretes a cocoon in which the eggs are deposited along with sperm from the other worm. Each cocoon hardens on contact with the air to form a short tubular case. The worm wriggles out of the cocoon, which then contracts at both ends to form an oval capsule about the size of a rice grain. Each worm produces cocoons for several weeks after mating; each cocoon contains one to 10 wormlings. The eggs develop within the cocoon and hatch as young worms. Recent observations indicate that all of the segments present in an adult are present in the young worm at birth.

26.4 The clam worm, *Nereis,* is an example of the largest class of annelids, the polychaetes.

Almost all polychaetes live in the sea. Upon external examination, these creatures seem to bear little resemblance to earthworms. True, they are segmented, round, and elongated; but unlike an earthworm they have eyes, jaws, tentacles, gills, swimming paddles, and other features which prepare them for an aggressive crawling and swimming existence.

The clam worm, *Nereis,* is a fairly typical polychaete (Figure 26-3). It grows up to 10 or 15 inches in length and is quite a fighter. It feeds on animal material, living or dead, and has powerful jaws for capturing and holding its food. These are connected directly to a protrusible **pharynx**, which can be everted through the mouth to pump food into the gut. The rest of the digestive system resembles that of an earthworm.

An animal that actively moves around in its environment needs some method for finding its food and detecting its predator. Four visual receptors ("eyes"), two tentacles, and four pairs of **touch receptors** on the anterior part of the worm enable it to sense its environment. In addition, small pits in the head are thought to be **chemoreceptors.**

Messages from these sensory organs are sent to two ganglia (groups of nerve cell bodies) in the head region. These are connected to two larger, fused ganglia on top of the esophagus. Although these structures receive most of the nerves from the sensory apparatus in the head region and are sometimes called a brain, most biologists prefer to call them **cerebral ganglia.** These are connected to the ventral nerve cord, which runs along the floor of the coelomic cavity of the worm.

Nereis uses the paddlelike **parapodia** extending from each segment both to swim and walk. These structures extend laterally in pairs and are actually continuations of the body wall which enclose fleshy, muscular tissues. The paddles can be flipped forward and backward to propel the animal through the water or along the bottom. While the worm walks or burrows, bristles extending from the ends of the parapodia are pressed against the substratum to provide traction. Many species have gills associated with the parapodia, but as in all polychaetes, respiratory activity also occurs directly across the membranes covering the body wall and parapodia.

The clam worm does not spend all of its time swimming and crawling. It often lies in a U-shaped, mucus-lined tube excavated in the mud with its proboscis. It obtains oxygen by forcing water through its mucus-lined tube with its parapodia.

26.5 Polychaetes sometimes have two ways of reproducing.

Spring and summer are the times when most clam worms reproduce. As with most other polychaetes, eggs and sperm develop in the gonads of dioecious (male or female) individuals and are released into the coelomic cavity to mature. Later, the body wall ruptures, and sperm and eggs are discharged into

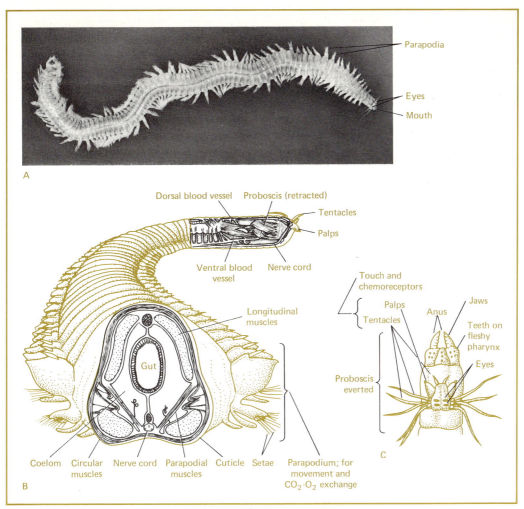

Labels in figure:
- Parapodia
- Eyes
- Mouth
- A
- Dorsal blood vessel
- Proboscis (retracted)
- Tentacles
- Palps
- Ventral blood vessel
- Nerve cord
- Longitudinal muscles
- Gut
- Touch and chemoreceptors
- Palps
- Tentacles
- Anus
- Jaws
- Teeth on fleshy pharynx
- Eyes
- Proboscis everted
- C
- Coelom
- Circular muscles
- Nerve cord
- Parapodial muscles
- Cuticle
- Setae
- Parapodium; for movement and CO_2-O_2 exchange
- B

26-3 Important features of *Nereis* (clam worm) anatomy: A. Photograph of whole animal. Exposure of $\frac{1}{250}$ second was required to stop its rapid motion. (Photograph by author.) B. Cross section and longitudinal section of the anterior region. C. Head region.

the sea. Fertilization occurs following a chance meeting of eggs and sperm.

Some worms that reproduce in this way also use an alternative method of releasing eggs and sperm. These worms are capable of budding off separate individuals devoted entirely to spawning. Prior to the spring spawning season, some of these worms begin to form new segments, and eventually long, many-segmented individuals with a complete head and enlarged gonads are produced at their posterior ends. The parapodia in this region enlarge to become very efficient swimming organs.

Suddenly, on a particular night, all of the reproductive half-worms break off from their "parents" and swim to the water surface and rupture. This releases eggs and sperm. The half-worm left in its tube at the ocean floor begins to produce a new posterior portion for the next season. This procedure protects the parent worm and insures that great numbers of eggs and sperm are released at the same time in the same vicinity.

The fertilized eggs of marine polychaetes develop into swimming **trochophore** larvae, which are shaped like spinning tops (Figure 26-4). Each has a band of cilia around its middle and tufts of cilia at each end for swimming. A rudimentary mouth, intestine, and anus are also present. After floating and swimming in the ocean waters for a short time, the lower end of the larva begins to elongate and segment. Tentacles emerge, and through a rather complicated series of changes, the adult worm forms. Some mollusks show similar trochophore stages of development and equally complex metamorphoses.

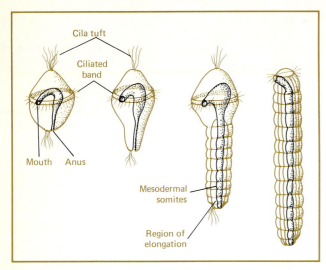

26-4 A diagram of a trochophore larva and its metamorphosis into a typical annelid.

26.6 The third group of annelids is the class Hirudinea, or leeches.

Contrary to popular opinion, most leeches are not bloodsucking parasites, and of those that are, few consume human blood. Many leeches are free-living carnivores; many others occasionally take blood meals from vertebrates or other invertebrates but should probably not be considered true para-sites. Only a few species live more or less permanently attached to a host animal, but their hosts are seldom human. Of the 300 species described, almost all live in fresh water.

Leeches appear segmented on the outside, but not on the inside. Instead of having a fluid-filled cavity separated into compartments by septal walls, the leech's body is filled with loose connective tissue and rather unspecialized cells. The internal organs are embedded in this material. Parapodia are also missing.

You can identify a leech externally by the presence of two suckers, one at each end of a flattened body that is composed almost invariably of 32 to 34 segments. The suckers are used in holding on to the substratum, attaching to predators, and moving in a series of looping motions. Leeches that consume blood have rasping teeth within their anterior sucker which are used for slitting the prey's skin so that blood can be sucked out. The anticoagulating effects of **hirudin**, a chemical extracted from the pharynx of leeches, are well known. This chemical is used by the leech to prevent blood clotting in the wound of the victim or in its own gut.

The stories associated with the bloodsucking ability of leeches are legend, and even today, in this country, leeches are used by some people to remove blood from a black eye or other bruised area. In the past, bloodletting by leeches was even more popular and was used for almost every ailment from chicken pox to cancer.

27

Mollusks

QUESTIONS TO KEEP IN MIND

In what ways are mollusks either useful or harmful to man?

To what other groups are mollusks closely related?

What are the major groups of mollusks, and in what principal ways do they differ?

Members of the phylum Mollusca, or mollusks as they are more commonly called, live in a variety of habitats, ranging from the ocean depths to the tops of trees in a tropical rain forest. Some 80,000 living species of this diverse group have been described and divided among six major classes. The most primitive of these is the class Amphineura, or chitions; the class Monoplacophora was thought to be extinct until the recent discovery of living representatives; Gastropoda is a diverse class and includes snails, slugs, and limpets; the class Pelecypoda includes the Bivalvia, such as clams and oysters; the class Scaphopoda includes tusk shells. The class Cephalopoda, squid and octopi, is considered by many to include the most advanced of all invertebrates (Figure 27-1).

A garden snail does not appear to have much in common with a giant squid or even with an oyster, yet the body plans of all groups in the phylum Mollusca are similar in many respects. Every mollusk possesses a **foot**, which is a muscular structure mainly involved in locomotion. (*Poda*, meaning "foot," is part of several class names.) A snail glides

on its foot, a clam extends its foot between its two valves (half-shells) to dig into the sand or mud. The squid and octopi are somewhat different, for their feet have been modified into tentacles, which they use for propulsion and other uses.

Mollusks also have a **mantle**, a tissue that covers most of the body and secretes the shell. Again the cephalopods seem to be the exceptions, for in squid the structure analogous to the shell is an internal skeletal structure called the **pen**. Inside the **mantle cavity** of the aquatic mollusks are featherlike gills; in most terrestrial ones there is a primitive lung.

Mollusks have a digestive system with mouth and anus, and all except the bivalves have a **radula** in their mouths to tear away bits of food, shell, or wood and bring them into their gut. The radula functions like a file mounted on a two-way conveyor belt. Mollusks also have circulatory and nervous systems that range from simple to complex. Most are dioecious, some hermaphroditic, and a few start life as males and later become females. Like annelids, mollusks typically begin life as trochophore larvae (Figure 26-4), but unlike annelids the adults are not segmented.

27.1 Gastropods are the largest, most diverse class of living mollusks.

Because of the attractiveness of their shells, mollusks have been collected, studied, and classified extensively for many years. As a consequence, more is known about the evolutionary relationships of gastropods than about any other invertebrate group.

27-1 Representatives of the six classes of mollusks.

Gastropods, the snails and their relatives, are the most versatile mollusks and are equally abundant in the sea, in fresh water, and on land. Many, but not all, are univalve (single-shelled) animals, and the shell is usually spirally coiled and beautifully colored. On land, snails are plentiful wherever vegetation is lush. In France, *Escargot* (garden snails) and in Japan, *Buccinum* (whelk) are valued as food.

Despite their beauty and delicacy of flavor, gastropod mollusks do a lot more harm than good as far as man is concerned. They are intermediate hosts to countless parasitic flatworms (section 24.16) which inflict misery on man, domestic animals, and other animals the world over. One carnivorous snail, *Urosalpinx,* the "oyster drill," drills into and consumes so many oysters that extensive oyster beds have been destroyed.

27.2 The common garden snail is a typical gastropod mollusk.

Snails have three body regions: the head-foot region, the visceral mass, and the mantle-shell complex.

Structures found in the head-foot part of the snail are involved in sensation, locomotion, and food ingestion. There is no sharp division between the head and foot. In the garden snail *Helix*, for example, the head-foot is the region of the body which protrudes from the shell when the animal is moving (Figure 27-2). The "sole" of the foot is ciliated and glides over the substratum by coordinated ciliary beats. Mucus is secreted from large glands just behind the mouth and serves as a lubricant for the gliding motion, especially over dry objects. Mucus also acts as an adhesive which allows snails to crawl vertically, or even upside down. *Helix* shows negative geotropism; that is, when placed on a stem or twig, it crawls upward, a reaction which presumably helps it reach the plant material it feeds on. The external head of *Helix* bears two pairs of tentacles, with two eyes located on the ends of the longest pair of tentacles; **statocysts** (organs of equilibrium); and the mouth. Inside the mouth is the radula, consisting of a hard chitinous band bearing teeth which is stretched over a rigid tongue. Mus-

cles attached to the two ends of this toothed band move it back and forth over the snail's food, and, like a rasp, the radula scrapes off small bits of food. The teeth point inward and carry the food into the esophagus. New teeth are formed at the posterior end of the belt as old ones wear off near the mouth. Subtle differences in radular structure provide the basis for classifying closely related molluscan groups.

The visceral mass contains the organs for reproduction, circulation, excretion, and digestion, respectively. *Helix* is hermaphroditic, and cross-fertilization occurs when the penis of one transfers sperm bundles into the vagina of its mate. Eggs covered by a gelatinous layer are deposited in damp places. The young emerge as minute snails.

Helix possesses a single kidney which drains the area around the heart, and filters wastes from the body fluids as they move through the vicinity. The snail also has what is known as an **open circulatory system.** The heart pumps blood via closed vessels to different parts of the body, and the blood runs freely through open spaces, or sinuses, in the tissues, eventually filtering back to the heart. The visceral mass is the most delicate part of the snail's body and remains inside, protected by the shell, at all times. Muscles connect the top of the mass to the upper portion of the shell and can retract the body into the shell when necessary.

27-2 Anatomy of *Helix*, the garden snail.

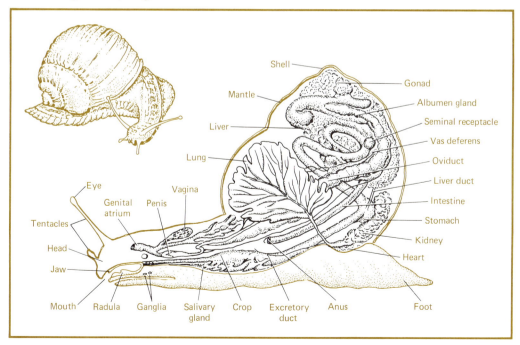

Certainly the most conspicuous portion of the mantle-shell region is the shell. It is a hard structure but can be easily dissolved in strong acid. It is essentially composed of calcium carbonate crystals embedded in a protein meshwork. The hardness of the shell depends on the relative amounts of calcium carbonate and protein laid down during its secretion. Just under the shell of *Helix*—and all other mollusks—is the mantle, a rather thin, fleshy fold of tissue. The edge of this mantle secretes new calcareous material at the free edge of the existing shell.

A shell obviously offers some protection from predators; it also protects terrestrial snails and intertidal marine species from desiccation. In dry times the animal remains tightly encased, with no tissues exposed to the air. Some snails have a lid which closes over the mouth of the shell when the snail is inside. Others secrete a mucus covering over the opening. Snails kept on laboratory and museum shelves for five years or more in dry bottles have resumed their normal life activities when placed again in a moist environment. The gastropod shell may well have contributed nearly as much as the "lung" in allowing some gastropods to become the only successful group of nonaquatic mollusks.

Garden snails belong to the gastropod order known as Pulmonata (from *pulmonis*, "lung") and can breathe air. Gills, which are present in the mantle cavity of most mollusks, are absent in pulmonates; instead part of the mantle has been modified to form a space which is heavily infiltrated with blood vessels. This semicavity is connected to the exterior by a small opening, and gases enter the cavity and pass into the blood via the specialized mantle surface. This is obviously a special adaptation for terrestrial existence.

27.3 *Mercenaria mercenaria,* variously called a quahog, cherrystone, or littleneck clam, is representative of the bivalve mollusks.

Mercenaira mercenaria is the most common of East Coast marine clams. As normally seen in the fish market, it is about two or three inches across, off-white or slate gray in color and if small is called a cherrystone clam; if considerably larger it may be called a quahog. Like most bivalves, *Mercenaria* is a marine species, but many clams do grow in fresh or brackish water.

The pattern of body construction in the clam (Figure 27-3) is quite different from that of the

27-3 Anatomy of the quahog (clam) *Merceniaria.* (*A*) surface view; (*B*) longitudinal section; (*C*) cross section.

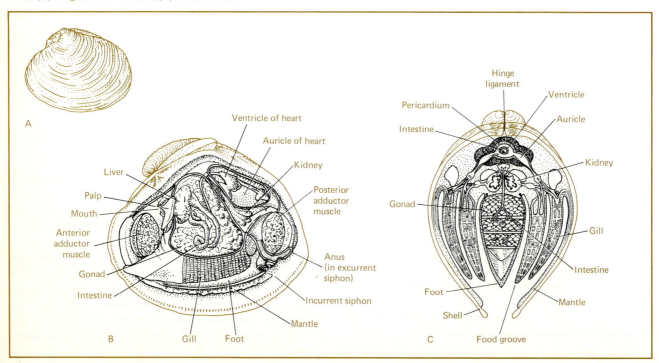

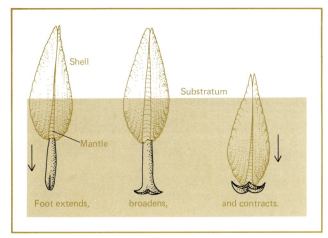

27-4 How a clam digs in the sand (rapidly!) The foot protrudes from the front portion of the shell.

gastropods. The clam *Mercenaria*, for example, resembles a laterally compressed gastropod lacking a head. Two valves, or half-shells, are hinged dorsally and enclose the soft portions of the animal, and in times of stress the heavy valves can be closed tightly and kept closed by contraction of the adductor muscles. Upon relaxation of these muscles, the valves open because of the elastic properties of the hinge.

The only parts of *Mercenaria* that extend beyond the limits of the shell are the **siphons**, which channel water to the gills, and a foot. Since most clams spend their lives buried in the mud or sand, they have not become cephalized with its many sensory and nervous structures for gaining information about the environment.

Lining the shell and clearly visible at its edges is the mantle. In the clam this tissue secretes the shell and also forms two siphons for circulating water through the mantle cavity which contains the gills and foot. A part of the mantle may also form a **brood pouch**, in which the young of some bivalves develop. (*Mercenaria* does not have this.)

A clam can quickly withdraw its foot, siphons, and other soft parts into the shell, but it extends these structures rather slowly, because these processes occur by two different mechanisms. The structures withdraw when their attached muscles contract and then extend when fluid pressure builds up in the foot or siphons.

The pelecypod's foot can be an extremely efficient digging tool, especially in moist sand and mud. *Ensis*, the razor clam, and its relatives can burrow much faster than you can dig in the sand with your hands, or even with a shovel. Their foot

is first extended into the mud as a sharp point; then the end fills with blood and spreads out to form a mushroom-shaped anchor (Figure 27-4). Next the muscles of the body contract, and the shell-enclosed part of the clam is drawn down to the anchored foot. The foot is again extended, and the cycle is repeated.

27.4 Except for some land snails, all mollusks have gills.

Mollusks have gills, which vary greatly both in structure and function from species to species. In *Mercenaria* the gills consist of two pairs of flaplike tissue which lie on either side of the foot in the mantle cavity. The dorsal edge of each is attached to the body, and the ventral edges hang free. Each gill is actually a W-shaped structure consisting of inner and outer **lamellae** (sheets). Between the two lamellae is an open space divided vertically into a series of narrow water tubes. Countless cilia beating on the surface of the gills in a constant and predicatable pattern draw water into the gill region and route it through the clam, in one siphon and out the other. As water passes over the gills, oxygen and carbon dioxide are exchanged.

The gills also act as food-gathering devices. The water passing over the gills contains small organisms and bits of organic solids. A sheet of mucus, produced continuously by the gill surface, flows by ciliary action ventrally along its face and traps this food. The mucus-bound food enters a small ciliated **food groove** and is carried to the mouth region and ingested. Thus we see in clams adaptations to a semisessile existence in which the environment is brought to the animal. *Mercenaria* can move around to some extent, but undoubtedly it spends most of its adult life in a very restricted locality, probably buried in the same patch of mud.

In *Mercenaria*, the sexes are separate, and once the clams are two years old they begin to produce gametes as the water warms up in the spring and summer. Gametes of both sexes are released simultaneously and pass directly via the siphon into the sea, where fertilization takes place.

As in almost all marine mollusks the fertilized eggs develop into motile larvae of so-called trochophore and **veliger** types. These ciliated larvae move about in surface waters, often drifting great distances from their parents and thus dispersing the species. After a period of time they metamorphose into adults.

BIOEPICUREAN DELIGHTS

Mussels (or Steaming Clams) Steamed in Wine

Mussels (Mytilus edulis), "steamers" (Mya arenaria), and a number of other mollusks found in the intertidal zone can be collected in sufficient numbers to make a delicious feast. (Caution: Check with knowledgable local residents to be sure the shellfish are not poisoned by "red tide" dinoflagellates. See Chapter 20.)

Wash mussels in fresh running water and drain. Place in a steaming vessel and pour about 300 milliliters of white or rosé wine (per kilogram of mussels) over the mussels. Steam for about 15 minutes or until mussels are open. Discard those that do not open. The main reason for steaming in wine is that the lower temperature of alcohol vapor leaves the mussels tender and succulent. Serve with a side dish of melted lemon butter and a cup of the steaming broth, to which has been added a gram of butter and several milligrams of dill weed and monosodium glutamate ("Accent").

27.5 For man, bivalve mollusks are a source of food and in the case of shipworms, of trouble.

The clam and its relatives are most useful to man as a source of food. Even in prehistoric times clams and oysters were eaten, and today, despite pollutants and overfishing that hold down the yield, reliable estimates indicate that over 3 million metric tons of edible mollusks are caught per year in the world. About 55,000 tons of oysters alone are caught in this country every year. Bivalves are also used for dyes and stains, roadbed materials, buttons, poultry feed supplement, and decorative paraphernalia.

A bivalve with important negative consequences for man is the shipworm, Toredo, which bores into wood. It riddles the submerged portions of rafts, wharves, and ships and causes billions of dollars worth of damage each year.

27.6 Loligo, a squid, is one of the more common cephalopods, or "head-footed" mollusks.

All cephalopod mollusks are marine. Many, such as the squid and the chambered nautilus, live in deep waters; the octopus prefers depressions or caves in relatively shallow water.

Cephalopods range in size from diminutive squid found in plankton to giant, deep-ocean squid which may be 60 feet long and weigh two tons. Scientists surmise that much larger ones existed and in addition to being the largest invertebrate ever known, may even be the largest animal.

The body of the squid Loligo (Figure 27-5) is elongate. There is no external shell; instead, an internal rod of cartilage called the pen serves as skeletal support.

The foot of the squid is partially modified into 10 sucker-bearing arms, two of which are called tentacles. Well-developed eyes are present in the head region and resemble vertebrate eyes in structure, function, and location (Figure 27-5). The squid's mouth lies within the circle of tentacles and is armed with a horny beak and radula. Saliva secreted by the mouth is poisonous in large amounts and is probably used to subdue prey. The beak is fantastically strong and can crack the shell of a crab or extract a bite from a fish.

In the squid, the mantle is a cone-shaped structure which completely encloses the visceral parts of the body. Fins along the sides of the body are formed from mantle tissue and aid in swimming and stability. A small, tubelike siphon, which, like the tentacles, is a modified portion of the foot, extends out from under one anterior edge of the mantle and is used in propelling the squid. Water is taken inside the mantle from around the neck or collar of the mantle. The collar is then closed, the mantle muscles contract, and a jet of water is shot through the siphon, pushing the squid. By pointing the siphon in various directions, the squid can move rapidly in whatever direction it wishes. Apparently it uses this method of movement only under stress and normally swims with its fins.

If irritated or agitated, both squid and octopi can eject a dark ink from a gland inside the mantle cavity. This puff of ink evidently disorients a would-be predator and may be toxic.

Squid and octopi can also try to avoid these confrontations with predators by blending in with their

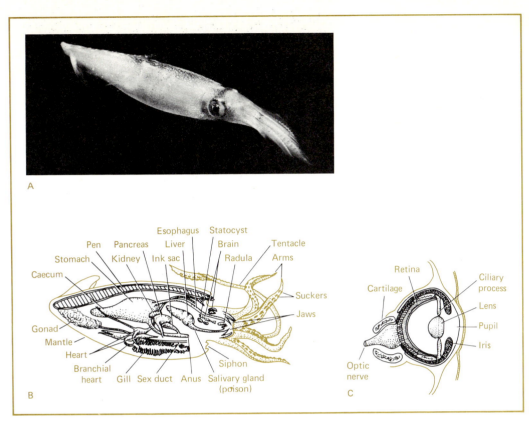

27-5 *A. A photograph of the squid* Loligo. *B. Squid anatomy. C. A section through the eye of a squid. (Note the similarity to the human eye.)*

environment. The animals have **chromatophores**, cells in their skins which can change color, so the skin can exhibit different color patterns. Under certain conditions, the animal blushes; under others, it pales. This system not only provides protective coloration; it is also used during courtship and defensive activities.

Courtship and mating among squid (and octopi) is quite an elaborate procedure. The male has one specially modified tentacle with which he transfers sperm bundles from his own mantle to that of the female. Fertilization takes place inside the female, and afterwards masses of eggs are passed out through her siphon. She grasps these with her tentacles and molds them into long pencil-shaped strings, or "dead man's fingers," which she attaches to a rock. This complete sequence of behavior is preset in the female and cannot be modified: She will enact the entire routine of egg-mass shaping and attachment even if her eggs are experimentally removed as soon as they come out of her siphon. This same type of behavioral rigidity is seen in many insects, birds, and mammals. Once a behavior pattern is started, these animals cannot stop or

change it, but must complete it even though it is unnecessary.

The nervous system of the cephalopods is amazingly well developed. Octopi can discriminate among certain visual and tactile (touch) stimuli and can be trained to learn responses. The similarity of the eyes of the squid and human is amazing (Figure 27-5).

In *Loligo*, there is a giant nerve axon (nerve process) which runs from the brain to the mantle muscles, and almost all we know today about how nerves transmit impulses was derived from experiments performed on this axon. The axon was discovered in the 1930s, and soon scientists were inserting electrodes into the nerve cells to study electrical changes in them as they conduct nerve impulses.

27.7 Annelids, mollusks, and arthropods have enough features in common to suggest that they descended from closely related organisms.

Although there are too few facts to ascertain the nature of a common ancestor for annelids, mollusks, and arthropods, or even to assert that members of

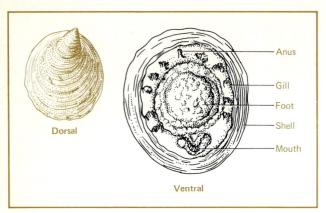

27-6 *Neopilina*, a representative of class Monoplacophora, once thought to be extinct.

these phyla descended from a single kind of animal, the three groups have a large number of characteristics in common.

For instance, the polychaete annelids and the mollusks both produce trochopore larvae, and embryologically, their coeloms are formed in about the same manner. Mollusks, however, do not seem to be segmented at any point in their lives. The mollusk *Neopilina* (Figure 27-6) belongs to a class thought to have been extinct for 400 million years; when first rediscovered by a Danish oceanographic expedition in 1952, it seemed to show that primitive mollusks had been segmented. A dozen living specimens of this small mollusk were taken from water two miles deep north of the Gulf of Panama. Each has a single shell, but underneath are five pairs of external gills and eight pairs of muscles for attaching the animal to its shell. At present, however, it is suspected that this segmentation was acquired secondarily and was not a primitive feature.

The relationship between annelids and arthropods is clearer. For one thing, arthropods are segmented, although this is not always so obvious as in the annelids. Secondly, they have an annelidlike nervous system, with a ventral nerve chord leading from a dorsal "brain" or ganglion. Also, coelomic development follows parallel paths in the two groups.

Still another link between these two phyla is the existence of the *Onychopora*, a small phylum of tropical wormlike creatures (Figure 27-7). Like annelids, these "walking worms" have soft, segmented bodies with repeating units of muscles and nephridia. Like arthropods, they have a hard chitinous cuticle, a tracheal respiratory system, and paired walking legs tipped with claws. These legs are not jointed as they are in arthropods but are operated by a combination of flexor muscles (for raising the legs) and hydraulic pressure (for extending them—there are no extensor muscles), as are the legs in such animals as spiders and millipedes. The fossil record of the Onychophora goes back over 500 million years into the Cambrian period, when arthropods were evolving rapidly, apparently at the expense of annelids with which they were successfully competing.

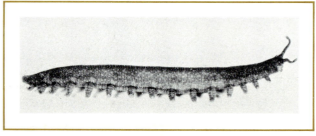

27-7 *Peripatoides*, a representative of the phylum Onychophora. (Photo courtesy Carolina Biological Supply Company.)

28

Arthropods

QUESTIONS TO KEEP IN MIND

Why are the arthropods, and especially insects, so successful?

What are the advantages and disadvantages of an exoskeleton?

What functions do the crayfish's many pairs of legs serve?

Why do insects rely more on instinctive behavior than on learning?

The phylum Arthropoda ("joint-footed") has more species alive today than all other phyla combined. Its one million or so known species go to make up such flourishing classes as Crustacea, Arachnida and, largest of all, Insecta. For every human being alive today, approximately 300 million insects exist. Paleontologists speak of the earth as having had an age of reptiles and an age of mammals. It seems no exaggeration to say we are now in the age of insects.

Considering the phylum as a whole there is almost no place on earth that is unoccupied by one of its members. Arthropods live in the deep ocean, in shallow bays and lakes, in dank caves, in scorching deserts, in steaming jungles, and in polar regions. An arthropod may swim, fly, hop, crawl, wriggle, drift, or, like the barnacle, stand on its head in a limestone house kicking food into its mouth (Table 9).

Why have the arthropods been so successful? For one thing they evolved the jointed foot or leg with antagonistic muscles. This characteristic, shared only with the vertebrates, gives them great agility

and makes them formidable hunters. They also developed a hard outer covering, or **exoskeleton.** This offers them protection from predators and desiccation and yet is hinged to allow freedom of movement. In addition, the insects were the first major group of animals to inhabit the newly formed forests of the Carboniferous period some 340 million years ago. With almost no competition, they diversified and were able to settle into almost every ecological niche. Subsequently they developed the ability to fly some 250 million years ago, and thus further

TABLE 9 THE RELATIVE ABUNDANCE OF ANIMAL SPECIES IN DIFFERENT PHYLA

Animal phylum	Approximate number of species (× 1000)		
Porifera	5		
Cnidaria	5		
Platyhelminthes	13		
Aschelminthes	10		
Annelida	9		
Mollusca	110		
Arthropoda	900	Insects	675
		Crustaceans	25
		Spiders	31
		Centipedes	2
		Millipedes	7
Echinodermata	6		
Chordata	45	Fishes	25
		Amphibians	3
		Reptiles	10
		Birds	9
		Mammals	4
"Minor phyla"	20		

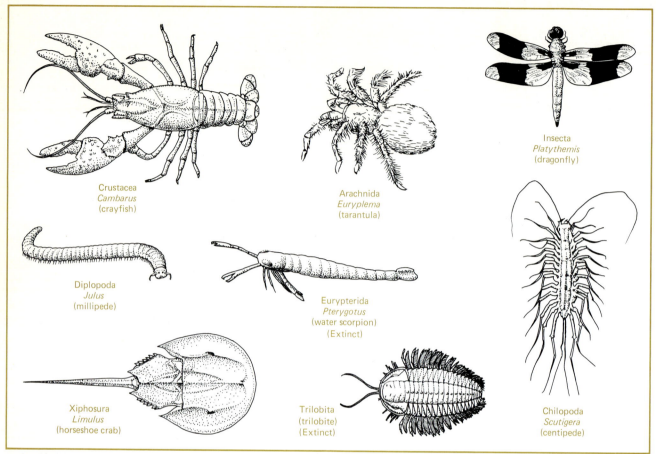

28-1 Representatives of the principal groups of arthropods.

expanded their territory. The various types of arthropods that have evolved are shown by examples in Figure 28-1.

28.1 Some primitive arthropods, such as the trilobite, have become extinct but a fossil record of their prodigious numbers was left behind.

The most primitive of animals that are recognizable as arthropods were the trilobites (Figure 28-2). These lived in the sea some 250 to 550 million years ago. They had heavy shells, and the segmentation of their bodies and appendages was simple and repetitive. Eurypterids, or water scorpions, which were sometimes 8 feet long, flourished and disappeared at about the same time as did the trilobites.

Another ancient, relatively primitive group has not become entirely extinct. A living representative of the class Xiphosura is *Limulus*, the horseshoe crab (Figure 28-3). These curious animals are common along the Atlantic coast of the United States

and a few other places, where they skim along the bottom hunting for worms and other organisms. The horseshoe crab has five pairs of legs and is similar in structure to fossil horseshoe crabs of 500 million years ago.

In comparing primitive arthropods with modern ones, several evolutionary trends are apparent. The legs become fewer and are modified into claws, mouth parts, or tubes for transferring sperm; the head region becomes the site of sensory organs and nervous tissue; and the original number of wormlike segments decreases, and the segments are combined into head, thorax, and abdomen, or some combination of these.

28.2 The class Crustacea is the only major group of arthropods that is primarily aquatic.

Among this large class of some 28,000 species are the familiar shrimps, lobsters, crabs, crayfish, and barnacles; the somewhat less familiar sowbugs or

28-2 Fossil remains of two extinct arthropods. *A.* The eurypterid, *Eurypterus lacustris*, in Bertie Waterlime, Erie County, New York; late Silurian age. *B.* A trilobite, *Isotelus gigas*, in Trenton limestone, Trenton Falls, New York; middle Ordovician age. (Courtesy of D. Fisher.)

28-3 Many specimens of the horseshoe crab, *Limulus polyphemus*, in a tank at the Marine Biological Laboratory, Woods Hole, Massachusetts, where they are extensively used for research on vision, nerve function, and blood chemistry. (Photo by author.)

pillbugs and sand fleas; and the inconspicuous copepods and their relatives, which make up the greatest part of the zooplankton.

Copepods are among the most abundant animals in the world. One of the more interesting groups of these microscopic or near-microscopic crustaceans is the cladocerans, which include *Daphnia* and *Leptodora* (Figure 28-4). *Leptodora* lives in clear northern lakes. Because it is almost completely transparent, it is possible to see each and every cell of its body while the animal is alive under the microscope. The head region contains a single compound eye connected to a simple brain by a long optic nerve. As in many cladocerans, the compound eye is made to "scan" or jiggle by rapid contractions of eye muscles. The striated muscles moving the appendages; the heartbeat; and the peristaltic contractions of intestinal muscles can all be seen. The circulatory system is a simple open type, and the blood contains amoeboid cells. *Leptodora* can either subsist on phytoplankton or capture smaller animals such as *Daphnia*.

28.3 The freshwater crayfish is a representative crustacean.

The crayfish *Cambarus* is abundant in the streams and lakes of this country. Its anatomy and physiology are almost identical to those of the lobster.

One of the most conspicuous features of the common crayfish is the large number of appendages which protrude from the body (Figure 28-5). Al-

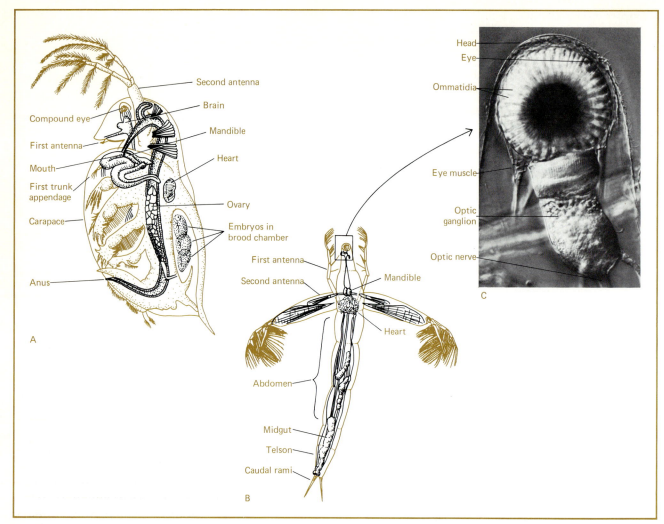

28-4 Two cladocerans, (A) the water flea, *Daphnia,* common in most lakes and ponds as far north as Point Barrow, Alaska; and (B) *Leptodora,* an almost completely transparent organism. (C) A low-power photomicrograph of the *Leptodora's* scanning compound eye, the muscles that cause it to scan, the optic ganglion, and optic nerve. Also visible are some amoebocytes. (Photo by author.)

though not entirely visible from the exterior, each pair of these appendages extends from a **segment,** or **somite,** of the animal's body. Segmentation is not as clearly marked in the crayfish as in the annelids, partly because of modifications which have taken place, resulting in fusion of some segments and rearrangement of others.

The primitive crayfish presumably had many pairs of identical appendages which evolved into the diverse structures found today. There are 18 paired appendages extending ventrolaterally in *Cambarus,* plus one fused pair at the tail. All are jointed, and most are subdivided into an outer and inner branch.

Covering all these appendages and the rest of the crayfish's body is a tough **exoskeleton,** or **integument.** The exoskeleton of crustaceans is tougher and heavier than that of insects or spiders because it contains calcium salts. This extra weight is not a problem for aquatic animals, for the water buoys them up, but it does limit the size of terrestrial forms.

28.4 The integument must be molted to allow the animal to grow.

In spite of its advantages as protection, because of its rigidity an exoskeleton has two major disadvantages: It cannot expand as the animal grows; and it prevents bending movements essential to locomotion. The many joints in the exoskeleton and an elaborate musculature allow for movement and

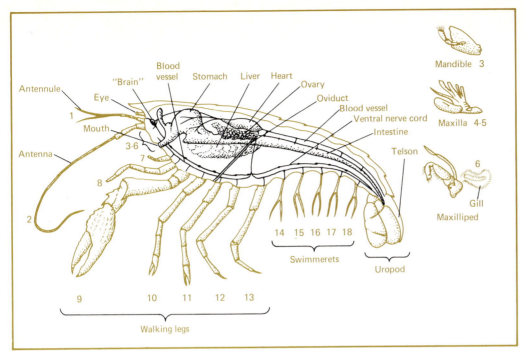

28-5 Crayfish anatomy. The names of the appendages and their functions are: *1.* antennule (touch, taste, and equilibrium); *2.* antennae (touch and taste); *3.* mandible (jaw); *4.* maxilla (food handling); *5.* maxilla (food handling and drawing water over gills); *6, 7, 8.* maxillipeds (food handling); *9.* cheleped (grasping); *10, 11.* periopods (grasping and walking); *12, 13.* periopods (walking); *14–18.* pleopods (males: sperm transfer; females: circulating water over eggs).

solve the locomotion problem. To overcome the growth limitation, crayfish and other arthropods molt, that is, they periodically shed their exoskeleton.

As an animal fills its exoskeleton, it begins to secrete a new one beneath it. When first synthesized, this new integument is relatively soft. At the appropriate time the old exoskeleton cracks open, and the animal crawls out. The newly exposed integument immediately begins to harden. The freshly molted crayfish imbibes large amounts of water and swells up. After the new integument hardens, the crayfish loses some of the water and is then protected by a loose-fitting exoskeleton that allows for growth. This molting cycle is repeated about twice a year until the animal reaches old age —several decades for some crustaceans.

For several days after the old exoskeleton is shed, and before the new one hardens, the crayfish is vulnerable. Its soft new integument can easily be torn and ripped by another animal. Since its pincers, mouth parts, and walking legs have no hard protective coverings, the crayfish cannot defend itself either. It also temporarily loses part of

its digestive system and its gravity-detecting ability. It therefore cannot eat or maintain its equilibrium. For all these reasons, the molting period is a dangerous time.

Between molts, when the exoskeleton is hard, it is sometimes advantageous for a crayfish to cast off a leg or pincer which is wounded or trapped by a predator. This process is known as **autotomy.** When this happens, the break is always at a predetermined fracture plane near the base of the appendage, and loss of blood is prevented by a diaphragm and valve located there.

The crayfish regenerates a new leg under the tissue of the wound and, at the time of the next molt, the new leg expands to the size of its predecessor and assumes its assigned function. In a lobster, a lost claw is replaced in the same way but in a large animal it takes years to catch up in size.

The crayfish is well equipped with muscles for moving parts of its body independently or in concert. Most of the muscles attach at one end to the immobile exoskeleton and at the other to a moveable part such as a leg. The abdomen is the only really flexible part of the body and in it the muscles are segmentally arranged. In both crayfish and lobster, this is the meaty part of the animal.

All muscles are arranged in antagonistic pairs and are usually designated **extensors** or **flexors.** Even the autotomy mechanism has its own set of muscles which causes the leg to break off under appropriate conditions.

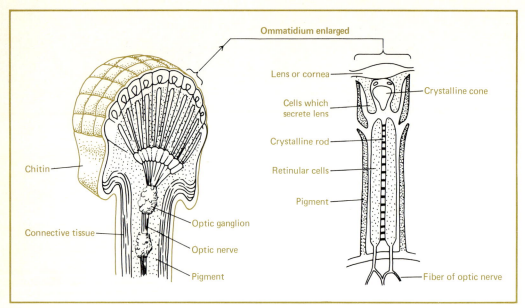

28-6 A diagram of the structure of a typical compound eye.

28.5 Both crustaceans and insecta have compound eyes which are excellent for detecting motion and judging distance.

Compound eyes are quite different from the simple cameralike eyes on your own head (Figure 28-6). Each compound eye is composed of many small light-receptive facets called **ommatidia**. These elongated structures have tiny lenses at their outermost edge. Each of these units is separately innervated. In each compound eye there are about 10,000 ommatidia.

Each of these photoreceptive units points in a slightly different direction and perceives the source of light, most often the sun, in a slightly different way. When objects come between a source of light and the eye, the eye integrates all the information from the individual units and derives a certain picture of the object. Such an eye can detect motion anywhere in the near vicinity. Since arthropods have two eyes, they can perceive distance.

28.6 Lobsters, close marine relatives of crayfish, produce a chemical which facilitates mating behavior.

Female lobsters usually mate soon after a molt. Recently it has been discovered that when ready to mate, they release a pheromone, **crustecdysone**, that modifies the behavior of male lobsters. There are many examples of such substances known in insects and mammals. Male lobsters are highly aggressive and will normally attack a lobster of either sex if it is placed in the same tank. However, pheromone from a sexually mature female makes the male docile; thus he can copulate with the newly molted female without harming her. The male approaches the female, gently turns her over on her back, and deposits sperm near her genital opening. After mating, the two animals separate and seem to have no further interest in each other.

28.7 No other group of organisms can compare in diversity with the class Insecta.

Insects are an enormously successful group, whose members have developed a myriad of forms as they have adapted to almost every ecological niche on the earth except the oceans. In spite of their diversity, insects have certain common characteristics. All have some kind of hard, waterproof exoskeleton and a tracheal breathing system which uses small tubes to carry air directly to the cells. An insect's body has three major parts—head, thorax, and abdomen; insects also have one pair of antennae (crustaceans have two), a pair of compound eyes, three mouth parts which are really modified legs, and three pairs of legs coming from the thorax.

BIOEPICUREAN DELIGHTS

Preparing and Consuming the American Lobster

Try to save enough money to purchase a live lobster before the Soviet fishing fleet fishes them to extinction along the North American continental shelf. Prepare it as the early American colonists did, and bear in mind that the treat you saved for was once so cheap and plentiful that household servants in colonial days complained bitterly that they had to eat it so often!

Bring a suitably large pot of salted water to a vigorous boil. (A 1:1 mixture of fresh water and sea water or a $1\frac{1}{2}$ percent solution by weight is best.) Cook the lobster for about 15 to 20 minutes, depending on size. Remove the lobster immediately when done, and allow it to cool a few moments before attacking it.

Every part is edible except for the stomach, gills, and shell. You must crack the large claws in order to reach the meat inside, but the smaller appendages can be crushed with the back teeth to force the meat out. Although most of the edible muscle is in the abdomen, the thorax is full of treats. Those who know prize the red or orange gonads and green digestive gland very highly. Don't forget the muscles in the base of the thorax. The stomach should be removed and discarded, as it may contain some un-killed bacteria.

Lobster meat can also be prepared after removal from the shell by sautéeing in butter. Add salt and much heavy cream. Heat but do not boil, and serve in a ring of white rice along with peas. The rice is essential to soak up the juices of the lobster.

Entomologists classify the one million or so known living insect species into about 26 orders, but the animals can be more simply divided into several grades which presumably reflect a progression up the evolutionary scale (Figure 28-7).

The most primitive insects are the wingless forms or **apterygotes.** Silverfish are common examples. All the rest are **pterygotes,** winged insects, and these may be subdivided into insects having wings that cannot be folded back against their bodies, such as dragonflies; and those having wings that can be compactly folded, such as beetles, moths, grasshoppers, flies, bees, ants, and so on. Insects that can fold their wings are sometimes further classified by the way they develop. Some emerge from their eggs as juvenile forms resembling adults, others progress from egg to a larval form which does not resemble the adult form, such as a maggot or caterpillar. This difference often means that the larvae are specialized efficient feeders, while the adults are specialized reproducers.

With even such a brief introduction to the diversity of insects you can appreciate the problem of trying to choose a single one that is representative of all the rest. It cannot be done. Instead, one animal has been chosen for its familiarity and accessibility: the housefly.

28.8 Flies have benefited to a large extent from the increased human population in the modern world.

The more people there are in the world, the more garbage is left around, and the more food there is for flies and their larval form, maggots. Consequently, there are a lot of flies in the world today.

Like other insects, the housefly is covered by a hard, chitinous integument which protects the underlying tissues from desiccation. The appendages of *Musca* are quite different from those of the crayfish, although both are jointed. The mouth parts of the fly are modified for sucking in food; the legs are used mainly for walking. One large pair of flying wings is present on the thoracic region of the body, and a second, greatly reduced pair act as small balancing organs. Many other insects have two pairs of functional flying wings.

Internally, insects are deceptively simple. There is a digestive tube with specialized regions for digestion and absorption. There are either male or female gonads, excretory organs in the form of long tubules which empty into the gut, an open circulatory system with a pumping heart, a double ventral nerve cord, various glands, special storage tissues,

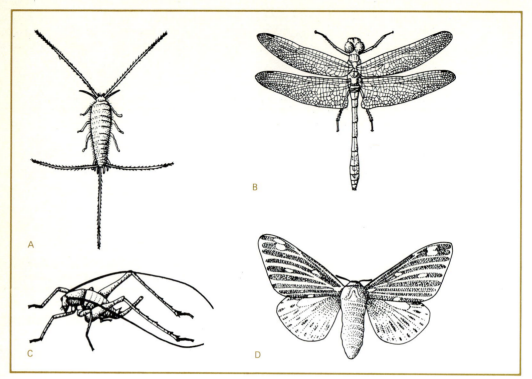

28-7 Examples of the four grades of insects: (A) silverfish; (B) dragonfly; (C) cricket; (D) moth.

and strong muscles, especially in the wing region. In addition, there is a tracheal respiratory system; that is, a system of air-filled tubes that penetrate all tissues of the insect's body.

Insects are the only successful flying invertebrates, and part of their success as flyers is due to their unique respiratory system. Chitin-lined tubes extend from exterior openings in the integument of the insect and branch into smaller and smaller tubes until they ultimately bring oxygen to individual cells or small groups of cells. Carbon dioxide produced in the tissues leaves through the same tracheal tubes. Thus we see a system which is the most efficient known for delivering large supplies of oxygen very quickly to such metabolically highly active tissues as the flight muscles.

Small valves, or **spiracles,** at the external openings regulate the entry and exit of air. Since the tubes are vapor filled, the prevention of water loss at the spiracles must be insured. The chitin in the trachea is spirally arranged, to allow linear expansion of the tubes when the animal moves and at the same time prevent compression and blockage of the otherwise thin-walled tubes. Besides moving gases into the animal efficiently, this type of system has another advantage. The fact that only a small part of the system, namely the spiracle openings, is ex-posed to the dehydrating effects of the terrestrial environment helps the animal to conserve its body water.

28.9 Flies use their feet to taste food and to adhere to vertical surfaces.

The fly has a long tubular **proboscis** associated with the mouth, which it unrolls and uses to suck up desirable food, but it has no sense of taste in this organ. Only recently has it been discovered that the animal uses its legs to taste. There are small hairs projecting from the front legs of the fly which are connected with nerves going to other parts of the body, mainly the proboscis. If the fly lights on something which its legs taste and the nervous system decides is of nutritional value, the proboscis lowers, and the animal eats.

In addition to these sensory hairs, there are tiny pads located at the ends of the fly's legs which contain adhesive glands and produce small amounts of a sticky substance. This attaches the lightweight insect to a wall, ceiling, branch or slippery garbage pail. The legs of some insects are equipped with various kinds of spines, pincers, and projections which allow them to cling to many different kinds of plants and animals.

28.10 The small size of insects and their preset behavioral patterns have contributed to their evolutionary success.

In the introduction to this chapter several reasons for the success of arthropods in general were advanced, namely a jointed outer skeleton and, for insects, the ability to fly. In addition, the small size of insects and their particular kind of nervous system seem to have given them an advantage over competing animals. Small animals can occupy numerous habitats not open to other animals. Most other terrestrial forms, which are almost exclusively vertebrate, are rather large and do not live in tiny cracks and crevices, under rocks, or under slivers of wood. Thus the territorial competition for small terrestrial animals is rather slight, although some insect species have to compete with other insect species for the same territory.

Another apparent evolutionary advantage associated with smallness involves geometrical considerations of muscle size and strength. Ants, wasps, and other insects can lift much more weight in relation to their size than can a rat, elephant, or man. The best human weight-lifters can only hoist about 2 or 3 times their own weight, whereas an ordinary ant can lift objects 10 to 15 times its own weight and carry them a great distance. This is not the result of the ant's especially strong muscles but is a function of the relationship between muscle size and strength.

A final characteristic of insects that seems to have equipped them remarkably well for evolutionary success is their behavior patterns. Behavior is defined here as the result—usually expressed through movements we can see—of the activities of the nervous system. For want of a better expression, insect behavior has been designated as "instinctive." It should be clearly stated that no one today understands the nervous actions underlying this type of behavior, but at least we can define it operationally.

Instinctive or **innate behavior** is based primarily on complex inherited or preset patterns of response which are triggered by certain stimuli. Often no learning or previous experience is required for this type of behavior to be enacted, but learning and

experience may play a role in some instinctive behavior. An example of instinctive behavior was given earlier in the case of the female squid's behavior during egg laying and attachment of the eggs to rocks; many other examples have been studied among arthropods. The shape and size of the nest built by different species of insects is determined instinctively, and an insect can build one without ever seeing another nest. Males of a certain species of mosquito will fly toward the sound of a tuning fork which corresponds in pitch to that of female mosquitoes of the same species even though they have never heard a similar sound in their lives. By the same token, they do not respond to a tuning fork of a slightly different pitch. Female moths and butterflies emit chemical attractants with a particular scent that evoke responses only from males of their own species, even though all males can detect a wide variety of scents.

A dramatic example of instinctive behavior concerns reproduction in the praying mantis. A male mantis placed beside a female will usually not make copulatory advances to the female until after the female has eaten the head of the male during premating activities. Evidently there are inhibitory centers in the brain of the male which prevent stimulation of copulatory nerves leading to other parts of the body involved in copulatory activity. In fact, sexually immature males if decapitated will attempt to copulate with a female. The fact that the male mantis can mount the female and mate even after its head and brain are removed, and that the female insures mating by eating the head of the male, and that these activities occur without prior learning or experience, suggests some of the complexity involved in instinctive activity.

A nervous system which triggers instinctive behavior certainly saves the animal from wasting time and energy learning by experience or making mistakes. It is much simpler to automatically follow a set of rules or preset instructions once the correct stimuli sets off the sequence. Insects are generally short-lived and cannot afford the luxury of learning. In the course of their evolutionary history, the ability to modify behavioral patterns or make corrections if unexpected hazards arise has been largely sacrificed. Their great success using instinct argues against the importance of learned behavior in an animal of this level of complexity, however. Most studies of instinctive behavior in insects have been done on adult forms, but larvae are undoubtedly equally endowed to handle certain types of stimuli and make suitable instinctive responses.

29

Echinoderms

QUESTIONS TO KEEP IN MIND

Where do echinoderms fit on the evolutionary tree?

What is unique about the way echinoderms move and respond to stimuli?

With the study of the phylum Echinodermata, containing the familiar starfish, brittle star, sea urchin, and others, we leave the annelid-mollusk-arthropod branch of the evolutionary tree, the **protostomes,** and approach the echinoderm-chordate branch, or **deuterostomes.** The differences between the two groups are apparent mostly during their development. As you may remember from Chapter 13, "Sexual Reproduction and Development in Animals," most multicellular animals develop from a hollow sphere of cells, the blastula, which during gastrulation, forms a primitive gut (section 13.6). At this point there is only one opening to the outside, the blastopore. If this opening eventually becomes the mouth, the animal is a protostome ("primary mouth"), but if, as in echinoderms and chordates (including vertebrates), the blastopore becomes the anus, the animals are said to be deuterostomes ("secondary mouth").

There are other differences in the embryonic development of these two broad groups. Among the protostomes, a typical zygote undergoes **spiral cleavage,** in which the new cells form a spiral pattern. The early cells formed by spirally cleaving animals are **determinate,** meaning that very early, even at the two-cell and four-cell stage, each cell is committed to developing into a specific tissue or organ. Deuterostome zygotes undergo **radial cleavage,** in which stacks of cells are built up. The cells remain in an **indeterminate** condition—that is, they are uncommitted—much longer than the cells of protostome zygotes. The larval forms are different too: Trochophore larvae are produced by many protostomes, and tornaria larvae by many deuterostomes. Taken together, these differences suggest that echinoderms and chordates are fairly close relatives which are evolutionarily distant from animals in the annelid-mollusk-arthropod line.

29.1 Echinoderms are entirely marine.

All of the 6000 or more known living species of echinoderms are marine and are widely scattered over the sea floor, from warm, shallow waters to the black abyssal depths. Their most striking characteristics are **radial symmetry** (although their larvae are bilaterally symmetrical), five-sidedness, prickly skin (*echino-* + *derm* literally means "hedgehog skin"), and a water-vascular system which will be discussed later. Because echinoderms have an internal skeleton or **endoskeleton,** of calcium-containing material, they fossilize well, and their remains can be traced back some 500 million years.

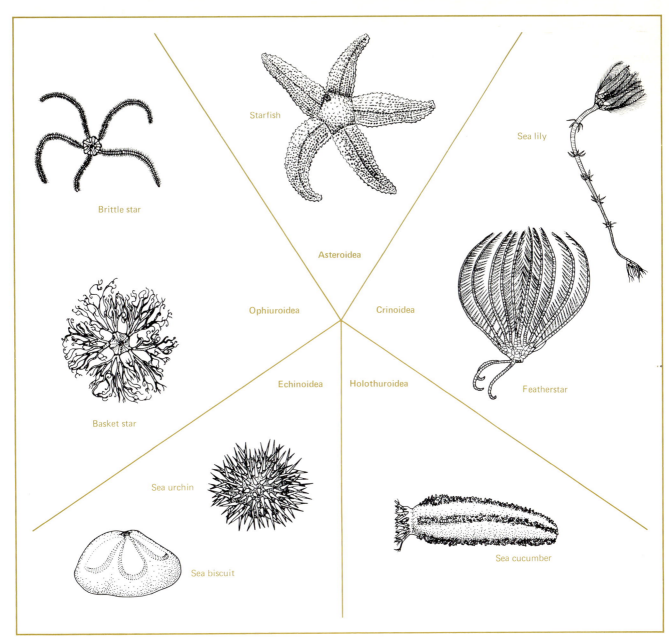

29-1 Representatives of the five classes of echinoderms.

More than 20,000 extinct species have been described.

Modern species are divided into five major classes (Figure 29-1): *Crinoidea*, sea lilies and feather stars; *Asteroidea*, starfish; *Ophiuroidea*, brittle stars and basket stars; *Echinoidea*, sea urchins and sand dollars; and *Holothuroidea*, sea cucumbers. All are true coelomates. They are not segmented, and their nervous system is not controlled by a brain.

29.2 Echinoderms have a distinctive water-vascular system which serves primarily as a means of locomotion for obtaining food.

Figure 29-2 shows the anatomy of the common Atlantic Coast starfish *Asterias*. In this figure you can see the sieve plate, through which seawater can enter and move, with the help of cilia, into a **ring canal.** From this, **radial canals** branch off and run

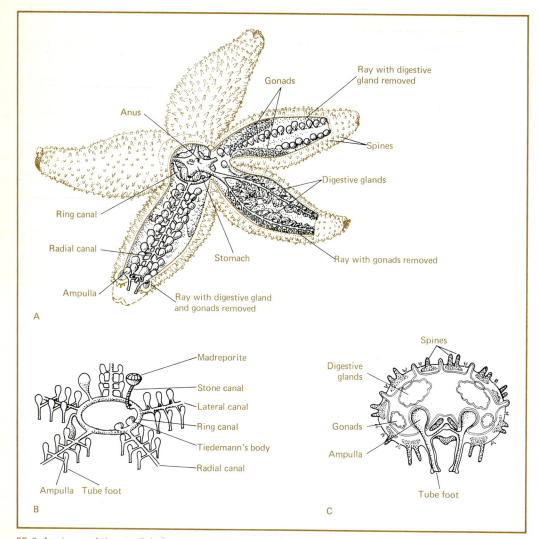

29-2 Anatomy of the starfish *Asterias. A.* Whole animal cut open to show the various organ systems. *B.* The central portion of the water-vascular system. *C.* A cross section through an arm.

down the arms of the starfish, leading from these smaller connections to hundreds of sac-like **ampullae** and **tube feet.** Water can be shifted about through this canal system. If a group of ampullae contract, each forces water into a tube foot, which is built like a cylinder with a suction cup at its end. Hydraulic pressure forces the tube foot to extend. There are muscles on either side of each tube foot, and by contracting these independently the animal can make its foot move in any direction. The foot's slightly cupped end sticks onto any smooth surface, such as a rock or a clam shell.

This unique hydraulic locomotor system appears to have originated as a feeding device and has since been modified to provide transportation as well. It is unique to the phylum Echinodermata. *Asterias*

uses its tube feet to pull open oysters and clams. Once its feet force a small opening, the starfish everts its stomach into the clam. The everted stomach secretes powerful enzymes and begins to digest the clam soft parts and muscles. After the tissues are digested into a broth, they are swept into the stomach by ciliary currents.

29.3 The starfish has no respiratory system but exchanges gases directly through thin-walled tissues.

Some of the body tissues of the starfish extend like small fingers through openings between the skeletal plates. Through the walls of these fingerlike structures, gases are exchanged. Additional oxygen and carbon dioxide exchange occurs through the thin walls of the tube feet. There is no well-developed circulatory system in these rather sluggish animals, and gases simply pass from the fluid surrounding the animal to that inside. The body fluid of echinoderms is similar to sea water.

The epidermis of echinoderms is ciliated, and any debris or sediment which settles on them is swept off by the cilia so that the areas with epidermal protrusions used in respiration do not become too dirty or clogged. The skin of the starfish is also equipped with numerous small, jawed appendages which are presumably used for protection against parasites and sessile animals which might settle, encrust, and generally "foul" the starfish.

29.4 In echinoderms the nervous system is typically scattered throughout the animal.

Unlike those of the annelids and the arthropods, the echinoderm's nervous system is not concentrated in a head. Instead, sense organs are distributed all over the body. It is not well understood how this nervous system functions, but clearly these animals can respond to environmental stimuli from many directions simultaneously. If food or danger appears from any direction, the animal is easily able to assess the situation and cope with it.

The starfish has light-sensitive eyespots at the top of each arm and tactile tentacles in the same regions. Receptor cells are also scattered abundantly over the surface of the animal. These are presumably sensitive to touch and may also aid in the reception of chemical and light stimuli.

Once a starfish receives a certain stimulus, it may decide to move in a particular direction. If, while gliding along with one arm in the lead and the others trailing it receives a stimulus from the side or back which causes it to change direction, the animal does not turn its body, but simply has another arm assume the lead. The tube feet on all the other arms change orientation, and the starfish goes off in the second direction.

29.5 Starfish and other echinoderms produce tremendous numbers of gametes that meet almost by chance in the open sea.

Most echinoderms are dioecious (have separate sexes), although occasional hermaphroditic animals are found. Each animal produces as many as several billion eggs or trillions of sperm at a time. Males

BIOEPICUREAN DELIGHTS

Sea Urchin Gonads, Two Ways

Gonads of most sea urchins are delicious, nutritious, and perfectly safe as long as they are taken from live animals (those with moving spines). Sea urchins can be collected from tidal pools in Maine, California, and the Pacific Northwest and can be dredged in deeper water off more southerly coasts. (Caution: Avoid poisonous spines of tropical species and the East Coast purple urchin, Arbacia, which is edible but not very delicious.) Cut around the external covering on the oral side. If the gonads are ripe, they will be swollen and either yellow-white (testes) or yellow-orange (ovaries); they should be drained of their visceral fluid (which is watery), but not of their gametes (milky). Each person can be expected to consume the gonads of 3 to 6 urchins, depending on size and season.

1. On toast: Apply butter thinly to bite-size pieces of toast; cover with gonads, and sprinkle with fresh lemon or lime juice and with black pepper. (Beverage: aquavit from the freezer or a dry white wine.)

2. As sushi: Boil rice in unsalted water containing about one percent vinegar. Cool the rice and mold into cylindrical cakes about 2 cm in diameter and the same height. Wrap with a sheet of Nori (japanese name for Porphyra, a red alga sold in stacks of dried green, red, or black sheets in most Oriental food stores). Wrap with extra Nori at the top, so as to form a container into which several sea urchin gonads can be placed. Served with shoyu, Japanese soy sauce, into which the sushi is dipped before eating (Beverage: Cold beer or hot sake.)

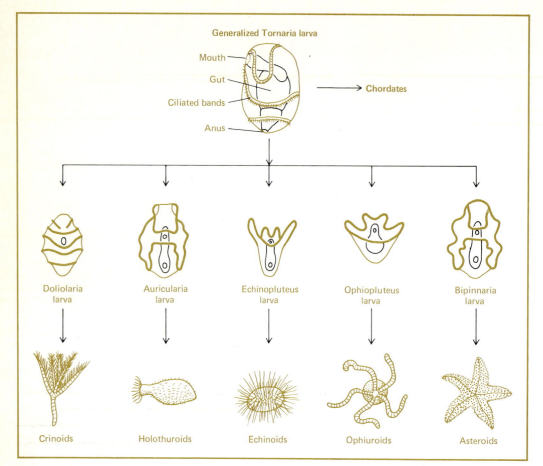

Generalized Tornaria larva

Mouth
Gut
Ciliated bands
Anus

→ Chordates

Doliolaria
larva

Auricularia
larva

Echinopluteus
larva

Ophiopluteus
larva

Bipinnaria
larva

Crinoids

Holothuroids

Echinoids

Ophiuroids

Asteroids

29-3 The tornaria is the basic larval type for both echinoderms and chordates. This diagram shows how the larval form is modified in each of the classes of echinoderms. All of these larvae can be found in marine plankton.

release semen in response to substances given off by eggs. Consequently, when females in an area spawn, it is likely that nearby males will release their sperm. Echinoderms occupy certain regions of the ocean bottom in such substantial numbers that the probability that some eggs will be fertilized is high. Only a fraction of the eggs produced by any female become zygotes, and only a tiny fraction of zygotes survive to metamorphose into adults.

The zygote undergoes radial cleavage and be-comes a ciliated blastula which swims and floats through the water as it develops further into one or another type of bilaterally symmetrical larva (Figure 29-3). These larvae exist as part of the oceanic zooplankton (the small animals that can be collected in a fine mesh net), and most are con-sumed by fish and filter-feeding invertebrates. In further transformations the echinoderm larvae lose their bilaterality, and the adults exhibit a modified radial symmetry. The symmetry of the larvae, how-ever, indicates that the animals are bilateral and that their five-sidedness is acquired secon-darily.

30

Chordates

What characteristics do all chordates have in common?

What adaptations enable amphibians to live on land?

What adaptations for flight have birds developed?

Are placental mammals more successful than others?

All the animals described so far, from sponges through annelids through arthropods and echinoderms, are **invertebrates** (animals without backbones). The phylum Chordata, composed of some 45,000 species, is usually divided into three subphyla; the members of two of these are also invertebrates. The third subphylum, Vertebrata, includes all the vertebrates, or animals with backbones. Members of the two invertebrate subphyla, the Urochordata ("tail chordates") and Cephalochordata ("head chordates"), are considered to be **lower chordates.**

30.1 All chordates have three distinctive characteristics: a notochord, a hollow nerve cord, and pharyngeal gill slits.

All chordates, be they sea squirts, sharks, toads, or princes, have these three features *at some time in their life cycle*—but not necessarily as adults.

The **notochord** (literally, "back cord") is a firm but flexible rod that extends the length of the back of the animal and gives it support for swimming or other locomotion. In vertebrates, the notochord becomes calcified to form the vertebral column.

The **hollow nerve cord,** or neural tube, also runs down the back of the animal and lies just dorsal to the notochord. This type of nervous system is markedly different from the double ventral nerve ganglia of annelids and arthropods or the diffuse system of cnidarians. In vertebrates, the dorsal tubular nerve cord develops into the brain and spinal cord.

Pharyngeal gill slits are found on either side of the pharyngeal (throat) region; they allow water entering through the mouth to pass over the gills and to leave the gill region. Some lower chordates retain the gill slits throughout adult life. In vertebrates they are evident only in larval or embryonic life, except in the case of fishes. The single, large water exit (*operculum*) on the sides of an adult fish are modified gill slits. Mammals, including man, have gill openings in the throat only as young embryos.

In addition to these three features, which are found only in chordates, members of the phylum also share certain important features with other groups. They are bilaterally symmetrical; they are partially segmented—branches of the spinal cord, vertebrae, and ribs are evidence of this segmentation in your own body; and they have a large, well-defined coelomic cavity in which the viscera are suspended.

30.2 The sea squirt *Molgula* is a typical example of the Urochordata.

The **urochordates** or **tunicates** are a group of marine animals which bear absolutely no external resemblance to their vertebrate relatives. There are three classes of tunicates, but the most familiar ones are sea squirts. These potato-shaped organisms grow subtidally on rocks or pilings and look for all the world like relatives of the sponge. But sea squirts are chordates. As adults they have only gill slits, but as **tadpole larvae** they have gill slits, a notochord, and a dorsal neural tube (Figure 30-1).

Molgula, a common, solitary sea squirt, will serve as a type specimen of urochordates. As an adult the animal is covered with a tough, rather thick **tunic** (covering) composed of a celluloselike material. The covering overlays the rest of the body, which is mainly filled by a **branchial sac** (gill chamber). Cilia on the gill slits create a current of water which flows in through an **incurrent siphon**, is filtered through the branchial sac, and is finally passed out through an **excurrent siphon**. Mucus secreted on the inner surface of the perforated branchial sac traps food particles as they pass through with the water. Small balls of food and mucus are propelled by cilia into the endostyle and on into the stomach. Undigested and unusable materials are shot out of the excurrent siphon along with the "used" water. Gases are also exchanged in the branchial sac.

Molgula lives attached to the substratum and is yet another example of a sessile animal making the environment move through its body instead of moving itself through the environment.

Molgula is hermaphroditic and the ovaries and testes lie along the inner side of the body wall. In the breeding season gametes are released into the sea. Fertilization occurs in the open water and tadpole larva form within 12 to 24 hours. The *Molgula* larva has a tail, tubular dorsal nerve cord, notochord, a few gill slits, and sensory organs for light reception and balance.

Within 10 hours the swimming larva attaches itself to a solid surface and metamorphosis begins. Metamorphosis is a complex process involving many changes. Tail reabsorption is one of the most dramatic, mainly because only a few minutes are required. The sense organs, presumably used by the larva to locate a homestead, also disappear as the animal matures into a sessile adult. The larval forms are much more chordatelike than the adults,

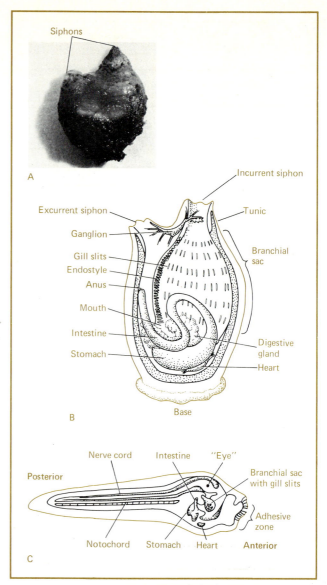

30-1 *A.* The common tunicate *Molgula. B.* A diagram of *Molgula* anatomy. *C.* The larva (tadpole).

and it is not surprising that the true status of adult tunicates was unknown for many years.

30.3 *Molgula*'s nervous system includes a neural gland which may be the forerunner of the vertebrate pituitary gland.

The tunicate nervous system is a fairly simple one, especially in the adult stage. Once the larva's sense organs and swimming muscles have disappeared and no longer require nervous connections, the adult is left with a small oval ganglion located between the siphons and a few auxiliary nerves which connect with it. These are enough to allow the animal to contract and draw in its siphons, expel unwanted particles, and make other simple behavioral responses.

Very near and in nervous contact with the cerebral ganglion of *Molgula* is a mass of tissues called the **neural gland.** This gland is believed to be a type of primitive pituitary. Extracts of the tunicate neural gland injected into vertebrates elicit a response similar to that resulting from the injection of vertebrate pituitary substances. For instance, the growth of ovaries, sperm discharge, uterine muscle contraction, and several other reactions have been noted.

Interestingly enough, extracts of both tunicate neural glands and vertebrate pituitary induce ovulation in the tunicate. The current theory is that the pituitary, which has a host of primary and secondary functions in growth and development of vertebrates (section 35.2), arose in the lower chordates as a structure to insure that many eggs and sperm would ripen and be released at the same time.

30.4 Vanadium is accumulated by tunicates. But why?

Many animals contain strange and unusual chemicals in their bodies which have no known function. There are at least two possible explanations. First, the substance in question, be it a rare inorganic chemical or a complex organic molecule, may actually have a function in the organism which investigators have not yet discovered. On the other hand, an unusual chemical may not play a role in the animal's physiology but may just happen to be present, either as a by-product of a reaction or because it was trapped in the tissues of the animal accidentally.

There is an example in the tunicates of such a substance. The element vanadium is found in very, very small quantities in sea water and cannot be detected at all in the bodies of most other organisms. Yet, in tunicates, it constitutes a rather large component of some cells. Its function has been the object of speculation for years but as yet no one has uncovered its function. Since vanadium is important in the manufacture of some types of steel, it seems likely that the tunicates' ability to extract it from sea water may some day become economically important.

30.5 *Amphioxus* is the most common member of the subphylum Cephalochordata.

Cephalochordates, or **lancelets,** as they are often called, are small, inconspicuous marine invertebrates. *Amphioxus* is a translucent animal two or three inches long, which spends most of its adult life embedded in the sand with only its head sticking out (Figure 30-2). It is uncommon in our own coastal waters and is abundant only in one area off the south China coast, where it is harvested by the ton for food. Larval stages of *Amphioxus* last only a few hours and are even more difficult to find and study than the adults.

In spite of its elusiveness *Amphioxus* is of considerable biological interest. Cephalochordates are believed to be similar to the primitive ancestors from which vertebrates evolved. They resemble superficially some fishes but structurally are much less complex. For example, they lack paired fins, jaws, and a well-defined brain.

Because the muscles in *Amphioxus* are arranged in discrete units, each innervated by segmentally arranged nerves, the animal can send waves of contractions along either of its sides. This produces undulations which propel the animal quite rapidly through the water. Many vertebrates which swim—for example, fish, snakes, and whales—utilize this same mechanism. In *Amphioxus* there is a single dorsal posterior extension of the skin and connective tissue which acts as a thin, flexible fin and facilitates movement. Gill slits, up to 200 of them, occupy almost half of the total length of *Amphioxus*. Small bits of food contained in the water are trapped in mucus produced in the gill region and are carried into the intestine by ciliary action. As the water passes over the gills, respiratory gases are exchanged; the water finally leaves the pharyngeal gill cavity via a ventral pore. As in tunicates, the mucus is produced in a ciliated groove, the **endostyle,** which may serve as a primitive thyroid.

The gill regions in *Amphioxus* are well supplied with arteries, and a large dorsal artery carries blood to the tail. Veins return blood from the front and back of the animal to the gill region. This closely resembles the vertebrate system.

There is no well-defined heart, but part of the arterial system is modified to pump blood at a

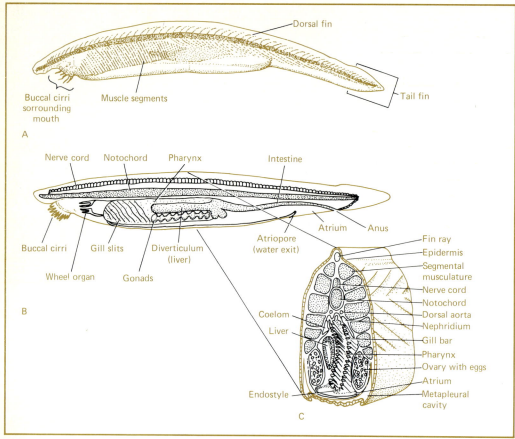

30-2 The lancelet *Amphioxus*. *A*. Whole animal. *B*. Anatomy as shown in longitudinal view. *C*. Cross section through the pharyngial basket.

rather high pressure throughout the animal. Capillary beds found along the intestine pick up digested food, just as those in the gills pick up oxygen and unload carbon dioxide.

30.6 There are seven major classes of living vertebrates.

The subphylum Vertebrata includes the agnatha, or jawless, fishes, cartilaginous fishes, bony fishes, amphibians, reptiles, birds, and mammals (Figure 30-3). The first four of these have retained a fishlike body form, and must live in water all or part of their lives.

Animals in all seven groups have the special characteristics of a vertebrate: a **cranium, gill slits, aortic arches** during part or all of the life history, **vertebrae,** and a well-developed **brain.** We will examine each group separately to show a few of the ways in which its members are specially adapted for their own particular mode of life.

30.7 *Petromyzon*, the lamprey, is a typical cyclostome, or modern jawless fish.

A great deal of research has been done on lampreys, because they are destructive of commercially valuable fish, especially in the Great Lakes. Adult lampreys fasten onto the bodies of other fish and feed on their body fluids, often causing the host's death.

The body of the lamprey is elongated and superficially resembles that of eels and snakes (Figure 30-4). The larval and adult stages differ somewhat in structure. The larvae, called **ammocoetes,** live in the mud, feeding on small bits of organic matter they sift out of the bottom ooze. The adults are free-swimming and have a single **median fin** which aids in movement.

The head region of the lamprey is modified for its particular way of life. At the anterior end there

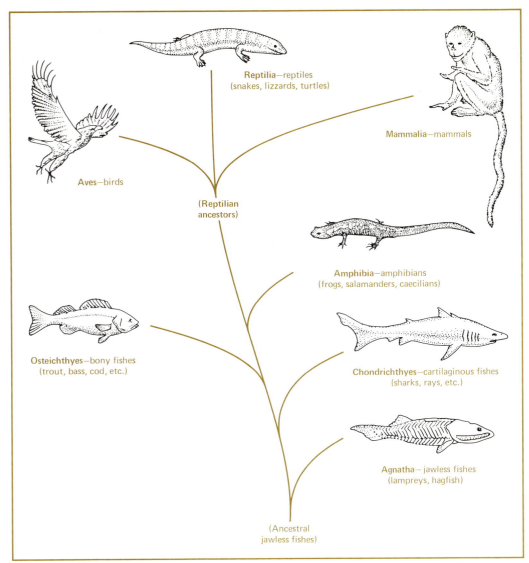

30-3 Representatives of the classes of vertebrates (subphylum Vertebrata).

is a round **sucking mouth**, which is used for attaching to stones while the animal rests or feeds. This mouth has teeth and a rasping tongue which tears a hole in the body of its prey. Salivary glands in the throat secrete an anticoagulant which prevents the host fluids from coagulating. Behind the round mouth there is a single nostril which sits high up on the head and allows the animal to breathe by passing water over its gills. It can thus keep its anterior end buried in the body of its host while breathing. The lamprey has seven pairs of gill slits for respiration.

Externally, the lamprey is covered with a tough skin rich in mucus-producing cells but no scales. There are no paired limbs or fins in either the embryo or the adult, and wriggling motions accomplish locomotion. (The adults are often carried about by host fish.) The lamprey has a noncalcified cartilaginous skeleton (including a skull) for protection and support.

30.8 The life cycle of the lamprey includes a long larval stage.

In the spring, lampreys migrate upstream from their ocean or freshwater habitat, depending on the species under consideration. When the migratory adults reach a suitable site in the stream, they make

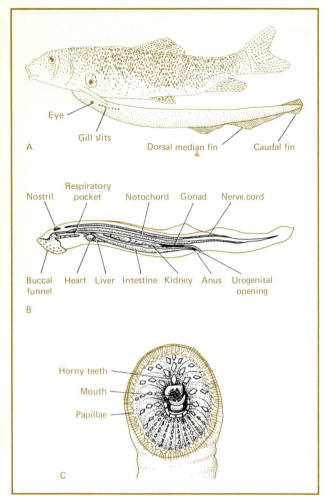

Eye

Gill slits

Dorsal median fin   Caudal fin

A

Respiratory
pocket

Nostril   Notochord   Gonad   Nerve cord

Buccal   Heart   Liver   Intestine   Kidney   Anus   Urogenital
funnel   opening

B

Horny teeth

Mouth

Papillae

C

30-4 The lamprey *Petromyzon*, its (*A*) external anatomy and (*B*) internal anatomy and (*C*) the structure of the mouth.

a depression in the sand or gravel bottom. The female fastens herself to a stone, and the male attaches to the female. Eggs are extruded by the female, and the male sheds sperm over the eggs as they fall into the sand depression. The adults die soon after spawning.

The young hatch in about one month and remain in the nest until they are 12 to 15 mm long. Then, as ammocoete larvae, they make their way to quiet waters in the stream and dig out a tunnel in the mud. Blind and toothless, they live in their tunnel and feed on materials filtered from the bottom mud. After three to seven years, the larvae metamorphose into adults which may or may not migrate farther downstream. They live one or more years, spawn, and expire. Conservationists are rapidly exterminating the lampreys from the Great Lakes by treating streams containing developing individuals with poisons specific for the ammocoetes.

30.9 The cartilaginous fishes, class Chondrichthyes, include sharks, skates, and rays.

Cartilaginous fishes evolved from a class now extinct, called Placoderma. This class had hard external plates, bony skeletons, paired fins, and hinged jaws. A mouth with jaws had an immense mechanical advantage over a jawless mouth and enabled the armored placoderms to become hunters of large prey.

Sharks, skates, and rays, the living representatives of the cartilaginous fishes, lost the armor and bony skeleton of the placoderms but retained paired fins, hinged jaws, and carnivorous feeding habits. Skates and rays have flattened bodies and cruise along the bottom feeding mainly on mollusks, whereas sharks live in the open sea and are predators of bony fish and, occasionally, mammals (Figure 30-5).

30.10 The spiny dogfish, *Squalus,* is a small but representative example of the cartilaginous fishes.

The dogfish, a small shark seldom exceeding three feet in length, is no danger to man and occurs in considerable numbers in our coastal waters. Its jaws are hinged, and in its mouth are rows of small razor-sharp teeth used to grasp the prey and tear its flesh into pieces small enough to swallow. The shark takes in water as well as food through its mouth. The water passes over gills in the neck region and leaves through the spiracles and gill slits located on each side of the body just in front of the **pectoral** (shoulder) **fins.** On the shark's head are two nostrils and two lidless eyes. The shark has pairs of pectoral and **pelvic fins** that are the forerunners of the legs of tetrapods (four-legged animals). The tail, or **caudal fin,** is asymmetrical, the dorsal lobe being larger than the ventral lobe (Figure 30-5). All of the fins aid in swimming and help direct the shark's movements. The skin of the dogfish shark is tough and abrasive, due to numerous small spinelike scales embedded in the epidermis. These scales undoubtedly help protect the shark from some predators.

The skeletal system of *Squalus* is composed entirely of cartilage. (The class name, Chondrichthyes, comes from *chondr-,* "cartilage," plus *ichthyes,* "fish,") There are some limy deposits in the skeleton but no real bone. The backbone, composed of a series of biconcave plates surrounding the notochord, bears **neural arches** through which passes a canal holding the spinal cord. The skull protects the brain and also forms the sensory capsules which overlay the olfactory, optic, and otic (equilibrium) centers of the nervous system. Cartilage also stiffens the fins and the gill arches.

The musculature of the dogfish is similar to that of other vertebrates, including man, in that the muscles are separated into **myotomes** (muscle segments). This arrangement makes it possible for the animal to contract individual segmental muscles successively. Bending waves pass down the length of the body and propel the animal.

In the dogfish and other sharks the sexes are separate, and fertilization is internal. Some sharks develop externally from a yolky egg covered with a tough case, others develop within the body of the mother but use nutrients stored in a yolky egg. Some sharks develop internally in the mother, and the mother's bloodstream supplies nutrients to and removes wastes from the bloodstream of the developing young.

30.11 The bony fishes, class Osteichythes, are by far the most abundant living fishes.

Over 20,000 species of bony fishes exist in a wide variety of body forms and have populated all parts of the sea and fresh water. They are herbivorous, carnivorous, and omnivorous; large and small; blind and stalk-eyed. All are complex.

They are divided into three subclasses, of which the **teleosts** are the most familiar. One of the other major subdivisions includes primitive fishes, such as sturgeons and paddlefish, and the other includes several specialized groups, such as the lungfish. In all of these the body plan is similar, as is the internal anatomy (Figure 30-6). All have skeletons of true bone, scales, and two pairs of lateral fins. The scales are formed within the mesoderm and are covered with a thin layer of living tissue.

Most bony fishes contain a **swim bladder** which is absent in cartilaginous fish. The swim bladder is a gas-filled baglike structure lying in the visceral region of the body just under the kidney. It helps regulate the density of the fish at different water levels so that the buoyancy of the animal can be regulated. Glands around the swim bladder and blood vessels in its walls absorb or secrete gases as needed, to match the density of the fish with the water around it. Some fish use the swim bladder for respiration or for receiving or sending sounds. In lung and gar fishes the swim bladder is modified into a kind of "lung."

In all fishes, the kidney is utilized to control the osmotic concentration of the blood. Kidneys vary in function according to the particular environment in which a species lives, but they tend to retain water in fish living in salt water and pump water out of fish living in fresh water. In addition, the kidneys also help the fish get rid of nitrogenous waste products.

Many species of bony fish, like the herring, cod, anchovy, trout, salmon, and perch, exist in great abundance. More than 100 billion pounds of fish are caught each year. They are of tremendous eco-

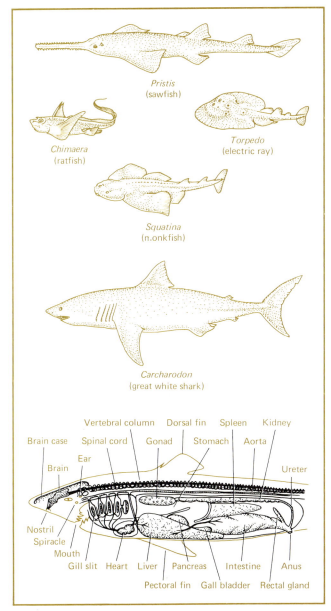

Pristis
(sawfish)

Chimaera
(ratfish)

Torpedo
(electric ray)

Squatina
(monkfish)

Carcharodon
(great white shark)

Vertebral column   Dorsal fin   Spleen   Kidney

Brain case   Spinal cord   Gonad   Stomach   Aorta

Ear

Brain

Ureter

Nostril
Spiracle
Mouth

Gill slit   Heart   Liver   Pancreas   Intestine   Anus

Pectoral fin   Gall bladder   Rectal gland

30-5 The internal anatomy of the dogfish shark, *Squalus*, and some other cartilaginous fish.

BIOEPICUREAN DELIGHTS

Haddock Stew

In a large skillet, cook together until brown $\frac{1}{4}$ lb finely diced salt pork and $\frac{1}{4}$ lb chopped onions. Add a layer of barely cooked diced potatoes, and add water to cover the potatoes. Add $\frac{1}{4}$ cup of Worcestershire sauce and a layer of haddock or other white fish fillets (freshwater perch is very good). Cook until the fish is done and starts to fall apart. At this point, the fish can be broken apart with a spoon and turned into the layer of potatoes in order to cook more rapidly. Add at least $\frac{3}{4}$ pt heavy (whipping) cream, and heat almost to a boil before serving. The stew is better if prepared a day before serving; however, reheat gradually to serving temperature, and do not boil.

---

nomic importance. Bony fishes provide protein-rich food for humans and livestock and serve as fertilizer. In the past, whole communities were so dependent on the annual arrival of herring or cod, for example, that if the fish did not come, a great proportion of the people starved to death. Even today sharp and unpredictable fluctuations in a fishery can have a devastating effect on many, many people and an indirect effect on millions. If the anchovy fishery off the west coast of South America has a poor year, for example, the price of beef rises sharply in the United States.

With overpopulation a reality in many countries and a threat to all countries, increasing attention is being given to the oceans as a source of food. So-called "trashfish" that are considered inedible are now ground up into fertilizer or made into pet food, but in the future more and more will be turned into high-protein fish flour for human consumption.

30.12 Amphibians were the first vertebrates to partially adapt to a terrestrial environment.

Amphibians probably evolved from fishes some 350 million years ago. Their ancestors were possibly lungfishlike animals possessing paired pectoral and pelvic fins which developed into legs. A mode of terrestrial locomotion, lungs, and rather large eggs gave the ancestral lungfish-amphibian a selective advantage over its completely aquatic neighbors under certain circumstances. It has been suggested that the amphibians' ancestors developed these characteristics not to "escape" from an aquatic environment, but to move across land from a drying waterhole to a larger pond in times of drought.

Today there are two major groups of amphibians —the frog and toad group, whose adults have no tails; and the salamander group. There is also one minor group, the tropical, wormlike **caecilians**, which lack legs (Figure 30-7). Most amphibians spend part of their lives in fresh water and part in moist regions on land. Typically the adults return to the water to reproduce. Fertilized eggs develop into larvae that live like fish—swimming, breathing with gills, and feeding on phytoplankton or small invertebrates. At some point the larvae undergo metamorphosis and develop into adults, with legs,

30-6 The striped bass, (A) external and (B) internal anatomy. (Photograph by author.)

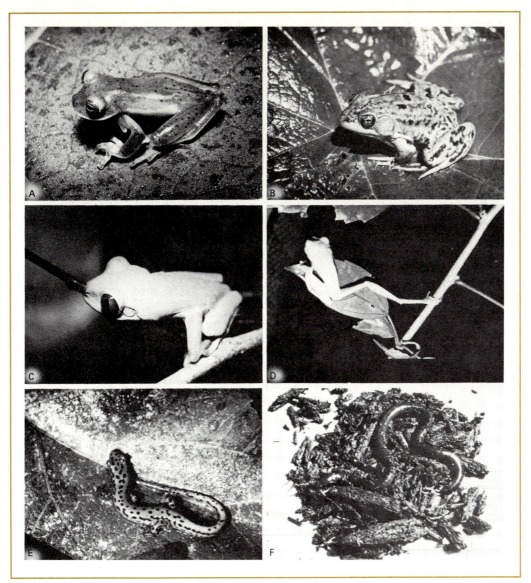

30-7 Some amphibians. *A.* Leaf frog, *Centrolenella prosoble-pon,* Costa Rica. *B.* Mink frog, *Rana septentrionalis,* Adirondack State Park, New York. *C* and *D.* Different views of a tree frog, *Agalychnis callidryas,* Costa Rica. *E.* Cave salamander, *Eurycea lucifuga,* Indiana. *F.* A caecilian, *Gymnophis multiplicata,* Costa Rica. (Photographs by M. Stewart.)

lungs, modified digestive system, and other features necessary for terrestrial life.

30.13 The life cycle of the leopard frog includes a typical but nonetheless amazing metamorphosis from tadpole to adult.

In early spring, adult frogs commence their croaking songs that attract mates from surrounding swamps and bogs. When the temperature and conditions are correct, the animals pair up, and the males climb on the backs of females, a process known as **amplexus.** The males squeeze the bodies of the females to help them lay eggs. Once the eggs are laid, the males discharge sperm over them. The eggs, which are in a jelly matrix, are left by the adults to develop into small embryos and later to hatch into larvae. The larvae, or tadpoles, have functional gills and gill slits, a lateral line system (a sensory apparatus also found in fish), and a tail fin. In addition, they excrete ammonia, which is characteristic of fishes; but not urea, as do adult frogs.

After a period of growth, metamorphosis begins. The process is triggered by the hormone **thyroxin,** which is released from the thyroid gland. Under the influence of thyroxin the tadpole resorbs its tail, grows legs, acquires increased pigmentation and mucus glands in the skin, develops lungs, absorbs its larval gills, undergoes structural changes in the mouth, switches from the excretion of ammonia to urea, and becomes altered structurally and functionally in a variety of less conspicuous ways. The product of all these changes—a frog—requires from one to several years to reach sexual maturity.

30.14 An amphibian's skin helps it resist dehydration, exchange respiratory gases, and blend in with its surroundings.

Before animals could emerge from the sea or fresh water and take up residence on land they had to develop a means of resisting dehydration. Insects and reptiles have dry waterproof skins, but amphibians, like many terrestrial worms, have moist permeable skins covered by a film of mucus. The mucus aids in preventing the frogs and salamanders from drying out; it also keeps the skin continually wet, to allow oxygen and carbon dioxide to pass in and out: Most amphibians respire extensively through their skin even though they have functional lungs.

This is especially true during winter, when frogs and other amphibians hibernate by burying themselves in the mud or moving below the frost line in some other way. There, as the temperature drops, their normal physiological processes are brought to a near-standstill. In this state, the animal can survive at a temperature just above that of the environment. For example, respiration normally requires a great deal of energy and oxygen, but in hibernation an amphibian does not feed, and gases are exchanged through its skin instead of its lungs. Sometimes in hot weather, frogs, toads, and salamanders will dig into the dirt or mud, apparently in an attempt to escape from conditions which may lead to desiccation. Such a temporary period of quiescence in warm, dry weather is called **aestivation** and should not be confused with hibernation, which lasts much longer and is more intense. The stimuli that trigger hibernation and aestivation are unknown, although day length and temperature have been suggested.

The skin of amphibians is scattered with pigment cells, **chromatophores,** which impart various colors to the organism. Most amphibians can change their colors to blend in with their environments. Some amphibians also have poison glands which secrete a white fluid with a burning taste. This helps repel enemies. In some species of toads this poison is dangerous, sometimes even lethal, to dogs and cats that catch them and take them into their mouth. The poison they secrete can also be fatal to other small mammals or birds.

Toads tend to remain in cool, damp areas and forage for small worms, snails, and insects. Insects are caught on their long, sticky tongue, which, like the frog's, is attached at the front of the mouth and can be flicked out a long way.

Salamanders and newts are tailed amphibians and, unlike most higher vertebrates, can regenerate a lost tail or leg. Usually these animals are small, but one species in Japan attains a length of five feet. Salamanders are often confused with lizards (reptiles), but the anatomy and life histories of the two are quite different.

30.15 Reptiles evolved from amphibianlike creatures some 300 million years ago and were the first truly land-dwelling vertebrates.

Reptiles became so well adapted to life on land that they displaced many of their amphibianlike ancestors and dominated the land during the Mesozoic era (70 to 225 million years ago). During the age of reptiles, as this era is sometimes called, there was a tremendous number and diversity of reptiles —dinosaurs and many others on land, ichthyosaurs in the sea, and pterosaurs in the air. Almost all of these had become extinct long before man evolved, and today only three significant groups of reptiles survive: the turtles, the lizard-snake group, and the crocodiles and alligators.

All reptiles are covered with scales formed of the protein **keratin.** These scales are similar in composition to the scales found on birds' legs and on the tails of rats and beavers. Reptilian scales are secreted along the surface of the epidermis and are derived from the ectodermal layer. Such completely external structures are a valuable protection against drying out, a major problem for terrestrial animals. Snakes and lizards periodically shed their skin; but turtles retain the old layers, and keratin piles up, especially in the regions overlying the bones of the hard shell. The armor of the turtle consists of bone overlaid with scales. Such a covering, combined with the ability to draw exposed parts of the body into the shell, makes for one of the most effective protective designs in the animal kingdom.

30.16 The anatomy of reptiles is intermediate between amphibians and mammals.

In some respects the internal body structure of a turtle, for example, is intermediate between a frog and a man. In other ways it is clearly either amphibianlike or mammallike. The inner skeleton, for instance, is very similar to that of a salamander but has fewer vertebrae. The muscular system resembles that of mammals.

The circulatory system has a three-chambered heart (two **atria,** or receiving chambers, and a single **ventricle,** or pumping chamber), but has the beginnings of a septum between the left and right valves of the ventricle, which if complete would make the heart four-chambered, like those of mammals and birds. Some turtles have a **hepatic portal system,** which is an advancement over the amphibians. It consists of a system of veins which carry blood from the intestine back through the liver.

Reptilian lungs are larger than those of air-breathing fishes and amphibians, and the air chambers within the lungs more extensive. Movements of the ribs by powerful muscles draw air into the lungs, much as happens in your own body. The excretory system also approaches the type found in mammals.

Although the higher centers of the nervous system are not nearly so large as in mammals, they resemble in size and architecture the mammalian system more than that of amphibians. There is one unusual feature in the nervous system of some reptiles—the presence of a third or **pineal eye** situated under the skin and higher on the head than the other two. Turtles lack this specialization.

The internal reproductive structures of reptiles are not very different from corresponding structures in amphibians. However, both sexes have copulatory organs since conditions on land clearly are not suitable for external fertilization. Sperm, released directly into the female genital tract, fertilize the eggs; only then is the leathery shell characteristic of all reptiles deposited on the eggs. To use the turtle as an example again, the female lays her eggs a few inches deep in the soil where they incubate during warm weather and hatch into miniature turtles. Even sea turtles come on land to lay their eggs, and the struggle of the clumsy female to pull her heavy body up the beach and the desperate race of the baby turtles to the sea just after hatching are classic tales. It is now known that the female is guided by white sand on the beach, whereas the young are able to follow ultraviolet light patterns in the sky to return instinctively to the water. Adult turtles do not take care of their offspring, so it is an advantage for the young to be large and well developed when they emerge from the egg. The eggs of all reptiles are richly supplied with yolk, which nourishes the developing young.

Like most other cold-blooded animals, reptiles cannot control their internal body temperature physiologically, and it fluctuates according to external conditions. Turtles, snakes, and lizards living in temperate regions are able to move about all year, and to some extent they can control their body temperature behaviorally, for example, by sunning themselves. In colder climates, however, they hibernate in the earth during the winter. Likewise, in hot seasons, they may aestivate to survive. Under either condition, physiological processes are slowed, and the animal utilizes very little energy and requires little oxygen.

30.17 Lizards and snakes make up the most plentiful and diverse group of living reptiles.

There are thousands of different species of lizards, living in many different environments. Most lizards are well equipped to run, climb, dig, and blend with their background. They are also known for their ability to autotomize (divide) their tails and regenerate new ones. If a predator grabs a lizard by its tail, the appendage will break off at a preset point but will continue to wriggle after it is detached. Thus the lizard can often escape while its predator is distracted by the moving tail.

There are many types of lizards (Figure 30-8). The gila monster of the American Southwest has a poisonous bite; the iguana, which may attain a length of six feet, lives in tropical America and is often used as food; true chameleons, abundant in Africa, Arabia, and India, are renowned for their ability to change colors; the flying dragon has thin membranes along its sides and can glide from tree to tree; horned toads (really reptiles and not toads) are sometimes kept as pets; swifts and skinks are the most common lizards in this country and are often found around old woodpiles, where they feed on insects and small larvae; and komodo dragon lizards of the Dutch East Indies, the largest of all lizards, may reach a length of nine feet and a weight

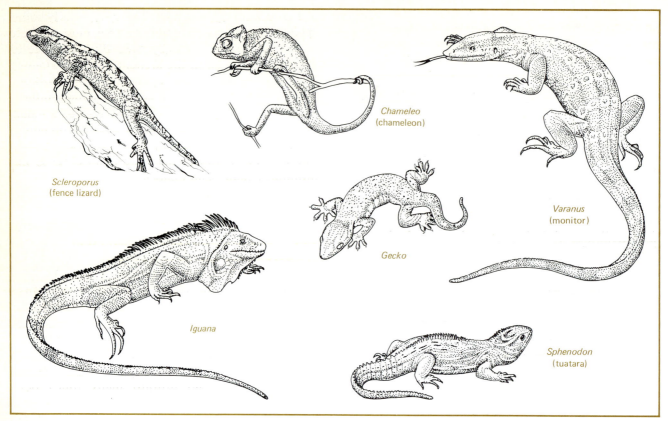

Scleroporus
(fence lizard)

Chameleo
(chameleon)

Varanus
(monitor)

Gecko

Iguana

Sphenodon
(tuatara)

30-8 An assemblage of modern lizards.

of 250 pounds. They are capable of capturing and ingesting a wild pig or other large mammals.

Snakes are closely related to lizards and apparently evolved from a group of burrowing lizards. Although poisonous snakes have given the entire group a bad reputation, only a small percentage of snakes are actually harmful to man. Those that are poisonous derive their poison (**venom**) from glands in the head which develop from salivary glands. The **fangs** of snakes are hollow front teeth that carry venom to the victim as the fangs penetrate the skin. There are two types of snake venoms: **hemotoxins**, which attack blood cell membranes, causing **hemolysis** (lysis of blood cells); and **neurotoxins**, which paralyze the nervous system. If the appropriate **antitoxin** is administered soon enough after a snakebite, the victim rarely dies. In some parts of the world, snakebites are still a major medical problem.

Snakes have some interesting sensory receptors and behavioral patterns involving their use. For example, when a snake repeatedly flicks its tongue out, it is actually "tasting" the air. The moist tongue picks up chemicals in the air and transfers them to **chemoreceptors** in the roof of the mouth. Pit vipers have special receptors in pits adjacent to their eyes and nostrils that can detect tiny amounts of infrared radiation given off from the warm bodies of small mammals which are prey to these snakes.

30.18 Crocodiles and alligators are the most mammallike of reptiles.

Like mammals, crocodiles and alligators have a bony palate in the upper part of the mouth and transport air from the nostrils to the back of the mouth. Their heart is almost four-chambered, and their tongue and mouth are hinged more like mammals than like reptiles.

With nostrils situated at the end of their snout, eyes protruding so they can see well while floating just under the surface of the water, and flaps which can be closed over the nostrils and ears when they dive under water, they are well adapted for an amphibious life. The tail of a crocodile or alligator is large and gives the animal balance, buoyancy, and a means of propulsion and defense.

30.19 Birds, the only animals with feathers, make up one of the most successful classes of vertebrates.

Most birds can fly, and it is this characteristic which most obviously sets them apart from other living vertebrates (except bats, the one type of flying mammal). The ability to invade an environment not utilized by any other large animal has been responsible to a large extent for the evolutionary success of birds. Living species number close to 9000.

Anatomical and developmental similarities can be seen between modern birds and reptiles, and the fossil record indicates without a doubt that birds arose from reptilian ancestors. Several intermediate forms such as *Archaeopteryx* (*archaeo*, "ancient" or "primitive"; *pteryx*, "wing") are known from the fossil record. These animals had feathers and wings, but also had teeth, a long reptilian tail, and a wingspread of over 40 feet!

Since the first birdlike reptiles some 150 million years ago, a great diversity of forms has arisen. Birds range in size from the tiny hummingbird to the ostrich. Some, such as the kiwi, penguin, and ostrich cannot fly, whereas others, especially some species of seabirds, are airborne for most of their lives except when rearing young. Birds are able to live far out at sea, in the arctic, in the tropics, and even in deserts. Several species, such as turkeys and chickens, exist in very disproportionate numbers because they are raised and protected by man for both eggs and flesh. Others, such as pigeons and sparrows, are not purposefully raised by man but have prospered by adapting themselves to his buildings, statues, and parks.

30.20 A bird is a bipedal organism with feathered wings and hollow bones.

In contrast to the terrestrial vertebrates studied thus far, birds possess only two legs; the anterior appendages have been modified into wings. Strong muscles (the "breast" of a fried chicken) making up one-fifth of the total weight of the body, connect the wings and the **sternum** (breastbone) and draw the wings downward and upward many times per second with great force (Figure 30-9). Feathers cover the wings as well as much of the rest of the body. They provide wind resistance during flight

30-9 The external (A) and internal (B) anatomy of the pigeon, *Columba.* (C) Diagram of the muscles that move the wings.

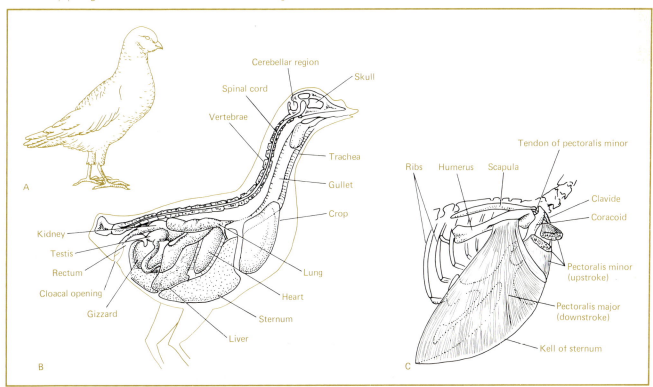

and insulate the animal from the environment. Tail feathers serve as a rudder for flying birds.

Feathers arise embryonically in much the same manner as the scales of reptiles and may be of three types: **contour feathers,** which are the large feathers of the wings, tail, and body; **down** feathers, typical of young birds and found under contour feathers of adults; and **filiplumes,** which have a slender, hair-like shaft and a small tuft of barbs at the terminus. Each year birds molt, losing their plumage to some extent, and a new set of several thousand feathers grows out.

The bones of birds are hollow, a condition obviously advantageous for flying animals. Air sacs extend from the lungs into the hollow portions of the larger bones. These bones are therefore much lighter than mammalian bones, which are filled with bone cells and blood tissue. The advantages of flight cannot be overemphasized: birds utilize it in escape, locating and capturing food, finding a mate, nesting, and migrating. Nonflying birds have poorly developed wings and a flat sternum, rather than a keeled, or ridged, sternum on which the pectoral flight muscles attach.

Because of the close relationship between reptiles and birds, the latter are sometimes jokingly referred to as "glorified reptiles" or "hot lizards." This points to a major difference between the two groups: Reptiles are cold-blooded, and birds are even more warm-blooded than humans. Most birds maintain an internal body temperature of about 40° C (104° F). This is a necessity for a flying animal, which must have a high rate of metabolism to supply its muscles with oxygen and energy for rapid muscular movements. As you might expect, birds have a rapid heartbeat and circulation of vascular fluids. They have a four-chambered heart that prevents mixing of oxygenated and unoxygenated blood; a type of lung system that allows air to flow in and out at a very rapid rate; and specialized contractile elements in the flight muscles which utilize energy-rich fuels (mainly ATP) at a faster rate than normal striated muscles.

To keep all this activity going and to maintain the body temperature at such a high level requires a great deal of energy, derived ultimately from food which the bird consumes. Birds feed on a wide variety of foods—seeds, insects, worms, mice, fish, fruit, vegetables, and so forth. They swallow their food whole or in large chunks and pass it into their **crop,** where it is stored and moistened. The food next enters the anterior part of the stomach which secretes gastric juices and then on into the posterior, or **gizzard,** portion of the stomach. This thick, muscular organ grinds and mixes the food, often with the aid of small pebbles or other hard objects swallowed by the bird. The partially digested food then enters the intestine for further breakdown and absorption. Such a system is valuable, since it allows a bird to gulp down large amounts of food and then retreat to a safe place to digest it.

30.21 The life cycle of the pigeon is typical of many birds.

Male and female pigeons mate in the spring, but they have no specialized copulatory organs. Sperm produced in the two internal testes of the male is transferred from the cloacal region to the **cloaca** of the female (*cloaca,* the Latin word for sewer, refers to the vent through which urine and feces pass together; gametes may move through it also). The adult female pigeon has one functional ovary. Fertilization occurs in the oviduct. Next the protein albumin (in the form of egg white), a two-ply membrane, and a hard calcareous shell are deposited around the egg as it passes through the oviduct.

30-10 Two familiar egg-laying mammals (subclass Monotremata): (*A*) The duck-billed platypus (*Ornithorhynchus*) and (*B*) the spiny anteater, *Tachyglossus* ("sticky tongue").

A   B

Pigeons lay two eggs per clutch, whereas some birds lay larger numbers. Birds which produce a fixed number of eggs once they begin to lay, are called **determinate layers.** Others, such as chickens, continue to lay if each egg is removed from the nest after being laid. These are called **indeterminate layers.**

Once the female pigeon has laid two eggs in the nest, the male and female take turns incubating them by sitting on the nest almost continually. After about 14 days, the young pigeons peck their way out of the egg and hatch as down-covered nestlings. During early life they are fed by both parents on "pigeon's milk," a secretion composed mainly of degenerating cells lining the crop of the parent and only incidentally of bits of half-digested food. The adult sticks its bill into the mouth of the young bird and ejects the "milk" directly into its throat. When the young pigeons have become proficient flyers, they leave the nest. The following spring, when they are approximately a year old, they mate and renew the cycle.

30.22 Birds exhibit a wide range of behavior including extensive migrations.

Humans have long been mystified by birds' instinctive or innate behavior—courting, nest building, homing, and seasonal migration. Research into many aspects of bird behavior is attempting to answer such questions as: How do birds recognize potential mates or competitors? How do they establish and maintain mating territories? How do ducks, geese, and other birds anticipate impending weather or temperature changes and undertake fall or spring migrations? How do owls catch prey in the semi-dark? How do young birds, raised in isolation, instinctively know how to build the nest characteristic of their species?

30.23 The three subclasses of living mammals are the monotremes, marsupials, and placentals, examples of which are the platypus, wombat, and man, respectively.

All mammals are characterized by mammary, or milk, glands; four limbs which may be modified in a number of ways; hair at some stage in their existence, and a fairly large brain. Except for the monotremes (section 30.24), all mammals bear their young alive, and most develop a **placenta** in the uterus between the maternal and embryonic tissues.

The young require parental care for a much longer time than do other animals. Mammals are warm-blooded and have a four-chambered heart.

Current evolutionary theory ascribes to mammals a reptilian ancestry. Mammals are thought to have originated from an advanced reptile that was distinctly mammalian in appearance some 150 million years ago. Their numbers were presumably kept down by the predation of reptiles dominant at that time. When the latter declined, mammals multiplied and diversified and became the dominant terrestrial vertebrates. Present-day mammals range in size from the pigmy shrew, less than one-sixth of an ounce, to large elephants weighing over 7 tons, or the blue whale, 105 feet long and weighing 100 tons.

All living mammals have been placed in three subclasses: **monotremes,** or egg layers; **marsupials,** pouched animals which lack a true placenta; and the dominant **placentals,** which includes approximately 15,000 species grouped into 16 orders.

30.24 Monotremes are aberrant mammals found only in Australia, New Guinea, and Tasmania.

The duck-billed platypus and the spiny anteater (Figure 30-10) are the only surviving representatives of the egg-laying mammals. In many ways they seem more closely related to the reptiles than to other mammals. For example, they produce eggs which hatch outside the body, and possess a reptilelike urogenital system.

Once the young hatch, they are nursed on milk that seeps from unspecialized **mammary glands** found along the ventral portion of the female's body. There are no breasts or nipples but, like the mammary glands of all mammals, those of the monotremes appear to be derived from sweat glands.

The platypus lives in a burrow dug near a stream or river pool and is well adapted to a semiaquatic mode of life. The body is streamlined, the front feet are completely webbed, the tail is flattened into a powerful swimming structure, and the upper jaw is modified into a flat beak which is used to feed on small animal and possibly plant food in the mud below the surface of the water.

The platypus has few natural enemies, but males possess sharp, hollow spurs on the inside of the hind legs which they can use to inject a poison into antagonists. The poison is painful to man and can rapidly incapacitate and kill smaller mammals.

As the name implies, the spiny anteater has spines covering the dorsal part of its body. It feeds on ants or other insects which are captured on its long sticky tongue. These animals live on land and have powerful claws with which to burrow.

Fertilized eggs are laid by the female directly into a ventral pouch on her belly region; the young hatch in the pouch and remain there until the spines developing on their bodies begin to irritate the mother.

30.25 The young of marsupials are born essentially as embryos and are often brooded in a pouch.

Marsupials lack a placenta and the young develop in the uterus only a short time. They are usually born before eyes, ears, hair, and many internal structures are functional. However, the organs involved with smell are well formed, as are the two front feet. In the Virginia opossum, for example, the young use their front feet to pull themselves along the hair on the mother's stomach from the vaginal opening to the pouch. Once inside the pouch they grab one of the mother's teats in their round mouths and the end of the teat swells up, thereby trapping the baby in a protected and well-fed position until the mouth grows large enough to allow the teat to slip out. The young of the kangaroo crawls into its mother's pouch and is fed in somewhat the same manner.

At one time, marsupials were scattered over much of the world, but they have been displaced by competing placental mammals. Today, they are abundant only in Australia and the neighboring islands, because until the advent of man and his livestock, marsupials were the only mammals there and existed without competition. Consequently they diversified and developed the form and life style similar to placental animals. The marsupial Tasmanian wolf, for example, is analogous to the placental wolf, the wombat to the groundhog, and the flying phalanger to the flying squirrel (Figure 45-2).

30.26 The dominant mammals, the placentals, are nourished within the mother's uterus by means of a placenta.

The placenta is an organ that develops from the fertilized egg and becomes attached to the wall of the uterus, where it effects an exchange of nutri-ents, gases, and wastes between the fetus and the mother. This allows the embryo to remain within the mother's body with an unlimited food supply for a long enough time to develop a large brain and other complex systems. A pronounced development of the brain is characteristic of most mammals, especially the primates (monkeys, men, and tree shrews).

There are 16 orders of placental mammals (Figure 30-11). It should be remembered that the groupings are somewhat arbitrarily set up by man in an attempt to make some sense out of the array of mammals that exist. Future observations, collections, and experiments may prove that some animals are misplaced.

The **insectivores,** or insect-eating mammals, include the moles, shrews, hedgehogs, and their relatives, and are believed to be among the more primitive placentals. Most moles and shrews are nocturnal and spend their lives in a burrow, which they can construct at a great speed. Other insectivores are aquatic or live in trees. Shrews have the highest metabolic rate of any mammal.

Bats (order Chiroptera) are the only flying mammals. Nearly all of the bats native to this country feed on insects that they catch at night. Bats navigate in the dark and manage to avoid hitting obstacles with a kind of radar system: Their sensitive ears pick up echo patterns from their own high-pitched squeaks as they bounce back from insects or other objects in their path. Bats feed mainly on moths, and some species of moths have evolved "bat radar detectors," special auditory receptors "tuned" to the frequency of the bat's squeaks. When the bat radar is on, the moths go into an erratic flight pattern to avoid being caught.

The wings of bats are composed of thin membranes extending between the elongated forelimb segments and stretching back to the hind limbs. The limbs all terminate in claws which makes hanging upside-down easy.

Sloths, anteaters, and armadillos belong to the order Edentata—"toothless ones." Sloths live in the tropics of Central and South America and spend most of their time suspended upside-down from a tree limb. Their claws are modified for holding on in this fashion, and their hair is sloped in the opposite direction from most other mammals so that it will shed rain and dirt while the animal is upside-down.

Anteaters have a snout fitted with a long, sticky tongue to capture ants. The front claws, curved and strong, are used to tear open anthills. Armadillos are protected by bony plates and live in Texas, its neighboring states, and South America.

Common mole (*Talpa*)—
order Insectivora

Big-eared bat (*Plecotus*)—
order Chiroptera

Beaver (*Castor*)—order Rodentia

Armadillo (*Dasypus*)—
order Edentata

Tiger (*Felis*)—order Carnivora

Rhinoceros (*Rhinoceros*)—
order Perissodactyla

African elephant (*Loxodonta*)—
order Proboscidea

Camel (*Camelus*)—
order Artiodactyla

Hare (*Lepus*)
order Lagomorpha

Gorilla (*Gorilla*)—order Primates

Blue whale (*Balaenoptera*)—order Cetacea

30-11 Familiar representatives of 10 orders of placental mammals.

Elephants, the largest land animals, are members of the order Proboscidea (*proboscis* means "nose") and are native to only two locations, India and Africa. The nose or trunk is greatly elongated and is used for smelling, feeding, breathing, and fighting. The tusks of elephants, actually their canine teeth, grow to great lengths and are prized as ivory. The size and weight of the elephant skull, especially when bearing tusks, is enormous, and even the oversized muscles in the neck are taxed to lift it. Elephants have evolved structural modifications to help solve this problem. Air spaces are found within the skull which cut down weight without reducing the effective size of the head.

It may be true that elephants are not very intelligent animals, but their slowness has probably been exaggerated. They can be trained to do civilian and military work or to perform in a circus. In the wild they usually live in herds, probably for protection and mating purposes. Gestation lasts from 18 to 21 months, and a single calf is born. Elephants rank with man in longevity, and 70 years is not an uncommon age. The big animals eat fantastic quantities, and a single large specimen in captivity may consume 500 pounds of hay per day.

"Odd-toed," that is, hoofed, animals belong to the order Perissodactyla. Horses, zebras, tapirs, and rhinoceroses all have legs which terminate in what is actually a modified third digit of the foot. The other digits are much smaller and do not help support the weight of the animal. True horns never occur in this group, although rhinos develop a "horn" of matted hair.

The evolution of the horse is probably better known than that of any other vertebrate and can be traced back to the Eocene period, some 60 million years ago. Many primitive horses were smaller than sheep and did not resemble those living today in the least. Modern breeds of horses are all of the same species and were bred by man to emphasize particular characteristics, such as size, strength, and speed. Domestic breeding in this order has also produced the mule, a sterile offspring of a mare (female horse) and a male donkey.

Even-toed or cloven-hoofed mammals (order Artiodactyla) include pigs, camels, hippos, deer, giraffes, cows, goats, and sheep. The weight of the animal is carried on the third and fourth digits of the foot, which are modified for this purpose. Man has domesticated nearly all the families of this group and uses them primarily for food. Oxen, camels, and llamas are sometimes used for transportation and for plowing or driving a pump or thresher.

Camels are well adapted for a desert life. When they go for long periods without drinking, they obtain metabolic water from oxidizing fat in their hump. They also have specialized kidneys, which conserve most of the water other animals would lose as urine. A camel's body temperature adjusts somewhat to external conditions, and the animal starts to sweat only at high temperatures. Its wooly hair is a good insulator against heat.

Cows, deer, sheep, goats, buffaloes, and giraffes are **ruminants,** even-toed mammals which have a three- or four-chambered stomach. Food, generally grass, is passed along with much saliva into the **rumen** portion of the stomach, where it lies until the animal finishes feeding. Then it is returned by regurgitation to the mouth, where it is chewed as **cud.** When fully masticated, it is swallowed again and this time passes into another part of the stomach for further enzymatic digestion. It goes on to the intestine, where absorption occurs. This process appears to be an evolutionary adaptation which allowed ancestral animals to eat in a great hurry and then retreat from exposed areas where predators might be present. The cud could be chewed in safety at a later time. (See section 30.20 for a similar adaptation in birds.)

Whales and dolphins (order Cetacea) are large mammals that returned to the sea. Their distant ancestors were land mammals, as can be seen in the bones of a cetacean's flippers, which still resemble a limb, not a fin. Cetaceans have strong tails for propulsion and steering; nostrils located on top of the head; small ear openings; smooth streamlined body surfaces with few hairs; and a thick layer of fat under the skin which adds buoyancy and protects them from cold temperatures. Some whales can dive to depths in excess of 3000 feet, but all must surface at intervals to breathe. Mating, birth, and nursing of young are accomplished in the water.

The blue whale is the largest living mammal and specimens have been taken which measured more than 105 feet long and weighed over 150 tons. The young may be 23 feet long at birth and consume half a ton of milk per day. Many small whales feed on fish and other sizable animals, but most larger species feed on plankton, which they strain from the water with whalebone (baleen) sieves along the jaws.

Members of the order Carnivora, the meat-eaters, include bears, skunks, cats, fur seals, and walruses. The tusks of the walrus are well-developed canine teeth and are used to dig up mollusks from the mud and for climbing over ice in northern habitats.

Many carnivores have different colored coats in summer and winter and animals in these "phases" were first thought to be separate species. The same species of fox, for example, may be gray, brownish-red, or white, depending on the time of the year.

Some carnivores living in temperate and subarctic regions hibernate during the winter months. Animals such as woodchucks and bears build up body fat in the autumn, then retire into their burrows or caves when the weather turns cold. The body metabolism decreases, the heart rate goes down, and the animals need little energy during the time of the year when feeding would present a major problem. Hibernation is broken intermittently until spring, when the animals emerge, remain active continuously, and begin to feed voraciously.

Rodents (order Rodentia) are the largest group of mammals in terms of numbers of individuals and have a great range in size and habitat. Most rats, mice, gophers, squirrels, and beavers, and so on feed on leaves, stems, seeds, or roots of plants, and many live in burrows in the ground. Unfortunately, rodents often compete with man for available food: It has been estimated that, along with insects, they destroy half of all stored grains in the world every year. In the major urban areas of this country, rats have become a threat to public health. Many rodents have prospered extremely well in association with humans, and their high reproductive rate makes them a major problem.

The hares and rabbits constitute the order Lagomorpha. Although sometimes grouped with the rodents, they are different enough in structure and ancestry to warrant separate status. These animals are known for their rapid rates of proliferation and their susceptibility to predation by many animals, including man. Both wild and domesticated species have been introduced into many countries of the world and have become pests in certain regions, such as New Zealand and Australia, where they compete with sheep for valuable grazing land.

Primates (the "number one" order) include lemurs, small night animals of India, the East In-

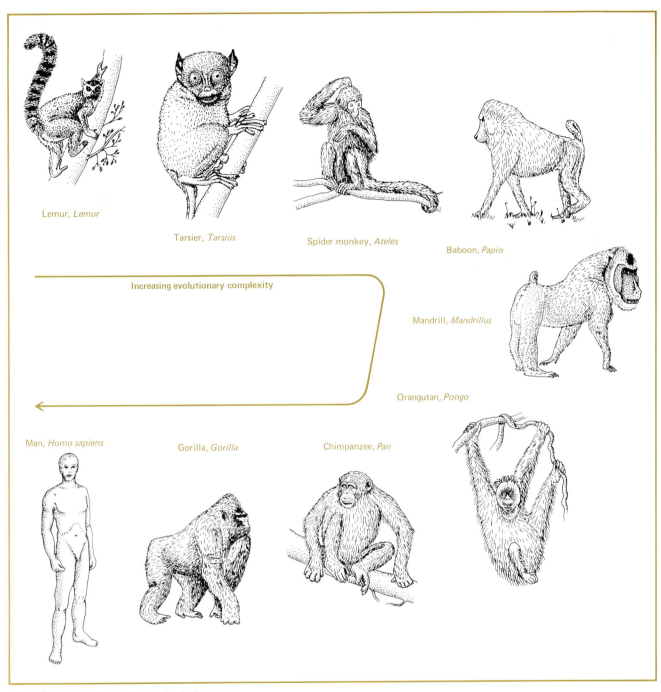

Lemur, *Lemur*

Tarsier, *Tarsius*

Spider monkey, *Ateles*

Baboon, *Papio*

Increasing evolutionary complexity

Mandrill, *Mandrillus*

Orangutan, *Pongo*

Man, *Homo sapiens*

Gorilla, *Gorilla*

Chimpanzee, *Pan*

30-12 Representative primates, including man.

dies, and Africa which live in trees and feed on fruits and insects; tarsoids, animals almost identical to lemurs but with huge eyes and a better-developed brain; and anthropoids (Figure 30-12).

The **anthropoids** are further subdivided into those with prehensile tails and flat noses, such as the spider monkey, the small marmoset monkey, and

the howler monkey; another group of Old World monkeys, with larger and more prominent noses, such as the baboon, macaque, and mandrill; apes, which include the gibbon, orangutan, chimpanzee, gorilla, and humans. Humans will be discussed in detail in Part V.

31

Behavior

What is behavior, how is it studied, and does it occur in all kinds of organisms?

What is a reflex? a tropism? a taxis?

What trends can be seen in the evolution of behavior?

What role do biological clocks play?

What types of behavior are inherited, learned, or a combination of both?

What are the simplest forms of learning?

What is memory?

How do motivational drives affect behavior?

In what ways do animals communicate?

31.1 Everyone knows what behavior is, but in very few examples can it be scientifically understood.

People, anteaters, tropical fish, paramecia, and even bacteria exhibit various kinds of complex **behavior.** All share in common the fact that the behavior causes an immediate change in the relationship of the organism to its environment. In most cases, behavior is **adaptive;** that is, it improves the chances for survival of the species, if not also of the individual.

It is not enough to know that behavior is adap-

tive. Biologists would like to know the **mechanism** by which a specific behavior occurs as well as the function that it serves in the life of the organism. In order to change in relationship to its environment, an organism must be able to sense—and perhaps even measure—features of the environment and respond to its sensations. What does the organism sense? How? What parts of the nervous system are involved? Are hormones or other chemicals involved? When a number of such questions can be answered, it will be possible to say that a particular type of behavior is **scientifically understood.** At present, we are a long way from understanding most types of behavior, especially in humans.

Perhaps because humans are complex, they sometimes feel intuitively that they "understand" simpler animal behavior in a subjective way. Explaining the singing of birds on a spring morning as due to their happy, carefree nature is an example of **anthropomorphism,** the tendency to explain the action of animals in human terms. This tendency is the very antithesis of science, and a sin to be rigorously avoided in trying to understand any aspect of animal behavior scientifically. It happens that birds sing as part of their territorial and sexual behavior. People exhibit territoriality too, of course, and might have selected that explanation as their anthropomorphic guess. An experiment can test which is correct, and it has been shown that a tape recorder playing bird songs is more than likely to be attacked, if brought into the forest, by the bird "in charge" of that territory. Clearly, however, it is difficult to describe animal behavior without using words that are full of human connotation.

31.2 Scientists have found more than one pathway leading to the understanding of behavioral phenomena.

The first step in understanding a particular type of behavior is to describe it in such a way that others, who may not have experienced it first hand, can visualize it. Thus descriptive studies always play a part in the early phase of trying to understand behavior. This is true regardless of which experimental approach is followed.

One experimental approach is to try to isolate the behavior of interest and study it under laboratory conditions. This is the approach of **psychology**; it has been a powerful tool in delineating the roles of sensation, learning, and memory in various types of behavior. Psychology has been moderately successful in the study of some aspects of human behavior and that of higher vertebrates. It has been somewhat less successful in the lower metazoans and protists, and in those higher organisms in which the ecological setting is of special importance.

A second approach is to study behavior in its ecological setting rather than in the laboratory. **Ethology** is the science of behavior in its evolutionary and ecological contexts. It attempts to concentrate primarily on the behavioral interactions of a species with its environment, living and non-living. An example of the ethological approach is the study of how bees communicate with other bees in the same hive about the location of food. The German ethologist Karl Von Frisch experimented by placing food for the bees in different locations around the hive. He then marked those bees that visited the food and observed their behavior after they had returned to the hive. By repeating the experiment with the food in different locations, and by analyzing the changes in behavior, he was able to decipher the "language" by which bees communicate both the direction and distance of the food from the hive. The language consisted of variations on two forms of simple "dance" in which both of these indispensable direction-finding parameters were communicated.

Studying behavior in the field is a valuable and important step, because it often reveals the roles of sensation, memory, learning, insight, social interactions, and so forth, in behavioral phenomena. Combined with laboratory studies it can clarify the relationships between **learned** and **inherited** ("instinctive" or innate) behavior. The limitation of ethology is that it does not always lead to the study of mechanisms.

Physiology is the attempt to sort out all the functional factors—chemical, physical, and neuronal—that underlie a given type of behavior. Good physiology starts with the information available from the classical ethological approach and goes deeper, with more penetrating tools. A simple example of the success of the physiological approach is the electrophysiological work that led to the understanding of why the protist *Paramecium* exhibits an "avoiding reaction" when stimulated. The stimulation causes the cell membrane to become suddenly permeable to calcium ions. This causes the direction of ciliary beating to reverse. Such simple results often have wide-ranging significance for other kinds of cells.

Physiologists interested in the mechanisms of behavior often study the action of neurons and their communication in collections of nerve cells called **ganglia**, or in larger groups of nerve cells, such as those that constitute a brain. The more complex the nervous system being studied, the more indirect the methods typically become. For obvious humanitarian reasons, experimentation on the human brain must be carried out using nondestructive methods, such as **electroencephalography**, the measurement of "brain waves"—the electrical signals emitted from the brain during nervous activity. Several decades of research using increasingly sophisticated equipment have resulted in much knowledge about what different kinds of electrical signals mean in terms of the normal and abnormal function of the brain.

Although no responsible physician or scientist would purposely damage a human brain, it often happens that accidents, gunshot wounds, tumors, and diseases damage or destroy a part of a brain. For a long time, the effects of accidental damage have been carefully studied and correlated with postmortem examination to obtain valuable clues on how the brain works. Moreover, since the brain is insensitive to pain during brain surgery, neurosurgeons have sometimes allowed patients to remain awake during operations so that they could report sensations, "experiences," emotions, and evoked memories as surgical tools are used to probe the brain for a tumor or foreign object.

31.3 Other organisms besides animals exhibit behavior.

Unicellular organisms respond to light, chemicals, oxygen, and nutrients by changing their pattern of locomotion. In unicells, however, there is only a fine line to be drawn between behavior and physiological adaptation. If a paramecium changes its direction repeatedly in response to some chemical in the medium, this is called behavior, because it causes an immediate change in the relationship between the organism and its environment. However, if the cell altered its membrane permeability or metabolism in response to a chemical, that would be called **physiological adaptation.**

It is often very risky to apply the same behavioral concept or term to organisms that are very different in their organization and complexity. For example, **habituation** is the term used to describe a simple learning situation in which the organism learns *not* to respond to innocuous stimuli. Amoebae subjected to repeated irradiation by strong white light gradually lose their ability to respond. However, it is doubtful that the mechanism of habituation in amoebae is at all similar to that in an animal that gradually learns not to respond to a loud noise.

Chapter 20 describes a number of simple behavioral responses among members of the Monera and Protista kingdoms. Among the unicells that have been most studied are *Amoeba, Paramecium,* and *Euglena.* These organisms exhibit a number of characteristic reactions to stimuli and in some cases show some form of **taxis,** a continuous movement toward or away from a stimulus. *Euglena,* for example, shows a **positive phototaxis,** meaning that it moves toward light. *Amoeba* has a **negative phototaxis,** since it moves away from light. Protists have a limited behavioral repertory, but do respond adaptively to the presence of food, oxygen, carbon dioxide, light, and pollutants of various kinds. Their simple behavior patterns allow them to select environments that best enable them to maintain **homeostasis,** or a constant internal environment.

Biologists have given comparatively little attention to the behavior of plants. Plants respond slowly, by growth movements, to certain environmental influences such as light, humidity, gravity, and nutrients. These slow responses are called **tropisms.** Plants also show a number of rhythmic movements that are light-mediated, or osmotic responses of cells that cause motion by shrinking or swelling. These movements are rarely of any demonstrated adaptive value, and they are slow in comparison to the responses and movements of protists and animals. Certain plants do have dramatically fast behavior, such as the sensitive plant *Mimosa,* whose leaves suddenly fold in response to being touched.

The cells of many plants respond to various kinds of stimulation. Electrical or mechanical stimulation of *Nitella* cells produces an electrical response in the plasma membrane that somehow turns off cytoplasmic streaming temporarily. Cells of the alga *Mougeotia* respond to light of various intensities by changing the orientation of chloroplasts so that they absorb more or less light.

Fungi also exhibit a number of rapid responses that alter their relationship to their environment. However, it is difficult to decide whether these responses are behavioral or physiological—if indeed there is a difference.

It is of course in the animal kingdom that behavior plays a major role in survival. In Part IV of this book, the behavioral repertories of several representative organisms were discussed in relation to their body plan and neuroanatomy. In the present chapter some generalizations will be introduced that may help explain a few salient facts about the evolution of behavior that has accompanied the structural and physiological evolution of the organisms that comprise the animal kingdom.

31.4 The lower invertebrates, the acoelomates and pseudocoelomates, have limited behavioral repertories due to the small number of neurons in their nervous systems.

Hydras, jellyfish, corals, and other cnidarians are the most limited of all metazoans in their behavioral repertory. The reasons are two: They have relatively few neurons, and these are connected mainly in a loosely organized **nerve net.**

Planarians and other flatworms (phylum Platyhelminthes) have a far more efficient nervous system; not only do they have more neurons, but their cell bodies are brought together in the cerebral ganglia of the head region. The concentration of nerve cells in the head (**cephalization**) is of obvious advantage to these organisms in searching for food, in feeding, in responding to various stimuli, and in mating. In fact, the flatworms are apparently the most primitive organisms in which learning and memory can be demonstrated. For this reason they have been used a great deal in behavior experiments.

Although the pseudocoelomate nematodes, rotifers, and gastrotrichs have a more advanced body plan and carry out some behavior with great efficiency, their behavioral repertories are also limited by the number of neurons in their nervous systems. Many have only a few hundred nerve cells, but these are adequate for the control of feeding (mostly grazing on microorganisms), mating, and defense. In the past few years these very simple nervous systems have become the object of study by scientists who hope to gain an understanding of one simple nervous system.

It is generally believed that the lower invertebrates are the most primitive organisms in which a **reflex** occurs. A reflex is an automatic response to a **stimulus;** and a stimulus is a sudden change in the environment that evokes a **response** that is mediated by neurons. In multicellular animals, reflexes involve at least two neurons, and usually more. In humans, the knee-jerk reflex requires two neurons in a circuit that carries the nerve impulse to the spinal cord. A sensory nerve transmits the impulse to a motor **neuron** leading to the extensor muscle of the thigh which causes the lower leg to swing forward. (See Chapter 32 for a more detailed discussion of reflexes in man and their relation to conscious control of the musculature.) Some human reflexes are remarkably complex and involve hundreds of neurons.

Although protists respond to sudden stimuli, no neurons are involved, so this behavior is not strictly considered a reflex. As an example of response in protists, the horn-shaped ciliate *Stentor* is normally extended when feeding, but can contract in milliseconds if its surface is touched by a predator. The cell membrane of *Stentor* responds in the same way as the cell membrane of a neuron in a multicellular animal. *Stentor* has no muscles but does contain contractile fibrils that respond directly to calcium ions that penetrate through the membrane when it is stimulated.

The lower invertebrates also show evidence of sexual behavior. Flatworms are **monoecious** (bisexual) and mate by **copulation,** the exchange of male gametes. This type of behavior requires that two worms of the same species recognize one another and maneuver into such a position that their genital openings are positioned properly for gamete exchange. In some rotifers the sexes are separate and have a different form (**sexual dimorphism**), each with a characteristic sexual or mating behavior. In the common rotifer *Brachionus*, for example, the smaller male swims in tight circles around the female before he literally punctures her body wall with his penis in order to introduce sperm into her body cavity.

major evolutionary trends toward more complex patterns of behavior can be discerned. First and foremost is the tendency on the part of the more advanced organisms to have their neurons concentrated in the head region (a trend called cephalization). This trend is visible chiefly in the phyla Annelida, Arthropoda, and Chordata, while the Echinodermata and most Mollusca have not become cephalized, and behavior of members has remained relatively simple. With these exceptions, the trend has been toward either more or larger ganglia, in which neurons can communicate in progressively larger **networks.**

A second trend has been toward increasing reliance on sensory input. Evolution has resulted in the perfection of such exquisite sensory organs as the eyes of humans, octopi, hawks, and owls; all are radically different variations on a common plan, and each highly specialized for the behavioral functions carried out by each organism. Special receptors have also evolved to detect tactile and vibrational stimuli, the direction of the gravitational field (statoliths), infrared radiation (in certain snakes), specific chemicals (e.g., taste and smell), and even magnetic fields (in birds).

A third trend is toward cyclic or rhythmic behavior, either **circadian** (from *circa diem,* "about a day") or **circannual** ("about a year"). Circadian rhythms are **endogenous;** that is, they are due to some kind of time-measuring activity inside the organism. This can be seen by isolating the organism as well as possible from environmental cues. Those organisms with endogenous daily rhythms then exhibit what is called a **free-running period** of rhythmic activity, showing remarkable precision in their ability to measure time.

The fourth and fifth trends are to some extent mutually exclusive, although they coexist to a degree. One is the trend toward increasingly stereotyped and rigid genetically determined behavior. The other trend is toward a complex nervous system allowing more flexible behavior and the capacity to acquire new forms of behavior through learning and by social interactions, including those that shape human culture.

31.5 Representatives of the coelomate phyla exhibit several trends toward more complicated and interesting patterns of behavior.

Among the coelomate phyla (Annelida, Mollusca, Echinodermata, Arthropoda, and Chordata), several

31.6 Rhythmic behavior, widespread among the eucaryotic organisms, plays an important role in the physiology, reproduction, behavior, and adaptation of higher animals.

It was once believed that only metazoan organisms could exhibit rhythmic activity—often called "**bio-**

logical clocks," as any rhythmic phenomenon obviously measures time. However, research of the past decade has shown that biological clocks exist in representatives of all four kingdoms of eucaryotic organisms. Thus rhythmic activity is not a property only of nervous systems, but of many kinds of biological material.

Among the protists, *Euglena* and *Gonyaulax* have been shown to have circadian rhythms. The latter organism exhibits cyclic changes in the frequency of **bioluminescent** flashes that it emits. Among the fungi, the geneticists' favorite bread mold, *Neurospora,* has certain mutant forms that show a characteristic growth period of about 24 hours. Plants have daily rhythms in the opening and closing of leaves and flowers. They also undergo changes in their sensitivity to light; this cycle is important in the induction of flowering.

A great many marine animals exhibit similar circadian rhythms. For example, many species of planktonic crustaceans undergo **vertical migration.** They spend the daylight hours in the ocean's deeper layers, but come to the surface at night. Oceanographers first detected these movements as sonar "echos" that moved. Many of these same organisms exhibit free-running vertical migration cycles of approximately 24 hours.

Lunar rhythms are also found in nature. An example is the once-a-month mating behavior seen in the polychaete annelid clam worm, *Nereis,* which lives in the soft, muddy bottoms of marine inlets. Females swim to the surface and begin to move in circles; males are attracted to them and swim in wider circles around them. Both sexes shed their gametes together in the same cubic meter of sea water, instead of randomly along the bottom as many marine organisms do.

Circannual rhythms are also found among the invertebrates. For example, in Japan there is a sea lily (ophiuroid echinoderm) that releases its gametes once each year. It is not known what external or internal influences cause this to happen in an entire population on the same day of the year.

Both circadian and circannual rhythms combine with the incredible sensory acuity of some birds to make possible migratory behavior and navigational accuracy that continue to defy human understanding after decades of study. It is known that many species of birds migrate from northern North America to winter grounds in the tropics or in the southern hemisphere. Many species of birds are known to use solar navigation; to do so, they must measure time, that is, possess a biological clock or circadian rhythm in order that they can interpret the position of the sun in terms of the time of day. Experiments on birds in which the clock has been reset by keeping them on an artificial light-dark schedule have shown that when they are returned to daylight, they interpret the position of the sun incorrectly and make navigational errors. The process of bird navigation is even more complicated, because birds can use information from the state of polarization of light, geographical cues, constellations of stars, and even the earth's magnetic field to navigate!

Humans also have circadian rhythms that many are only dimly aware of. The ability of some people to awaken themselves at a predetermined time is well known. Physicians have found in addition that the effects of a given dosage of a drug depends very much on the time in the daily cycle when it is administered. It is also becoming clear to jet-age travelers that it takes at least several days to reset a human biological clock that is suddenly forced to change, for example by a flight from New York to Paris or Tokyo. The individual suffers what is popularly known as "jet lag" and is characteristically hungry, sleepy, or wide awake at all the wrong times. It is now possible to purchase certain prescription drugs which are reputed to assist individuals in overcoming jet lag. Jet lag is not an insignificant complaint; insurance company analysts noticed that executives traveling abroad experienced an unusually high mortality rate during their first few days of foreign travel. The standard advice given to travelers today is to preadapt to the new schedule if possible, and in any case, to avoid too strenuous a schedule just after a long trip abroad that involves resetting the biological clock.

31.7 In the invertebrates, most patterns of behavior are innate or inborn, although the capacity for learning can be seen in animals as primitive as the flatworm.

If a given pattern of behavior is innate, then it should be observed in a young organism of a given species that has been reared in isolation. Often it is found that certain reflexes and more complicated reactions involved in feeding, defense, or mating are carried out correctly the first time by an animal in isolation, showing that the neural mechanisms specifying that behavior have been inherited from the parents. This presumably means that natural selection, working over extremely long periods of time, has favored the survival of certain species-specific patterns of behavior. In other words, the nervous system comes "all wired" to perform certain kinds of behavior.

Insects are the most remarkable of all animals for their great variety of complex species-specific behavior. Their ganglia are few and small, but are wired for some tasks so complex that many biologists long believed that they must be intelligent. This they are not, by human standards at least. However, they are capable of some learning: Ants can learn to run a maze, and bees, as we have seen, can remember the location of a source of food at least long enough to communicate it to other bees in the same hive. The arthropod side of the animal evolutionary tree in general stresses rigid and stereotyped behavioral mechanisms that are inherited; however, even in the most stereotyped cases there may be room for modification by learning.

The same can be said about many of the organisms on the chordate side of the evolutionary tree of animals: The echinoderms and primitive chordates are, if anything, simpler behaviorally than the arthropods. However, evolution among the chordates has proceeded along the lines of increasing brain size and complexity. While innate species-specific patterns of behavior are widespread, there is a greater potential for learning. The increased brain size of higher chordates means that there are greater numbers of neurons to interact in the nervous system, and a greater possible number of neuronal circuits. In other words, higher chordates, especially the vertebrates, are "wired" for behavioral adaptability. A newborn human possesses a number of highly adaptive innate responses, including the coughing, sneezing, and gag reflexes, as well as what appear to be innate fears of height, loud noises, "bad" tastes, and foul smells. Beyond these inborn patterns of behavior, humans can develop enormously diverse patterns of behavior, depending on learning and culture. Consider such behavior patterns as throwing a javelin, diving for pearl oysters, and orbiting the earth in a spacecraft; these are determined culturally as opposed to genetically and are subject to modification through learning and communication.

31.8 Conditioned response, operant conditioning, habituation, and imprinting are considered to be among the simplest learning phenomena.

The great Russian physiologist Ivan Pavlov showed in a now-classical experiment that dogs fed meat at the sound of a bell for several trials would learn to salivate at the sound of the bell. Dogs require only a dozen or so trials to learn this conditioned response. If the bell is sounded a few times without the presentation of food, the response persists, but if continued a sufficient number of times without food, the response gradually becomes extinguished; that is, the amount of saliva secreted gradually decreases to zero.

Another form of conditioning, called operant conditioning, is a simple form of learning in which a response is reinforced when it leads to a reward. B. F. Skinner showed that a rat in a box would press a lever repeatedly after it had learned to associate pressing the lever with the automatic presentation of a food pellet. Operant conditioning can also be extinguished after a number of trials without a reward. However, after a while the response may reappear, indicating that the animal remembers the reward.

Another simple form of learning is habituation, a gradual decline in the response to a stimulus that is "not significant," that is, neither leads to reward nor is threatening. An example of habituation is the decline in the **agonistic** (aggressive) behavior of a kitten brought into a house with a friendly dog.

Imprinting is another form of learning, but one that can occur only during a very short period in development. Newly hatched ducklings, for example, become imprinted on the first moving object they see during their first day or two of life. They then follow that object wherever it goes for weeks. The famous German ethologist, Konrad Lorenz, found that ducklings would imprint on him or on a mechanical toy. By studying the phenomenon of imprinting systematically, he was able to show that there was a brief **critical time** over which it could occur. Evidently some animals, especially birds, inherit a time-limited capacity to learn a form of behavior that has an obvious adaptive value.

31.9 More complex learning apparently involves some form of memory. Just what is memory?

If a dog manages to avoid a second encounter with a skunk or porcupine, it is more than likely that its natural aggressive behavior has been inhibited by memory of past unpleasant associations with these animals. Memory is not easy to study in animals other than man, because animals nearly always lack the communication skill required to report their memories. However, there are certain situations in which animals can demonstrate their memory in a way that can be made quantitative. Many animals, including people and flatworms, can learn to run a maze in which the "correct" pathway leads to a reward. The time required to reach the reward decreases as the maze is learned, and serves as a quantitative estimate of how well the maze is remembered. Of course, the maze has to be designed with the characteristics of the particular animal in mind. Maze running is often used as an experimental tool in order to study memory. An animal that has learned a maze can then be subjected to some experimental treatment, such as drugs or electrical stimulation, to see if there is any effect on its memory.

Human memory has some fascinating subjective aspects that all of us, as humans, would like to understand. Why is it so difficult for most of us to remember telephone numbers? What happens when we "memorize" a poem or a piece of music? How is it possible for some people to have a "photographic memory," which enables them somehow to summon from long in their past experience a scene, a sonata, or a particular page of a book and describe it correctly in its every detail?

Memory is the storage and recall of some kind of information. Computer specialists store mathematical information on **magnetic memory,** and polymer chemists speak of **molecular memory.** In either case, memory consists of the maintenance of a nonrandom pattern—either in the distribution of magnetic particles on tape or in the order of monomers in a polymer. In a sense, molecules of deoxyribonucleic acid (DNA) in the cell nuclei of every human cell "remember" the genetic traits of that individual's ancestors.

Human experience has suggested that there may be at least two kinds of memory: short- and long-term. Anyone can remember a telephone number long enough to dial it, but then it is forgotten unless memorized. In committing such information to long-term memory, it is necessary to expend conscious effort by reciting the number several times, by writing it down and staring at it, or by some other repetitive method.

The ways in which we manage to remember information that we wish to learn have suggested one of the major theories of memory: that it consists of pathways among the neurons of the nervous system that are somehow reinforced by continued use. The difficulty with this theory of memory lies in the fact that one of its major predictions is not correct. It would be expected that if learning a particular task was associated with a particular pathway or "memory trace" in the brain, that trace would be wiped out if a portion of the brain containing that trace was removed. Experiments have shown that animals trained to do a certain task perform worse if a portion of the cerebral cortex is removed, but it does not matter so much which region of the brain is excised. This result has led to an alternative theory of memory: that a chemical substance may be deposited in various cells scattered over the brain, and that these substances in some way evoke memories when "activated" in some way. There is no experimental evidence that clearly discriminates between these theories. It is quite likely that memory is more complicated than we think.

31.10 Motivational drives determine what behavior an animal will exhibit at a particular time.

Protists that react to the presence of food by ingesting it probably do not do so because they are motivated by hunger. It is more likely (and less anthropomorphic to say) that they have genetically determined responses (or reflexes, in the case of metazoans) for feeding. Higher on the evolutionary scale, however, something more akin to "hunger" probably occurs. What we see and describe in a grasshopper, brook trout and raccoon, however, is **appetitive behavior** that is characterized by the search for food. The sight or smell of food then serves as a releaser for feeding, which is the **consummatory act.** Once the animal has fed, its appetitive behavior ceases.

Motivation is best defined as the set of internal factors that determine what kind of behavior an animal will exhibit at a given time. Motivation consists of a combination of **drives** which seek to satisfy the objective needs of the organism for survival. The basic drives for animals are the same as those for people: hunger, sex, thirst, and defense or self-preservation. Higher animals exhibit behavior that appears to be under the control or influence of one of these drives at a time. That is, a hungry animal is preoccupied with its search for food, and only when this drive is satisfied will it respond to sexual stimuli or thirst.

In man and other vertebrates, the **hypothalamus,** in the basal portion of the brain, is the seat of nervous activity corresponding to the interplay of the various drives. This part of the brain receives stimuli from receptors that monitor the internal environment in order to achieve **homeostasis,** that is, a stable, regulated internal environment. The hypothalamus is also acted upon by hormones that elicit certain kinds of behavior. For example, sex hormones acting upon the hypothalamus and perhaps other parts of the brain, induce sexual or mating behavior. Male (or even female) rats that have been castrated early in life and will therefore never exhibit sexual behavior have been shown to exhibit characteristic mounting (sexual) behavior when sex hormone pellets are directly injected into the hypothalamus.

31.11 Certain types of behavior are triggered by specific stimuli called releasers.

Male mosquitoes are attracted by the sound of the female's beating wings. This can be demonstrated by the fact that male mosquitoes will congregate around a tuning fork vibrating at the same frequency as the sound of the female's wings. Male fireflies are attracted to females by the frequency at which they flash in the night.

Sexual behavior in insects is sometimes initiated in peculiar ways. The male praying mantis only copulates when the female has removed his inhibitions by removing his head. Such stimuli that trigger specific patterns of behavior are called **releasing stimuli,** or simply **releasers.**

Often the releaser is a visual signal of some kind, such as the red coloration on the underbelly of a fish called the stickleback. This red coloration serves both as a releaser for female sexual behavior during mating and as a releaser for male aggressive behavior.

In many species, sexual behavior is turned on only by a chain of several releasing stimuli in which the partners alternate in the releaser role. This situation is a little like keeping sexual behavior under several locks, each unlocked in turn by a specific releasing stimulus. Such patterns of behavior are adaptively important, because they limit very strictly any tendency there might otherwise be for cross-mating among related species. In other words, they constitute behavioral mechanisms for genetic isolation.

31.12 Animals have developed a wide variety of methods by which they communicate information about themselves and about their environment.

Many organisms communicate by means of vibrations transmitted through air or water. The antennae of mosquitoes are tuned to vibrate at the frequency of wingbeats of other mosquitoes to aid in mating. Crickets and cicadas have species-specific calls that serve both for defense and sexual recognition. Bird songs play primarily a territoriality role, although they serve also to synchronize the sex drives of mating birds.

An interesting example of both "autocommunication" and interspecies communication is found in the **echolocation** of bats. Bats emit bursts of high-frequency sounds that sound to human ears like clicks. The ears of bats are exquisitely tuned to hear the reflections of their own clicks from insects on which they prey. While the clicks are not communication between bats, because there is no transfer of information from one bat to another, they do serve as communication to some of the moths that the bat is trying to capture, which are sensitive to the bat clicks and respond to them by evasive maneuvers that are more often successful than not. Some moths even emit sounds that confuse the bats' echolocation mechanism.

One of the most interesting and unexpected forms of communication by sound is that of whales. Underwater microphones have picked up veritable concerts by singing whales. The songs are remarkable both for their variety and beauty, and some samples have been published as commercial recordings and have appeared on concert programs. It is both ironic and encouraging that man, who has driven whales to the very brink of extinction, should now take both a scientific and an aesthetic interest in whale songs. There is now considerable interest in whale communication, and comparative studies of whale songs in different areas along the Pacific coast of North America have shown that the whales in different regions sing in slightly different "dialects." Very little is yet known about what information whale songs convey (if any), or what effects they have on other whales.

Animals also communicate by means of chemical substances released into the environment. J. Atema of Woods Hole, Mass. has recently discovered that an injured snail releases a chemical substance which causes snails in the region to bury themselves in the sand. The chemical is not only active at extremely low concentrations, but is remembered by snails that have been exposed to it.

Chemicals produced by animals that affect the behavior of other members of the same species are called **pheromones.** Pheromones play a role in the mating of lobsters (section 28.6), silkworm moths, queen honey bees, and cockroaches. Ants that locate food communicate their find to other ants by releasing a pheromone called a **trail substance.** Other ants follow it to the source of food. Dead, decomposing ants release a **death pheromone** that causes other ants to discard the dead body. If some of this pheromone is applied to a living ant, it too will be discarded, even though it is perfectly alive.

Animals also communicate by **visual displays** of many kinds. Aggressive behavior often takes the form of a visual display, such as that exhibited by a cat or dog; often it is answered by submissive behavior, as shown by the dog that slinks away from an aggressor with its head down and tail between its legs. Such displays serve an adaptive function because they avoid combat that might result in the death of one or both of the combatants.

31.13 Recently, sociobiologists have been attempting to explore the roots of human behavior by comparative studies on other primates and on primitive human societies.

Much has been written about "human nature" that is both contradictory and controversial. Some claim that humans are naturally aggressive, territorial, and warlike. Others think humans are fundamentally peace-loving.

The real key to human nature probably lies hidden in prehistory. Some of the questions sociobiologists would like to answer are: What factors launched humans into the rapid evolution of a large brain, intelligence, communication, social cooperation, agriculture, and finally industrialization. What roles were played by tool-making, social organization, and so forth, in the evolution of humans as we are today? In the years ahead we can hope for the growth of a science that will help us to understand our own roots, and in the process, understand ourselves, our limitations, and what our future may be.

SUGGESTED READING

Alexopoulos, C. T. and H. C. Bold. *Algae and Fungi.* New York: Macmillan, 1967.

Bayer, F. M. and H. B. Owre. *The Free-Living Lower Invertebrates.* New York: Macmillan, 1968.

Bell, P. R. and C. L. F. Woodcock. *The Diversity of Green Plants.* Reading, Mass.: Addison-Wesley, 1968.

Bonner, J. T. *The Cellular Slime Molds.* Princeton, N.J.: Princeton University Press, 1959.

Borror, D. J. and D. M. DeLong. *An Introduction to the Study of Insects* (third edition). New York: Holt, Rinehart and Winston, 1971.

Brock, T. D. *Biology of Microorganisms.* Englewood Cliffs, N.J.: Prentice-Hall, 1970.

Buchsbaum, R. *Animals Without Backbones.* Rev. ed. Chicago: University of Chicago Press, 1948.

Chandler, A. C. and C. P. Read. *Introduction to Parasitology.* 10th ed. New York: Wiley, 1961.

Cronquist, A. *The Evolution and Classifications of Flowering Plants.* Boston: Houghton Mifflin, 1968.

Dawson, E. Y. *Marine Botany: An Introduction.* New York: Holt, Rinehart and Winston, 1966.

Fingerman, M. *Animal Diversity.* New York: Holt, Rinehart and Winston, 1969.

Gamow, G. *The Creation of the Universe.* New York: New American Library, 1965.

Gardiner, M. S. *The Biology of Invertebrates.* New York: McGraw-Hill, 1972.

Grasse, P. P., ed. *Traité de Zoologie.* Paris: Masson et Cie, 1948.

Grell, K. *Protozoologie.* Berlin: Springer, 1956.

Hanson, E. D. *Animal Diversity.* 2nd ed. Englewood Cliffs, N.J.: Prentice-Hall, 1964.

Heywood, V. H. *Plant Taxonomy.* London: Edward Arnold Publishers Ltd., 1967.

Hyman, L. *The Invertebrates.* Vols. 1–6. New York: McGraw-Hill, 1940–1967.

Jahn, T. L. and F. F. Jahn. *How to Know the Protozoa.* Dubuque, Iowa: Brown, 1949.

Klopfer, P. N. and J. P. Hailman, *An Introduction to Animal Behavior: Ethology's First Century.* Englewood Cliffs, N.J.: Prentice-Hall, 1967.

Kudo, R. R. *Protozoology.* 5th ed. Springfield, Ill.: Thomas, 1966.

Kummel, B. *History of the Earth: An Introduction to Historical Geology.* San Francisco: Freeman, 1961.

Laurence, G. H. M. *Taxonomy of Vascular Plants.* New York: Macmillan, 1967.

Lorenz, K. *Evolution and Modification of Behavior.* University of Chicago Press, 1965.

Mackinnon, D. L. and R. S. J. Hawes. *An Introduction to the Study of Protozoa.* Oxford: University Press, 1961.

Margulis, L. *Origin of Eukaryotic Cells.* New Haven, Conn.: Yale University Press, 1970.

Mayr, E. *Principles of Systematic Zoology.* New York: McGraw-Hill, 1969.

Nester, E., C. E. Roberts, Jr., B. McCarthy, and N. Pearsall. *Microbiology.* New York: Holt, Rinehart and Winston, 1973.

Pennak, R. W. *Fresh-water Invertebrates of the United States.* New York: Ronald Press, 1953.

Prosser, C. L. and F. A. Brown, Jr. *Comparative Animal Physiology.* 2nd ed. Philadelphia: W. B. Saunders Company, 1961.

Romer, A. S. *The Vertebrate Story.* 4th ed. Chicago: University of Chicago Press, 1959.

Russell-Hunter, W. D. *A Biology of Lower Invertebrates.* London: Macmillan, 1968.

Russell-Hunter, W. D. *A Biology of Higher Invertebrates.* London: Macmillan, 1969.

Scagel, R. et al. *An Evolutionary Survey of The Plant Kingdom.* Belmont, Calif.: Wadsworth, 1966.

Sistrom, W. R. *Microbial Life.* New York: Holt, Rinehart and Winston, 1969.

Sleigh, M. *The Biology of Protozoa.* London: Edward Arnold, 1973.

Stanier, R. Y., M. Doudoroff, and E. A. Adelberg. *The Microbial World.* 3rd ed. Englewood Cliffs, N.J.: Prentice-Hall, 1970.

Tinbergen, N. *Social Behavior in Animals.* London: Methuen, 1953.

Ward, H. G. and G. C. Whipple. *Fresh Water Biology.* Edited by W. T. Edmondson. New York: Wiley, 1959.

Wells, M. J. *The Brain and Behavior in Cephalopods.* Stanford, Calif.: Stanford University Press, 1962.

Whitaker, R. H. "New Concepts of Kingdoms of Organisms." *Science* 163(1969):150–160.

Wilson, C. L., W. E. Loomis, and T. A. Steeves, *Botany.* 5th ed. New York: Holt, Rinehart and Winston, 1971.

Wilson, E. O. *Sociobiology: The New Synthesis.* Cambridge, Mass.: Harvard University Press, 1975.

Young, J. Z. *The Life of Vertebrates.* 2nd ed. New York: Oxford University Press, 1962.

V MAN

Any elementary course in biology should include a fairly detailed consideration of the relationships between structure, function, and behavior in one organism. Part IV treated these relationships in all plants and animals in a preliminary way that will serve as a useful background for more advanced studies in microbiology, protozoology, botany, and zoology. In selecting an organism for a more detailed study, we could have selected almost any higher vertebrate, such as the frog, cat, dog, pig, gorilla, or cow. Although the pictures would have been different, the principles would have been similar. The main advantage of studying humans is that students have a natural interest in the human body and its functions and a natural curiosity about human behavior.

Human biology, including anatomy, physiology, biochemistry, genetics, behavior, and of course the study of human health and disease, has become vast and complex, and new knowledge is being acquired at a dazzling rate. Out of this immense body of knowledge, however, it is possible to select some basic information that is interesting, provides a general concept about total body function, and indicates some of the malfunctions that can occur.

Although the human organism is convenient to study because so much is known, there is also a great deal that still remains unknown, often because the relationships between function and behavior on the one hand and human psychological and social processes on the other are not always clear.

32

Support and Locomotion

QUESTIONS TO KEEP IN MIND

What is the skeleton for, and how does it grow?

What are the properties and functions of the body's three different kinds of muscle?

What is involved in making a motion such as reaching for a book?

In Chapters 24–31 we have briefly examined members of the animal kingdom. Some types of animals have been on this earth for hundreds of millions of years. Some have populated every nook and cranny on earth with billions upon billions of their numbers. Compared to these organisms humans are extremely late arrivals and have adapted themselves to live only in fairly restricted regions. And yet, given our natural bias, we are more interested in the single species *Homo sapiens*, how its body functions, and how it relates to the rest of the world, than we are in all the other organisms combined.

The next 10 chapters will deal with two broad topics in the study of humans—the construction and function of the human body.

32.1 A human body is supported by a skeleton consisting of 206 bones.

Like other terrestrial animals, humans are constantly subjected to a gravitational acceleration of 980 cm/sec^{-2} pulling them toward the earth's cen-

ter. The human body, therefore, needs some form of support to prevent collapse. This support is provided by an internal bony **skeleton.** Furthermore, to allow an animal to move, a supporting system must be flexible or jointed, and in humans this condition is accomplished by the joining together of 206 bones in a specific manner. These bones are of different sizes and shapes and move varying amounts, depending upon their position and the muscles and ligaments attached to them. The major bones of the human skeleton are depicted in Figure 32-1. Parts of the skeleton also protect the brain and spinal cord.

Each bone has a specific structure. A long bone, such as the **femur** or thigh bone, for example, consists of an outer cylinder of hard, **compact bone** (Figure 32-2) and an inner hollow region or **marrow cavity.** Microscopically, compact bone is organized into **haversian systems,** which are sets of concentric microcylinders of bone with a blood vessel at the center of each. This arrangement contributes to the structural rigidity of the bone. Within the marrow cavity, red and white blood cells and platelets are produced constantly and are released into the bloodstream as the body requires them.

32.2 Bones grow by two different processes, which are controlled by several factors.

Growth in the length of a bone occurs by the addition of bony material to each end. This process occurs in zones of dividing cartilage cells called the **epiphyseal plates. Cartilage** is a soft tissue which,

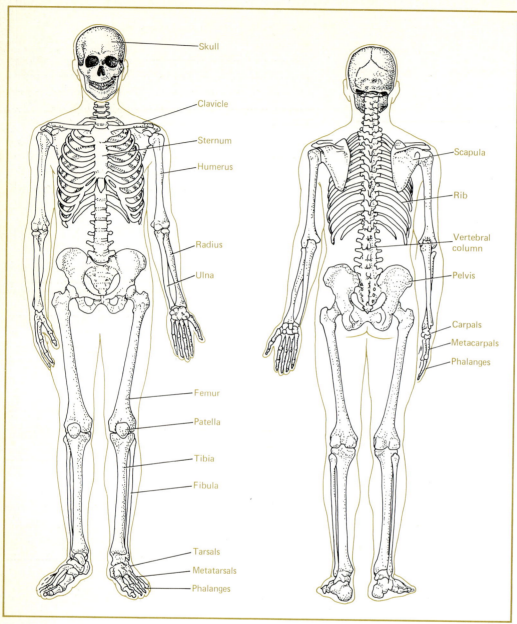

32-1 The human skeleton, anterior and posterior views.

for the most part, undergoes calcification and subsequent invasion by blood vessels and connective tissue cells. As the cartilage cells are replaced, bone is formed.

A continuous remodeling process also occurs, ensuring proper shaping as the bone lengthens. Growth in length ceases when the growing ends become completely converted to bone, usually by about age 20 in the human.

Growth in circumference occurs beneath the **periosteum**, the connective tissue sheath that cov-ers the external surface of a bone. This growth ensures that the bone will increase in strength as it increases in length.

The length a bone will attain is influenced mostly by an individual's genetic makeup and by hormones. Genes partially determine a person's height; **growth hormone**, a substance liberated from the pituitary gland (section 35.2), stimulates bone growth; and vitamin D, because it promotes the absorption of calcium from the small intestine, also influences the growth process. In a healthy body the calcium thus absorbed is transported via the bloodstream to growing bones where it is incorporated and is responsible for the hardness of the bone. When vitamin D is deficient in the diet, however, calcium is poorly absorbed from the gut. Bones continue to grow but are relatively soft. A child with this deficiency, known as **rickets,** may become bow-legged as the long bones become bent under the body weight.

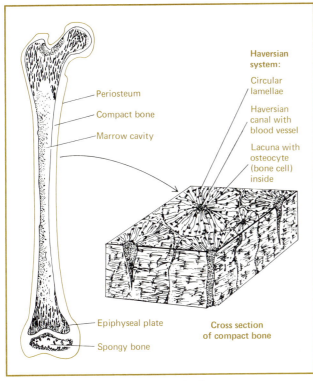

32-2 The structure of a long bone, the femur.

32.3 Joints are formed where bones meet.

A joint is formed where two or more bones come into contact with one another. Whereas some joints (such as the joints or suture lines of the skull) permit no motion of the bones that form them, other joints permit a fair degree of motion (joints of the spine and ribs) and others, in the extremities, allow great freedom of movement.

The hip joint (Figure 32-3), is formed where the femur connects with the pelvis. The bones are held together by **ligaments,** tough connective tissue fibers which completely surround the joint and form its **articular** (or joint) **capsule.** Within this capsule is a space called the **joint cavity,** filled with **synovial fluid,** a lubricant which eases the motion of one bone against the other. To further facilitate motion, the portions of the bone in contact with each other are covered by a layer of smooth, hard **articular cartilage.** Degeneration of this smooth surface, as in osteoarthritis, may result in rough, grinding, and painful joint motion.

32.4 The skeleton is susceptible to different kinds of injuries.

From time to time, the skeleton is subjected to unusual stress, and injuries may result. Ligaments may be torn slightly, as in a "sprained ankle," the cartilaginous disks between two bones (such as those of the knee joints) may be torn or split, or a bone itself may break. Fortunately, bone has extraordi-

32-3 The hip joint, (A) external and (B) cross-sectional views.

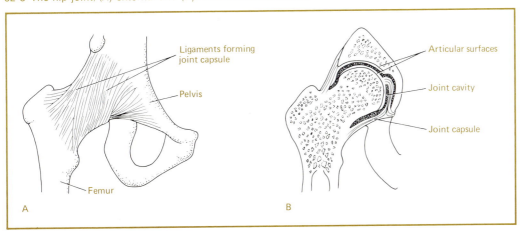

nary healing powers and, if set properly, usually heals in a number of weeks.

Man is particularly susceptible to back injuries because of his erect posture. By picking up a heavy weight while bending over a person may crush an **intervertebral disk** (Figure 32-4). This is a ring-shaped cartilaginous structure between the vertebrae of the spinal column. A crushed disk may partially slip out of place and compress or pinch a spinal nerve as it exits from the spinal cord, producing pain and eventually nerve injury. (This condition is known as a "slipped disk.")

32.5 Calcium and phosphorus are stored in bone.

Bone serves as a storage depot for various important minerals. The bone substance itself is composed of a tough matrix containing an orderly array of **collagen** fibers. This matrix gives a bone a certain resilience so that it can bend slightly. At even intervals in the collagen matrix are crystals of hydroxyapatite—$Ca_{10}(PO_4)_6(OH)_2$—a substance which makes the bone hard. As can be seen from the chemical formula, the crystal contains calcium (Ca) and phosphorus (P). Calcium in particular is important for maintaining many normal body functions, including muscle contraction (section 8.4).

Two hormones affect bone stores of calcium and phosphorus. **Parathyroid** hormone causes bone to be resorbed, thus freeing calcium into the bloodstream. **Thyrocalcitonin** promotes deposition of calcium in bone and thus removes this element from the bloodstream. These two hormones regulate the blood level of calcium at a near constancy. If insufficient calcium is absorbed by the intestine, as in rickets, the calcium in bone is sacrificed to maintain a constant blood calcium level.

32.6 The human body has three types of muscle, each with distinct properties and functions.

Muscles permit motion; they also give the body some support, much of its shape, and a means of partially regulating its temperature. Shivering, a form of muscular activity, generates heat and is an important and automatic response to reduced body temperature. The muscles recognizable from the outside of a body are **skeletal muscles**. The other two types are **smooth muscle** and **cardiac** or **heart muscle** (Figure 32-5).

Skeletal muscles, also called **striated** or **volun-**

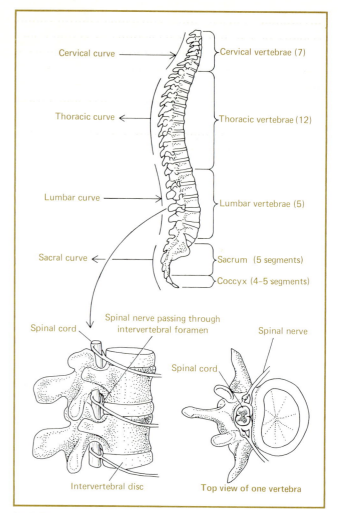

32-4 The vertebral column and its relationship to the spinal cord and nerves.

tary muscles, are attached to the bones of the skeleton by cartilaginous **tendons** (Figure 32-6) and enable the body to move in different ways. The cells of skeletal muscles are multinucleate and are larger and longer than those of smooth muscle. Under a microscope they appear banded or striated due to the arrangement of actin and myosin filaments (Figure 8-1).

Skeletal muscle is termed "voluntary" because a person voluntarily and consciously controls its action. This is possible because each muscle fiber is innervated by a nerve cell from some part of the somatic portion of the central nervous system.

Smooth muscle is found in the walls of the hollow organs of the body. In the gastrointestinal tract it helps propel food along its course. Smooth muscle is also found in the walls of blood vessels, ureters, and other structures. It is sometimes called

involuntary muscle since conscious control cannot usually be exerted over its actions. It is innervated by the autonomic nervous system. Smooth muscle cells are tapered at each end, have a single nucleus, and contain relatively few myofibrils (section 8.1). This kind of muscle contracts slowly and powerfully in response to stretch.

Cardiac muscle, like skeletal muscle, is striated, has several nuclei per cell, and contracts rapidly.

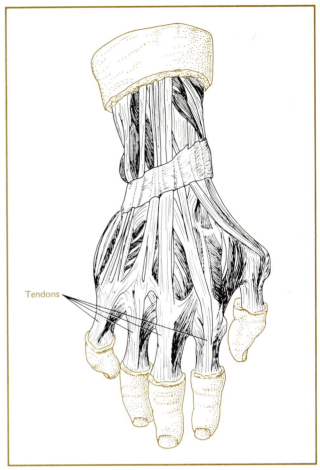

Tendons

32-6 The long tendons of the forearm muscles that move the fingers insert into the various bones of the hand, enabling finger mobility while permitting fingers to remain slender.

Unlike skeletal muscle, however, cardiac muscle cells are branched and are joined end to end by junctions called **intercalated disks.** Moreover, cardiac muscle is involuntary and contracts rhythmically in response to electrical stimuli produced by specialized "pacemaker" cells. Heart muscle does not fatigue as does skeletal muscle and is therefore able to maintain a regular beat throughout the life of the individual.

32.7 Controlled, coordinated movement is a complex process which involves both the activation of some muscle groups and the inhibition of others.

A step, a handshake, a nod of the head all require the perfectly timed and graduated buildup of tension in one set of muscles and the equally precise

32-5 The microscopic appearance of striated muscle (*top*), smooth muscle (*middle*), and cardiac muscle (*bottom*). (Photomicrographs by the author.)

relaxation of another set. Voluntary muscle action is controlled by nerve impulses from the brain. Some of these impulses activate **alpha motor neurons** in the spinal cord to produce contraction in the skeletal muscle fibers they innervate. Other impulses from the brain regulate the activity of specialized **gamma motor neurons,** and by so doing permit relaxation of muscle groups that would prevent the desired movement. The interaction of both mechanisms, one for excitation and the other for relaxation, can be seen in the **stretch reflex.**

When a muscle or its tendon is tapped, the muscle quickly contracts. This is what happens when a physician tests the knee jerk reflex by tapping the tendon just beneath the knee cap. The tap on the tendon stretches its muscle in the thigh. This brings about reflex contraction, causing the lower leg to jerk forward. Such a reflex action is caused by the presence among the skeletal muscle fibers that move the parts of the limb of striated fibers of another kind, called **intrafusal fibers.** These fibers have not only a contractile striated portion but also a central sensory portion (Figure 32-7) which houses a specialized stretch receptor. The latter is the terminus of a sensory neuron. When a mus-

cle is stretched, receptors in the sensory portion of the intrafusal fibers sense the change in tension and feed back nerve impulses along their processes to the spinal cord. Once in the spinal cord, these impulses excite motor neurons to send impulses away from the spinal cord to activate the main mass of the muscle. In this way the reflex muscle contraction is brought about. The sensory and motor neurons involved in this action form a **two-neuron reflex arc.** The same sequence of events occurs when a person touches a hot stove and jerks his hand away, except that in this case, the signal to the spinal cord comes from temperature receptors in the skin, not stretch receptors in muscle. These reflex actions are important for survival and are among the simplest elements in the behavior of animals.

If a reflex action occurred every time a muscle were stretched, however, each movement would be a spastic jerk. This does not occur because specialized gamma motor neurons adjust the sensitivity of a muscle to stretch by regulating the tension of the intrafusal fibers. When the contractile portions of the intrafusal fibers are relaxed, for example, there is no tension on the sensory portions. If a muscle is stretched slightly under these circumstances, the stretch will be taken up by the relaxed contractile portions, and the sensory portions will not react. If the contractile portions were already

32-7 The stretch reflex arc involving sensory nerve cell (*1*) and motor nerve cell (*2*), and its regulation by gamma motor nerve cell (*3*). Arrows indicate direction of impulse conduction. See text for full details.

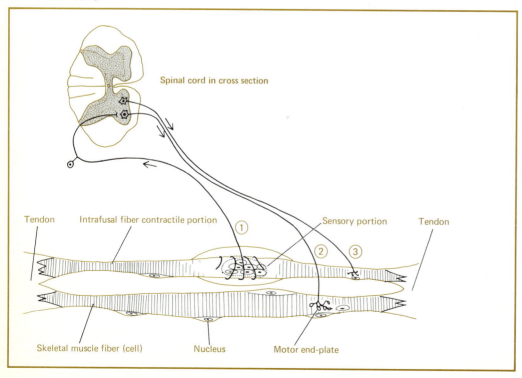

taut, however, any further stretch of the muscle (such as occurs when a tendon is tapped) would stretch and stimulate the sensory portions, causing them to generate nerve impulses.

The gamma regulatory system is operative in most if not all body movements. While taking a book down from a high shelf, for example, various muscles contract to bring about the necessary arm motions, while others that could prevent your reaching up (if they were stimulated) are stretched. If there were no gamma regulatory system, the muscles for downward arm motion would contract and pull your arm away from the shelf as soon as they were initially stretched. Impulses from the brain, however, "order" gamma nerves to permit relaxation of intrafusal fibers in this latter group of muscles. When this happens, the intrafusal sensory re-ceptors are no longer stretched and therefore do not signal for reflex contraction. Continued motion of the arm in reaching up for the book is made possible.

The successful completion of virtually any voluntary motion of the body requires the coordinated combination of these two distinct but simultaneously occurring mechanisms, each controlled by impulses from the brain. One mechanism, contraction of skeletal muscle fibers activated by alpha motor neurons, accounts for the forcible contraction of the muscle groups that move the body in the intended direction, while the other mechanism, regulation of intrafusal fiber sensitivity to stretch by gamma motor neurons, accounts for the relaxation of muscle groups that could, if activated, oppose the intended motion.

33

Integration and Regulation: The Neuron

QUESTIONS TO KEEP IN MIND

What are the characteristics of the three basic types of neurons?

How does a nerve impulse travel along a neuron and from one neuron to another?

Control and coordination of body functions is effected by the nervous and endocrine systems. Although considered separately in this chapter and the two following it, the two systems work together.

33.1 The neuron is the functional unit of the nervous system.

There are many kinds of **neurons** or nerve cells in the human body and together they help regulate such diverse functions as heart rate, respiration, and stomach secretion. Also, by stimulating muscular contraction, they help control the movement of limbs, blood pressure, and intestinal motility.

In spite of their varied functions, most neurons have certain basic structural features in common (Figure 33-1). These include (1) a highly branched collection of thin processes called **dendrites**; (2) the **cell body**, which contains one or more nuclei; and (3) a single **axon**, a process usually unbranched over most of its length and thinner and longer than any dendrite. The dendrites receive nerve impulses from other neurons and conduct them to the cell body. The axon conducts impulses away from the cell body toward another cell, which may be micrometers or even meters away.

Three types of neurons will be considered here: **motor, sensory,** and **association** neurons (Figure 33-2). One class of motor neurons in the spinal cord has relatively large cell bodies with large nuclei, prominent nucleoli, a well-developed dendritic tree, and a very long axon. These axons leave the spinal cord, and may extend a considerable distance before reaching their muscle targets. At its terminus, the axon branches, sending one branch to each of some 10 to 1000 muscle cells. Death of a single such motor neuron results in loss of function of every muscle cell it innervates. Other types of motor neurons make contact with glands and stimulate glandular secretion.

Sensory neurons have many of the same features as motor neurons, with one or two differences. The axon of sensory neuron divides into two major branches, one of which has at its terminus a specially modified sensory receptor for receiving information from the external environment. Depending on the type of sensory neuron, these receptors may react to changes in temperature, pressure, stretch, and so on, and may be located in the skin, tendons, or other tissues. Impulses from a receptor on this branch of the axon travel *toward* the cell body, bypass it, and travel away along the other major axon branch into the dorsal portion of the spinal cord. Impulses arriving at the spinal cord in this manner may be transmitted to an association neuron or directly to a motor neuron.

Association neurons are small neurons found within the spinal cord. They may connect sensory neurons with motor neurons and may be excitatory or inhibitory. If excitatory, they increase the probability that the motor neuron will fire an electrical impulse; if inhibitory, they diminish the likelihood that the neuron will fire (section 33.3).

33.2 Schwann cells are closely associated with neurons and affect the conduction of impulses.

Schwann cells (Figure 33-2) are distinct from neurons but are closely associated with them. Schwann cells wrap around axons, forming a multilayered, spirally arranged insulating sheath called a **myelin sheath** (Figure 33-3). An axon wrapped in this manner is said to be **myelinated.** Myelinated axons conduct impulses faster than unmyelinated axons because the myelin sheath alters the electrical properties of an axon. In myelinated axons, there are short unwrapped regions, or **nodes of Ranvier** (Figure 33-2), where Schwann cells meet. A nerve impulse jumps rapidly from node to node in the myelinated portion of a myelinated axon, covering a given distance in a shorter time than it would in an unmyelinated axon. Rapid conduction of nerve impulses, up to 120 meters per second, minimizes delay in voluntary action. It permits the driver of a speeding automobile to make split-second adjustments in steering or allows a baseball batter to hit a fast ball, to cite two simple examples.

A class of cells similar to and closely associated with neurons are the **glial cells,** found in the brain and spinal cord. The function of these cells is not clearly understood, but some evidence suggests they play a role in the regulation of the external environment of neurons.

33.3 Neurons are connected with one another and with muscle cells at synapses.

An axon of one neuron may functionally connect with the dendrites, cell body, or even axon of another neuron or with a muscle cell at a small connecting region called a **synapses** (Figure 33-4). Synapses are of two kinds, **chemical** and **electrotonic.** At a so-called chemical synapse, the first

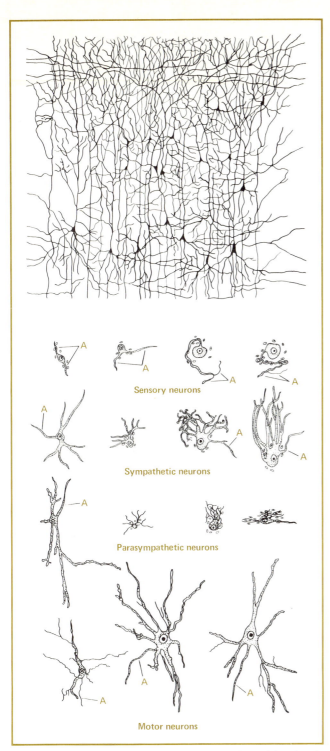

Sensory neurons

Sympathetic neurons

Parasympathetic neurons

Motor neurons

33-1 *Upper:* Arrangement of neurons in a portion of the visual cortex of a child's brain. Intricate connections among these neurons (and others not shown) make human three-dimensional color vision possible. *Lower:* Scale drawings of other types of neurons outside the brain. (*A* denotes axons.)

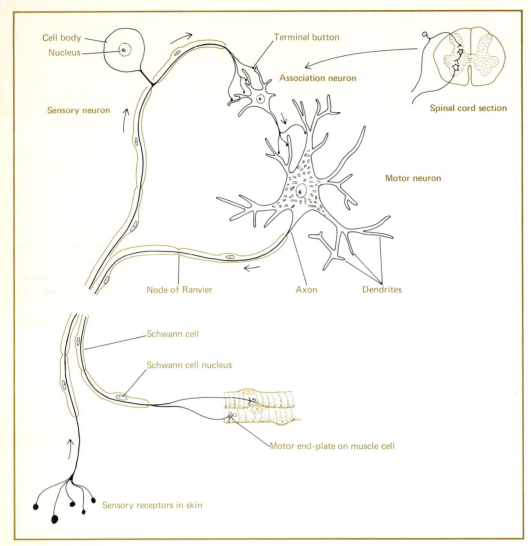

33-2 The structure of the neurons in a reflex arc. Arrows indicate direction of impulse conduction.

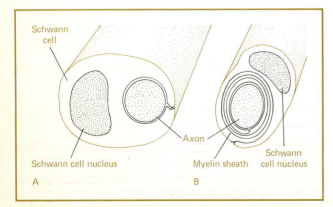

33-3 Schwann cells envelope axons and wrap their membrane many times around it to form a myelin sheath: (A) early stage, (B) later stage in the formation of the myelin sheath.

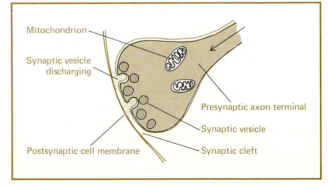

33-4 A chemical synapse. Arrow indicates direction of the nerve impulse.

neuron, the **presynaptic cell,** is separated from the second neuron, called the **postsynaptic cell,** by a gap of approximately 200 Å (Ångstroms), which is termed a **synaptic cleft.** When an impulse arrives at the end of the axon of a presynaptic cell, a chemical substance, or **neurotransmitter,** is released into the synaptic cleft, diffuses across it, and transmits the effect of the impulse to the postsynaptic cell membrane. Transmission occurs one way only, from pre- to postsynaptic cell, and the small, round **synaptic vesicles,** where neurotransmitter substance is stored, are found only in the presynaptic nerve terminal.

The neurotransmitter substances at the chemical synapses are of various kinds and have varying effect on the postsynaptic cell. Transmitters which excite a postsynaptic cell to "fire" (transmit an impulse) are released at **excitatory synapses** (and some inhibitory synapses); they include acetylcholine and norepinephrine. Transmitters which decrease the likelihood that a postsynaptic cell will fire are released at **inhibitory** synapses and may include glycine and gamma-amino butyric acid. Much research is directed toward the definitive identification of various neurotransmitters.

Transmitters exert their effects by altering the permeability of postsynaptic cell membrances, as will be discussed in section 33.4. Excitatory neurotransmitters **depolarize** (i.e., lower the electrical potential across) a membrane, bringing it nearer to its firing threshold. If this activity is sustained, or if there are many such excitatory synaptic inputs on the same postsynaptic cell, the cell may be caused to fire. Inhibitory transmitters function to **hyperpolarize** (increase the electrical potential across) a membrane, moving the resting membrane potential further away from threshold and thus rendering the cell more difficult to excite. Whether a cell fires or not is determined by the net effect of all the excitatory and inhibitory inputs impinging upon it. Commonly, hundreds of presynaptic nerve endings, both excitatory and inhibitory, converge on any one postsynaptic cell.

Various enzymes degrade the neurotransmitters once they have acted, thus clearing the synaptic cleft for transmission of a new message. In certain animal species special **electrotonic synapses** between neurons which do not involve chemicals have been discovered. Here there is no synaptic cleft, and transmission occurs directly from neuron to neuron.

We have briefly examined the events which occur at a synapse, but how does a nerve impulse actually travel along an axon?

33.4 A nerve impulse is a wave of transitory electrochemical disturbance that travels along an axon.

The unusually large neurons of the squid allow investigators to measure the electrical changes that occur as an impulse travels along the axons. Certain chemical analyses can also be performed to gain an idea of the chemical changes that occur at the same time.

By inserting a microelectrode inside an unexcited neuron and placing a second microelectrode just outside the cell it can be seen that the neuron is **polarized,** that is, it has a net negative charge inside and a net positive charge outside its cell membrane (Figure 33-5). This separation of charges by the membrane accounts for and defines the **electrical potential difference** (measured in volts) across it. This potential difference in a resting polarized neuron is termed the **resting membrane potential.**

The separation of charges and creation of the resting membrane potential is due to the **selective permeability** of the membrane to various charged particles or ions. Sodium ions (Na^+), potassium ions (K^+), and large negative organic ions or **polyelectrolytes** (mostly proteins) are principally involved. In a resting state the membrane of a neuron is relatively impermeable to sodium ions. Even though these ions are in much higher concentration outside the cell, they do not cross the membrane into the cell. Potassium ions are more concentrated inside but not enough so to overbalance the cell's net negative charge, caused largely by the negatively charged polyelectrolytes.

When a neuron conducts an impulse, its resting membrane potential is temporarily disturbed as an electrochemical change sweeps down the axon. The electrical manifestation of this disturbance (measured at the small region of the membrane where stimuli occur) is called the **action potential.** The process is initiated when a nerve cell receives a sufficiently strong stimulus of some kind which brings about a sudden change in the permeability of the membrane. In the area of the stimulus, sodium channels of atomic dimensions are opened and positively charged sodium ions rapidly diffuse into the cell (Figure 33-5). They enter so fast that the **sodium pump** (section 4.6) that normally pumps them out of the cell cannot keep up with their influx, and the interior of the cell in this region becomes positively charged relative to the outside of the cell. The action potential of the cell is thus cre-

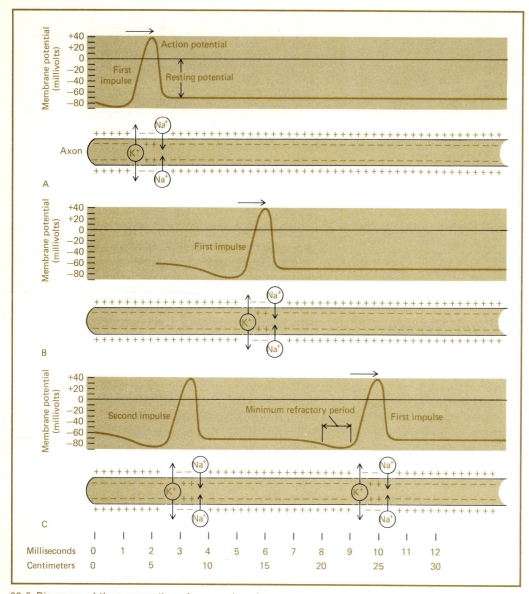

33-5 Diagrams of the propagation of a nerve impulse, or action potential, and the ionic fluxes across an axon's membranes at different phases during the passage of one impulse. *A*, *B*, and *C* represent sequential times.

ated with the reversal of the polarized state at this site.

The initial area of disturbed membrane alters the permeability of the neighboring portion of membrane and sodium ions pour in there as well, disturbing the resting membrane potential in that region. This sequence continues, and the action potential sweeps along an axon (Figure 33-5).

An action potential is only a temporary disturbance. When the impulse has passed, the cell membrane returns to its normal state and is impermeable to sodium ions. The initial negative charge inside the cell is partially restored by a transient increase in the permeability of the membrane to potassium atoms. Potassium channels are opened, and potassium ions flow out, leaving large negative polyelectrolytes inside. The negative charge of a cell is also restored by the sodium pump, which removes sodium ions that entered during passage of the action potential. Energy to fuel this active transport mechanism comes from ATP.

A short time (roughly one millisecond) must elapse after the passage of one action potential before an axon membrane can be stimulated to conduct a second impulse. This time interval is called the **refractory period.**

A nerve impulse is an **all-or-none phenomenon:**

it is either sent at a uniform strength or not sent at all. To get a neuron to fire, a stimulus of a certain minimum intensity must be applied to the membrane. The stimulus must reach or exceed the membrane's **threshold.** If the stimulus is not sufficient to bring the membrane to its threshold excitation level, no action potential results. Thus an action potential either is or is not generated.

33.5 All nerve impulses are essentially the same, and yet qualitatively different information can be conveyed by them.

When viewed on an oscilloscope, an action potential has a characteristic shape, and action potentials of all nerve axons of one type look alike. There are no "strong" impulses or "weak" impulses. How then can information of different sorts be transmitted down the same axon?

One way to vary a signal is to increase or decrease its frequency. Indeed, a stimulus that causes more impulses to arrive per unit time at a presynaptic terminal is defined as being strong. For example, if your finger rests lightly on a thumbtack, sensory nerve cells will relay the sharp sensation by transmitting impulses at a certain frequency. If the thumbtack penetrates your finger, impulses will

travel at a much greater frequency along sensory nerves, indicating a more severe sensation of sharpness. Here, degree of sharpness may have been coded in the frequency of impulses.

A nerve impulse can carry a great deal more abstract information than that, however. Suppose you are climbing a cliff with ropes, pitons, and full gear. As you climb over a ledge jutting out from the cliff, your body is partially upside down. There is a 300-foot drop beneath you. As you reach around to get a critical handhold to pull yourself over the ledge, you grab a razor-sharp rock. The sharp edge of the rock slices deeply into your hand. What happens in the nervous system at that point is an example of the remarkable control of which it is capable.

The events which might occur are depicted in Figure 33-6. First, impulses from the sensory nerves in the hand arrive at the spinal cord and signal immediately for the hand to be withdrawn. The sensory nerve releases an excitatory transmitter at its excitatory synapses, 1 and 2. The excitatory influence at synapse 1 activates an association neuron which inhibits motor neuron A and in this way tends to prevent motor neuron A from activating muscles that would keep the hand where it is. The excitatory influence at synapse 2 tends to activate motor neuron B, which if it fired would help activate the muscles for hand removal.

There are, however, other influences acting on motor neuron B. As you hang on the ledge, impulses from retinal cells in your eyes inform higher centers in the brain that there is a 300-foot drop beneath you. Impulses from your balance system in-

33-6 Decision making at the level of the spinal cord. Whether or not an action potential is fired by motor neuron A or B is decided by the net effect of all excitatory (*color*) and inhibitory (*black*) impulses impinging on each. See text for details and explanation.

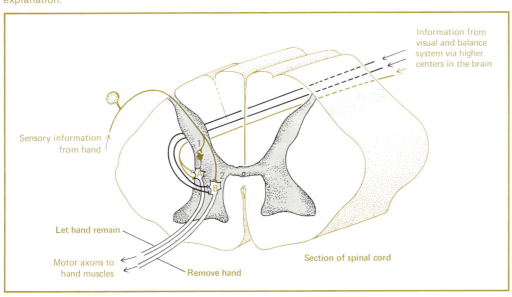

Information from visual and balance system via higher centers in the brain

Sensory information from hand

Let hand remain

Motor axons to hand muscles

Remove hand

Section of spinal cord

form the higher centers that you are partially upside down. This information is instantly computed in the brain, and appropriate impulse traffic is initiated. The net result is that impulses from the brain traveling down the spinal cord to prevent your hand from being removed outnumber those trying to excite the neurons for hand removal. Hand removal is prevented by strongly inhibiting motor neuron B. This inhibition overcomes the excitatory influence from the sensory nerve at synapse 2. Keeping the hand where it is, despite the pain, is accomplished by strongly exciting motor neuron A, thus overcoming the inhibitory influence from the sensory neuron at synapse 1.

The point to remember is that the action potentials involved in reacting to crisis situations are no different from those in shaking hands or holding a cup of coffee. The information that such action potentials carry can be qualitatively different, however, depending on the abstract significance of the summed excitatory and inhibitory influences that resulted in their generation.

34

Integration and Regulation: The Central and Autonomic Nervous Systems

QUESTIONS TO KEEP IN MIND

What are the major components of the central nervous system?

How does man keep in touch with the environment using special senses?

What are the functions of the autonomic nervous system?

Neurons operate in much the same way whether they belong to a frog, a bird, or a human being, but man's high degree of neural organization is unique.

The nervous system is somewhat arbitrarily divided into two major parts, the **central nervous system** (CNS) which is composed of an estimated 10^{12} or 1 trillion (!) neurons and the **autonomic nervous system** (ANS). In reality, both form one single system, but there are some bases for distinction which will become apparent.

34.1 The central nervous system is composed of brain and spinal cord, sensory and motor nerves leading to and from the spinal cord, and peripheral receptors.

The central nervous system (Figure 34-1) may be thought of as a central clearinghouse that integrates and controls all nerve impulses. Impulses coming from **peripheral receptors**, for example, include those coming from the special senses—sight, taste,

smell, balance, hearing, touch, temperature, and pain. Sensory nerves together with their peripheral receptors are sometimes referred to as the **peripheral nervous system.**

The **cerebrum**, consisting of two **cerebral hemispheres**, is the most highly developed portion of man's brain. Its surface is thrown into a series of convolutions or folds called **gyri** (singular, *gyrus*), with deep crevices in between called **sulci** (singular, *sulcus*), both of which first become apparent during fetal development. The outermost zone of a gyrus, containing the cell bodies of brain neurons, appears darker than the inner zone because the cell bodies are unmyelinated, and is called the **gray matter** or **cortex** (Figure 34-2). The inner zone, called **white matter**, consists of bundles of myelinated axons and/or dendrites.

Each cerebral hemisphere is divided into five regions or **lobes**, each with distinct functions (Figure 34-3): (1) the **olfactory lobe**, which is chiefly concerned with the sense of smell; (2) the **temporal lobe**, involved in hearing, speech, and memory; (3) the **occipital lobe**, concerned with vision; (4) the **frontal lobe**, which has a specialized region called the **motor cortex**, where voluntary body motions are initiated; and (5) the **parietal lobe** which receives sensory impulses from a variety of receptors.

The various lobes are connected, and their functions coordinated. Information from muscle stretch and tactile (peripheral) receptors, for example, is relayed to the **sensory cortex** of the parietal lobe, the site of conscious recognition of movement and position. This area keeps the motor cortex of the frontal lobe "informed" of the position and progress of the body. Thus, the motor cortex receives infor-

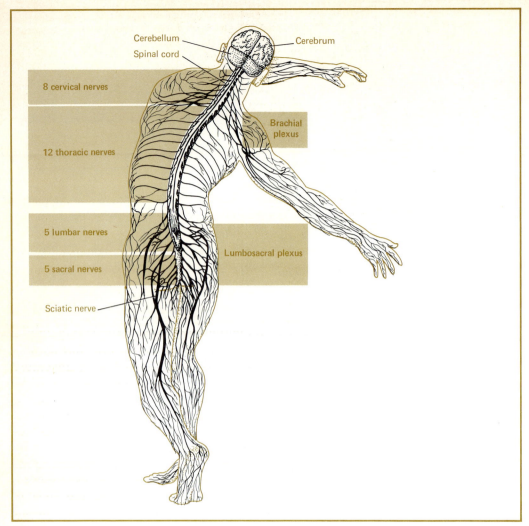

Cerebellum

Cerebrum

Spinal cord

8 cervical nerves

Brachial plexus

12 thoracic nerves

5 lumbar nerves

Lumbosacral plexus

5 sacral nerves

Sciatic nerve

34-1 The nervous system of man. The central nervous system is shown in color, the peripheral nervous system in black and white.

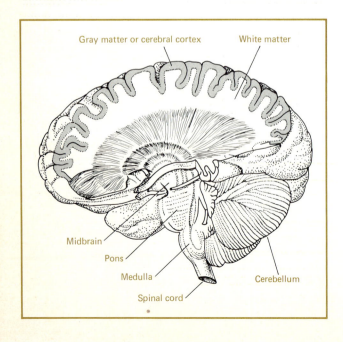

Gray matter or cerebral cortex

White matter

Midbrain

Pons

Medulla

Spinal cord

Cerebellum

34-2 A view of the brain from the side with a portion of the left cerebral hemisphere cut away to show the gray matter or cerebral cortex (made up of neuronal cell bodies) and the white matter (neuronal processes).

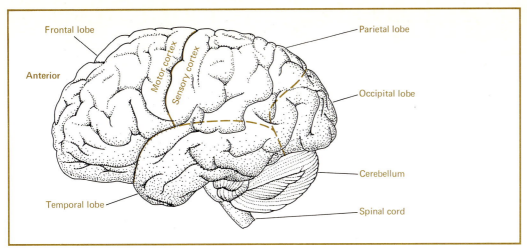

Frontal lobe

Anterior

Motor cortex

Sensory cortex

Parietal lobe

Occipital lobe

Cerebellum

Temporal lobe

Spinal cord

34-3 A surface view of the left side of the brain, showing the different lobes. (The olfactory lobe, not shown, lies beneath the temporal lobe.)

mation on the extent to which its commands to the body musculature have been carried out.

The **cerebellum** (Figure 34-2) is situated posterior to and beneath the cerebral hemispheres and makes smooth coordination of body motion possible. In reaching for a glass of water, a number of muscles in the shoulder, upper arm, forearm, and hand are brought into play. In simplified form, the action is carried out as follows:

Impulses from the motor cortex signal your desire to reach for the glass, and muscles are activated. As the motion progresses, impulses from stretch and other receptors in the shoulder and arm relay information to the cerebellum concerning the degree of contraction and relaxation of the various muscles and the position of the limb. This con-

tinuous stream of changing information is correlated in the neuronal network of the cerebellum, and new impulses are automatically, and without conscious control, sent out to the muscles involved in the movement, controlling their rates and degrees of contraction. While you are conscious of what you are doing, you are not aware of the heavy impulse traffic in the cerebellum which makes the action smooth and coordinated.

The **brain stem**, including the **midbrain, pons,** and **medulla,** connects the cerebral hemispheres and cerebellum to the spinal cord. (Figure 34-2). It is also the location of various centers of the autonomic system, such as the respiratory center (section 38.4), and the cardioaccelerator and cardioinhibitory centers.

The **spinal cord** is situated within the **vertebral canal** and extends from the brain stem to the lumbar region of the back. From it, at the level of each

BIOEPICUREAN DELIGHTS

Calf Brain Purée Viennese

Calf brain is a rich source of the protein tubulin and also contains actin, myosin, and great quantities of phospholipids. Soak two fresh calf brains in cold water for 2 hours, then soak for 1 hour in water to which 1 tablespoon of vinegar has been added per quart. Simmer for 20 minutes in an enamel saucepan with enough water to cover, containing a teaspoon of salt and a tablespoon of lemon juice per quart. Remove the brains, drain, and chop into very small pieces. Preheat the oven to 400°, and melt a half-stick of butter in a large skillet. Sautée the chopped brain in butter for a few minutes, then mix in two tablespoons of flour, and continue to stir for a few more minutes. Then add a can of chicken broth, and stir until the mixture forms a smooth paste; season to taste. To serve, pour the purée into separate baking dishes or scallop shells, sprinkle croutons or bread crumbs on top, and bake 10 minutes. Add fresh parsley and lemon wedges before serving.

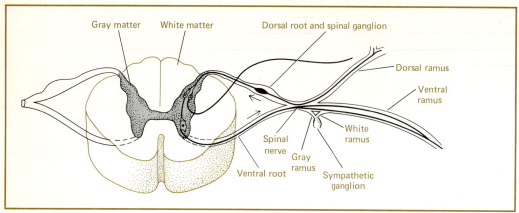

34-4 A cross section of spinal cord showing the directions of nerve impulses in spinal reflexes.

vertebra, spring the spinal nerves. These nerves are a composite of motor axons coming *from* the spinal cord in its **ventral root** and of sensory axons going *to* the spinal cord via its **dorsal root** (Figure 34-4).

In cross section, the spinal cord is composed of an inner core of gray matter, somewhat H-shaped, surrounded by an outer rim of white matter. Gray matter is characterized by the presence of large numbers of unmyelinated axons, dendrites, and nerve cell bodies. White matter derives its color from the myelin sheaths surrounding the axons in the spinal cord.

Both the spinal cord and the brain are protected both by bones (vertebrae and skull) and by a cushion of liquid, the **cerebrospinal fluid.** Cerebrospinal fluid (CSF) is a clear, slightly viscous ultrafiltrate of blood plasma produced in a series of hollow passages and chambers within the brain called **ventricles.** CSF is produced constantly and flows throughout the ventricles, bathing various brain structures and eventually returning to the bloodstream. Aside from its protective role in forming a kind of water jacket around the brain and spinal column, thus insulating these neural structures against mechanical shock, CSF may also function in a nutritive capacity. The hydrogen ion content of the CSF is monitored in the regulation of breathing.

34.2 Knowledge of CNS function is gained by experimentation and from accidents and diseases.

The function of the CNS is studied by performing experiments on animals, and by observing the be-

havior of people who have suffered accidents to or diseases of the CNS. In addition, microelectrode studies have been carried out recently as part of brain surgery. To give an example of how knowledge about the CNS is gained, if the cerebellum of an animal is experimentally removed, the animal loses all muscular coordination and jerks crazily about as it tries to move. As a further example, an accidental blow to the back of a person's head, such as might be sustained in a motorcycle accident, may cause damage to the occipital lobes, with concomitant loss of vision. Should the victim of such an accident die several weeks later, damage to the occipital lobes can be confirmed at autopsy.

A hard blow to the head may result in a **concussion,** a temporary disorder of instantaneous onset characterized by widespread paralysis. In a more severe head injury, a person may sustain a **contusion,** an actual laceration of the surface of the brain, with rupture of blood vessels and hemorrhage. The nerves in the damaged portion become nonfunctional, and blood hemorrhaging from torn vessels forms an enlarging pool or **hematoma** inside the skull. This pool of blood forms a clot which, as it expands, compresses other parts of the brain, causing a multitude of nervous symptoms. Even without injury, a blood vessel in the brain may burst due to high blood pressure. The same sequence of events (clot formation and the compression of other brain structures) may occur. This situation is called a **cerebrovascular accident** or **stroke,** and may lead to sudden paralysis, coma, or death.

Paralysis due to circulatory failure may also be produced by a thrombus, or blood clot which forms within a blood vessel in the brain, or by an embolus, a clot formed elsewhere which travels in the bloodstream and lodges in the brain. As for diseases, a severe infection of *Herpes simplex*, the virus that causes cold sores, for example, can selectively destroy the temporal lobes of a young child. Such a child is unable to hear and remember. If the child dies and the brain is examined at autopsy, the temporal lobes are found to be shriveled and greatly distorted. Thus, by correlating observed mental or behavioral defects and sensory and motor losses with structural abnormalities, physiologists and physicians gain an understanding of how various parts of the brain function.

Metabolic abnormalities and insults to the brain are just beginning to be explored. Blood glucose (sugar) concentration, for instance, must be maintained within certain limits for normal brain function. Markedly low blood glucose levels such as may occur in persons producing excessive amounts of insulin due to a pancreatic tumor may result in periods of frankly psychotic behavior. This behavior may completely disappear if the tumor is removed so that the normal glucose concentration in the blood is restored.

34.3 Specialized receptors such as the ear inform the brain about changes in the environment.

The body is constantly beset by chemical and physical events. It is bombarded by light waves, cosmic rays, and sound waves, and it is subjected to changes in atmospheric pressure, temperature, and position with respect to the ground.

The ear is designed to detect alternating waves of compression and rarefaction, which humans interpret as **sound**. Its structure is organized into

three regions (Figure 34-5). There is the **outer ear**, which includes the portion outside the skull as well as the ear canal that leads to the **tympanic membrane**, or eardrum; the **middle ear**, which houses the three ossicles or ear bones—the **malleus** (hammer), the **incus** (anvil), and the **stapes** (stirrup); and finally the inner ear itself, which contains a number of fluid-filled chambers, one of which, the **cochlea**, houses the actual sound-detecting receptors. The middle ear is connected to the throat by the **auditory** or **eustachian tube.**

In order to hear, sound waves must be converted or transduced into nerve action potentials in the auditory nerves. Sound vibrations cause the **tympanic membrane** to vibrate. This vibration is amplified 22 times as it is transmitted from the eardrum through the three ossicles to the fluid in the cochlea. Vibration of this fluid is eventually transmitted to a membrane lined with receptor cells called the **organ of Corti**. These cells send impulses via the **auditory nerve** to the temporal lobe of the brain. In this way the ear transmits information on sounds ranging in frequency from 20 to 20,000 hertz (vibration cycles per second).

Amplitude or loudness is also detected by the receptor cells. It is measured in **decibels**, the lowest audible sound being rated at 1 decibel. The decibel scale of measurement is a logarithmic one such that a 20-decibel sound is not twice but 10 times louder than a 10-decibel sound and 100 times louder than a 1-decibel sound. Sustained exposure to loud sounds (above 90 decibels) can damage groups of receptor cells and cause an irreversible loss of hearing. Such noises also foster tension which can lead to a wide variety of physical and emotional problems. A list of common sounds, with average

34-5 *A*. Parts of the human ear. *B*. Structure of the ossicles of the middle ear that transmit amplified sounds to the cochlea.

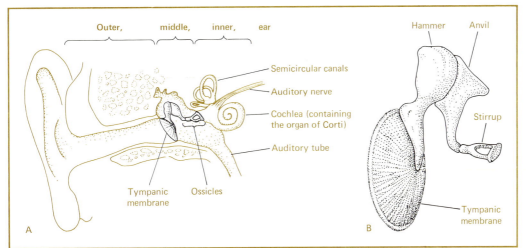

TABLE 10 WHEN NOISE BECOMES A NUISANCE
(Intensity of different levels of sound measured in decibels*)

		Decibels	
Discomfort and danger area	Pain threshold ⟶	180	Rocket engine
		150	Jet takeoff
		120	Machine gun (close range)
		120	Pneumatic chipper
		115	Party, with 4-piece rock band
		111	Motorcycle
		108	Pneumatic hammer (6 ft away)
		107	Power mower
		104	Nearby helicopter
		102	Outboard motor
		102	Outside, jet taking off at airport
		100	Heavy traffic or jet aircraft overhead
		100	Train stopping in station
		95	Subway train
		95	Inside a subway train, windows open
		94	Inside a jet airplane on take-off
		93	Food blender
		92	Screaming child
		90	Bus idling
	Hearing damage after 8 hrs. ⟶	90	Niagara Falls at base
		86	Sports car in street
		85	Garbage truck (200 ft away)
	U.S.A.F. recommended maximum noise level ⟶	85	Inside a city bus
		82	Traffic at a residential intersection
		75	Average traffic
	Telephone use difficult ⟶	70	Automobile
		60	Conversational speech
		60	Large office
		50	Quiet restaurant
		45	Homes
		40	Quiet office
		35	Library
		30	Whispering
		20	Leaves rustling in breeze
		20	Broadcasting studio
		10	Normal breathing
		0	Audibility threshold (weakest sound that can be heard by a young person with excellent hearing)

* The decibel is a logarithmic not a linear unit. 10 decibels equals 10 times the power of 1 decibel. 20 decibels equals 10 times the power of 10 decibels or 100 times the power of 1 decibel.

decibel levels, is shown in Table 10. The number of common sounds to which we are exposed which far exceed the danger level testifies to the increasing noise pollution of our environment.

The **vestibular system** of the inner ear is concerned with the sense of balance and helps in orientation. It consists of three **semicircular canals** (Figure 34-5), oriented roughly at right angles to one another. This arrangement permits the body to detect motion in the three directions of space: up and down, right and left, forward and backward. Special sensory cells detect motion of the fluid in the semi-

circular canals as the head is moved. When stimulated, these receptors generate action potentials that convey information about changes in body position to the brain.

34.4 Chemical receptors in taste buds on the tongue permit humans to taste.

The sensation of "taste" is a complex phenomenon involving both the tongue and the nose. Hold your

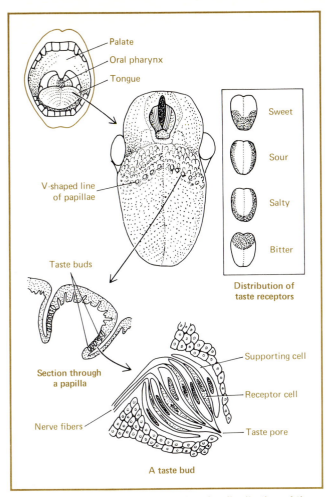

34-6 Diagrams of the tongue, showing the distribution of the four basic taste receptors, and the cellular structure of a taste bud.

nose while eating a delicious steak, and the role of your sense of smell in tasting will become immediately apparent. Although the role of the tongue in tasting is only partially understood at present, some features of its anatomy and physiology are known (Figure 34-6). The surface of the tongue itself is covered by small projections called **papillae,** on the sides of which are numerous **taste buds,** which contain **taste receptor cells.** The type of taste these receptor organs are sensitive to varies with their location on the tongue. Discrete zones on the tongue for detecting four basic tastes—sweet, sour, salty, and bitter—have been mapped out. Exactly how taste receptor cells interact with certain molecules and carry out their function has yet to be determined.

34.5 Olfactory cells are the odor receptors of the nose.

Man's sense of smell, although not nearly as highly developed as that of lower animals such as the dog, is an important means of detecting changes in the environment around him. The smell of gas in the bilge of a boat warns a skipper not to start his engine and risk an explosion, and the odor of other fuels in the home or laboratory dictate similar precautions. However, some gases that may prove fatal if breathed in sufficient concentration for any length of time, such as carbon monoxide from automobile exhaust, cannot be detected by smell.

The mechanism of smell is only partially understood at present (Figure 34-7). It involves the in-

34-7 Diagram showing the location of the olfactory receptors used in the sense of smell.

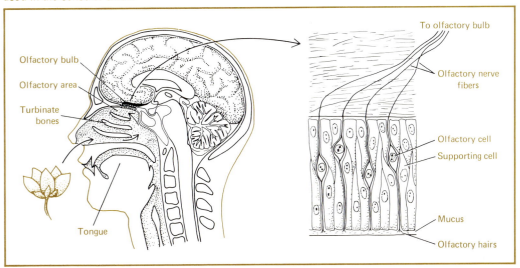

halation of a certain number of odorous molecules which react in some manner with specialized **olfactory cells** in the olfactory area. This reaction triggers impulses which are conducted to higher centers of the brain.

34.6 The eyes are the most highly developed human special sense organs, and in conjunction with parts of the brain, provide three-dimensional color vision.

A fine view of the environment is possible with the movable eyeball. Situated in its bony orbit in the skull, it is moved by six **extrinsic muscles.** Each of four **rectus muscles** moves the eye in one specific direction: up, down, to the right, or to the left. The **superior oblique muscle** moves the eye downward and outward and also rotates it slightly; and the **inferior oblique muscle** moves the eye upward and outward and also rotates it slightly. Three different nerves control these muscles, coordinating the movement of both eyes in all directions.

The human eye is more than just a light detector. It is like a camera, with an automatic iris and a photosensitive surface recording images anologous to those of high-speed, high-resolution instantaneously processed color movie film. Moreover, information from both eyes is integrated by the brain into a three-dimensional image.

The transparent **cornea** of this remarkable receptor forms the outermost window (Figure 34-8). It merges with the opaque white **sclera** which surrounds the rest of the eyeball. Behind the cornea lies the **iris,** the colored portion of the eye which responds automatically to light, like the diaphragm of a modern automatic camera. When strong light reaches the retina, the muscles in the iris react to decrease the diameter of the **pupil.** Similarly when the light is weak, the pupil widens.

Immediately behind the iris is the **lens.** Small **ciliary muscles** attached to its circumference can, by contracting or relaxing, control the lens shape and thereby regulate its focus. Where an image is focused depends on the distance of an object from the lens and on the shape of the lens. In the eye, the lens focuses images on the **retina,** the light-sensitive layer of cells at the back of the eye. The changeable shape of the lens makes it possible to sharply focus on the retina images of near and distant objects, but not simultaneously.

Between the cornea and the lens is a chamber filled with a fluid called **aqueous humor,** and the rest of the eye behind the lens is filled with a jelly-like substance, **vitreous humor.** These substances are both transparent. They may help nourish cells of the lens and cornea.

At a point slightly removed from the center of the retina is the **fovea,** a region where images are centered and focused most of the time. At another point, the **optic nerve** leaves the eye. Light receptor cells are not present in the small, round zone where

34-8 A diagram of the human eye (above), and the cellular structure of the retina (receptor layer).

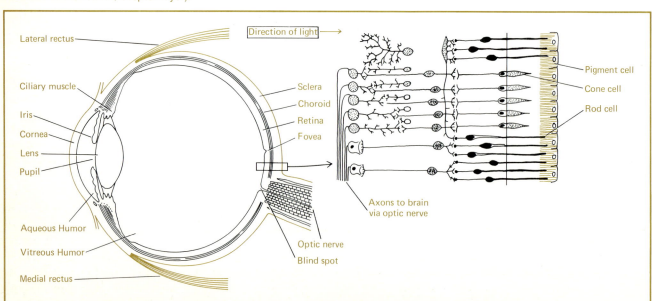

it leaves, called the **optic disk**; consequently, no vision is possible here. You can detect a **blind spot** in your own visual field by making a black dot on a sheet of paper and an *X* 2 inches to the right. Close your left eye and focus on the dot with your right. As you bring the paper closer to your face, the image of the *X* will be brought over your optic disk and will disappear.

The neuronal structure of the retina itself is elaborate (Figure 34-8). Light must pass through various layers of nerve cells before it reaches the sensitive **rod** and **cone receptor cells.** Rod cells, of which there are about 120 million in each eye, are extremely sensitive to light and are adapted to night vision. They can neither resolve details or boundaries of objects nor determine color. The six million cone cells in each eye are largely concentrated in the fovea. They have a much greater acuity and are responsible for color vision in bright light. Behind these rods and cones is the dark **pigment cell layer** of the retina, which participates in the various chemical reactions that take place in the visual process.

34.7 The autonomic nervous system regulates internal body functions.

The second major division of the nervous system, the **autonomic nervous system** (ANS), regulates the motions and activity of internal organs (Figure 34-9). The ANS is subdivided into the **parasympathetic system** and the **sympathetic system.** The nerves of the parasympathetic system arise from the base of the brain and from the very bottom or **sacral** region of the spinal cord. The nerves of the sympathetic system arise from regions of the spinal cord in between.

In the simple reflex arcs we have so far examined (section 32.7), a single motor neuron extends from the spinal cord to the muscle it activates. In the ANS, however, there are two motor neurons in each pathway, the **preganglionic neuron** leading from the spinal cord to a **ganglion** (mass of nerve cell bodies) outside the cord, and the **postganglionic neuron** leading from the ganglion to the organ it activates. The ganglia of the sympathetic nervous system lie on either side of the spinal cord and are linked together into the **sympathetic chain.** The ganglia of the parasympathetic system are farther from the spinal cord, often in or near the organs they innervate. Both sympathetic and parasympathetic preganglionic axons synapse with eight to nine postganglionic neurons, thus spreading and diffusing autonomic output.

34.8 The parasympathetic system regulates the normal daily activities of internal organs.

The parasympathetic nerves that arise from the brain stem include fibers which produce pupillary constriction, salivation, lacrimation or tearing (which keeps the surface of the eye moist and prevents it from drying), and slowing of heart rate. Fibers which produce slowing of heart action travel via the **vagus nerve,** the tenth cranial nerve, about which a good deal more will be said in subsequent chapters. Other fibers in the vagus nerve innervate the stomach and intestines and aid digestion by increasing stomach secretions and promoting gastrointestinal motility. Parasympathetic nerves that arise from the sacral portion of the spinal cord innervate the pelvic viscera, including the kidney, bladder, sex organs, and rectum.

34.9 The sympathetic system prepares the body for emergencies.

Sympathetic nerves arise from the thoracic and upper lumbar portions of the spinal cord. Preganglionic sympathetic axons may synapse with postganglionic neurons in the ganglia of the sympathetic chain as indicated previously, or may pass through the sympathetic chain without forming synapses, only to form synapses with postganglionic neurons in the so-called **collateral ganglia** closer to the organs being innervated.

Unlike postganglionic parasympathetic neurons, most postganglionic sympathetic neurons secrete norepinephrine at their synaptic endings, and their effect on the body is quite different (section 33.3). Under sympathetic stimulation, for example when a person is frightened or angry, heart rate and blood pressure are increased, and blood vessels in muscles dilate, while those in the abdominal viscera constrict. The net effect is to shunt more oxygen-rich blood to muscle where it may be needed (to run away or fight, etc.). At the same time, blood vessels in the skin constrict (the person may become pale), thus minimizing blood loss should a surface wound be inflicted. In this manner, the sympathetic system automatically and appropriately prepares the body to meet emergency situations.

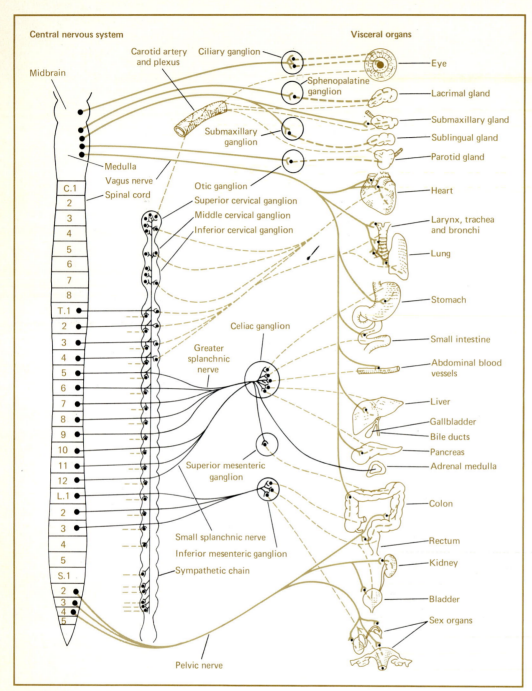

Central nervous system

Midbrain

Carotid artery and plexus

Ciliary ganglion

Sphenopalatine ganglion

Submaxillary ganglion

Medulla

Vagus nerve

Spinal cord

Otic ganglion

Superior cervical ganglion

Middle cervical ganglion

Inferior cervical ganglion

C.1
2
3
4
5
6
7
8
T.1
2
3
4
5
6
7
8
9
10
11
12
L.1
2
3
4
5
S.1
2
3
4
5

Celiac ganglion

Greater splanchnic nerve

Superior mesenteric ganglion

Small splanchnic nerve

Inferior mesenteric ganglion

Sympathetic chain

Pelvic nerve

Visceral organs

Eye

Lacrimal gland

Submaxillary gland

Sublingual gland

Parotid gland

Heart

Larynx, trachea and bronchi

Lung

Stomach

Small intestine

Abdominal blood vessels

Liver

Gallbladder

Bile ducts

Pancreas

Adrenal medulla

Colon

Rectum

Kidney

Bladder

Sex organs

34-9 The autonomic nervous system consists of the parasympathetic (*color*) and the sympathetic (*black*) systems. All postganglionic neurons are indicated by dashed lines, preganglionic neurons by solid lines.

35

Integration and Regulation: The Endocrine System

What are the general characteristics of endocrine glands?

How is the thyroid gland stimulated? What effect does this have on the body?

How is the level of blood glucose regulated?

The **endocrine system** works with the nervous system to control and regulate the functions of the body, but whereas the nervous system governs by means of action potentials, the endocrine system uses **hormones.** A hormone has been defined as a chemical substance which is produced in one place and has its effect somewhere else. In humans, various glands secrete hormones into the bloodstream, which carries them to the **target cell** tissue or organ.

35.1 Endocrine, or ductless glands, are scattered widely throughout the body and, with rare exceptions, secrete their hormones directly into the bloodstream.

The location of the dozen or so major endocrine glands in the human body are shown in Figure 35-1, and the two dozen or more hormones they produce and the effects these have on other parts of the body are listed in Table 11. Except in the case of certain hypothalamic neurons, secretory cells in all these glands are in direct contact with blood vessels and secrete their hormones directly into the bloodstream (Figure 35-2A). The cells are not arranged in clusters around a duct as are the secretory cells of other glands, such as the salivary glands.

A common characteristic of hormones is the relative slowness with which they act. Compared to neural regulation, which is measured in seconds or fractions of a second, hormonal control is usually measured in seconds, minutes, hours, days, weeks, or even months.

35.2 The hypothalamus, a portion of the brain, is situated directly above the pituitary and controls the release of hormones from it.

The close proximity of the **hypothalamus** to the **pituitary** (Figure 35-1) is not coincidental. Neurons in one portion of the hypothalamus transport **releasing factors** down their relatively short axons and secrete them directly into the bloodstream, where they are carried via a network of **portal veins** to the **anterior lobe,** or front portion, of the pituitary. Upon arrival the releasing factors stimulate the release of anterior lobe hormones into the bloodstream and in turn affect other endocrine glands. As you can see from Figure 35-3B this is not the usual mechanism of endocrine secretion. A wide variety of releasing factors stimulate the release of different hormones from the pituitary. These pituitary hormones in turn stimulate the secretion of hormones from most of the other endocrine glands.

A good example of the interaction between hypothalamus, anterior lobe, and target endocrine gland involves the **thyroid.** Figure 35-3 outlines three steps in this interaction: (1) Thyrotropin re-

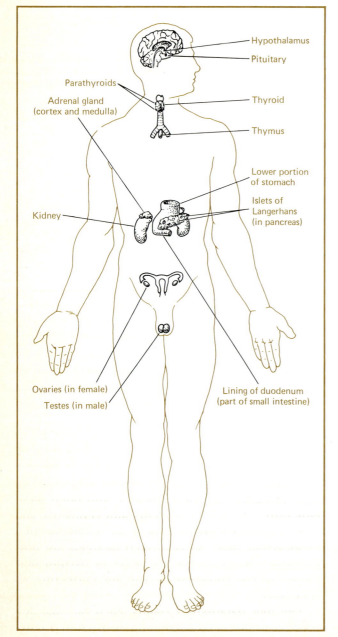

35-1 The location of the major endocrine glands in the human body.

leasing factor (TRF) is secreted by hypothalamic neurons into the network of portal veins and is swept along in the bloodstream to the anterior lobe of the pituitary. There it triggers the release of thyroid stimulating hormone (TSH), which in turn travels (2) in the bloodstream to the thyroid gland in the neck and stimulates it to secrete its thyroid hormones. The amount of circulating thyroid hormones is controlled by a **negative feedback system.** When the thyroid hormone concentration in the blood reaches a normal level, the hypothalamic secretion of TRF is shut off (3), thereby curtailing the secretion of TSH and the production of more thyroid hormone. In this manner, the level of thyroid hormone is controlled.

In the absence of iodine, an essential component of thyroid hormones, insufficient thyroid hormone is produced, and the system fails to be shut off. The secretion of TRF and TSH continues, and the thyroid gland becomes progressively enlarged, forming a visible swelling or **goiter** in the neck as its iodine-depleted tissues respond to the TSH. Goiters are common in inland regions of the world where the diet may be deficient in iodine.

The relationship between the hypothalamus and the **posterior lobe** of the pituitary is unique, in that hypothalamic neurons which send axons to the posterior lobe secrete their hormones *directly* into posterior pituitary cells, which store them and subsequently release them into the bloodstream (Figure 35-2C). Two hormones, about which more will be said below, are transported to and stored in the posterior lobe in this fashion.

Despite its relatively unimpressive size (about that of a pea) the pituitary secretes six hormones from its anterior lobe in addition to the two released from its posterior lobe. Five of the former are **trophic,** that is, they stimulate growth and secretion in other glands. The sixth, **growth hormone** (GH), exerts its action directly on the body and promotes development of long bones and statural growth (Figure 35-4).

The two hormones released by the posterior lobe of the pituitary are **vasopressin,** which promotes water retention by the kidney (section 40.6); and **oxytocin,** which causes milk ejection from the breast of a mother who breast-feeds her baby. The function of another hormone, **melanocyte-stimulating hormone** (MSH), produced in the middle zone of the pituitary formerly referred to as the middle lobe, is not fully understood. It stimulates the development of melanocytes, or dark-pigment-producing cells, in human skin and in some fish, reptiles, and amphibians.

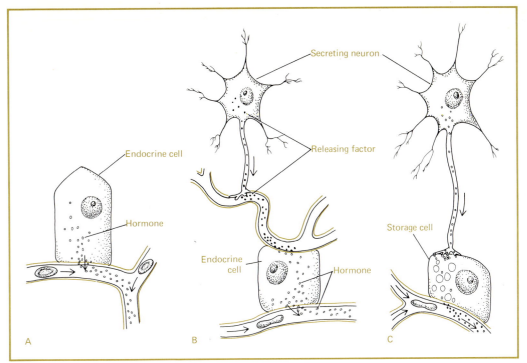

35-2 Different forms of endocrine activity. *A.* Secretion of a hormone into the blood from an endocrine cell. *B.* Secretion of a specific releasing factor from a neuron into the blood. The releasing factor induces hormone secretion by receptor endocrine cells. *C.* Storage by a second type of cell of a neurosecretion for later release.

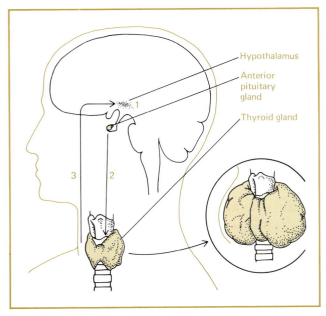

35-3 The feedback loop that normally regulates the level of thyroid hormones in the blood: (*1*) neurosecretion from the hypothalamus stimulates the pituitary, which (*2*) stimulates the thyroid gland by releasing thyrotropic hormone; (*3*) thyroxin shuts off hypothalamic neurosecretion. At the right is a swollen thyroid as seen in cases of goiter.

35.3 The thyroid produces two hormones, *thyroxine* and *tri-iodothyronine,* which help regulate metabolism.

The thyroid is a two-lobed gland located in the neck. The two hormones it secretes bring about an increase in cellular oxygen consumption, help regulate body temperature, and affect growth and development. A person with excessive amounts may be restless and irritable and have an elevated body temperature and heart rate, and protruding eyes. These are manifestations of a **hyperthyroid condition.** In contrast, a person with a lack of thyroid hormones becomes lethargic and often mentally dull. The skin appears dry and yellowish, and cold is tolerated poorly. These are symptoms of a **hypothyroid** state. Sophisticated methods are now available for measuring the level of thyroid hormones in the bloodstream and for correcting an imbalance.

The four **parathyroid glands,** each the size of a match head, are situated behind the thyroid (Figure 35-5). Parathyroid hormone helps regulate blood calcium levels (section 32.5).

TABLE 11 THE PRINCIPAL HUMAN HORMONES, THEIR ORIGIN AND EFFECTS

Hormone	Secreted by	Principal effects
Releasing factors	Brain (hypothalamus)	Individual ones cause release of ACTH, TSH, LH, FSH, prolactin, and growth hormone by pituitary
Adrenocorticotropic hormone (ACTH)	Pituitary	Stimulates the adrenal cortex
Follicle-stimulating hormone (FSH)	Pituitary	Stimulates production of gametes: egg cells by the ovary, spermatozoa by the testes
Luteinizing hormone (LH)	Pituitary	Helps maturation of ovarian follicles or stimulates production of testosterone
Thyrotropin (TSH)	Pituitary	Stimulates the thyroid gland
Prolactin	Pituitary	Regulates breast development and milk production
Growth hormone (GH)	Pituitary	Regulates growth and metabolism
Oxytocin	Neurohypophysis (part of the pituitary)	Facilitates movement of sperm in fallopian tube, stimulates uterine muscle in childbirth, stimulates secretion of milk in mammary glands
Antidiuretic hormone (ADH), or vasopressin	Neurohypophysis	Regulates absorption of water in the kidney
Melanocyte-stimulating hormone (MSH)	Posterior pituitary	Regulates pigment cell activity
Thyroid hormones	Thyroid	Regulates rate of body's metabolism
Thyroxine, triiodothyronine	Thyroid	Regulates rate of body's metabolism
Calcitonin	Thyroid	Promotes calcium deposition
Parathyroid hormone	Parathyroid glands	Mobilizes calcium from bones, teeth
Adrenaline	Adrenal medulla (inner layer of adrenal gland)	Stimulates brain and heart rate, mobilizes sugar and fat for metabolism
Noradrenalin	Adrenal medulla	Increases force of heart contraction and constricts arterioles
Aldosterone	Adrenal cortex	Regulates excretion and retention of minerals, particularly sodium and potassium, by kidneys
Hydrocortisone	Adrenal cortex	Regulates excretion and retention of minerals, particularly sodium and potassium, by kidneys
Estradiol 17-B ("Estrogen")	Ovaries	Regulates development of female characteristics and the menstrual-ovulatory cycle
Progesterone	Ovaries	Works with estrogen to regulate ovulation cycle and pregnancy
Testosterone	Testes	Regulates development of male characteristics and reproductive system
Insulin	Pancreas (islets of Langerhans)	Regulates utilization of sugar, proteins, and fats
Glucagon	Pancreas (islets of Langerhans)	Helps to regulate utilization of sugar, antagonizes effect of insulin
Renin	Kidney	Regulates blood pressure
Erythropoietin	Kidney	Stimulates red blood cell production
Gastrin	Lining of part of stomach and intestine	Stimulates stomach to secrete hydrochloric acid
Secretin	Lining or part of intestine	Stimulates pancreas to secrete chemicals needed for digestion of food
Cholecystokinin	Lining of part of intestine	Stimulates liver and pancreas to secrete chemicals needed in digestion, causes gallbladder to discharge

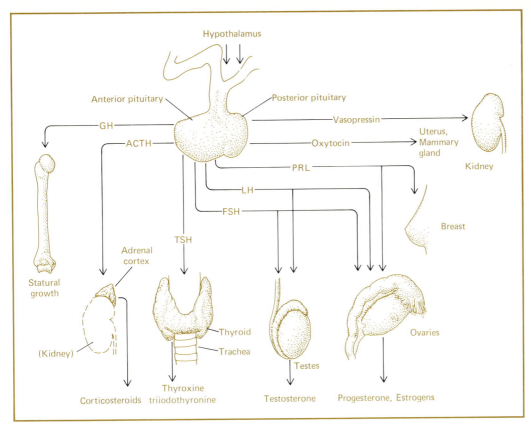

35-4 A diagram showing the influence of the pituitary gland (often called the "master endocrine gland") on other endocrine organs. Most pituitary hormone secretion is under hypothalamic control, often involving a negative feedback system similar to that described for the thyroid.

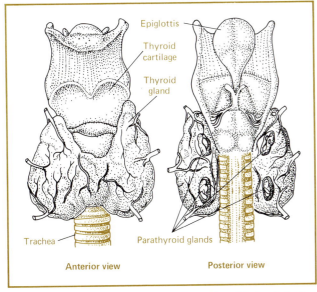

35-5 The location of the thyroid and parathyroid glands in the neck.

35.4 Each adrenal gland is composed of two separately functioning parts—an inner medulla and an outer cortex—each of which secretes several hormones.

There are two pyramid-shaped adrenal glands, each situated on the top of a kidney (Figure 35-6). The **adrenal medulla** or inner portion of each gland contains cells which secrete **adrenaline** and **noradrenaline** upon stimulation by the sympathetic nervous system (section 34.9). These hormones produce a quickening of pulse, a rise in blood pressure, diminished bowel muscle activity, and shunting of oxygen-rich blood to muscles—all preparing the body for "fight or flight."

The **adrenal cortex** (outer region) secretes a different set of hormones, called **corticosteroids,** in response to pituitary **adrenocorticotropic hormone,** or hormone ACTH. The two most important of these essential hormones are **aldosterone** and **hydrocortisone.** Aldosterone functions to promote salt retention by the kidney and thereby helps maintain blood volume and blood pressure. Hydrocortisone helps regulate carbohydrate, fat, and protein me-

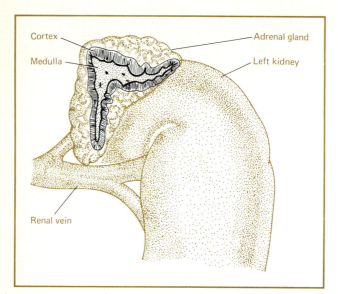

35-6 The location and structure of the adrenal glands. Both the outer cortex and inner medulla of the adrenal gland are endocrine glands.

tabolism and also helps the body meet stressful situations. A lack of these two hormones such as is seen in **Addison's disease** is characterized by low blood pressure and a feeling of weakness and fatigue, as well as low blood sugar, wasting away of tissues, and vulnerability to various stressful situations, which might cause collapse. An excess of corticosteroids, such as is seen in **Cushing's syndrome**, is characterized by high blood pressure (too much salt retained), an elevated blood sugar, and fat deposits. As with the case of thyroid hormones, adrenal cortex hormones can be measured and an imbalance corrected.

35.5 The ovaries in the female and testes in the male respond to pituitary FSH and LH in a number of ways.

In the female, **follicle-stimulating hormone** (FSH) makes its way from the pituitary via the bloodstream to the ovaries, the female reproductive glands, or gonads, on either side of the uterus. There it stimulates growth of a **follicle**, which contains an ovum, or egg cell. It also stimulates the production within the ovary of **estrogen** which promotes the development and maintenance of the female reproductive tract and of secondary sexual characteristics. In the male, FSH stimulates the production of sperm cells in the testes, the male gonads.

Another hormone from the pituitary, **luteinizing hormone** (LH), stimulates **ovulation**, the rupture of the follicle and release of the egg, and **progesterone** production in the female and **testosterone** production in the male. These hormones and their actions are taken up in greater detail in Chapter 41.

35.6 Growth hormone (GH), liberated by the anterior lobe of the pituitary, has a marked effect on skeletal development.

Too much growth hormone can cause bones to grow so much that a person may attain a height

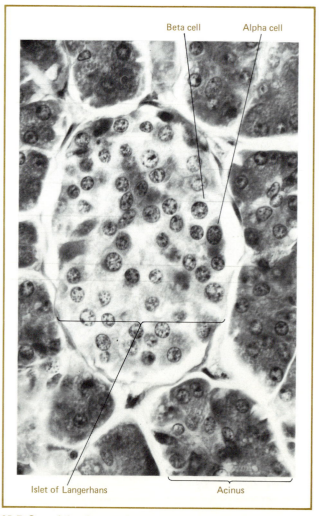

35-7 One of the thousands of islets of Langerhans scattered throughout the pancreatic tissue. Alpha cells secrete glucagon, and beta cells secrete insulin. See Figure 35-1 for the location of the pancreas relative to other organs in the body.

greater than eight feet, a condition known as gigantism. Conversely, a lack of growth hormone may produce a midget. Growth hormone also promotes protein synthesis and has other metabolic effects.

35.7 Islets of Langerhans, small groups of cells in the pancreas, produce insulin and glucagon, two hormones which help regulate blood sugar.

The maintenance of a certain concentration of glucose in the blood is essential for life. All cells in the body need glucose for energy. Brain cells, in particular are sensitive to a lack of glucose. As you might expect, there are various mechanisms to ensure that the glucose concentration in the blood remains fairly constant.

Glucose is absorbed from the intestinal tract into the bloodstream. When its concentration in the blood reaches a certain level, **beta cells** in the **islets of Langerhans** (part of the pancreas) are stimulated to release the hormone **insulin**. The islets (Figure 35-7) form the endocrine portion of the pancreas, as opposed to its exocrine portion, which plays a role in digestion.

Once released, insulin promotes absorption of amino acids from the bloodstream and the synthesis of cellular protein. Its main effect, however, is on the glucose transport system in cell membranes. Insulin permits cells to absorb glucose from the blood. Any disease that destroys the beta cells in the pancreatic islets can produce a condition known as **diabetes mellitus.**

In 1923, C. F. Banting and J. J. R. Macleod received the Nobel Prize for work culminating in the isolation and purification of insulin. As a result of their work and subsequent refinements, insulin is now available for administration to those with diabetes. Injections temporarily enable the cells to absorb sugar.

Under normal conditions, as glucose is used up and its concentration in the bloodstream falls, the amount of insulin secreted decreases. Furthermore, as blood sugar decreases, a second hormone is liberated by the pancreatic islets. This hormone, **glucagon**, is secreted by **alpha cells.** Glucagon stimulates the liver to release glucose back into the bloodstream. In this manner, the level of blood sugar is maintained between meals.

Insulin and glucagon, then, help regulate blood sugar level and cellular access to it. Critical control over cell metabolism is thus exerted.

35.8 Other hormones have great influence on the body.

There are several other hormones which profoundly influence body functions (Table 11). Some special cells in the kidney, for example, secrete **renin**, a hormone that indirectly produces a rise in blood pressure. Another hormone produced by the kidney is **erythropoietin**, which stimulates red blood cell production in bone marrow. In the intestinal tract, certain cells secrete **gastrin**, a hormone which stimulates stomach acid production, and others produce **secretin**, which stimulates the secretion of a watery, alkaline fluid needed for digestion from the exocrine portion of the pancreas. Other examples will be discussed in later chapters.

36

Transport Systems

How do the various cells and plasma that make up the blood function?

What are the vessels that transport blood to and from the heart, and how do they differ in structure and function?

What route would a red blood cell follow if it passed from head to heart and back to the head again?

How is blood pressure controlled?

What purpose does the lymphatic system serve?

The human **circulatory system** provides a suitable environment for the trillions of cells that make up the body. It delivers nutrients and regulatory substances to the cells and removes metabolic wastes. A second transport system, the **lymphatic system,** is primarily concerned with regulating the amount of fluid present between cells and limiting the spread of bacteria.

The circulatory system is a closed circuit of blood vessels with one-way valves and a pump—the heart. Blood is pumped by the heart into **arteries,** which branch into smaller **arterioles** and lead ultimately to microscopic thin-walled **capillaries.** The exchange of nutrients and wastes occurs between blood in capillaries and cells of the body, and the blood is then returned to the heart via the **veins.** To understand how this system functions we must first learn something about blood.

36.1 Whole blood is composed of formed elements —cells and platelets—and the fluid plasma.

Formed elements in blood include **erythrocytes** (red cells), **leukocytes** (white cells), and **platelets** (Figure 36-1).

Erythrocytes contain the red pigment **hemoglobin,** which enables them to transport oxygen (section 38.6). The red cells are produced in the bone marrow and have a life span of approximately 120 days. A mature human erythrocyte has the shape of a flat, biconcave disk (Figure 36-1A), has no nucleus, and can bend considerably to facilitate passage around corners in the smallest capillaries. As the cell approaches the end of its life span, irreversible alterations in its shape cause it to be sequestered in the spleen and destroyed there. Iron, which forms an essential part of hemoglobin, is conserved and reutilized in new red cells. Red cells are constantly being destroyed or lost and are replaced by new ones. Only when bone marrow production fails to keep pace with red cell destruction does a person become **anemic.**

The principal function of leukocytes is to protect the body against microorganisms. There are several types of leukocytes, two of which are shown in Figure 36-1B and D. All differ considerably from red blood cells. Leukocytes lack hemoglobin and are colorless, have a nucleus, are relatively large, and move like amoebae. They engulf bacteria and other foreign particles by the process of phagocytosis (section 4.7). Several types of **granulocytes** are produced in the bone marrow. **Lymphocytes** are made there also, as well as in the lymph nodes,

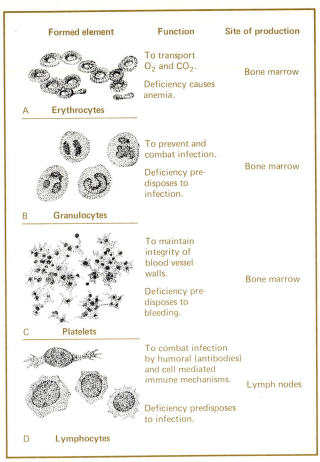

Formed element	Function	Site of production
A Erythrocytes	To transport O$_2$ and CO$_2$. Deficiency causes anemia.	Bone marrow
B Granulocytes	To prevent and combat infection. Deficiency predisposes to infection.	Bone marrow
C Platelets	To maintain integrity of blood vessel walls. Deficiency predisposes to bleeding.	Bone marrow
D Lymphocytes	To combat infection by humoral (antibodies) and cell mediated immune mechanisms. Deficiency predisposes to infection.	Lymph nodes

36-1 The formed elements of the blood, their function, and origin.

thymus, and spleen. Their role in defending the body is discussed in Chapter 37.

Platelets are small bits of cytoplasm which break off special cells in the bone marrow (Figure 36-1C). They form part of the blood-clotting mechanism.

About 55 percent of blood is plasma. Plasma is mostly water carrying dissolved inorganic ions, nutrients, waste products, and a number of proteins. Plasma has buffering capacity and helps regulate blood acidity.

36.2 The vascular circuit is composed of arteries, veins, and capillaries; blood is driven through it in a one-way circuit by the heart.

Arteries are thick-walled blood vessels that carry blood away from the heart. The major ones are shown in Figure 36-2, but there are many thousands of miles of smaller ones, arterioles, that do not appear.

Far from being inert tubes, arteries play an active role in promoting blood flow and regulating blood pressure. The elastic properties of their walls transmit pressure waves or pulses from the heart. These pulses help to squeeze blood along its course. The pulses felt at the wrist are those of the radial artery.

Smooth muscles in the arterial walls enable arteries to regulate blood pressure. When these muscles contract, they cause the arteries to constrict. This increases resistance to blood flow and thereby raises the pressure.

Veins are thin walled and distensible; they carry blood back toward the heart (Figure 36-2B). Because of their expandability, they can serve as temporary reservoirs of blood. Furthermore, in medium-sized veins, such as those in the arms and legs, there are a series of one-way valves. These ensure that blood continually flows toward the heart and not away from it (Figure 36-3). The smallest veins, venules, receive blood directly from capillaries.

Veins lack the thick walls and pulsatile properties of arteries but manage to effectively return blood to the heart with the help of muscle and thoracic pumping mechanisms. In the former, veins located near muscles (or pulsing arteries) are periodically compressed as the muscles alternately bulge during contraction and then relax (or as the arteries pulse). This intermittent compression helps squeeze blood toward the heart (Figure 36-3B).

The second mechanism, the thoracic pump, is related to breathing movements. During inspiration, the reduction of pressure inside the thorax and the increase in pressure in the abdomen forces venous blood upward into the thoracic (chest) cavity.

Capillaries are the smallest blood vessels and connect the arterial and venous sides of the vascular circuit (Figure 36-4). They are just large enough to permit passage of one erythrocyte or leukocyte at a time. Through their walls, which are composed of a single layer of cells, nutrients and oxygen diffuse into the tissues. Waste products diffuse in the opposite direction, to be carried away by the bloodstream. Blood flow through capillaries can be regulated by tiny precapillary sphincter muscles located at the beginning of each capillary vessel. When contracted, the sphincters effectively close the capillary. Blood loss can thus be minimized in the case of injury, or blood can be shunted to an area where it is needed more, such as shunting from skin to muscle when fight or flight seems imminent.

The heart contracts with sufficient force to propel

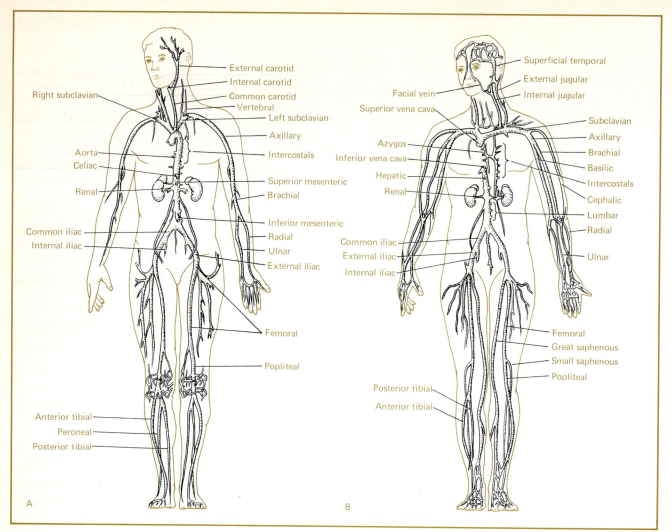

External carotid
Internal carotid
Common carotid
Vertebral
Right subclavian
Left subclavian
Axillary
Aorta
Celiac
Intercostals
Renal
Superior mesenteric
Brachial
Inferior mesenteric
Common iliac
Radial
Internal iliac
Ulnar
External iliac
Femoral
Popliteal
Anterior tibial
Peroneal
Posterior tibial
A

Superficial temporal
External jugular
Facial vein
Internal jugular
Superior vena cava
Subclavian
Axillary
Azygos
Brachial
Inferior vena cava
Basilic
Hepatic
Intercostals
Renal
Cephalic
Lumbar
Common iliac
Radial
External iliac
Internal iliac
Ulnar
Femoral
Great saphenous
Small saphenous
Popliteal
Posterior tibial
Anterior tibial
B

36-2 *A.* Arteries (*in color*) carry blood away from the heart to all parts of the body. *B.* Veins carry blood toward the heart from all parts of the body. Only major arteries and veins are shown.

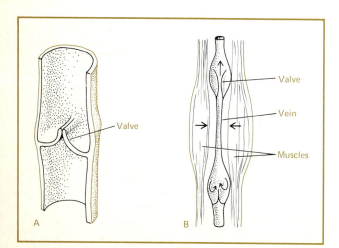

Valve
Valve
Vein
Muscles
A
B

36-3 *A.* A one-way valve in a medium-sized vein (e.g., in the arm or leg). *B.* The venous pumping mechanism which helps pump blood back toward the heart is shown schematically. Adjacent bulging muscles compress the vein (*arrows*) and passively force blood through the one-way valves.

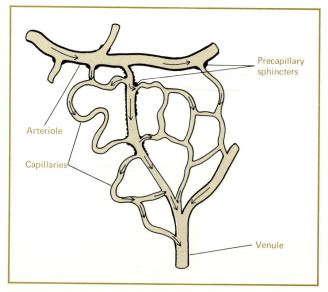

Precapillary sphincters
Arteriole
Capillaries
Venule

36-4 A portion of a capillary bed, demonstrating how capillaries connect the arterial and venous sides of the vascular circuit. Flow through the capillary bed is regulated by precapillary sphincter muscles. Vessel walls are thickened by muscle on the arterial side of the circuit.

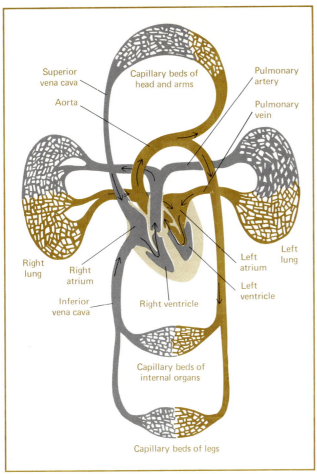

Superior
vena cava

Capillary beds of
head and arms

Pulmonary
artery

Aorta

Pulmonary
vein

Right
lung

Right
atrium

Inferior
vena cava

Right ventricle

Left
atrium

Left
lung

Left
ventricle

Capillary beds of
internal organs

Capillary beds of legs

36-5 The human heart is a four-chambered muscular pump that sends oxygen-depleted blood through the pulmonary circuit and oxygen-rich blood through the systemic circuit from the left side of the heart.

blood through the 60,000-odd miles of blood vessels. It is situated in the thorax or chest between the sternum and the vertebral column, just to the left of the midline, and is composed of four chambers, each of which has a one-way valve to insure that blood flows in one direction only.

Oxygen-poor venous blood returns to the heart via the largest veins, the **inferior** and **superior vena cavae,** enters the first chamber, the **right atrium,** and as this chamber contracts, is pumped into the second chamber, the **right ventricle** (Figure 36-5). From this chamber, blood is pumped through the pulmonic (pulmonary) valve and via the pulmonary artery into the lungs, where it loses its carbon dioxide and takes on oxygen. Next, blood returns to the heart via the pulmonary veins and enters the third chamber, the **left atrium.** From there it is pumped into the **left ventricle,** from which it is forcefully ejected through the aortic valve into the

aorta. The walls of the left ventricle are thickest, for it is from this chamber that blood is pumped to the entire body.

The left side of the heart and the vessels which conduct blood to the tissues are called the **systemic circuit.** The right side of the heart is concerned with pumping blood to the lungs; together, it and the pulmonary (lung) vessels are known as the **pulmonary circuit.**

The one-way valves of the heart, which normally prevent backflow of blood and allow forward pumping action in both circuits, may be damaged by infection, especially rheumatic fever. Blood flow through such a damaged valve is turbulent and sets up an audible sound, called a **heart murmur.** In severe cases of valvular damage, open heart surgery must be performed, the diseased valve removed, and a synthetic valve sewn into place.

36.3 The heart is an extraordinary organ with its own blood supply and impulse-conducting system.

Heart muscle differs from skeletal muscle in a number of ways. Unlike skeletal muscle, which is continuously supplied with a steady stream of arterial blood, heart muscle is intermittently supplied and drained by its own vessels, the **coronary arteries and veins** (Figure 36-6B). Only during the relaxation phase of the cardiac cycle (**diastole**) can blood move through these. During the contractile phase (**systole**), the vessels are temporarily squeezed shut by the bulging muscle around them.

Another difference is that heart muscle extracts a higher percentage of the total oxygen carried by the blood in its vessels than does any other tissue. If blood supply to a skeletal muscle is restricted, the muscle can compensate in part by extracting more oxygen than usual from the blood it is receiving. The heart, already taking so much, has almost no margin of safety. Should one of the two coronary arteries become permanently blocked (by a blood clot, for instance), the heart is rapidly damaged by a lack of oxygen. When this occurs a person experiences a "coronary," better known as a heart attack. This is accompanied by severe chest pain and poses a serious threat to life. As can be seen in Figure 36-6A, heart attacks are the number-one killer in the United States today. Research has identified several major risk factors which increase any individual's chances of having a heart attack. The four most important of these are high cholesterol

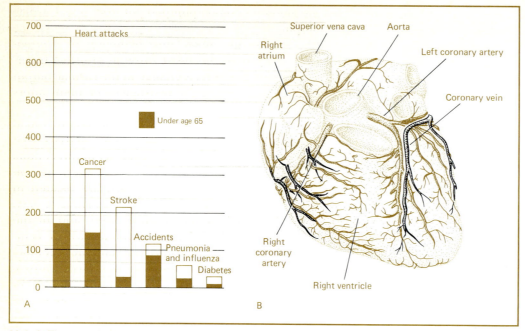

36-6 *A.* The mortality rates from heart attacks and other causes in the United States (figures are in thousands). *B.* The coronary circulation serving the heart muscle itself is partly visible on the surface of the heart. (Chart from the American Heart Association.)

level in the blood, obesity, high blood pressure, and smoking.

Still another way in which heart muscle differs from skeletal muscle is in the inherent rhythm of its contraction. Heart muscle is not under voluntary control and beats at its own rate. The area which normally initiates periodic bursts of impulses (about 72 per minute) is called the pacemaker, or **sinoatrial (SA) node** (Figure 36-7). It is located at the junction of the superior vena cava and the right atrium. Impulses arising from this node pass through and activate atrial muscle on their way to the **atrioventricular (AV) node,** which is situated between the right atrium and right ventricle. Normally the SA node governs heart rate, but should it fail to fire for some reason, the AV node then causes the ventricles to beat rhythmically, but at a slower rate.

Impulses originating in the AV node are conducted down a specialized group of muscle cells called the **bundle of His** and out along its finer branches to activate the ventricles to contract. The heart continues to beat 65 to 80 times per minute (or faster or slower, depending on the individual and the type of activity engaged in). This amounts to roughly 100,000 beats per day at average heart rates, with rests only between beats!

The various electrical events in the heart set up small currents throughout the body which can be monitored by an **electrocardiograph (EKG) machine.** Aberrations in the electrical pattern can be used in the diagnosis of several kinds of heart disease.

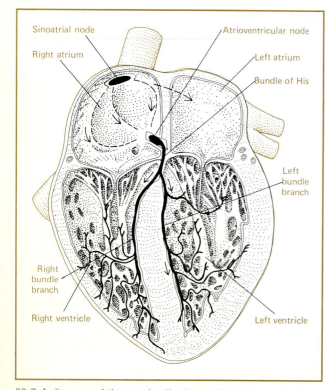

36-7 A diagram of the conductile tissue of the heart, showing the sinoatrial and atrioventricular nodes and the bundle of His (see text).

Should the heartbeat become very slow and irregular, as it does in a number of diseases, it is now possible to implant a small package of batteries just under the skin of the chest. This is connected by wires to the heart muscle itself. The circuitry of the batteries is arranged to emit periodic impulses which cause the heart to contract and thus serves as an artificial "pacemaker." Thousands of people who would otherwise be hopeless invalids now lead nearly normal lives because of these devices.

The rate of the heart's beat is increased and decreased by two sets of nerves. The ones that increase heart rate are part of the sympathetic nervous system and secrete noradrenalin at their endings. The nerves that decrease heart rate are the pair of **vagus nerves,** part of the **parasympathetic system,** that liberate acetylcholine from their endings (section 34.7).

36.4 Blood flow and pressure, which must respond to the demands of the body, are regulated by several factors.

The **heart rate** and the volume of each stroke (**stroke volume**) together determine **cardiac output,** that is, the amount of blood pumped per unit time. Cardiac output, together with **total peripheral resistance** (the friction the blood must overcome as it circulates away from the heart through arteries) determine **blood pressure.** A change in the cardiac output or the total peripheral resistance, or both, alters blood pressure.

Three important mechanisms which control blood flow and pressure are **baroreceptor activity,** the **Starling mechanism,** and **adrenaline.** Baroreceptors (from *baro-*, "pressure") are specialized neuron endings sensitive to stretch. Most are located in two areas: in the walls of the **carotid sinus** of each internal carotid artery (Figure 36-8), and in the walls of the **aortic arch.** When blood pressure is increased, the walls of these vessels and the receptors in them are stretched, generating nervous impulses. Some of these impulses bring about a dilation of blood vessels, thus decreasing total peripheral resistance and hence lowering blood pressure. Other impulses excite the cardioinhibitory center in the brain stem, causing a decrease in heart rate, again lowering blood pressure. As blood pressure drops, baroreceptors are stretched less, and fewer impulses are generated, so that blood pressure is not lowered below normal.

The Starling mechanism raises blood pressure by increasing stroke volume without affecting heart rate. Ernest Starling (1866–1927) formulated a law that states that the force of contraction of a cardiac muscle fiber is proportional to its initial length.

36-8 Blood pressure is regulated in part by baroreceptors (*in color*) in the aortic arch and carotid sinus of the internal carotid artery. The carotid and aortic bodies are chemoreceptors that regulate breathing (Chapter 38).

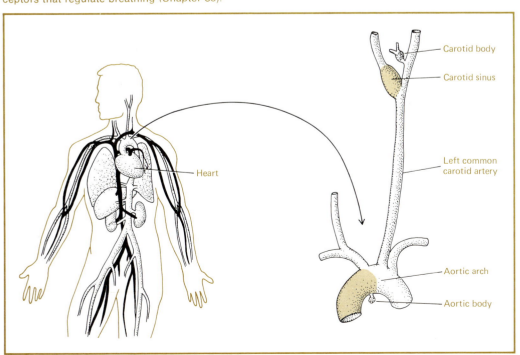

Carotid body

Carotid sinus

Heart

Left common carotid artery

Aortic arch

Aortic body

That is, the more stretched a heart muscle fiber is at the beginning of contraction, the greater will be the force of contraction that it develops. This is a property heart muscle shares with skeletal muscle.

In practical terms, the Starling mechanism means that when you exercise and thus return more blood to the heart, the chambers become dilated, and the heart muscle fibers are stretched and elongated. This causes the heart's contractions to become more forceful, so that more than the usual volume of blood can be ejected during a single heart stroke. A relatively slow heart rate and large stroke volume are signs of "being in shape." A trained athlete can achieve a significant increase in cardiac output by increasing stroke volume and therefore, during mild exercise, need not increase his or her heart rate significantly. In contrast, the flabby spectator must increase his heart rate much more to achieve a similar increase in cardiac output.

Adrenaline increases cardiac output and alters total peripheral resistance. Liberated directly into the bloodstream from the adrenal medulla in response to anger, excitement, or fright (section 35.4), adrenaline acts directly on the heart to increase heart rate and force of contraction and thereby to increase blood pressure. In addition it alters the peripheral resistance of blood vessels so that vessels supplying the skin and internal organs (e.g., stomach and intestines) constrict, while vessels supplying muscle beds dilate. There is a net decrease in the total peripheral resistance, with the overall effect that blood is shunted to muscle capillary beds, where it may be needed to support increased muscle activity. Adrenaline also liberates stored sugar into the bloodstream, further aiding muscles to meet an emergency.

Drugs also affect blood flow and pressure. Some are used to lower blood pressure in people with **hypertension** (high blood pressure), while others are used to raise blood pressure in individuals in a state of **shock** (inadequate blood pressure). **Nicotine,** the vasoactive substance in tobacco smoke, causes blood vessels to constrict, a consequence which may be extremely harmful to persons with poor circulation.

36.5 Another transport system, the lymphatic system, conducts fluid that has leaked out of blood capillaries back into the circulatory system.

The lymphatic system consists of a network of thin-walled vessels which transport intercellular fluid in a one-way direction to lymph nodes and subsequently empty it into veins just below the base of the neck (Figure 36-9). The one-way flow of the intercellular fluid, or **lymph,** is assured by the presence of one-way valves in the larger lymph vessels. The propulsive force which moves the lymph is similar to that of the venous system: The bulging of contracted muscles in which the vessels are embedded squeezes the fluid along its course.

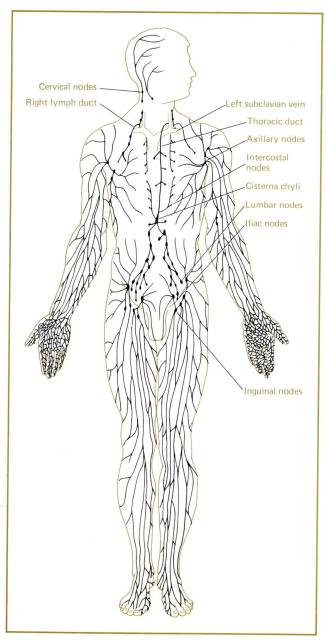

36-9 A diagram of the lymphatic system, which collects fluid from the extracellular space and returns it via the thoracic duct to the heart.

Before reaching the major lymph vessels, lymph is conducted to and passes through lymph nodes, small, rounded structures several millimeters in diameter.

36.6 Located at intervals along the various channels, lymph nodes act as filters and "purify" lymph by removing particulate matter, such as bacteria.

Phagocytes in the lymph nodes engulf bacteria and other particles as they are carried through the cen-tral **reticular zone** of the node (Figure 36-10). Bacteria traveling from a site of infection are thereby prevented from spreading further into the body. In such a case, the active lymph nodes tend to swell, as, for example, do the ones in the neck when there is an infection causing a sore throat. Lymph nodes also help defend the body by producing lympho-cytes (Figure 36-1), a class of white cells. Some of these are produced in the **germinal centers** of **lymphatic nodules,** which are specialized regions within a node. When mature, many leave the node and travel with lymph into the venous system. Some differentiate into antibody-producing plasma cells. Others differentiate into specialized lymphocytes that are responsible for cellular immunity. Both cell types play critical roles in defending the body against infection.

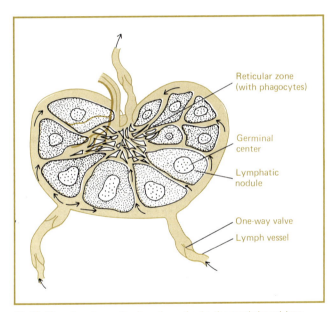

Reticular zone
(with phagocytes)

Germinal
center

Lymphatic
nodule

One-way valve

Lymph vessel

36-10 The structure of a lymph node. In the peripheral lymphatic nodules, lymphocytes can differentiate into antibody-producing plasma cells (Chapter 37). The reticular zone is made up of a network of fibers, among which are large phagocytes that ingest foreign particles, thus purifying lymph before it passes via the thoracic duct back into the bloodstream.

36.7 Specialized lymphatic vessels, lacteals, located in the intestinal wall, absorb long-chain fats and transport them into the bloodstream.

The **lacteals,** so called because of their white, milky appearance when engorged with fats, are situated in the intestinal villi, the minute, fingerlike projections of the internal intestinal wall. After a fatty meal, lipids (fats) are digested within the intestine and then absorbed by the cells lining the villi. Here they are hydrolyzed. Only the long-chain fats (those with a long molecular structure) pass into lacteals and are transported via the lymphatic system into the bloodstream. All other fats, the medium- and short-chain ones, as well as other nutrients absorbed from the gut, travel via the venous portal system to the liver, where they must first be processed before being emptied into the systemic circulation.

37

Defense Mechanisms

QUESTIONS TO KEEP IN MIND

What protection do the skin and other epithelial surfaces provide?

How do phagocytic cells defend the tissues of the body against procaryotic invaders?

What is involved in an immune response?

How does a tetanus shot offer protection against infection from tetanus bacteria?

What defense is there against rival infection?

How does the liver function as a defense mechanism?

Humans have a number of mechanisms for protection against the harmful substances that abound in the environment and against invasion by harmful microorganisms. These protective devices include skin, the surface linings of the lungs and intestine, phagocytic cells, the immune response, interferons, and the detoxification mechanisms of the liver.

Although the body has these defenses, they are being increasingly overridden in our polluted world. It is becoming clear, for instance, that a person can suffer serious heart and lung damage by breathing the polluted air of our cities or by polluting otherwise clean air by drawing it through a lighted cigarette. It has even been said—only partly in jest—that mother's milk, because of its concentration of the insecticide DDT, could not legally be shipped across state lines in any other container!

37.1 The skin is man's first line of defense.

Human skin is composed of two major layers, the outer **epidermis** and the inner **dermis.** The epidermis is further subdivided into the innermost, or **basal layer** (Figure 37-1), of cells that multiply and are pushed slowly toward the surface as cells beneath them divide, and the outer **horny layer.** As cells from the basal layer approach the surface, they become progressively flattened, are filled with a tough fibrous structural protein called **keratin,** and die. These dead, keratinized cells form the outermost layer.

The skin forms a protective covering over the entire body which prevents tissues from drying out and blocks entry to most microorganisms. Some microorganisms do penetrate the skin through scratches, open wounds, or in an insect bite. Should the skin be injured, blood clotting and scab formation close over the torn surface until the skin regenerates.

Skin perpetually renews itself every 7 to 21 days. Old cells slough off the surface of the horny layer, and new cells grow up from beneath. The epithelial lining of the digestive tract likewise renews itself at a rapid rate—every 4 days. This continual shedding and replacement of surface layers makes it more difficult for harmful substances and organisms to enter the body.

Although the respiratory and digestive tracts are not conventionally thought of as part of man's exterior, in a very real sense they are. In a worm, built on a simple tube-within-a-tube plan, it is easier

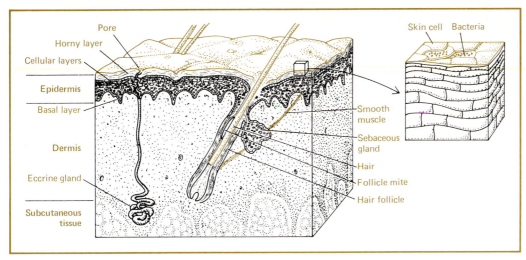

37-1 A block of human skin, showing a hair follicle with an associated sebaceous (oil) gland and smooth muscle that causes the hair to "stand on end", an eccrine (sweat) gland, and the layers of skin cells. Note the follicle mite (an arthropod) on the hair shaft. The insert shows the size of bacteria always present in the skin surface.

to see that the animal really consists of the tissues between those two tubes. The worm's "environment" exists both outside the outer tube and inside the inner one. So it is with humans, and protective epithelial cells not only cover their outside surfaces, but line their inside "tubes" as well. The latter are often lined with special mucus-secreting epithelial cells.

In the respiratory tract, mucus secretions of epithelial layers provide a measure of protection against potentially harmful agents. Mucus, which traps smoke particles or bacteria, is continually swept out of the respiratory tract by ciliated cells lining the various air-conducting tubes. However, not all the particles taken into the lung can be removed by this mechanism.

Epithelial surfaces, then, provide a first line of defense against various microorganisms invading the body. The skin also plays a role in temperature regulation, salt excretion, and the prevention of desiccation.

37.2 Phagocytic cells constitute a second line of defense.

Phagocytosis is the process by which cells engulf bacteria and foreign particles (section 4.7 and Figure 4-9). Phagocytic cells are found in various parts of the body, including the reticular zone of lymph nodes, where the phagocytic cells help purify lymph; the sinusoids of the liver, where they do the same for blood; the alveoli of the lung; and the glomeruli of the kidneys. Leukocytes and **tissue macrophages** also engulf foreign particles. Unlike the phagocytes just mentioned, however, these cells move about extensively. Leukocytes circulate in the bloodstream, and tissue macrophages move about between cells. When bacteria enter the body, these cells are attracted to the area and engulf the intruders.

37.3 The immune response forms a third line of defense.

Recent investigations show that there are two main categories of immune phenomena—**cellular immunity** and **humoral immunity.**

Cellular immunity involves specialized lymphocytes (section 37.4) and is responsible for the rejection of skin and organ grafts, for reactions to microorganisms which live inside tissue cells (such as tubercle bacilli), and perhaps the destruction of cancer cells.

Humoral immunity involves the production of antibodies and is responsible for protection against most bacteria and their products. Briefly, **antibodies** are specific proteins that are produced by plasma cells in the blood in response to the introduction of foreign cell products (mostly protein) called **antigens** (section 37.4). The latter are defined as any substances that will induce the production of an antibody. An antibody combines with antigen, undergoes a shape change, and as a result, proteins called **complement factors** bond onto the antigen-antibody complex and help to destroy the antigen (section 37.5).

37.4 Two types of lymphocytes, T-cells and B-cells, are involved in cellular immunity and humoral immunity, respectively.

In the bone marrow are **stem cells,** which give rise to lymphocytes that either pass through the thymus and become **T-cells** or pass through other lymphoid tissue and become **B-cells** (Figure 37-2). T-cells go into action when they encounter tissue (an antigen) that has been processed by a macrophage. The T-cells respond by dividing many times and differentiating still further into **T-cell effectors.** These attach to and destroy the foreign (transplanted) cells. In organ transplants, drugs must be used to repress this immune reaction.

37-2 A diagram showing the origin and interactions of the two types of lymphocytes involved in cellular and humoral immunity.

In addition to reacting to unnatural situations such as organ transplants, it has been proposed by Dr. L. Thomas of New York University that cellular immunity may play a role in combating cancer. His concept is that when a cell in the body changes (for as-yet-unknown reasons) and becomes cancerous, the antigenic determinants on its cell surface also change, and the cellular immune system then recognizes the cancerous cell as foreign and attempts to destroy it. In this manner, the cellular immune mechanism acts as a continual **immunological surveillance system,** weeding out cancerous cells as they appear. Only when the immune system is defective or begins to wear out with age are such cancer cells permitted to survive and multiply. The increased incidence of cancer in persons with rare immunological deficiency diseases and in older persons could be explained on this basis.

B-cells, like T-cells, must come in contact with an antigen processed by a macrophage before becoming effective. In this situation, B-cells multiply into **B-cell effectors,** or **plasma cells.** These produce antibodies that circulate in the bloodstream, where they combine with and neutralize the specific antigens that stimulated their production.

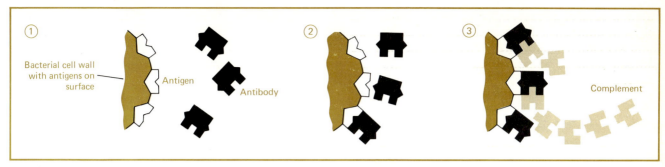

37-3 A schematic diagram of the interactions of bacterial cell wall antigens with plasma antibodies and complement factors.

37.5 Antibodies combine with antigens in a specific "lock-and-key" fashion.

When a bacterium enters the body, unique molecular structures on its wall are recognized as being different from every other molecular configuration found in the human body. The humoral immune system is stimulated to produce antibodies which circulate in the bloodstream and react with the specific antigens, that is, foreign proteins wherever they are found (Figure 37-3). At this point, complement factors, a series of nine proteins found in the plasma, attach to the antigen-antibody complex in a specified sequence to complete the process of destruction. When all nine complement factors are attached, the bacterium ruptures and is destroyed.

37.6 The memory response of the humoral immune system forms the basis of current immunization procedures.

When a person is vaccinated or innoculated with small doses of live or dead viruses or bacteria (or other substances, such as pollen grains in the case of hay-fever shots) there is a lag of several days before significant quantities of antibodies are measurable in the recipient's blood serum. Upon re-exposure to the same antigen at a later date the lag is much shortened, and a greater number of antibodies are produced. The rationale for immunization against various foreign agents, therefore, is to prepare the immune system for a possible exposure to the same antigen by a natural route and in a large disease-producing dose. Primed by a primary exposure, the immune system mounts a full-scale battle, effectively destroying the invaders and preventing disease. Protective immunization against such diseases as cholera, diphtheria, bubonic plague, whooping cough, smallpox, rabies, measles, and yellow fever is available.

In some cases the "memory" of the immune response can cause grave problems, as is the case for persons allergic to bee stings. In these people, high levels of a certain class of antibodies are produced in response to a first sting. For some reason, these adhere to special cells in human tissue called **mast cells.** Some time later, a second bee sting introduces more antigen, which combines with the antibodies affixed to mast cells, causing the mast cells to release **histamine.** This substance has drastic physiological effects on the body when released in large amounts. When the mast cells release their histamine, blood pressure falls rapidly, and various tissues swell, including those of the larynx, causing the individual to choke. Medical intervention must be swift. Adrenaline, found in most bee-sting kits, is injected to restore normal blood pressure. In severe cases, a **tracheotomy** (an operation to cut a hole into the trachea) may have to be performed to permit the person to breathe.

37.7 Interferons defend against viral infections.

Diseases such as measles, chicken pox, smallpox, mumps, and poliomyelitis, to mention but a few, are all caused by viruses. Virus particles live and multiply inside host cells (section 2.6) and cannot therefore always be attacked effectively by antibodies circulating outside the cells. The body has a special defense mechanism against them, namely **interferons.** The interferons, a class of protein molecules produced only by cells infected with a virus, function to block replication of viral particles within the cell. Interferons are quite different from antibodies. They are species specific—that is, they will not work if transferred to another animal species—whereas antibodies can be transferred from horse to man, for example, with no loss of effectiveness. Furthermore, interferons are *not virus-type specific* and will inhibit the multiplication of many types of virus particles. Antibodies, on the other hand, are very specific for a given antigen.

Relatively little is known about interferons, and much work is currently being done to further elucidate the mechanism of their action. They are, however, the main defense against the common cold and other virus diseases.

37.8 The liver detoxifies potentially harmful substances by changing them chemically.

Ingested substances, after absorption in the intestinal tract, are carried to the liver by the portal venous system (Figure 37-4), where they are processed before reaching the rest of the body. Some potentially harmful substances are altered chemically in the liver and thus rendered harmless. Ethyl alcohol, for example, undergoes **oxidation** to acetaldehyde, which is further oxidized to acetate. In this way the effects a person notes after drinking a cocktail or two gradually wear off, as the alcohol level of the blood is constantly lowered by the action of the liver. Narcotic drugs undergo **hydrolysis**, which renders them inert, thus limiting the time they can affect the body. Two additional kinds of chemical reactions, **reduction** and **conjugation**, are also carried out in liver cells.

Should a poisonous substance escape biochemical processing in the liver, the last remaining barrier to its harmful action is the selectively permeable membrane of the cell itself. Although the defense mechanisms of the body together form a formidable system of protection, they are nevertheless being

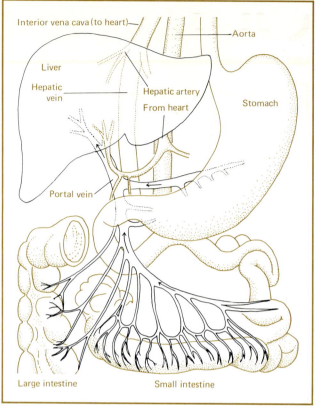

37-4 Most compounds absorbed in the stomach and intestines are transported in the bloodstream to the liver, where they are processed before passing on to the other parts of the body.

stressed in an increasingly polluted environment. The fact that the U.S. Food and Drug Administration has a tremendous backlog of chemicals (drugs, food additives, insecticides, and so on) to test and that these substances cannot be kept off the market until and unless they are fully tested and *proved to be unsafe*, does not improve the situation. Chemicals such as cyclamates (one type of artificial sweetener) were sold for several years before being banned because of evidence that they caused cancer in experimental animals.

38

Respiration

QUESTIONS TO KEEP IN MIND

How are gases exchanged in the lung?

How is oxygen carried from lungs to tissues?

How is carbon dioxide carried from tissues to lungs?

Resting or working, the human body is in constant need of energy. The cells utilize this energy packaged in the form of ATP to perform their various functions. As was seen in Chapter 6, small amounts of ATP can be synthesized in the absence of oxygen, but nearly 20 times as much can be produced from the same amount of glucose *in the presence of oxygen*. In an organism as complex and demanding as a human being, the more efficient process of aerobic glycolosis is absolutely essential. Consequently, the human body has elaborate mechanisms to ensure that all cells are supplied with adequate amounts of oxygen. In its absence, life can be sustained only for a few minutes. The process by which cells acquire an adequate supply of oxygen and eliminate carbon dioxide is called **respiration.** This process may be subdivided into four steps: the mechanics of breathing, the exchange of gases in the lungs, gas transport via hemoglobin in red blood cells (erythrocytes), and cellular respiration and oxidation.

38.1 Air is filtered, moistened, warmed to body temperature, and monitored for harmful gases as it passes through the nasal passages.

The earth's atmosphere consists of 21 percent oxygen, 78 percent nitrogen, and fractions of a percent of several other gases, including argon, carbon dioxide, and hydrogen. An elaborate system of air-conducting tubes and other structures brings atmospheric oxygen inside the body and gets it into contact with a moist lung surface of approximately 90 square meters.

On its way to the lungs, air first passes through the nasal passages (Figures 34-7 and 38-1), where it is filtered, moistened, warmed, and checked for noxious gases. As it enters the passages, which are situated above the mouth and separated from it by the **hard** and **soft palates,** the larger dust particles are trapped by hairs in the **nostrils** and by mucus. Then, deflected by the **turbinate bones,** the air swirls around and becomes both moistened and warmed. Olfactory receptor cells (section 34.5) are exposed to this air and alert the brain to halt inhalation if a noxious or harmful odor is detected. In these ways air is processed and checked for purity before passing further into the body. Unfortunately, humans have no olfactory receptors sensitive to some poisonous gases, such as carbon monoxide.

38.2 Branching tubes conduct air into the innermost regions of the lungs.

The **pharynx,** the region at the back of the mouth where the nasal passages and oral (mouth) cavity join, conducts air to the junction of the **esophagus,** through which food is passed to the stomach, and the **larynx.** Food is prevented from entering the larynx by a special flaplike structure, the **epiglottis,** which together with the tongue, closes off the larynx in a reflex action during swallowing. The layrnx houses the **vocal cords,** thin membranous structures

323

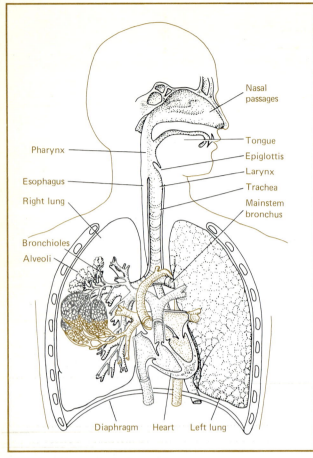

Nasal passages

Tongue
Epiglottis
Larynx
Trachea
Mainstem bronchus

Pharynx

Esophagus

Right lung

Bronchioles
Alveoli

Diaphragm   Heart   Left lung

38-1 The human respiratory system.

that vibrate when air is forced between them, producing speech, song, groans, and other vocal sounds.

The larynx leads to the **trachea,** a large hollow tube in the anterior portion of the neck, which divides into two **mainstem bronchi.** Each of these conducts air to one of the two **lungs.** The trachea and bronchi together form the major air passages and are supported by tough, ringlike structures of cartilage which prevent the airways from collapsing. The two mainstem bronchi subdivide into secondary bronchi, which in turn subdivide, and so on, until the smallest air-conducting passages—the **respiratory bronchioles** and the **alveolar ducts**—are reached. It is these passages that lead to approximately 300 million tiny air sacs, or **alveoli** (Figure 38-2).

The lungs themselves are each divided into several **lobes** or sections. The right lung has three, and the left two. Each lobe is surrounded by a membrane, the **pleura,** which helps separate them from one another. The millions of tiny alveoli give

lung tissue its spongy texture. They are richly supplied with capillaries because the alveoli are the site of gas exchange. In order for oxygen to leave the alveoli and enter the bloodstream and for carbon dioxide to leave the bloodstream and enter the alveoli, these gases must diffuse across a number of membranes. These membranes include those of the alveolar wall, the capillary wall, and that of the red blood cell itself. If capillary or alveolar membranes become abnormally thickened, as in the conditions called "farmer's lung" and "coal miner's lung," the diffusion of gases is impaired. Likewise, if alveoli become inflamed and plugged with mucus or other secretions, as in pneumonia, gas exchange is impaired. All three conditions have dire effects on health.

38.3 Air is sucked into the lungs by the expansion of the chest cavity, whereas expulsion of air from the lungs is essentially a passive process.

The chest, or **thoracic cage,** is formed by the vertebral column in back, the sternum in front, the ribs at the sides and top, and the diaphragm below (Figure 38-3). Both lungs are thus situated inside an airtight container and connected to the outside by the trachea and bronchi. During **inhalation,** contraction of the muscles attached to ribs causes the ribs to move upward and outward, and contraction of the **diaphragm,** the thin, flat muscle separating the thorax from the abdomen, causes the floor of the chest cavity to move downward. The net result is an expansion of the thoracic cage. The lungs, wrapped in their pleural membranes, are separated from the chest wall by only a thin layer of fluid and expand as the chest expands. The pressure of the alveolar gas drops below that of the atmosphere, and air flows in.

During **expiration** the rib muscles relax, causing the rib cage to lower to its resting position. The diaphragm also relaxes and arches upward to its resting position. The volume of the chest cavity decreases, and air is expelled. In addition, the elastic recoil of the lung tissue itself, together with alveolar surface tension, help to expel air.

Obviously, the normal expulsion of air requires that the air-conducting tubes remain open. In healthy lungs these passageways are held open by supporting tissues. **Emphysema,** a condition common among habitual smokers, is characterized by

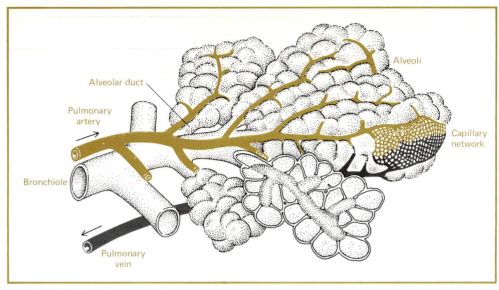

38-2 The alveoli of the lung and their blood supply.

progressive destruction of these supportive elements, so that expiration, more than inspiration, is seriously impaired.

38.4 The rate and depth of breathing are regulated by several mechanisms.

Normally people at rest breathe 10 to 14 times per minute and move half a liter of air in and out with

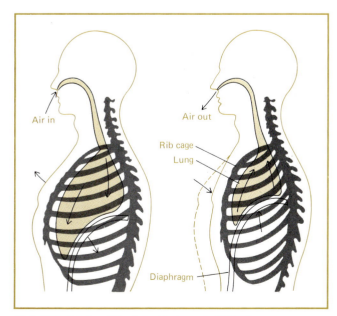

38-3 Movements of the thoracic cage and diaphragm cause the flow of air into and out of the lungs.

each breath. This resting volume of ventilation may be increased up to 30-fold during periods of vigorous activity. The actual rate and depth of breathing is geared to meet the metabolic demands of the body for oxygen and adjusts appropriately for any level of exercise or altitude change.

Spontaneous breathing is maintained by periodic neural discharges from the **respiratory center** in the medulla (section 34.1) to the respiratory muscles of the chest wall and diaphragm. This center automatically insures the next inspiratory effort. It also has cells which can respond to increases in carbon dioxide or acidity in the cerebrospinal fluid by increasing both the rate and depth of breathing.

A neural mechanism, initiated in the lung itself, also regulates breathing. The Herring-Breuer reflex is initiated by stretch receptors in the lung tissue. Deflation of the lung causes these receptors to discharge nerve impulses to the brain stem, resulting in an inspiratory effort. Inflation of the lung with consequent stretching of the receptors results in a generation of impulses which bring about an inhibition of inspiratory movements. Thus, when the lungs are full of air, neural signals for further inspiration cease.

Breathing is also controlled by special chemoreceptor cells located in the carotid and aortic bodies (Figure 36-8), which automatically analyze blood chemistry. These cells are sensitive to changes in blood acidity and to changes in the relative amounts of oxygen and carbon dioxide. When these parameters deviate from the norm, the chemoreceptor cells relay impulses to the brain stem for appropriate adjustments in rate and depth of breathing. Other factors, such as brain signals reflecting the state of awareness or wakefulness and muscle activity itself, also affect rate and depth of breathing.

38.5 During gas exchange, oxygen from the alveolar air moves into the capillary bloodstream, and CO₂ from the blood moves into alveolar air.

The exchange of gases in the lungs occurs by diffusion and is a rapid process. Oxygen, in higher concentration in the alveolar air, diffuses into the capillaries, where its concentration is lower. Carbon dioxide does the reverse. From 200 to 250 milliliters (ml) of oxygen per minute diffuses into the bloodstream per minute in a person at rest.

There are several factors that influence the amount of gases exchanged. These include the thickness of the membranes the gases must traverse, the area of alveolar surface in contact with blood circulating in the capillaries, the length of time the diffusion is allowed to take place, and the difference in concentrations of the gases in the alveoli and the blood.

38.6 Oxygen is transported from the lungs to the tissues by the hemoglobin in red blood cells.

The hemoglobin molecule (Hb, molecular weight, 68,000) consists of a protein portion, **globin**, and four **heme** groups. Each heme group consists of a porphyrin ring with an iron atom at its center (Figure 38-4). It is to this iron atom that oxygen binds. Since each Hb molecule has four heme groups, it is capable of combining successively to four oxygen molecules. As each heme group binds an oxygen molecule, the capability of the other heme groups to bind oxygen is actually enhanced, and it is this heme-heme interaction that confers upon hemoglobin its remarkable properties as a respiratory pigment.

In lung capillaries, Hb takes up oxygen readily and releases it "reluctantly." As oxygen joins a heme group, the affinity of the other three heme groups in that molecule for oxygen is increased, thereby insuring that the Hb molecule is loaded with oxygen before leaving the lung. In tissue capillaries,

however, where oxygen is in demand, the release of each oxygen molecule from a heme group weakens the binding of the remaining oxygen molecules to the other heme groups, and thus promotes the release of other oxygen as needed before the Hb molecule returns to the lungs.

38.7 Carbon dioxide is transported from the tissues to the lungs in the plasma, but hemoglobin also plays a critical role.

Carbon dioxide diffuses out of tissue cells as a waste product of their metabolism at a rate of 200 cm³ per minute in a person at rest. A small amount, 10 percent, travels to the lungs dissolved in the plasma, and a slightly larger amount, 20 percent, reacts with proteins in both the erythrocytes and plasma to form **carbamino compounds.** The majority of the remaining 70 percent enters erythrocytes and, in the presence of the enzyme carbonic anhydrase, combines with water to form carbonic acid ($CO_2 + H_2O \rightleftharpoons H_2CO_3$). The carbonic acid then dissociates (splits), forming hydrogen ions and bicarbonate ions ($H_2CO_3 \rightleftharpoons H^+ + HCO_3^-$), which diffuse into the plasma and are swept along by the bloodstream to the lungs.

The dissociation of carbonic acid with the production of hydrogen ions would considerably increase the acidity of venous blood were it not for the buffering action of unoxygenated (reduced) Hb. A **buffer** is a substance which can reversibly bind hydrogen ions and which thereby acts to minimize large changes in acidity. Within certain limits, by binding hydrogen ions as fast as they are produced, such a buffer tends to hold the concentration of free hydrogen ions constant. Oxygen-free hemoglobin acts in this manner.

Carbon dioxide, then, is carried in the plasma principally in the form of bicarbonate until it reaches the lungs. There it is reconverted to gaseous Co₂ and exits into the alveolar air to be subsequently exhaled.

38.8 In cellular respiration, oxygen is necessary to efficiently convert the energy in glucose molecules to chemical energy in the form of ATP.

As was discussed in Chapter 6 the breakdown of sugar for the production of energy occurs first along a common pathway (glucose to pyruvate), then

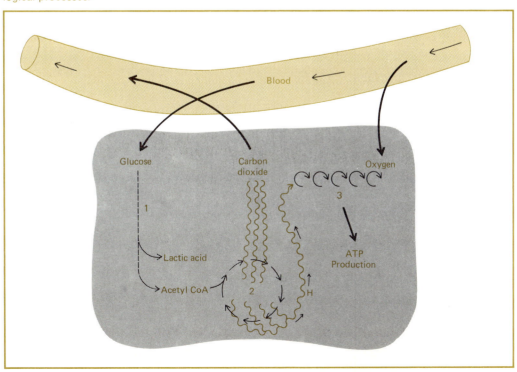

38-4 Each of the four heme groups in each molecule of hemoglobin in red blood cells undergoes a change in configuration,
as shown above, when it changes from the reduced (*left*) to
oxidized (*right*) state. An oxygen molecule becomes bound to
the iron atom at the center of each heme group.

38-5 A diagram summarizing the processes in cellular respiration in tissue cells. Glucose entering tissue cells from nearby
capillaries is broken down by the reactions of oxidative glycolysis (1) to acetyl coenzyme A, which enters the Krebs cycle
(2). The major products of the Krebs cycle are carbon dioxide,
which is returned to the bloodstream to be exhaled from the
lungs, and hydrogen in the form of H^+ ions and reduced coenzymes. This hydrogen is oxidized in the electron transport
chain (3), and the electrons it loses are passed on to the oxygen carried to tissues by hemoglobin molecules in the red
blood cells. The mitochondria of the tissue cell couple the
oxidation of reduced coenzymes to the synthesis of ATP,
which then serves as a ready source of energy for many biological processes.

through the Krebs cycle (pyruvate to CO_2 and H),
unless the body is strenuously exercising and has
insufficient oxygen for this and subsequent processes (section 6.3), and finally along an electron
transport chain oxydizing H to H_2O. The oxygen
that we have followed from the atmosphere to the
individual cell is the final hydrogen and electron
acceptor in this chain. From this entire process
one molecule of glucose yields the energy to produce 38 molecules of adenosone *tri*phosphate from
inorganic phosphate and adenosine *di*phosphate.
The chemical energy in ATP is released within a
cell as it is needed to synthesize compounds, actively transport substances across a membrane, and
drive many other reactions.

38.9 Normal breathing mechanisms can be overcome by grossly polluted air.

Although man has elaborate mechanisms to ensure that all his cells have the oxygen they need, nature has simply not endowed people with the apparatus necessary to cope with grossly polluted air. Toxic substances such as carbon monoxide, sulfur and nitrogen oxides, hydrocarbons, and particulate matter are spewed into the atmosphere of the United States alone by automobiles (the major offenders), industrial processes, solid waste disposal practices, and other sources at a rate of 600,000 tons every 24 hours. Under certain atmospheric conditions these substances combine to produce smog, the dense haze that hangs over cities. In 1930 smog hung over the Meuse Valley in Belgium and before it lifted caused 60 deaths and 6000 illnesses. In 1948 20 people died of a similar condition in Donora, Pennsylvania, and in 1952 the "killer smog" settled over London and took the lives of 4000 persons.

The National Air Pollution Control Administration has said that at least 300 American cities are suffering from severe air pollution, and more will suffer until and unless an educated citizenry moves to protect its own health.

More horrifying are the statistics compiled by the U.S. Public Health Service on cigarette smoking, the voluntary pollution of one's own atmosphere. Extensive evidence has clearly shown cigarette smoking to be a major risk factor for heart attack and lung cancer. Lung cancer alone now claims more than 50,000 American lives each year.

39

Digestion

QUESTIONS TO KEEP IN MIND

What are the nutritional requirements, both in terms of kinds of foods and calories, of an average adult?

How is food digested in the mouth, stomach, and small intestine?

What role do enzymes and hormones, respectively, play in digestive processes?

What function does the large intestine have?

The food we eat provides us with the substances we need for growth and energy. Although starvation and malnutrition are not major problems in the United States and Europe, they are in many other areas. *Over half the world's population is underfed and/or malnourished.* When malnutrition affects growing children or unborn infants, the effects are irreversible, both on bodies and on minds. There seems little hope that the world's food supply will suddenly explode to catch up to or even keep pace with the population explosion. Starvation is becoming more and more common and will exert its own grim pressure against overpopulation until man chooses to effect a more enlightened method of control. Even in the richest countries, where today at least, there is enough food, man seems determined to contaminate, overpreserve, artificially color and flavor, stabilize, emulsify, chemically supplement, and in other ways overprocess foods until many are, in all likelihood, rendered harmful for long-term consumption.

39.1 Three different kinds of foodstuffs—carbohydrates, proteins, and fats—are necessary to supply the body's needs.

Carbohydrates, sugars, and starch found in bread, potatoes, rice, bananas, and other foods, are made up of chains or polymers of sugar molecules linked together in either straight or branched chains (section 3.3). Some carbohydrates, such as cellulose found in plant cell walls (e.g., in vegetables), cannot be digested or broken down by humans. Others, such as starch, made up of many glucose molecules, can be broken down with the help of enzymes.

Proteins, another major class of food found in meat, fish, eggs, and cheese, are polymers of amino acids (section 3.4). Protein foods are broken down into these basic components in the intestinal tract by enzymatic action. Once digested, the protein's amino acids are used by cells to make their own proteins. Of the 28 amino acids found in human proteins, 20 can be manufactured within human cells, but 8 cannot. These **essential amino acids** must be provided in the diet.

Fats and other lipids are the third major class of food substances. They are found in chocolate, butter, bacon, and so on, and most are composed of three fatty acid molecules linked together by a glycerol molecule, forming a so-called triglyceride (section 3.8). Lipases, the enzymes that break down fats, catalyze the removal of fatty-acid molecules from triglycerides. Several fatty acids, phospholipids, and steroids (Chapter 3) essential to health must be provided in the diet.

Other substances, such as minerals and vitamins,

TABLE 12 VITAMINS REQUIRED BY HUMANS

Common name	Chemical nature or function	Deficiency disease	Known effects of excess ingestion
Vitamin A	β carotene	Night blindness, skin and growth	Nerve, skin, liver, and bone disorders
Vitamin B$_1$	Thiamine (a coenzyme)	Beriberi (polyneuritis), circulatory defects	–
Vitamin B$_2$	Riboflavin (a component of flavoprotein, part of cellular respiration system)	Eye inflammation, dermatitis	–
Vitamin B$_5$	Niacin (nicotinic acid, or its amide), a part of the cellular oxidation system	Pellagra, digestive disturbances	–
Vitamin B$_6$	Pyridoxine	Nervous disorders	–
Vitamin B$_{12}$	A cobalt-porphyrin compound essential to red cell maturation	Pernicious anemia	–
Vitamin C	Ascorbic acid involved in oxidation-reduction reactions in cells	Scurvy (swollen, bleeding gums), mental disturbances	–
Vitamin D	Several sterols related to ergasterol, calciferol	Rickets in children, osteomalacia in adults, disturbance in calcium and phosphorus metabolism	Hypercalcemia with central nervous system, neuromuscular, cardiac, renal, and other abnormalities
Vitamin E	Tocopherols (α, β, and γ)	Poor red cell survival, low fertility	–
Vitamin K	(A group of naphthoquinal derivatives)	Defects in the blood-clotting mechanism	–
Folic acid		Anemia, growth disorders	

are also needed. Vitamins are small molecules that participate in biochemical reactions within cells. Vitamin A, for instance, is needed for production of substances active in the photochemical process in retinal cells in the eye. The B group of vitamins are water-soluble substances essential for proper cellular function. Many are coenzymes in biochemical reactions. Other examples of necessary vitamins are shown in Table 12.

39.2 Food energy is measured in Calories. The number of calories taken in must equal the number used up if body weight is to be held constant.

A small calorie (or calorie with a small c) is defined as the amount of heat energy needed to raise the temperature of one gram of water one degree centigrade (C)—from 15° C to 16° C, for example. The unit commonly used in physiology and dietetics is the Calorie (with a capital C) or kilocalorie (kcal), which equals 1000 calories. Measured in kilocalories, the energy stored in the different classes of food substances varies, there being 4 kcal per gram in protein, 4 kcal per gram in carbohydrate, and 9 kcal per gram in fats.

The average adult must take in from 1600 to 3000 kcal per day, depending upon the type of activity he or she undertakes.

When caloric intake from absorbed nutrients exceeds caloric expenditure, weight gain is the inescapable consequence. How to combat the problem of excess weight, a problem faced by 80 million Americans, is a complex subject. A great deal of fact and fiction have been written about it, but whatever may be promised, there are only two ways to lose weight: by increasing energy expenditure and/or by decreasing food intake. Special "eat all you want of blank and get thin" diets may result in weight loss in some cases but also may be dangerous. Such diets should be followed only on the advice of a physician.

39.3 Food is processed in the alimentary tube, which extends from the mouth to the anus.

The gut or **alimentary tract**, a muscular tube some eight meters in length, lined with mucous mem-brane, includes the mouth, esophagus, stomach, small intestine, and large intestine (Figure 39-1). The basic muscular structure of this tube (Figure 39-1B) includes an outer layer of longitudinal smooth muscle, a middle layer of circular muscle, and a thin inner layer of submucosal muscle usually longitudinally arranged.

These muscle layers contract both to churn food about so that it mixes well with digestive fluids and to propel it along its course. **Nerve plexuses**, in two layers, are the locations of clusters of neurons from which axons extend to innervate the smooth muscles. The innermost layer of the tube, the **mucosa**, contains the structures and cells which process and absorb food from the gut's **lumen**, or hollow interior. The outer covering, or **serosa**, is continuous with the **mesentery**, which contains the nerves,

39-1 A. The human digestive system. B. The layers of the alimentary canal are similar along its entire length, except that the esophagus lacks a mesentery.

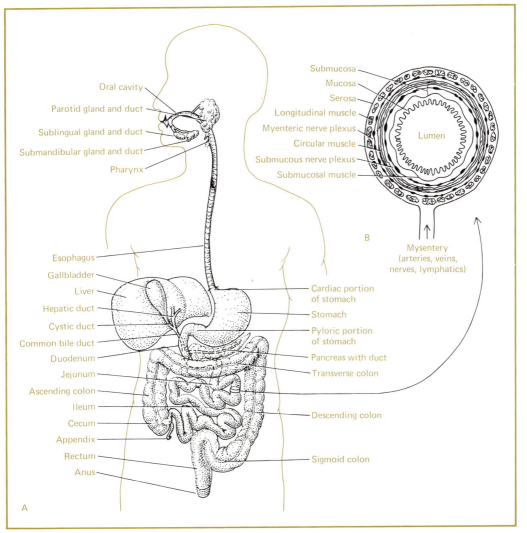

lymphatics, and blood vessels supplying the intestinal tract. At various points along this tube the processes of **digestion** (the breakdown of food into usable molecules), **absorption** (the taking of these molecules into the body proper), and **egestion** (the elimination of waste material, most of which has never been inside cells of the body, from the gut) occur.

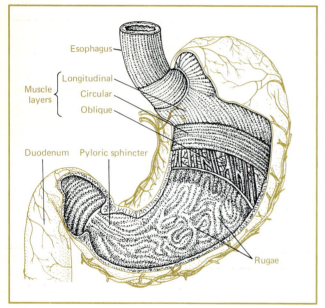

39-2 A cutaway diagram showing the layers of the stomach.

39-3 *A.* Gastric glands (the openings of which are the gastric pits), as seen in a section of the stomach. *B.* Diagram of a parietal cell, which secretes hydrochloric acid (HCl).

39.4 The progressive breakdown of food begins in the mouth, where food is chewed.

As food is chewed it is mixed with secretions from three pairs of **salivary glands.** One secretion is a mucuslike substance which lubricates the food and facilitates its subsequent passage down the **esophagus** to the stomach. Another secretion, the enzyme **salivary amylase,** initiates the breakdown of starches into their constituent glucose molecules.

When sufficiently chewed, food is swallowed and moved to the stomach by **peristaltic waves,** a series of involuntary constrictions of the alimentary tube which pass down the length of the esophagus one after the next, squeezing food ahead of each constricting portion. This process occurs in all parts of the digestive tract.

39.5 Once in the stomach, food is further digested by the action of stomach acid and pepsin.

The stomach, the largest dilatation of the alimentary tube, has well-developed muscle layers that work mechanically to mix the food with digestive secretions. Its mucosa is thrown up into ridges, or **rugae,** which increase its surface area (Figure 39-2). **Gastric** (stomach) **glands** or **pits** in the mucosa are lined with various types of cells which secrete **hydrochloric acid, pepsinogen, mucous,** and **intrinsic factor** (Figure 39-3). Hydrochloric acid,

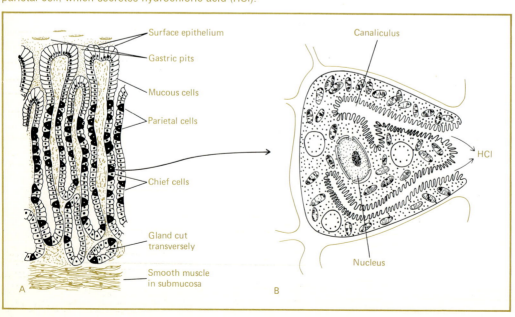

secreted by **parietal cells**, turns pepsinogen into its active form, **pepsin**. This enzyme is a **proteinase**; that is, it degrades proteins to polypeptides, smaller aggregates of amino-acid molecules.

Sometimes the acid and pepsin digest the stomach wall itself, and a hole or "peptic ulcer" results. Normally ulcers do not form, but how the stomach protects itself so well from its own juices is not clearly understood.

39.6 Gastric secretion is under neural and hormonal control.

The sight, smell, or even thought of food can cause an increase in the flow of saliva and gastric secretions. Impulses from higher brain centers that exert this influence travel down the vagus nerves to the stomach, where they stimulate the output of the gastric pits. Emotional states such as fear, anger, hostility, and depression may also stimulate gastric acid secretion.

Hormones regulate gastric acid secretion as well. The presence of proteins or alcohol in the lower part of the stomach, stomach distention, or stimulation from the vagus nerves can each cause the release of another substance from the gastric mucosa, the hormone **gastrin**. Like other hormones, it is released into the bloodstream, which carries it throughout the tissues of the stomach, where it increases the secretion of acid and pepsinogen.

Enterogastrone, another hormone, is liberated by cells in the upper portion of the small intestine. The stimuli for its release include the presence of fats, carbohydrates, and acid in the lumen of the intestine. Its action on the stomach is to decrease gastric secretion and motility. The rate at which acid secretions and food are passed into the intestines is thereby regulated.

39.7 Digestion is continued in the first portion of the small intestine, the duodenum, where enzymes from the pancreas and bile from the liver are added.

When food enters the **duodenum**, a series of enzymes from the pancreas and **bile** from the liver are added to it to assist in digestion.

The **pancreas**, situated beneath and behind the stomach, consists of an endocrine portion; the islets of Langerhans (section 35.7); and an exocrine portion that secretes pancreatic juice, a mixture of many enzymes in a watery alkaline secretion. The enzymes are secreted in their inactive form so that the pancreas does not digest itself, and only become active upon reaching the lumen of the small intestine. Secretion of pancreatic juice is stimulated by hormones elaborated in the upper portion of the intestine which are themselves stimulated by the presence of certain foods.

The **liver**, situated in the right upper quadrant of the abdomen is the largest abdominal organ, weighing nearly four pounds in the adult. It aids digestion by manufacturing bile, which empties into the duodenum.

Bile is a greenish-yellow fluid composed of bile pigments, bile salts, and various other substances. Bile pigments are the breakdown products of hemoglobin which are transported to the liver from the spleen (section 36.1). They give feces its brown color. The bile salts emulsify fats; that is, they break up large fat globules into minute particles which can then be degraded by fat-splitting enzymes (lipases) from the pancreas.

As bile is produced by the liver, it flows into **hepatic ducts** and from there it drains either into the duodenum, during meals; or, between meals, into the **gallbladder**, a small reservoir which stores the bile (Figure 39-4). Bile release is stimulated by the hormone **cholecystokinin**, made in the mucosa of the duodenum and released in response to the presence of fats in its lumen. This hormone causes the gallbladder to contract, squeezing its content of bile into the duodenum. Thus the flow of bile is also geared to the presence of food to be digested.

The liver performs a great number of other functions as well. All blood from the intestinal tract carrying absorbed food molecules is circulated to the liver via the portal veins and filtered before being returned to the heart. Blood moving via this portal system traverses a series of passages or **sinusoids**, where it is processed both by **phagocytic cells** which engulf particulate matter and by **hepatic (liver) cells** which remove certain compounds and detoxify other potentially harmful substances.

39.8 Digestive enzymes, like other enzymes, catalyze specific reactions.

Proteolytic enzymes, for example, catalyze (promote a chemical reaction without being used up in the process) the breakdown of protein molecules. Three of these proteolytic enzymes—**trypsin, chy-**

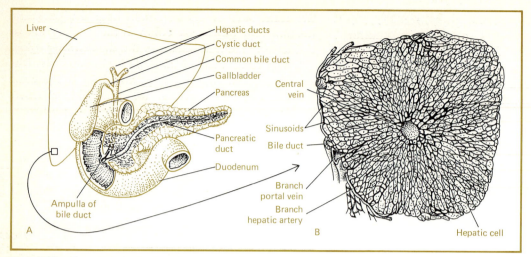

39-4 *A.* The ducts connecting the liver, gallbladder, and pancreas with the small intestine. *B.* Diagram of the cellular structure of the liver, showing its sinusoids, hepatic cells, and phagocytic cells.

motrypsin, and **carboxypeptidase**—may all work to help digest a single protein molecule, but each enzyme will only break one particular kind of bond. Trypsin, for example, will only act on the peptide bonds situated adjacent to the amino acid arginine.

39-5 *A.* The wall of the small intestine. *Plicae circulares* (circular folds) and the villi projecting from them greatly increase the surface area for the absorption of nutrients. *B.* Each villus is lined with epithelial cells.

39.9 Once broken down into small molecules, most food is absorbed by cells lining the small intestine.

After traversing the duodenum, food passes through the **jejunum** and **ileum**. The small intestine is about three meters long, without its convolutions and projections, it would not provide sufficient surface area for absorption. To increase its surface area folds of its wall project into the lumen of the gut (Figure 39-5). These are covered in turn by millions of tiny

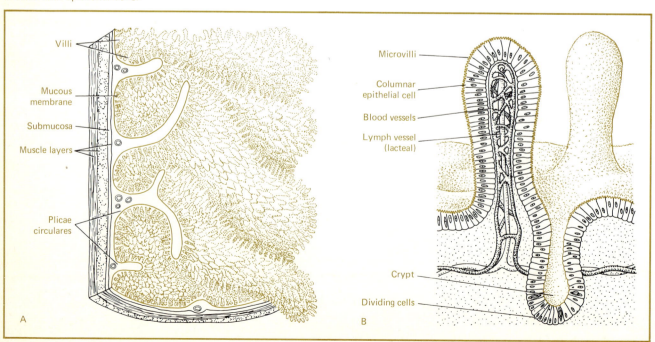

projections called **villi** (singular **villus**). Finally, the epithelial cells lining each villus have even smaller microscopic projections, **microvilli,** on their surfaces.

With these modifications providing tremendous surface area, the small intestine absorbs food molecules in several different ways, including active and passive transport and pinocytosis (section 4.7). Most nutrients are absorbed through the epithelial layer into the blood. Small droplets of emulsified lipids are absorbed into the lacteals, lymphatic vessels that lead eventually to the venous system via the thoracic duct (Figure 36-9).

39.10 Water and salts are absorbed in the large intestine.

Like the small intestine, the **colon,** or large intestine, is divided into several regions (Figure 39-1). At its junction with the small intestine, there is a blind, saclike portion called the **caecum,** from which the **appendix,** a blind tube about the thickness of a pencil arises. Occasionally food becomes caught in this structure, causing inflammation and pain in the lower right region of the abdomen—appendicitis. The caecum is continuous with the **ascending colon,** so called because it ascends along the right side of the abdomen near the back and becomes the **transverse colon.** It then swings across to the left side and becomes the **descending colon.** Near its lowest part it takes an S-shaped twist (the **sigmoid colon**) before emptying into the **rectum** with its external opening, the anus. This structure is surrounded by bands of muscles, or **sphincters,** which permit voluntary control of bowel movement.

As it enters the colon, digested food or **chyme** contains a large amount of water. A main function of the colon is to reabsorb this water and return it, together with various salts, to the bloodstream. Some of the bacteria living in the colon produce substances useful to the body, such as vitamins, and these are absorbed along with the water and salts.

Chyme gradually becomes less watery as it passes along the large intestine, and a residue of **fecal matter** or **feces** is formed. This consists largely of undigested food such as cellulose from plant tissue, sloughed-off cells from the lining of the intestine, large numbers of bacteria (which may constitute one-third of the solid matter), and mucus. This last substance is produced along the length of the intestinal tract and lubricates and facilitates the passage of chyme along it.

Chyme passes through the small intestine in from 2 to 6 hours, whereas it may remain in the large bowel as feces from less than 12 to more than 48 hours.

40

Excretion

QUESTIONS TO KEEP IN MIND

How is urine formed in the kidneys?

How do the kidneys function to keep the composition of body fluids stable?

Excretion is the elimination of metabolic wastes from cells. Carbon dioxide is eliminated through the lungs; bile pigments, the breakdown products of hemoglobin, are excreted by the liver into the gut; and some water and certain salts are excreted by sweat glands in the skin. Another most important excretory system is the urinary system, with its principal excretory organs, the kidneys.

40.1 The kidneys dispose of excess water and nitrogenous wastes such as urea.

The two kidneys in the human body are situated behind the abdominal cavity, one on either side of the spinal column at about waist level. The function of the kidneys is to rid the body of excess water and certain metabolic wastes which together comprise urine. Urine passes from each kidney through a tube called a **ureter** (Figure 40-1), into the **bladder,** where it is stored until it can be voided through the **urethra.** Bands of muscles (sphincters) at the junction of bladder and urethra allow for the voluntary control of urination.

If a kidney is sliced longitudinally, it can be seen that an outer **capsule** surrounds the entire organ. Within the kidney itself are two zones, an outer **cortex** and an inner **medulla.** Both regions are packed with **nephrons,** of which there are about 1.5 million in each kidney.

40.2 The nephron is the functional unit of a kidney.

Each microscopic nephron consists in part of a cup-shaped structure, **Bowman's capsule,** which surrounds a highly coiled capillary loop or tuft, the **glomerulus.** It is here that fluids and waste products leave the bloodstream and filter into the first part of the nephron. The filtrate then flows through the rest of the nephron, which consists of the **proximal convoluted tubule,** the **loop of Henle,** and the **distal convoluted tubule.** The long, thin tubule known as the loop of Henle is composed of a descending and an ascending limb. It projects into the medulla of the kidney, as do looped blood vessels, the **vasa rectae.**

From the ascending limb of the loop of Henle, the filtrate passes into the distal convoluted tubule and then into a **collecting duct,** from which it drains into the ureter. The loops of Henle, vasa rectae, and collecting ducts form most of the substance of the medulla of the kidney. The cortex is characterized mainly by the glomerular structures and the convoluted portions of the tubules.

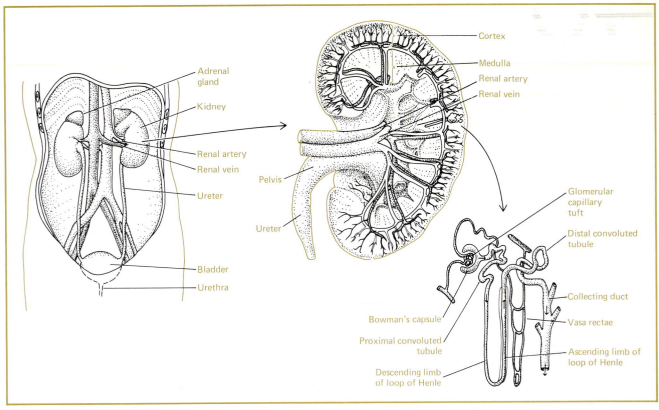

40-1 The human excretory system, showing the structure of the kidney and a single nephron with its blood supply.

40.3 The kidneys process approximately 1.3 liters of blood per minute.

As blood passes through the glomerular capillary tuft, its plasma, minus most of its protein, is filtered into Bowman's capsule. The glomerular filtration rate is about 125 milliliters (ml) per minute, which means that the kidneys filter out 180 liters per day—four times the total body water. Fortunately, about 99 percent of this is reabsorbed, and normal urine production is only one to two liters per day!

The glomerular filtration rate is influenced both by the permeability of the membranes and by blood

---

BIOEPICUREAN DELIGHTS

Rognons en Casserole

Slice fresh beef kidneys or lamb kidneys in half, and soak for half an hour in fresh water. (Four veal kidneys or 12 lamb kidneys serve 4.) Sauté in $\frac{1}{4}$ lb of butter for 10 minutes for beef, 5 minutes for lamb, until the kidneys are brown. Transfer to a heated platter to keep warm while making the sauce. Add $\frac{1}{4}$ cup of chopped scallions or shallots to the butter (adding more if needed) and sauté for 1 minute. Add $\frac{3}{4}$ cup of Chablis or other dry white wine, and boil while stirring, long enough to reduce the volume of the fluid to less than half the original volume. Cream half a stick of butter, softened by leaving it at room temperature, then mix in 2 tablespoons of Dijon (French) mustard, salt, and black pepper. Place the kidneys in a casserole, and cut them into slices; pour the sauce over them, sprinkle with lemon, and decorate with parsley. (Beverage: Burgundy wine.)

pressure, which is higher in the glomerular capillaries than in capillaries elsewhere in the body. This higher pressure helps force or squeeze the plasma filtrate into the nephrons. By changing the volume of urine or its concentration of solutes ever so slightly, the kidneys exert exquisite control over the volume and composition of body fluids.

40.4 The glomerular filtrate is modified by reabsorption and secretion in the tubules.

Although a great deal of fluid is filtered, little is finally excreted, because much water and some other substances are reabsorbed. On the other hand, some substances are added to the filtrate by secretion. Both processes occur in the tubules. Water and various ions are reabsorbed passively by diffusion, whereas other substances, primarily glucose, are reabsorbed by active transport (section 4.6).

Glucose is reabsorbed in the proximal convoluted tubules. Normally, almost all the glucose filtered through the glomeruli is reabsorbed for use by the body. However, the active transport mechanism can pump only so much glucose per minute, and if the glucose in the nephron is excessive, some will be lost in the urine. This occurs in persons with diabetes mellitus.

In addition to transporting glucose and other substances out of the tubule and back into the blood, the cells of the tubules pass some compounds such as certain dyes, and drugs in the other direction and increase their concentration in the urine.

40.5 Nitrogen is excreted mostly in the form of urea.

Urea, an important constituent of urine, is generated in the **Krebs-Henseleit cycle** (not to be confused with the Krebs cycle of intermediary metabolism). Reactions of the cycle are depicted in Figure 40-2. As can be seen in the figure, each urea molecule contains two nitrogen (N) atoms. The excretion of this compound thus provides the principal means for the removal of nitrogen from the body. Some nitrogen is also excreted as ammonia or uric acid.

40.6 Normal renal function is essential for water, acid-base, and salt balance.

By helping maintain the composition of the blood within narrow limits, the kidneys play a large part in keeping the body's internal environment stable. They do this by regulating the amount of water to be excreted, maintaining the proper acid-base balance, eliminating excess inorganic salts, and as we have already seen, removing waste products of nitrogen metabolism.

The amount of water reabsorbed by the tubules of the nephron must be just enough to keep the **osmolarity** of the body fluid stable; that is, the concentration of solutes must remain constant. One way in which the amount of water in the body is regulated is by the antidiuretic hormone, **vasopressin.** When **osmoreceptors,** special receptor cells, in the hypothalamus of the brain, detect an increased concentration of solutes in the blood (increased fluid osmolarity), nerve cells in this region cause vasopressin to be released from the pituitary gland (section 35.2 and Table 11).

Vasopressin makes the walls of the distal tubule and collecting duct more permeable to water. More water is therefore absorbed back into the collecting duct (Figure 40-1), and a small volume of highly concentrated urine is formed. The reabsorbed water dilutes the solutes in the blood and lowers body fluid osmolarity, and vasopressin release is shut off. When vasopressin is absent, less water is reabsorbed by the distal tubule and collecting duct, and a large volume of dilute urine is excreted.

40.7 Blood acidity is regulated by hydrogen and bicarbonate ion excretion.

When blood is too acid, hydrogen ions are secreted by the tubules into the tubule lumen, in exchange for sodium ions. These sodium ions pass through the tubule cells and into the bloodstream. Bicarbonate ions formed during the dissociation of carbonic acid which led to hydrogen ion secretion also pass into the bloodstream. The net result is that a hydrogen ion is secreted into the urine and a bicarbonate ion is returned to the blood along with a sodium ion.

If the blood is very acid (a condition termed severe **acidosis**) many hydrogen ions must be secreted to correct the disturbance. Were there no provision for the buffering of hydrogen ions in

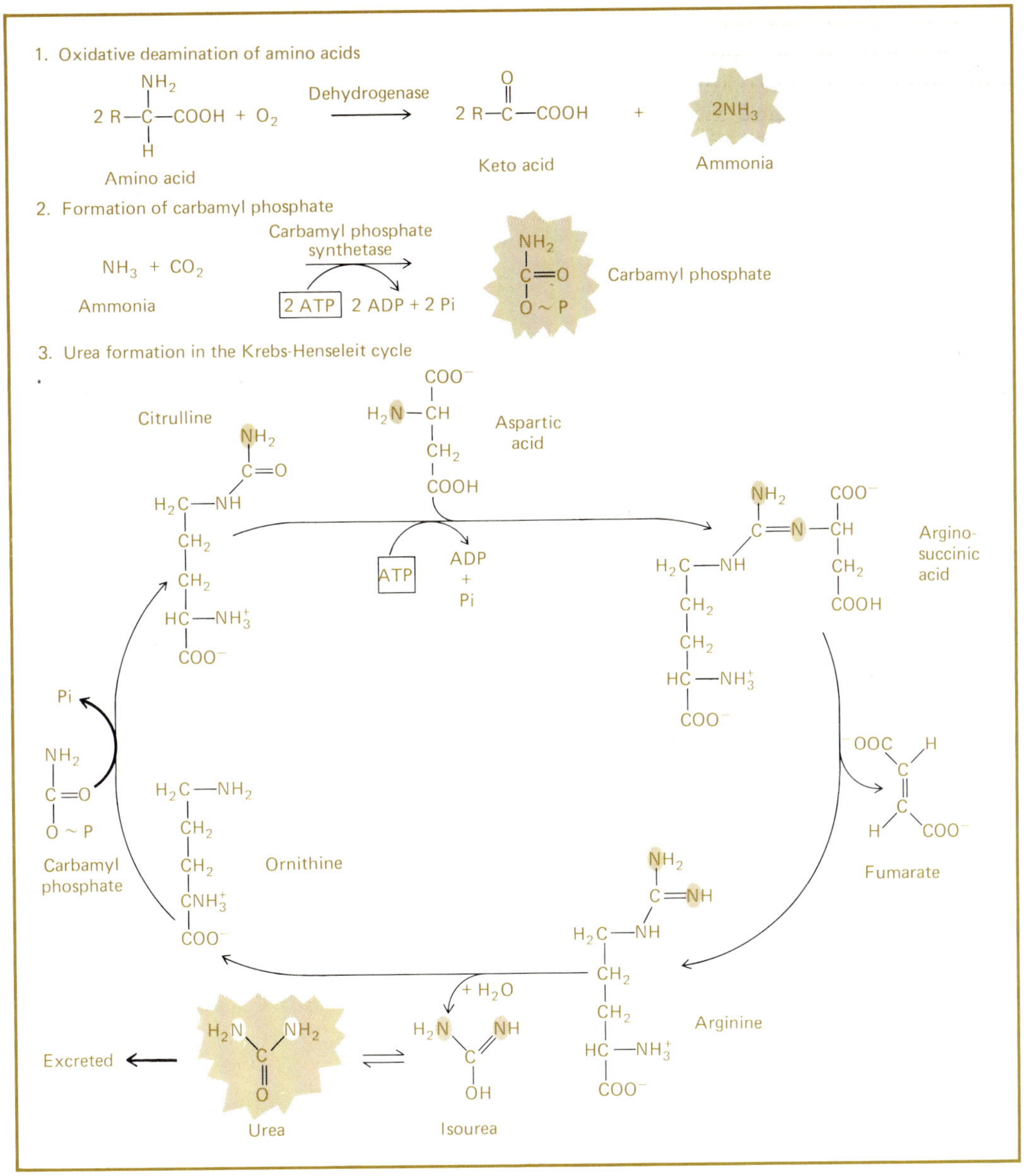

1. Oxidative deamination of amino acids

$$2\ R-\underset{\underset{H}{|}}{\overset{\overset{NH_2}{|}}{C}}-COOH + O_2 \xrightarrow{\text{Dehydrogenase}} 2\ R-\overset{\overset{O}{\|}}{C}-COOH\ +\ 2NH_3$$

Amino acid   Keto acid   Ammonia

2. Formation of carbamyl phosphate

$$NH_3\ +\ CO_2 \xrightarrow[\boxed{2\ ATP}\ 2\ ADP + 2\ Pi]{\text{Carbamyl phosphate synthetase}}$$

Ammonia

$\underset{O \sim P}{\overset{NH_2}{\underset{|}{\overset{|}{C=O}}}}$ Carbamyl phosphate

3. Urea formation in the Krebs-Henseleit cycle

Citrulline

Aspartic acid

ATP

ADP + Pi

Arginosuccinic acid

Carbamyl phosphate

Ornithine

Fumarate

Arginine

Pi

Excreted ←

Urea

Isourea

$+ H_2O$

40-2 The principal chemical reactions of nitrogen excretion: (1) oxidative deamination; (2) formation of carbamyl phosphate; and (3) the Krebs-Henseleit cycle in which urea is produced.

urine, secretion of these ions would soon stop. It would cease because the concentration of hydrogen ions in the lumen of the tubule would become too great to pump against. However, there are a number of ways hydrogen ions, once secreted, are buffered, thereby "making room" for further ion secretion. They may combine with a number of anions such as hydrogen phosphate (HPO_4^{-2}), or with bicarbonate ions.

When the blood is too basic or alkaline, a different situation obtains. In this condition, **alkalosis**, hydrogen ions are conserved and not secreted. Furthermore, bicarbonate ions in high concentration in the glomerular filtrate are excreted.

40.8 Sodium is reabsorbed to maintain body fluid volume.

Sodium plays an important role in the reabsorption of water and in acid-base balance. Being the most abundant cation in the **extracellular fluid** (plasma plus interstitial fluid), its concentration regulates fluid volume by osmosis. The maintenance of extracellular fluid volume is so important that there are a number of mechanisms for carefully regulating sodium balance. These mechanisms operate to insure that the amount of sodium excreted is adjusted to equal the amount ingested. Variations in the amount of sodium excreted are brought about by changes in glomerular filtration rate, changes in tubular reabsorption processes, or both. The tubular reabsorption of sodium is controlled by a hormonal mechanism.

41

Reproduction

QUESTIONS TO KEEP IN MIND

How are sperm and egg cells produced?

By what mechanisms are sperm and egg cells brought together for fertilization to occur?

How do the menstrual and uterine cycles fit into the broader scheme of the reproductive cycle in the human female?

How is a fetus nourished inside its mother?

How does a baby adapt to independent existence at birth?

Thus far we have considered the structure and function of organ systems that are essential to the survival of individuals. Although reproductive hormones have various physiological and behavioral effects on an individual, the reproductive system is not essential to the survival of individuals. Sexual reproduction *is* essential to the survival of the species, or perhaps it would be more accurate to say that *controlled* reproduction is essential. Unchecked population increase, with its attendant problems of crowding, altered social behavior, depletion of natural resources, food shortages, and pollution, poses a very real threat to the survival of the human species.

Sexual reproduction is characteristic of the majority of living organisms, from unicellular organisms to man, and is a very successful method of ensuring not only survival of the species, but especially genetic innovation (see Chapter 11).

In man as in other organisms, the reproductive systems of two sexes produce gametes (eggs and sperm) and deliver them to an environment where fertilization can occur and where the zygote (fertilized egg) can grow and develop.

41.1 In the male reproductive tract, sperm are produced in the testes.

The reproductive tract of the male (Figure 41-1) consists of two testes suspended in a saclike pouch, the **scrotum**; several glands; and conducting tubes. The testes house tiny coiled **seminiferous tubules**. **Spermatogenesis**, the production of sperm, takes place in the **spermatogenic epithelium**, a specialized lining of the seminiferous tubules (Figure 41-2). For sperm to reach the outside, they must pass through the **epididymis**, a collection of tubules where they may be stored temporarily; the **vas deferens**, a thick-walled conducting tube; and the urethra in the **penis**.

Various substances are added to the sperm as they traverse this path. The **seminal vesicles, prostate gland**, and **Cowper's gland** add secretions that nourish and activate the sperm. Together, the sperm and secretions comprise the **semen**, a white, opalescent, viscous fluid.

Sperm cannot develop normally at internal body temperature—it is too high. By being suspended in the scrotum, the testes are kept at a lower temperature suitable for sperm production and maturation. Should this temperature fall too low the testes can

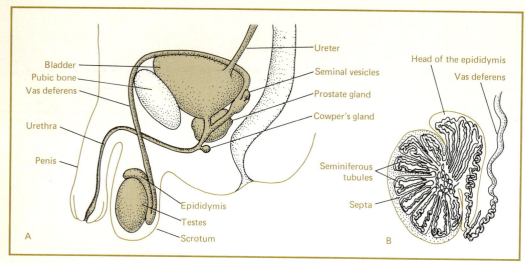

41-1 *A*. The male reproductive system. *B*. A longitudinal section of one testis, the epididymis, and vas deferens.

be drawn closer to the body by the involuntary action of the **cremaster muscle.**

41.2 Sperm cells have a head, which contains the genetic material, and a flagellum, which propels the sperm.

When activated by secretions from the glands mentioned above, sperm cells travel at a rate of approximately three millimeters per minute. It is the motile mechanism of the tail, as well as contractions of the female reproductive tract, that permits the sperm to reach an ovum when introduced into a female.

Semen is released from the male reproductive tract after sufficient sexual stimulation and spurts outward from the urethra in the process of **ejaculation.** After several days of sexual abstinence, the volume of an average ejaculate is between 2 and 4 milliliters. Although it only takes one sperm cell to fertilize an ovum, there are approximately 60 to 120 million sperm per milliliter of semen, and sperm counts below 20 million per milliliter of semen are associated with apparent inability to reproduce.

41.3 Testosterone, the male sex hormone, is produced by specialized interstitial cells between the seminiferous tubules in the testes.

Testosterone, produced under the influence of luteinizing hormone (LH) released by the anterior

pituitary, is responsible for the development of male secondary sex characteristics such as growth of body hair in the characteristic male pattern, deepening of the voice, enlargement of the penis and body musculature. Testosterone also promotes the male balding pattern in individuals with appropriate genes and fosters the development of characteristic aggressive male behavior by an as-yet-unknown mechanism.

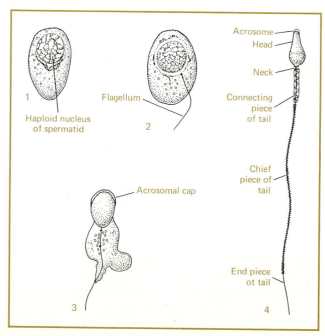

41-2 Stages in the differentiation of a human sperm cell.

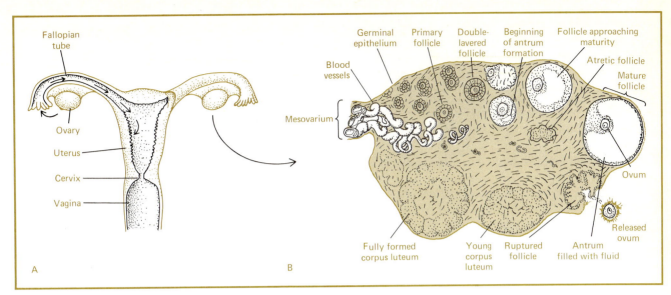

41-3 *A.* The reproductive organs of the human female. *B.* Diagram of the progressive stages in the maturation of an ovarian follicle before ovulation, and the development of a ruptured follicle into a corpus luteum. The stages are shown in clockwise sequence, beginning near the mesovarium.

41.4 In the female reproductive system, egg cells are produced in the ovaries.

Female reproductive cells, egg cells or ova, are produced in the ovaries (Figure 41-3) and are released in a process called ovulation. One ovum is usually released about every 28 days. Once released, the ovum is swept by the ciliary action of cells lining the funnel-shaped opening of one of the **fallopian tubes** down this tube to the **uterus.** Fertilization of the ovum usually takes place in a fallopian tube but may take place in the uterus itself. The uterus opens through the **cervix** into the **vagina,** where semen is deposited during the reproductive act. Above and outside the opening of the vagina is the **clitoris,** a small, highly innervated structure which becomes swollen and erect during sexual stimulation. The only known function of this sensitive structure is to transmit pleasurable impulses to the brain when stimulated by any kind of friction.

41.5 In man and many other animals, semen is discharged directly into the reproductive tract of the female in a process called copulation.

In humans, copulation, usually called **sexual intercourse** or **coitus,** is a somewhat more complicated process than in lower animals, not only because the biological factors are complex (many hormonal and neural mechanisms take part), but also because psychological and social factors play an important role as well. Lovemaking usually begins with a wide range of tactile and other sensory stimulation, or **foreplay,** which readies both partners for coitus. The senses of sight, smell, taste, and especially touch are stimulated, with the result that the woman's heart rate and blood pressure rise, her nipples become erect, her vagina and clitoris become swollen with blood, and secretions from glands in her cervix lubricate the vagina in preparation for insertion of the penis.

The male usually responds to foreplay more quickly, by a doubling in heart rate, rise in blood pressure, and erection of the penis. Erection is a neurally mediated process initiated by dilation of arterioles causing the erectile tissue to fill with blood. Veins compress, blocking the outflow of blood and thus adding to the turgor of the penis. Introduction of the erect penis into the vagina assures that sperm cells, when ejaculated, are deposited near the opening of the cervix and are constantly in contact with moist surfaces at physiologic temperatures. Ejaculation occurs almost simultaneously with the male's **orgasm,** a climax of excitement characterized by spasms of sexual gratification. In females sexual excitement may develop more slowly than in the male; orgasm is sometimes brought on by the stimulating action of the penis on either the clitoris, the vaginal walls, or both during sexual intercourse. Many men fail to learn to delay ejaculation long enough for their partners to reach orgasm; on the other hand, those who learn patience and restraint are often gratified to learn that many women are capable of multiple orgasms.

The achievement of orgasm by self-manipulation of either the penis or the clitoris is known as **masturbation.** Recent evidence suggests that it is practiced at some time in life by some 90 percent of the male population and by only a slightly smaller percentage of females. It is particularly common among adolescents who often have no other safe or attractive alternative for sexual release. Also fairly common among this age group are experiments with homosexuality, a sexual urge for a relationship with a member of the same sex. Homosexuality is no longer considered a psychic disorder, and some scientists and sociologists have pointed out that juvenile homosexuality is often a stage in the development of heterosexual adjustment and have suggested that without the strong social taboos enforced in many societies, a good number of humans might naturally be bisexual—sexually attracted to members of both sexes.

41.6 The menstrual cycle is one manifestation of the total reproductive cycle of the human female.

The human female has a **menstrual cycle** lasting approximately one lunar month, or 28 days. Throughout this period there are cyclical changes in the ovaries and in the uterus which are coordinated to prepare the female reproductive tract for possible insemination and fertilization. Although intimately related, these changes are discussed below separately.

The process of ovulation is shown in Figure 41-3B. Cells that are destined to become ova originate in the **germinal epithelium** of the ovaries. At birth a female possesses all such cells that she will ever have. These grow and develop in response to follicle-stimulating hormone (FSH) from the anterior pituitary, and on about the sixth day of the cycle, one follicle in one of the ovaries begins to grow faster than the rest. For unknown reasons, the other follicles regress, thus assuring that only one follicle, containing a single egg develops each lunar month. As the "chosen" follicle grows, a fluid-filled cavity or **antrum** develops around the ovum. On about the fourteenth day, a burst of luteinizing hormone (LH) from the anterior pituitary brings about **ovulation.** The follicle ruptures, and the egg is re-

leased from the ovary, to be swept by ciliary currents into the fallopian tube and travel to the uterus.

The ruptured follicle fills with blood. In a few days, yellow **lutein cells** replace the blood, and the structure becomes a **corpus luteum** (yellow body), the development of which is fostered by LH. The corpus luteum itself becomes an endocrine organ and secretes the two hormones **estrogen** and **progesterone**, which affect many structures, including the uterus. If an ovum is not fertilized, the corpus luteum gradually shrinks and ceases to function.

41.7 The uterine cycle is concerned with changes brought about by estrogen and progesterone in the *endometrium,* the spongy inner lining of the uterine wall.

During the first 14 days or so of the menstrual cycle, called the **proliferative phase,** the uterine wall grows under the stimulation of estrogen from the developing follicle in the ovary (Figure 41-4). On about day 14, ovulation occurs and shortly thereafter the corpus luteum begins to secrete both estrogen and progesterone. Under stimulation from these hormones, glands in the uterine wall begin to secrete various substances, and the uterus enters into the **secretory phase** of its cycle.

All these changes prepare the uterus to receive a fertilized ovum. If the ovum is not fertilized, it degenerates, as does the corpus luteum. With less and less estrogen and progesterone, the secretory uterine wall begins to degenerate and slough off, producing the **menses,** or menstrual flow, during a woman's **monthly period.** Bits of tissue and blood pass out through the cervix and vagina. A special anticoagulant factor present in the uterine wall prevents the blood from clotting, while constriction of blood vessels in the secretory endometrium prevents excessive bleeding. The uterus is thus cleaned out, and the whole system is readied for the beginning of a new cycle.

If fertilization of an ovum does occur, it may then **implant** on the uterine wall. As it develops there, specialized cells arise which secrete a hormone, **human chorionic gonadotrophin** (HCG), which furthers the development of the corpus luteum and its hormones. Hormonal support (estrogen and progesterone) from the corpus luteum is thereby ensured, the monthly period does not ensue, and development of the fertilized ovum is permitted to continue.

41.8 Conception, or fertilization, occurs when a sperm penetrates an ovum.

After deposition at the opening of the cervix, sperm cells, aided by contractions of the female reproductive organs, make their way through the uterus and up the fallopian tubes. It is there that fertilization takes place. The acrosome in the head portion of the sperm plays a role in this penetration process (section 13.4). After penetration, the male and female genes join in directing the development of the fertilized egg.

The fertilized ovum begins to divide before it reaches the uterus, and by the time it arrives there consists of a small ball of cells, or blastula (section 13.5). The blastula attaches to the side of the uterus and becomes embedded in the uterine wall, which has been prepared to receive it. As the embryo enlarges and develops organs, it is called a **fetus** and is protected by a surrounding series of saclike membranes. The fetus itself is suspended in the fluid interior of the innermost sac, the **amnion**.

41.9 As development proceeds, the fetus gains nourishment via its umbilical cord, which conducts fetal blood to and from the placenta.

The **placenta**, a flat, disk-shaped structure closely applied to the wall of the uterus, enables the fetus to derive nourishment from the mother. Blood coming from the fetus enters placental capillary networks and by so doing, passes close to maternal capillary networks in the wall of the uterus. The two circulatory systems normally do not mix, however. Oxygen, nutritional substances, and waste products pass back and forth by various mechanisms across the placental membranes.

As it grows (Figure 41-5), the fetus stretches the uterus and compresses other organs in the mother's abdominal cavity, especially the bladder and colon. A normal pregnancy lasts nine months.

41.10 An infant's blood circulation changes at birth.

As the end of the nine-month period approaches, various hormonal changes occur, and **labor** begins.

41-4 The related changes during the ovarian and uterine cycles are shown diagrammatically in an ordinary menstrual cycle and one that ends in pregnancy.

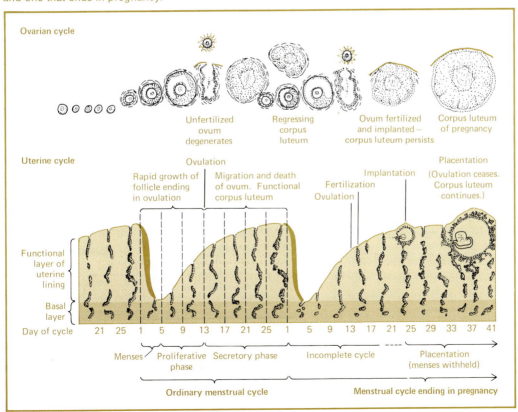

Ovarian cycle

Unfertilized ovum degenerates

Regressing corpus luteum

Ovum fertilized and implanted – corpus luteum persists

Corpus luteum of pregnancy

Uterine cycle

Ovulation

Placentation (Ovulation ceases. Corpus luteum continues.)

Rapid growth of follicle ending in ovulation

Migration and death of ovum. Functional corpus luteum

Implantation

Fertilization

Ovulation

Functional layer of uterine lining

Basal layer

Day of cycle   21 25 1 5 9 13 17 21 25 1 5 9 13 17 21 25 29 33 37 41

Menses   Proliferative phase   Secretory phase   Incomplete cycle   Placentation (menses withheld)

Ordinary menstrual cycle   Menstrual cycle ending in pregnancy

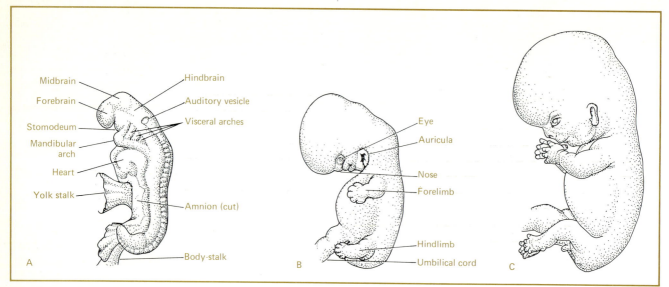

Midbrain
Forebrain
Stomodeum
Mandibular arch
Heart
Yolk stalk
Hindbrain
Auditory vesicle
Visceral arches
Amnion (cut)
Body-stalk
A

Eye
Auricula
Nose
Forelimb
Hindlimb
Umbilical cord
B

C

41-5 Stages in the development of human embryos: (A) fourth week; (B) sixth week; (C) eight and one-half weeks.

The uterus begins to contract at shorter and shorter intervals, and eventually, the child is expelled down through the birth canal, traversing the cervix and vagina, which have greatly dilated to allow its passage. What happens then is truly remarkable.

Up to this point the fetus has never used its lungs. Before birth, the circulation of the heart is ingeniously arranged to pump blood from the umbilical cord directly to the tissues, entirely bypassing the lungs. At the moment of birth, however, as air is taken into the lungs, the circulation in the heart reorganizes to pump venous blood into the pulmonary circuit—an amazing adaptation that is effective within seconds after birth.

41.11 An infant's initial nutritional needs can be entirely met by a healthy mother's breast milk.

The female breast develops under the stimulation of both estrogen and progesterone. Two hormones are directly involved in milk production and secretion. Estrogen stimulates proliferation of the **lactiferous ducts,** while progesterone stimulates the secretion of milk by the **alveolar cells** in the mammary glands. Milk is carried to the nipples by the lactiferous ducts (Figure 41-6). In humans, full-scale milk production is not achieved until one to three days following delivery. Suckling of an infant at the mother's breast evokes reflex oxytocin release from the mother's posterior pituitary (section 35.2) and consequent milk ejection. Along with milk,

the breast-feeding infant obtains maternal antibodies which afford it some protection from infections during its first few months of life.

41.12 Human reproduction may be controlled by various means.

A healthy human female has the biological capacity to produce about 20 children. The vast majority, however, have far fewer children and for one reason or another take active measures to limit the size of their families. Various approaches have been used.

The only 100-percent-effective ways to prevent pregnancy are to avoid sexual intercourse entirely or to have one of the sexual partners sterilized. Since it is known that a woman ovulates approximately in the middle of her menstrual cycle, it is theoretically possible to avoid conception by abstaining from intercourse during that time. This "rhythm method" of contraception is highly unreliable, because the time of ovulation is both variable and difficult to predict. Pregnancies have resulted from isolated intercourse on every day of the cycle.

A more effective way to prevent conception is for a woman to take "the pill," a mixture of synthetic hormones, estrogen- and progesteronelike substances that interfere directly with ovulation. This method is not 100 percent effective, probably because some women forget to take a pill each day as prescribed.

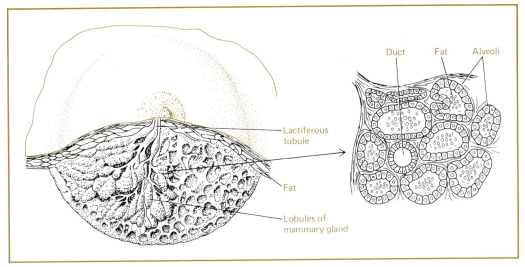

41-6 The structure of the human breast, showing a mammary gland in cross section (*right*).

Mechanical devices that are reasonably effective in preventing conception include the **condom**, a rubber membrane that fits over the penis, preventing semen from entering the female reproductive tract; the **diaphragm**, a rubber cap that is slipped over the entrance to the cervix deep inside the vagina; and the more permanent **intrauterine device** (IUD). An IUD, such as a loop or coil, is inserted in the cervical opening by a physician to prevent an egg from implanting in the uterine wall.

More recently, surgical sterilization techniques have beem employed. **Vasectomy,** the removal of a portion of the vas deferens of the male, interrupts transport of sperm from the testes. In women, fallopian tube ligation (tying) or removal prevents ova from reaching the uterus.

Because of the very real danger that the human population will outstrip the world's food supply, exhaust its resources, and irreversibly pollute its environment, there is much active research into methods of contraception. The problem is to find a method that is inexpensive, acceptable on religious and aesthetic grounds, and does not require the services of a physician. The development of an effective worldwide contraceptive program may turn out to be the most important problem facing mankind today.

SUGGESTED READING

Bloom, W. and D. W. Fawcett. *A Textbook of Histology.* 8th ed. Philadelphia: Saunders, 1968.

Comroe, J. H. *Physiology of Respiration.* Chicago: Year Book Medical Publishers, 1966.

Davenport, H. W. *Physiology of the Digestive Tract.* 2nd ed. Chicago: Year Book Medical Publishers, 1966.

Ganong, W. F. *Review of Medical Physiology.* 6th ed. Los Altos, Calif.: Lange Medical Publications, 1971.

Gatz, A. J. *Essentials of Clinical Neuroanatomy and Neurophysiology.* 4th ed. Philadelphia: Davis, 1970.

Katz, B. *Nerve, Muscle, and Synapse.* New York: McGraw-Hill, 1966.

Kennedy, D., intro. to "The Living Cell" in *Readings from Scientific American.* San Francisco: Freeman, 1965.

Pansky, B. and E. L. House. *Review of Gross Anatomy.* 2nd ed. New York: Macmillan, 1969.

Pitts, R. F. *Physiology of the Kidney and Body Fluids.* Chicago: Year Book Medical Publishers, 1963.

Sidman, R. L. and M. Sidman. *Neuroanatomy—A Programmed Text.* Vol. I. Boston: Little, Brown, 1965.

THE MODERN THEORY OF THE DESCENT OF MAN.

VI EVOLUTION

There is more in this book about evolution than it may appear from the Table of Contents. In Part I, the characterizations of procaryotic and eucaryotic cells and their reproduction showed in a general way what cellular mechanisms have evolved for reproduction. These mechanisms have in turn established the limits on the further evolution of their respective organisms. Part II showed, among other things, that developing organisms have a tendency to pass through stages that retrace their evolutionary history. Part III dealt with the principles and mechanisms of inheritance, heredity, and variation. These are the ''raw materials'' of evolutionary processes. Part IV presented the five kingdoms of organisms as an evolutionary fait accompli and organized the discussion of organismal diversity in terms of evolutionary sequence.

Part VI deals not with the details but with the mechanisms of evolution and the history of ideas about how evolution took place.

It is difficult for young people today to imagine the turmoil, both scientific and social, that Charles Darwin's ideas generated, not only a century ago, but even up to the present time. The turmoil now seems to have subsided almost everywhere,* but Darwin's theory of evolution by natural selection has left an indelible mark on the entire culture. Thanks to him and others who followed, we now have a clear picture of how the amazingly complex web of life on earth evolved and how it continues to evolve, as the earth and the genetic constitution of its creatures change and interact.

* Except for some new laws in some states requiring that in public school courses where the theory of evolution is taught, ''equal time'' be devoted to the biblical account of creation.

42

Evolution Has Occured

QUESTIONS TO KEEP IN MIND

What is evolution, and what evidence do we have that the process has occurred?

How and from what has man evolved?

Why should one particular species have evolved rapidly within the last century or so?

In 1831, Charles Darwin shipped out on H.M.S. *Beagle* as a guest naturalist. When he left, he supported the prevalent belief that each living species had been created in a unique and unchanging form. Global catastrophes had demolished previous populations—evidence of their existence remained as fossils—and new, unrelated species had been created to take their places.

When Darwin returned almost five years later he had changed his thinking. He was convinced that organisms evolved very slowly and that fossils, ancestors of present forms, represented a partial record of this process.

What did Darwin see that so altered his view of creation? Certainly he did not find so much evidence of the evolution of any species of plant or animal compared to the powerful and convincing examples that evolutionists have discovered in the past 100 or more years. Yet in a sense, Darwin saw a great deal, and he did a great deal with what he saw, as will be discussed in this and following chapters.

42.1 Evolution is a change in the inherited phenotypes (the heritable appearances) of members of a population.

Evolution is a special kind of change that can occur only in a group of organisms. An individual does not evolve. Evolution occurs within a **population,** which may be defined as a group of organisms of the same species living in a more or less restricted place; the process consists of a change in the inherited **phenotype** (section 15.3). The phenotype of any organism, as you may remember, is all of its measurable features: its color, size, biochemical constituents, rate of development, behavior—everything, in fact, that one can tell about the organism from examining it directly. Darwin used "similar phenotype" and "same species" interchangeably and this is still an adequate approximation.

It is possible for a population to evolve even though evolutionary changes do not develop within a given individual. An adult gray moth does not become black nor does a bacterium become resistant to a drug, but one of the gray moth's offspring can turn out to be black, and so on. A population is made up of different individuals at different times, and therefore it reflects the net changes that have occurred over many generations. If a population is examined at two widely separated times, and new phenotypes of a sort that can be passed on to future generations are present at the later date, then we say that the population has evolved (Figure 42-1).

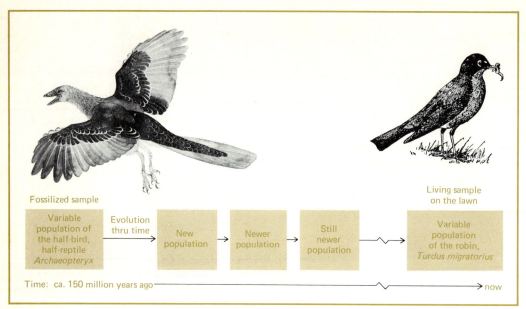

Fossilized sample

Living sample
on the lawn

| Variable population of the half-bird, half-reptile *Archaeopteryx* | → Evolution thru time → | New population | Newer population | Still newer population | Variable population of the robin, *Turdus migratorius* |

Time: ca. 150 million years ago ──────────────────→ now

42-1 Diagram of the definition of evolution. Evolution is the modification of heredity in a population's membership. Although *Archaeopteryx* (*left*) may not have been the exact ancestor of modern birds, it must have been fairly close to that ancestor. It retained many reptilian features: teeth, bones in the tail, and digits on the forward edge of the wing. But its feathers are completely birdlike. (*Archaeopteryx* restoration by Maurice Wilson. Courtesy of the Trustees of the British Museum (Natural History). Robin by G. M. Sutton, *An Introduction to the Birds of Pennsylvania*, 1928. From *And Replenish the Earth*, © 1974 by Michael L. Rosenzweig.)

42.2 The record we need of past populations is usually only available as fossils.

Because evolution usually takes thousands or millions of years to produce a noticeable change, it is usually observed by comparing modern populations to ancient ones, which exist only incompletely, as fossil remains. We cannot be sure that the fossils we do find are typical examples of their population, but from what we know about the process of fossilization this would seem to be true. The close correspondence between individual fossils and the populations they represent is dramatically demonstrated by the occasional discovery of a living "fossil" — a living representative of a supposedly extinct fossil group.

For example, *Latimeria*, the coelacanth, is a member of an ancient subfamily of fish known for a long time only from fossils. Scientists used to believe that coelacanths of all types became extinct 75 million years ago. But in 1939, a living coelacanth was caught in the deep waters off the Mala-

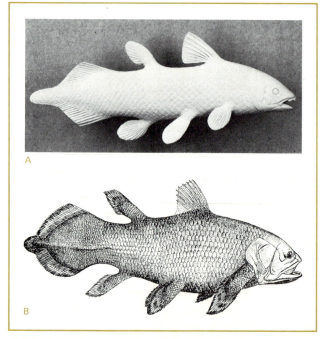

42-2 (*A*) A fossil coelacanth, *Diplurus* and (*B*) its only close living relative, *Latimeria*. (*Diplurus* courtesy of the American Museum of Natural History. *Latimeria* from J. F. Crow, *Ionizing Radiation and Evolution*. Copyright © 1959 by Scientific American, Inc. All rights reserved.)

gasy Republic, and others were soon taken. Figure 42-2 shows that the phenotype of the fish recreated from fossil remains is remarkably similar to its modern relatives. Instances like this enable scientists to use the fossil record with confidence.

42.3 Fossils are most often dated by an examination of the radioactive substances they contain.

Radioactive substances change or **decay** into other substances. Radioactive uranium, for instance, decays into lead and helium (a stable gas); radioactive potassium becomes argon (also a stable gas) and ordinary calcium; radioactive carbon becomes nitrogen; and so on.

Some radioactive transformations take hours, others years, still others eons. In 4.56 billion years, only half of a sample of ^{238}U (a uranium isotope) will have become lead and helium. The number of years it takes for half a sample to is called the **half-life** of the isotope, and each kind of radioactive substance has a characteristic half-life. Once we know these half-lives we can use them to determine the ages of rocks and the fossils found in them. For instance, when a mass of 1.0 grams (g) of ^{238}U decays to 0.5 g in 4.56 billion years, it produces over 0.4 g of lead. (The rest of the mass has become helium and nuclear energy). In another 4.56 billion years, only 0.25 g of the ^{238}U will be left, but the lead will have increased to over 0.6 g. To tell any age, the ratio of uranium to lead is measured. The more uranium in proportion to lead, the younger the rock.

^{238}U has too long a half-life to make it useful in dating younger fossils. But ^{235}U has a half-life of 713 million years. An isotope of potassium, ^{40}K, breaks down into an isotope of argon (^{40}A), with a half-life of 1.3 billion years. These half-lives are just right for dating many fossils.

Another useful isotope is ^{14}C. It occurs with regular carbon in all living things as a certain small but constant fraction of living tissue. Like all radioactive substances, it is constantly decaying. But as long as an organism lives, its radiocarbon is constantly being replenished as it decays. After death, the ^{14}C begins to become a smaller and smaller proportion of the total carbon of the dead tissues. In fact, in 5570 years it will be only half as common as it was when the organism died. Therefore, comparing the amount of regular carbon to radiocarbon allows us to date the youngest fossils, as well as teeth, bones, wood fragments, and the charcoal of campfires dead for 10,000 years.

In general, the repertoire of radioactive tests now encompasses the ages of life on the earth. Thus most fossils can be dated more precisely than ever before.

42.4 To study the evolution of man, that is, the divergence of the *hominids* (humans) from the *pongids* (great apes), we must first consider the present differences between men and apes.

Because many people are reluctant to admit that the process of evolution encompasses humans, we have chosen this species as the major example of evolution. Many other organisms would serve as well or better, especially those whose remains were preserved in areas where bacterial decay was minimal.

We must begin our reconstruction of human evolution with an examination of the differences between mankind and "apekind." Once these are known we will have an idea of what to look for in terms of common ancestors, or "missing links." There are few anatomical differences between apes and men. Our brains are much larger, and our foreheads are higher. We have shorter jaws and flatter faces from which our noses protrude. Our teeth are different, too. They are set in our jaws in a gracefully curved arch called a **rounded dental arcade.**

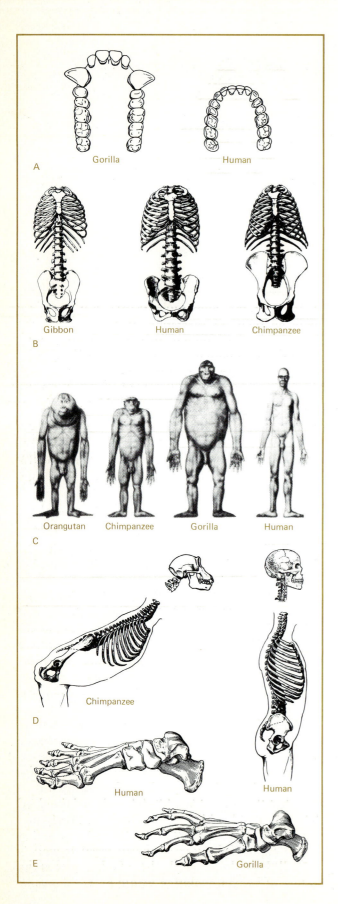

A Gorilla   Human

B Gibbon   Human   Chimpanzee

C Orangutan   Chimpanzee   Gorilla   Human

D Chimpanzee   Human   Human

E   Gorilla

The ape's arcade is more rectangular than arched. Apes have large spaces between some of their teeth; human teeth touch each other. Also, our canines, or eyeteeth, are no longer than our other teeth; an ape's are longer, resembling fangs.

Humans are **bipedal**; we walk. Apes **brachiate**; they swing along in trees, hanging by their arms. Because we are bipedal, we differ from apes in having (1) a broad, cup-shaped pelvis; (2) large, muscular buttocks; (3) a stout heel; (4) long legs; (5) arched feet; (6) a backbone curved into an S-shape; and (7) a **foramen magnum** (the large hole at the base of the skull through which the spinal cord enters) which points downward instead of backward (Figure 42-3). There are a few other differences, such as our relative hairlessness and our lack of a **baculum** (penis bone).

Since bones fossilize easily we should expect the fossil history of the foregoing skeletal differences to be traced successfully. There are, however, significant differences between us and apes which do not fossilize well: (1) We take longer to reach sexual maturity (humans take about 17 years, apes 8 to 10 years); (2) we are left- or right-handed; (3) we associate in large groups and have intricate means of communicating ideas, symbols, and abstractions to each other; (4) we are aggressive, even warlike (apes are quite peaceable); (5) we are fertile throughout the year, while ape fertility is seasonal. However, one nonskeletal difference is eminently "fossilizable." Humans make abundant tools, which reflect, because they shape, our complex cultures.

There are more similarities than differences between apes and humans. We share most features of anatomy and many biochemical peculiarities. For example, neither apes nor humans can synthesize vitamin C, and neither have tails.

42.5 Extinct forest apes living about 15 to 30 million years ago may have been the common ancestor of apes and man.

Fifteen million years ago, apparently, neither modern apes nor humans existed. Fossil remains of apelike primates which might be ancestors of both have been found that date from this time back to

42-3 Anatomical differences between apes and man: (A) upper jaw; (B) pelvis and trunk skeleton; (C) body proportions (Courtesy of Prof. A. H. Schultz); (D) Curvature of spine and pelvis and direction of foramen magnum; (E) Foot skeleton.

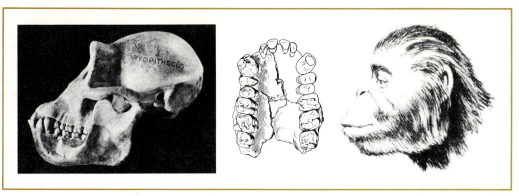

42-4 Fossils of *Dryopithecus* and a reconstruction. (Photo courtesy of the American Museum of Natural History, New York.)

30 million years ago. However, these older fossils are scanty remains, often only a part of the jaw, perhaps only a tooth, rarely anything approaching a complete skeleton. For our discussion, the most interesting of these fossils are derived from the group called dryopithecines, forest apes (Figure 42-4). These fossils have been found in Africa, India, and Europe. They clearly represent reasonable candidates for ancestors of apes like the gorilla and the chimpanzee, and are quite closely related to what must have been our ancestors, too.

The dryopithecines had a pelvis adapted to a four-footed gait, but not the large pelvis of modern chimps and gorillas. Their legs were not long like ours, but then their arms were not long like an orangutan's or a chimp's. Some dryopithecines had larger canine teeth ("eyeteeth") than we do, but smaller ones than those of the great apes. Indeed, the roots of our canine teeth are larger than they seem to need to be, suggesting that our ancestors had bigger ones. There are also similarities between human and dryopithecine molars (grinding teeth).

Sets of dryopithecine teeth are not all the same, for the forest apes belonged to a number of different families, genera, and species. Most had apelike teeth (though not as extreme as modern apes), but some are known with slightly rounded dental arcades, relatively small canines, and other details suggesting human teeth. Elwyn Simons has united the humanlike forms under the name *Ramapithecus punjabicus*. These fossils come from Africa and India and probably parts in between. They lived roughly 14 million years ago, as determined by potassium-argon dating of the site where the late Lewis Leakey collected one.

Drs. Leakey and Simons disagreed on the names of some of the humanlike fossils, but they did agree on their interpretation: that some 12 to 14 million years ago, animals that showed signs of developing the apelike features we find in modern Pongidae roamed the Old World's warmer regions. Coexisting with them and probably appearing quite similar to anyone but an expert primate anatomist was a group of primates with dental features that are decidedly humanlike (Dr. Simon's *Ramapithecus*). Dr. Leakey, in fact, has formally removed these possessors of humanlike jawbones from the dryopithecine group and classified them into the family of humans, Hominidae. Dr. Simons agrees.

The *Ramapithecus* fossil, known as the "Calcutta mandible," provides a piece of information which is remarkable: It shows that *Ramapithecus* individuals took a very long time to mature—like man, but unlike pongids. The mandible, or lower jaw, contains all three molars, but these molars show very uneven wear: The first is badly worn; the second only moderately worn; the third hardly worn at all. Similar differential molar wear is known among humans and fossil humans (including *Australopithecus*, which is discussed next), but it is never found in apes. According to Simons the third molar (or wisdom tooth) is a sign of maturity in all humans and apes. It erupts as skeletal and sexual maturity are attained. In quickly maturing apes, the molars erupt in quick succession, and there is not time for them to wear differently. In humans, the first molar erupts at about the same chronological age as in apes, but the second is a bit delayed, and the third delayed considerably longer. Hence, a newly mature human has a fresh third molar and a worn first molar—traits of the *Ramapithecus* fossil as well.

If all this is borne out by later finds, a part of the picture of human evolution would be as follows:

1. The first apes evolved from Old World monkeys that gradually lost their tails. Soon these apes diverged into forms which appear to be dryopithecine ancestors and gibbon ancestors (gibbons are a separate ape family).
2. Dryopithecines then diverged 15 to 20 million years ago into (a) forms that would later become humans (*Ramapithecus*) and (b) forms that would evolve into the modern Pongidae (*Dryopithecus*).

42.6 Man's more immediate ancestor may well have been *Australopithecus.*

By about 2 million years ago, probably even 3 or 4 million years ago, the Hominidae not only existed, but had developed most of the anatomy of the human being except for the head and even the head had many human characteristics. The teeth were almost human (except for the large size of the molars), and the jaws, though large, were smaller than those of *Dryopithecus.*

The original discoverer of these hominids, R. A. Dart, did not immediately conceive of the small-brained skull that he found as a true hominid, even though he recognized the many hominid features of its teeth and jaws (Figure 42-5B and C), so he named his find *Australopithecus africanus.*

In 1936, a decade after Dart's find, Robert B. Broom discovered the pelvic bones of *Australopithecus* (Figure 42-5A). Except for small details, the familiar human form was evident, proving that *Australopithecus* had walked upright. This was not a complete surprise, however, since the foramen magnum of Dart's find was directed downward,

42-5 Remains and a reconstruction of *Australopithecus.* Notice especially the modern pelvis, indicating an upright posture (A); the rounded dentition (B); and the small canine (C). (Dental arcade used by permission of the Trustees of the British Museum (Natural History)).

which had already indicated an upright posture. Many other details of the skeletal anatomy showed that *Australopithecus* was far more like a mini-brained man than anything else.

In Africa in the late 1950s, Louis Leakey's wife, Dr. Mary Leakey, made the most exciting discovery of all: skeletal remains of *Australopithecus* lying with stone tools of the earliest known type. The remains were dated by the potassium-argon radioactive decay method at 1.75 million years—*A. africanus* was a toolmaker.

42.7 *A. africanus* gradually changed into a form called *A. habilis,* which in turn, changed into *Homo erectus* about 1 million years ago.

Although the record of change from *Australopithecus africanus* to *Homo erectus* is most complete at Olduvai gorge in Tanzania (thanks to the Leakeys and the Tanzanian climate), *Homo erectus* was first discovered by a Dutch physician, Eugene DuBois, in Java in 1891. DuBois had decided Java was the place to look for the "missing link." So he went there and found it! The species he discovered has now been found in most parts of the tropical and warm-temperate Old World. Still, his luck remains amazing. For 40 years other expeditions tried to duplicate his find in Java without success. Originally DuBois's discovery was called *Pithecanthropus erectus* (erect ape-man), but now it is considered *Homo erectus* (erect man).

The anatomical advances in *Homo erectus* were mostly in the skull: He had a brain size approaching modern man's. In fact some larger-brained members of *H. erectus* had brains as large as some of today's smaller-brained *H. sapiens.* A cautionary word about small brains: The most famous small-

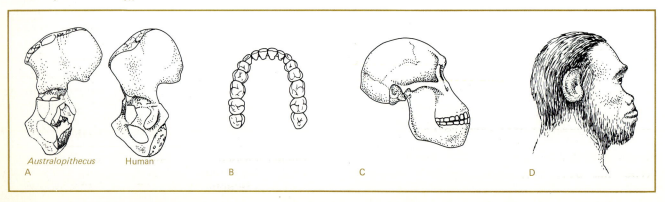

Australopithecus   Human
A   B   C   D

brained *H. sapiens* on record was the French author Anatole France, no illiterate, whose cranial capacity was only 1017 cm³, compared to the average 1350 cm³. So *H. erectus* was not necessarily a stupid species. Tools that he made are evidence of extreme dexterity and technical skill. Apparently *H. erectus* also shared other aspects of man's modern behavior: Some *H. erectus* skulls have been found neatly opened as if their contents had been eaten in a cannibalistic feast or ritual.

42.8 Human brain enlargement in the past 2 million years is one of the fastest-known evolutionary changes.

There is now a series of fossil skulls which take us from mini-brained *A. africanus* to *H. sapiens* in tiny steps. Despite the relatively small steps, this brain enlargement is one of the fastest known evolutionary changes in the history of life on earth. In less than 2 million years, the average hominid brain volume has more than doubled. That is blinding speed compared to more ordinary rates of evolution. For example, it has taken the horse some 60 million years to evolve from a dog-sized ancestor to its present form.

Human brain size is no longer increasing, and apparently it has not been for about 250,000 years. In fact, *H. sapiens neanderthalensis* (Neanderthal man, a race of our species which flourished during the most recent glacial period) had a brain whose volume averaged about 100 cm³ more than ours. Perhaps brains are no longer increasing because the already large size of the newborn's head is just barely able to squeeze through its mother's pelvis, which must actually separate slightly during delivery to allow the baby passage. Other reasons may be more important.

42.9 *Homo sapiens* evolved from *Homo erectus* over a period of time some 300,000 years ago.

Paleontologists say that *H. erectus* evolved into *H. sapiens* about 300,000 years ago, but they know that is an arbitrary division. The evolution of human anatomy, behavior, and physiology—man's phenotype—has been a gradual process. It continues today.

42.10 The process of evolution has actually been observed in one species over the past 100 or more years.

The first well-documented observation of evolution concerned moths which evolved a black coloration as their woodland environment became increasingly sooty.

When Darwin was a young man, almost all British peppered moths, *Biston betularia*, were a mottled pale-gray and white—they looked peppered. The black form of the peppered moth existed but was rare. We know this because moth collectors made quite a to-do about them. Yet today the woods about Birmingham, England are full of them, and they are as common as they once were rare. Evolution has actually occurred in our own time.

Contemporary biologists noticed that the black form was common in areas to the east of big industrial centers like Birmingham, and knowing that the winds in England usually blow from west to east, they guessed that the smoke and soot from factories had something to do with the success of the black form. British biologist H. B. D. Kettlewell found that in forests in which the blacks were common, trees were black and sooty, whereas forests that still had lots of the gray and white peppered moths—the old "typical form"—were relatively clean. The tree trunks in these forests were covered with speckled gray and white lichens. He further determined that blackness is due to natural moth pigmentation and is inherited. The typical peppered form is also inherited.

Kettlewell guessed that the moths' greatest enemies were birds and that the more conspicuous a moth was sitting on a tree trunk, the more likely it was to be seen and eaten. A peppered moth would be relatively safe on a lichen-covered trunk, a black one on a sooty trunk (Figure 42-6).

To test his hypothesis Kettlewell raised moths himself to release in both clean and sooty forests. He painted a dot under the wing of each moth before he released it so that he would know later which animals were his.

Kettlewell released 799 moths into the lichen woods and 11 days later recaptured 73 with his marking on them. The moth types were as follows:

	Black moths	Peppered moths
Number released	406	393
Number recaptured	19	54
Percentage recaptured	4.7	13.7

42-6 Both forms of peppered (peppered and black) moths on both types of background. Can you find the peppered form on the lichen-covered tree trunk (A)? (From the experiments of Dr. H. B. D. Kettlewell, University of Oxford.)

Thus the peppereds had a better chance of survival in the lichen-covered woods. In those 11 days, each peppered moth was about 2.9 times as likely to last as a black one.

In the sooty woods, it was the black form which held the advantage. Here the experiment was done in two different years. In 1953, in 11 days, 27.5 percent of the blacks, but only 13 percent of the peppereds, were recaptured. For this period, a black was 2.1 times as likely as a peppered to survive. In 1955 a black was again 2.1 times as like to survive.

In the next chapter, we shall learn that such inherited differences related to survival are intimately connected with a measurement called fitness. Darwin predicted, in fact, that differences like these are just the things that cause evolution by natural selection.

Kettlewell also used motion pictures to record the actions of birds being given the chance to eat moths when they had at least one moth of each type on the tree trunk before them. In Birmingham, the blacks were seen much less often by the birds. For example, redstarts ate 43 peppered but only 15 blacks in two days. In the clean woods, the opposite was true. The spotted flycatcher ate 81 blacks and 9 peppereds, and so on. It is clear from the films that the birds had trouble seeing the peppered moths against a peppered background of lichens, and the black ones against a black background of soot. It is not surprising to learn that some 100 species of moths have begun to evolve blackness in sooty environments.

There are other cases of observed evolution known to science, many of them caused by our radical interference with nature. One is the development of resistance to DDT in mosquito populations. Another is the development of tolerance to antibiotics by disease-causing bacteria.

These examples, as well as those taken from the fossil record, solidly support the occurrence of evolution. We are led, then, to a second question: What causes biological evolution?

43

The Theory of Evolution by Natural Selection

QUESTIONS TO KEEP IN MIND

What major assumptions and deductions are made by the theory of evolution by natural selection?

What is the Hardy-Weinberg law?

What forces can bring about evolution?

What are the four ways in which alternate phenotypes can be related to each other in terms of relative fitness?

In 1836 Charles Darwin returned from his voyage on the *Beagle*. In South America he had seen large numbers of fossilized bones of extinct vertebrates; in the Galapagos Islands he had observed many unusual species of animals that resembled more prosaic mainland species in their choices of places to live and their ways of finding food. And at sea he had read a book on geology by Sir Charles Lyell that argued for an earth that was slowly and continually changed by geologic processes rather than by a series of catastrophic events. These, plus a great deal of other information, combined to convince Darwin that evolution had occurred, and a few years after his return he stated (although not publicly) that the mechanism behind the process of evolution was natural selection. In modern terms we say that Darwin had formulated the theory of evolution by natural selection. To understand this simple but profound formulation we must first define two words: *heritability* and *fitness*.

Because of the **heritability** of certain traits, offspring resemble their parents more than if they were picked at random from all the offspring of their species. The phenotype consists of a traits inherited genetically, by tradition, or any other way. For this discussion, it is important that the phenotype is partly inherited. Because it is, fitness is partly inherited too.

Fitness is the ability of an individual to reproduce itself in its own environment. This involves **viability** (the ability to stay alive) and **fertility** (the ability to produce offspring). Furthermore, it involves producing offspring that will themselves survive and produce offspring.

It does the species no good for the individual to survive and be sterile; it is of no benefit for an individual to inherit the potential for bearing a thousand young without the potential for surviving for the time necessary to bear them. Bearing inviable or sterile young is similarly worthless.

Fitness has been quantified both in terms of the number of descendants left after a period of time and the number of descendants compared to an arbitrary standard.

43.1 Assume that different phenotypes have different fitnesses and that fitness is at least partly inherited; then, in a stable environment, the average fitness of the members of a population is always increasing.

This is the theory of Charles Darwin that revolutionized biology in 1859, and today it still stands, with only minor improvements. It subtly predicts that evolution will occur. If each phenotype has a different fitness, then an increase in the average fitness can only occur by a change in the mixture of phenotypes. In other words, the mixture of phenotypes we see now in the population is not the same (on the average) as that we would have seen had we looked last year or 10 years ago. This is the change of phenotypic make-up in a population that we have called evolution.

This theory not only predicts evolution but provides a mechanism to explain it: The heritability of fitness causes an increase of organisms with higher fitnesses in the population. This mechanism is called **natural selection**.

43.2 The assumption that different phenotypes have different fitnesses, subject to debate in the past, is continually being verified.

Darwin's critics were not disposed to agree with his sweeping generalization regarding phenotypes and fitness. They conceded that certain heritable traits such as wing color in a moth had survival value, but they maintained that most alternate phenotypes are selectively neutral; that is, each has the same fitness. How could a person's fitness be influenced by hair color or blood type? And why, they asked, do we have vestigial organs such as the thymus and appendix? If populations were evolving toward greater fitness, then organisms could not afford to waste energy on unnecessary structures. These complaints were based on ignorance. Just because the function of an organ or effect of a characteristic is unknown does not mean it is lacking. Each must be carefully tested. The effect of human hair color on fitness has not been tested, but that of blood types has. Researchers have now found that people with blood-type O get duodenal ulcers 1.38 times more often and stomach ulcers 1.19 times more often than other types. People with blood-type A are 1.19 times more likely than those with B or O to get stomach cancer; 1.26 times more likely than those with O to have pernicious anemia; and 1.16 times more likely than O's to get diabetes mellitus. Blood types are anything but selectively neutral.

In the cases of vestigial organs it seems they either have a function or are on their way out. The effect, if any, of the appendix is still unknown, but the thymus, which is no longer considered vestigial,

TABLE 13 SIZES OF FAMILIES IN TWO GENERATIONS

Number of children born to mother	Number of such mothers	Average number of children born to their daughters
1	35	2.97
2	67	3.54
3	111	3.14
4	136	3.41
5	138	3.99
6	132	3.89
7	114	3.93
8	87	4.07
9	74	4.23
10	50	4.38
11	24	5.08
12	19	5.21
13	9	8.56
14	1	10.00
15	3	6.67

Source: R. A. Fisher, *The Genetical Theory of Natural Selection*, 2nd rev. ed., New York: Dover, 1958.

has been found to be an important component of the body's defense system (section 37.4).

This is not to say that there are no such things as selectively neutral characteristics, but they are probably due to miniscule differences in the chemical make-up of different organisms. When we can readily sense differences in phenotype, as when the same organ is noticeably different in size or shape in different individuals, or a protein is missing in certain individuals or behavior is variable, then the differences probably cause important differences in fitness.

43.3 That fitness is at least partly inherited has also been validated.

The heritability of fitness is the second assumption Darwin made. We have already seen that several characteristics important to fitness, such as wing color in moths and blood type, are inherited. More direct substantiation was offered by a British researcher who tried to see if the number of children raised by a mother influenced the number of children her daughters bore. It did (Table 13). The numbers of children born in the next generation were also recorded, and, as expected, the influence was half as great from grandmother to granddaughter as from mother to daughter.

43.4 The assumption that the environment is stable is not always valid.

The assumption of a stable environment is made to simplify the theory and to examine its functioning during those periods of time in which an environment is relatively stable. The fact that environments do change does not challenge the theory of evolution by natural selection, but the changes do alter the relative fitness of the organisms in that environment. In other words, while a stable environment reigns, average fitnesses will be increasing. A change in environment may well cause average fitnesses to plummet, but selection will never stop. Fitnesses will immediately begin rising again, but the fittest phenotypes may be very different from what they were before. Thus, the direction of evolution may change radically as a result of environmental change.

43.5 Evolution usually proceeds so slowly that observations can only partially substantiate the process.

There are two ways in which deductions from Darwin's theory may be examined: one is by observation, the other by the construction of mathematical models. Using the former to tell if the average fitness of the members of a population is increasing requires either extensive use of the fossil record or examination of an environment that has radically changed. In the case of the peppered moth population, for example, pollution changed an environment much more quickly than natural forces would do, the process of natural selection was speeded up, and the moths evolved blackness which made them more fit than if they had remained light-colored. Sickle-cell anemia (section 18.3) is a case in which humans have been measurably improving in average fitness—in some environments, at least.

43.6 Evolution can be studied in mathematical terms.

A fuller understanding of natural selection has been derived from mathematical studies. In the example that follows, evolution will be considered as a change in the frequencies or proportions of genes in a gene pool. The gene pool is an abstraction. It means the imaginary barrel containing all the genes of all the individuals of a population. Thus in the case of the peppered moths we would say that evolution has occurred if the genes for color occur with different frequencies (more for blackness, less for peppered) in a modern population than in one studied 100 years ago.

In studying the make-up and possible changes in a gene pool we must first ascertain what happens in a nonevolving population. It used to be thought that dominant genes were always on the increase and recessive ones on the decrease, but this is not so.

The Hardy-Weinberg law, formulated about 1908, states that under certain ideal conditions, relative frequencies of genes in a population will tend to remain constant from generation to generation. This constancy of gene frequencies is referred to as genetic equilibrium. The "ideal conditions" refer to a large, randomly mating population undisturbed by mutation, selection, or migration, any or all of which tend to change gene frequencies.

This will make more sense if we examine the law as it expresses the equilibrium established among persons with the dominant gene (T) for the ability to taste the chemical phenylthiocarbamide (PTC) and those with the recessive gene (t), nontasters.

In this example, individuals who are TT or Tt will detect a bitter taste and are classified as tasters. Individuals who are homozygous tt will be nontasters. Suppose we are following the frequency of the T and t genes in successive generations of an island population where there is no migration, no mutation of that gene, and no advantage in having T or t that would lead to selection (different survival or reproduction rates). We must further assume that the population is infinitely large, and that mating is random; that is, people do not choose their mates on the basis of whether or not they can taste PTC.

Let us say that the gene pool contains eggs and sperm, 60 percent of which carry t and 40 percent of which carry T. The recessive gene occurs in greater frequency than does the dominant gene. We now want to find out what kinds of genotypes will result from this gene pool and whether this next generation will create a similar or dissimilar gene pool. Since mating is at random, and union of gametes is at random, we can construct a Punnett square (section 15.4) to answer these questions.

	Proportions of female gametes	
	0.4 T	0.6 t
Proportions of male gametes 0.4 T	0.16 TT taster	0.24 Tt taster
0.6 t	0.24 Tt taster	0.36 tt nontaster

We can see from this that one generation of random matings will result in the following proportion of genotypes and phenotypes: $F_1 = 0.16$ TT tasters; $0.24 + 0.24 = 0.48$ Tt tasters; and 0.36 nontasters (tt). Consequently, 16 percent of the total gametes in the gene pool produced by the F_1 generation will come from TT individuals; 48 percent will come from Tt individuals; and 36 percent will come from tt individuals. The composition of the gene pool will be: $T = 0.16$ (from TT) + 0.24 (from Tt) = 0.40; $t = 0.36$ (from tt) + 0.24 (from Tt) = 0.60.

The frequency of T and t in the new gene pool is exactly the same as the frequency we started with in the parental gene pool. It should be obvious that if we continue to simulate successive generations, we will continue to get exactly the same relative proportions of T and t in the gene pool.

The Hardy-Weinberg law thus predicts the stability of gene frequencies and serves as a standard against which evolutionary change can be quantitatively measured.

43.7 Migration, mutation, and chance are some of the forces that can upset genetic equilibrium.

Changes in the frequency of genes in a gene pool may result because organisms with different alleles are migrating in or out of a population at different rates. To pick an extreme example, if only TT tasters move onto our hypothetical island and only nontasters, tt, leave, the 40:60 ratio will be constantly changing. It will also change if, due to unequal rates of mutation, more T's change to t's than vice versa. Actually, mutation rates for most genes are very low, and mutation by itself is not a major cause of changes in gene frequency. However, mutation is the ultimate source of inherited variation and is necessary if evolution is to occur.

If a population is very small, changes in gene frequency may occur simply as a result of chance. Genetic drift is the term used to describe these random fluctuations in gene frequencies. If, for example, we have a breeding population that has only 10 parents each generation, the chance of getting a true expression of random matings would be the same as getting a 1:2:1 ratio of head-heads, head-tails, and tails-tails in 10 tosses of two coins.

43.8 Natural selection is generally the most important factor causing changes in gene frequencies.

In cases of genetic equilibrium all the genotypes are equally fit, that is, are equally viable and equally productive of fertile young. Among real organisms, this is rarely the case. One genotype usually contributes more to the gene pool than another, thus disturbing the equilibrium. Often the reproductive advantage is only slight. Sometimes, however, it is a dramatic one caused by such factors as a longer reproductive life, greater sexual activity, greater fertility, greater resistance to disease, greater beauty, and so on.

In considering any characteristic such as taster versus nontaster, there are only four possible relationships of relative fitness among the different genotypes. In one, homozygous tasters (TT) are the fittest, homozygous nontasters (tt) are the least fit, and heterozygous tasters (Tt) are intermediate. The second case simply switches the positions of TT and tt so that the latter is the fittest. Third, the heterozygous Tt could be least fit, a situation known as **heterozygote inferiority**; and fourth, the heterozygous Tt could be the fittest, **heterozygote superiority,** or **heterosis.** (In the last two cases it doesn't matter whether TT is better or worse than tt.)

43.9 Heterosis results in *balanced polymorphism,* the maintenance of many phenotypes.

In all the cases of relative fitness except heterosis, natural selection works to eliminate one of the genes and produce a homogeneous population. In heterosis, however, all three combinations of the two alleles (TT, Tt, and tt) are left, even though each form has a different fitness. A dynamic balance is struck such that if one gene accidentally becomes rare, it has a greater chance of being included in the heterozygous form than in the homozygous form. It will therefore be increased by natural selection—which favors the fittest (heterozygous) form—until equilibrium is reached. In other words, where heterosis exists, natural selection has produced evolution, and then, after an equilibrium has been reached, selection works against further changes. Heterozygote superiority appears to be a major force for maintaining the variety of individuals that make up a population at one place and one time. Heterosis also seems to be the basis for the superiority of some hybrid offspring, whether a special variety of corn or an intelligent mutt.

An interesting example of balanced poly-morphism is the sickling trait, which among black Africans conferred some measure of protection from the ravages of malaria. Homozygotes for the sickling trait died at a young age because their red blood cells were deformed and blocked their blood vessels. Heterozygotes, on the other hand, enjoyed a selective advantage in their resistance to malaria. Because of this heterozygote superiority, the gene persisted. Among black Americans, who are not challenged by malaria as their ancestors were, the sickling trait is better known as **sickle-cell anemia** (section 18.3).

44

The Role of Mutation in Evolution

What do mutations contribute to the process of evolution by natural selection?

Are most mutations beneficial? Are they helpfully directed?

How do mutations and natural selection strike a balance?

The late Sir Ronald Fisher gave the best definition of mutation: "Mutation is the initiation of any heritable novelty." He wrote this in 1930, more than two decades before scientists learned that genes are composed of DNA and before they began making distinctions between mutations that result from changes in the internal structure of genes and those that result from changes in the structure or number of chromosomes (section 17.7 and 18.10). In spite of these advances Fisher's definition is still a good one, because it is not dependent upon any mechanism.

Mutations are the events which provide the raw material for natural selection. They are, in other words, the source of variety in living organisms. Without mutations, natural selection would not exist: There would be no alternate phenotypes to choose among. Thus, in the process of evolution by natural selection, mutations produce variations and natural selection chooses the ones that will increase fitness and "discards" the others. Fisher used this concept to deduce that the greater the

heritable variety in a population, the faster selection improves the average fitness. This is what we believe today, but mutation has not always been linked to natural selection to explain evolution.

44.1 Many early evolutionists believed that evolution proceeded by mutation only.

Since the late 1700s there has been quite a variety of evolutionary theories. Scientists have argued that evolution occurs by mutation alone, either as **adaptive evolution,** which increases fitness, or as **random evolution,** which is change by accident and unrelated to fitness. In the former theory, mutations were believed to occur *because* they were needed to improve fitness. Adaptive evolution itself had two variations depending on whether or not one believed that acquired characteristics could be inherited.

Lamarckism, a theory of evolution named after the French paleontologist Jean Baptiste de Lamarck (1744–1829), rested upon the belief that acquired traits were inherited. This assumes a kind of evolution by mutation, because the acquisition of a new heritable phenotype is tantamount to mutation. Lamarckians believed these kinds of changes increased fitness, a point they liked to illustrate with the giraffe. Each generation of giraffes, they believed, stretched their necks a little in the effort of reaching the leaves high in the treetops. Each successive generation was therefore born with slightly longer necks. This process continued until stretching was no longer necessary.

This contention infuriated many scientists, and one raised generation after generation of mice, cutting off the tails of every one of them; yet, as he predicted, he never obtained a naturally tail-less mouse. None of the characteristics transmitted by genes can be modified in an offspring by its parents' acquired traits. However, acquired traits—an extra mouth grafted onto a paramecium, the celebration of Christmas on December 25—can be inherited nongenetically.

44.2 Are mutations helpfully directed?

The other variation of adaptive evolution by mutation was more sophisticated than Lamarckism and survived well into the twentieth century. Let us use the evolution of antibiotic-resistant strains of bacteria as an example. Believers in evolution by mutation stated that there were no resistant bacteria before the use of antibiotics, but that when such a drug was introduced to their environment, it caused mutations to occur that rendered some bacteria resistant.

Darwinists did not agree. They claimed that the mutations already existed before the penicillin was added. They were just rare. Putting them in this new environment simply changed relative fitnesses around. Penicillin-resistant bacteria then became quite common.

In an elegant experiment, S. Luria and M. Delbrück proved that the Darwinists were absolutely right. Later, by developing a technique they called "replica plating," J. and E. Lederberg were able to confirm these results. Both experiments are impressive, but the second is infinitely easier to explain.

There are two principal ways to grow bacteria in the lab—on a disk of jellied medium called agar and in a liquid medium in a test tube. Each medium contains nourishment, but the bacteria on the jellied disk grow on the surface and remain in a fixed position. Penicillin, streptomycin, or any antibiotic can be added to either medium, and all nonresistant bacteria will then die.

Usually, bacteriologists transfer bacteria from one agar disk to another by means of a single platinum loop. The loop picks up some cells and carries them to the new surface for continued growth.

The Lederbergs' innovation was to use a piece of velvet the size of the disc, instead of a platinum loop. The pile of velvet behaved like thousands of tiny needles, each picking up a few cells so they can be moved to the new surface.

Since the bacteria do not move around on an agar plate, each cluster of cells is a genetically pure colony which arose from a single bacterial cell. When the velvet touched the surface of the new agar plate, it made a replica of the old one, picking up cells from each cluster and keeping them in the same position relative to other clusters. The Lederbergs could locate the relatives of bacteria anywhere on the new plate by looking at the same spot on the old one.

The Lederbergs proposed proving that the mutation for antibiotic resistance occurred before exposure to the antibiotic. They grew some *E. coli* on a plain agar plate. Then they made a replica on a plate full of streptomycin. Only a few bacteria lived. These bacteria were discarded, but the Lederbergs sought out their families by looping some bacteria off the spots on the original plates where they had obtained the resistant ones. These loopfuls were placed in a liquid medium for rapid growth. Then they were plated onto new but harmless plates. Remember that these bacteria had never been exposed to streptomycin. Nor had any of their direct ancestors. The exposed relatives had all been discarded. Replicas of these new plates were then produced on plates full of streptomycin. Sure enough, far fewer bacteria died than in the first replica. The unexposed relatives of the resistant bacteria did contain the mutation (Figure 44-1).

The Lederbergs then looped out parts of the second harmless plate that corresponded to the resistant spots in the second replica plate. These second transferred loops contained a far higher proportion of resistant relatives than the first did. The Lederbergs grew these second transfers in liquid, and replated them. When they exposed this third round to a toxic replica plate, an even larger number of resistant bacteria were present than in the preceding replicas: The third replica plate was dense with resistant bacteria, yet none of their predecessors had ever been in contact with streptomycin. The mutation had occurred before the new environmental pressure. By the fourth round of replica plating, every single bacterium was resistant.

Each step of this experiment enriched the population of bacteria with resistant alleles. In effect, the Lederbergs exchanged natural selection for artificial or man-made selection.

The Lederbergs tested other antibiotics and obtained similar results.

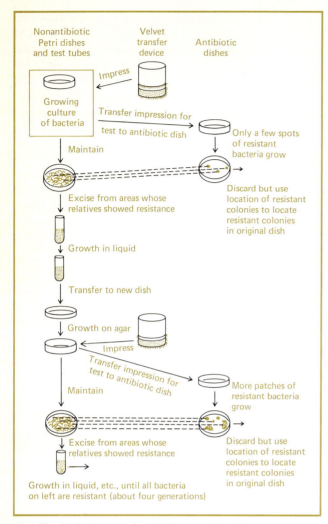

Nonantibiotic
Petri dishes
and test tubes

Velvet
transfer
device

Antibiotic
dishes

Impress

Growing
culture
of bacteria

Transfer impression for
test to antibiotic dish

Only a few spots
of resistant
bacteria grow

Maintain

Excise from areas whose
relatives showed resistance

Discard but use
location of resistant
colonies to locate
resistant colonies
in original dish

Growth in liquid

Transfer to new dish

Growth on agar

Impress

Transfer impression for
test to antibiotic dish

Maintain

More patches of
resistant bacteria
grow

Excise from areas whose
relatives showed resistance

Discard but use
location of resistant
colonies to locate
resistant colonies
in original dish

Growth in liquid, etc., until all bacteria
on left are resistant (about four generations)

44-1 The Lederberg replica-plating experiment.

44.3 Mutations are rarely beneficial.

Not only do mutations fail to occur because of an organism's need for fitness, but when they do occur, they are usually harmful. Less than 1 percent of the hundreds of observed mutations of barley have been helpful; and of the 10 kinds of human mutations where the mutation rates are known, 8 produce low relative fitness and the effects of the other two, albinism and aniridia (no iris in the eye), are unknown. In an experiment on laboratory populations of *Drosophila melanogaster* in which the rate of mutation was speeded up by exposure to X-rays, an estimated 95 percent of the flies did not even survive to adulthood, much less produce fertile offspring.

Although mutations are usually harmful, when one does occur that helps the organism out-produce the others in its generation, then natural selection will increase the proportion of the population that has this allele. The same mechanism tends to curtail the spread of deleterious alleles. In fact, it is because natural selection has increased fitness and produced populations of well-adapted individuals that a situation exists in which most changes are harmful. If you are playing tennis almost perfectly, then trying a new grip or stance is almost sure to impair your performance.

44.4 Random mutations cause evolution, but this is a weak force compared to natural selection.

All mutations occur without regard to whether they increase or decrease fitness. However, they often change the phenotype and hence cause evolution. What, then, is the evolutionary pressure generated by mutation? (We will talk about genic mutation, but the same principles can be applied to molecular mutation, chromosomal mutation, or any other type.)

Suppose there are two alleles at a locus, Y and B. There is a chance that a detrimental B will become a beneficial Y, and another chance that a Y will mutate to a B. These chances may be small, even near zero, but they do exist. What is the net result of evolution by these mutations (assuming no selection)?

Imagine a whimsical experiment involving two barrels: one for B and one for Y alleles. Imagine also the participation of two mutator genies (of *Arabian Nights* fame): One picks out B's, changes them to Y's, and throws them in the Y barrel. The other does the reverse. The only rules are that the $B \rightarrow Y$ genie must change exactly x percent of the B genes per generation; whereas the $Y \rightarrow B$ genie changes z percent of the Y genes in that time.

Which genie wins? Neither. When the $Y \rightarrow B$ genie is winning, he will be depleting his own stock of Y genes. That gives the $B \rightarrow Y$ genie the advantage in the next round. In fact, when the number of Y genes converted equals the number of B genes converted we have reached equilibrium.

Geneticists have found that the actual rates of genic mutation are very, very small: five per million or eight per hundred thousand or so. Since mutations tend to be harmful, these low rates are themselves beneficial. In fact, many evolutionists believe they have evolved to be low, since organisms with high rates tend to be less fit. (There are genes known to influence mutation rate.)

In section 43.9 we noted that in all cases but

44-2 Three types of abnormally small adults. The two dwarfs to the immediate left of the normal adult are chondrodystrophic. All these individuals were older than 20 when the photo was taken. (From *And Replenish the Earth.* © 1974 by Michael L. Rosenzweig.)

heterosis, one of two alleles for a given trait would become extinct, owing to selection against it. From the previous section, however, we see that extinction of an allele is not final. It is an abyss from which return is *predictable*. When an allele is absolutely gone, it is produced from another allele at the prevailing mutation rate. Although the forces of selection and mutation are unequal, a balance is struck

between them. Selection usually dominates the picture, and mutation is just barely able to prevent deleterious alleles from becoming totally extinct.

An example is the detrimental human allele for dwarfism. Persons with one normal allele and one harmful one are **chondrodystrophic** (Figure 44-2). (Persons with two detrimental alleles do not live.) Although chondrodystrophics should disappear if only selection were at work, mutations keep renewing the allele.

E. T. Morch, a Danish investigator, interviewed the families of 108 dwarfs and found that whereas the dwarfs had produced only 27 living children, their 457 normal brothers and sisters had produced 582 children and were more than five times as fit. Clearly selection worked against dwarfism. Next Morch estimated the mutation rates by collecting the records of 94,075 live births in a Copenhagen hospital. Of all these babies, only 10 were chondrodystrophic dwarfs. But 2 of the 10 had been born to dwarf parents, so they were not produced by mutation. Thus, out of 94,073 potentially normal births, 8 mutations occurred causing dwarfism. Since each of the 94,073 have two genes for this locus, there were 188,146 loci. Eight mutated, so the mutation rate (x) is 8/188,146, or 42.5 mutations per million genes. This low rate isn't at all unusual. No mutations to normality were observed. A balance has been struck between selection, which tends to eliminate dwarfism, and mutation, which tends to produce 8 dwarf alleles per 188,146 normal ones. The equilibrium that Morch observed was that the frequency of the dwarf allele is 10/188,150.

45

Adaptation:The Main Product of Evolution

QUESTIONS TO KEEP IN MIND

How does genetic drift produce nonadaptive evolution?

What evidence supports the contention that evolution is primarily adaptive?

In spite of some very strong evidence that evolution is primarily adaptive and has only rarely been influenced by nonadaptive mechanisms, some scientists believe that nonadaptive mechanisms such as mutation, migration, and genetic drift are relatively important forces producing nonadaptive evolution. In consequence, we will discuss genetic drift before going on to examine some of the evidence for adaptive evolution. (The evolutionary pressure of mutation has been discussed in Chapter 44.)

45.1 Genetic drift is a change in the frequency of alleles resulting from random-sample accidents.

Genetic drift (section 43.7) is more likely to occur in small populations where there are not enough members for the law of averages to work out. The frequencies of the alleles undergo unpredictable and nonadaptive changes; in other words, they appear to drift aimlessly. There are cases, especially among isolated island species, where genetic drift has been a strong force of evolution (section 46.8), but usually it is not.

There is a more extreme form of genetic drift described by the **founder principle.** Suppose, for instance, that an island is free of birds. Perhaps once in a hundred years a storm of just the right size and direction, at just the right season, blows a few birds to the island. If the birds survive their voyage, they are Fletcher Christians (*Mutiny on the Bounty*). They have no better chance of returning to the mainland than they had of arriving in the first place. And they certainly waste no time waiting for the return storm. If the habitat is at all suitable, they found a new colony, like the shipwrecked souls of Pitcairn Island.

Soon the island is teeming with this bird species. But all the members of the population can trace their ancestry back to those first founders, whose tiny gene pool began the whole population. These founders couldn't carry all the variety of their parental mainland population with them. They had only their own few genes. These genes were necessarily a small sample of those available. So the new colony may look very different from the mainland population. If there were, for example, 100 alleles at a particular locus in the mainland population and there are only five Fletcher Christians, then immediately at least 90 percent of the mainland alleles are extinct on the island. The fact that genetic drift results when a few members of a large population establish a new population in some isolated spot is known as the founder principle.

Several examples of the founder principle at work in isolated human populations are known. Some have been noted in the strict Amish communities, because these people tend to intermarry. One such example, a genetic disease, was first described in 1860. Since then, only 100 cases are known to have occurred, 55 of them were among a small group (8,000) of Amish in Pennsylvania. The ancestors of all 55 were a couple who came to America in 1774. The frequency of the allele in this group is estimated to be about 0.065. This is far higher than in any other known human population. Naturally, the average fitness of these people must be lower than it would be if the drift had not occurred.

45.2 Parallelism and convergence support the concept that adaptive evolution is so much more important than nonadaptive evolution that in any environment the direction of evolution will be predictable.

To find out if natural selection does indeed dominate the nonadaptive forces, the ideal experiment would be to place several populations in the same new environment, separately. If evolution is truly adaptive, then each population should evolve to be the same as all the others, because presumably there is some *best* way to occupy this environment and adaptive evolution should achieve it. Nonadaptive evolution, especially random evolution, should not yield the same result in all the separate replicas of the environment.

Nature has performed such experiments for us innumerable times. When she has, sometimes the results are observable as parallelism, or even more spectacularly as convergence. In **parallelism**, separate populations begin with similar phenotypes and change markedly, but always resemble each other. In **convergence**, they start out quite dissimilar and evolve to look (often extraordinarily) alike (Figure 45-1).

Nature's "experiments" have occurred because similar environments tend to develop in many different parts of the world; but the populations that are forced to adapt to them from place to place are often from different species. For example, we might have a population of roses forced to adapt to a new desert in Mexico and a population of euphorbias forced to adapt to a similar new desert in Tanzania. If the two populations converge, it must be because evolution is adaptive and will act to change plants and animals to the predictably fittest forms for that environment. The fact of their convergence is preserved, however, because the two species were so dissimilar to begin with that each is likely to have kept some fundamental traits which express their true origin.

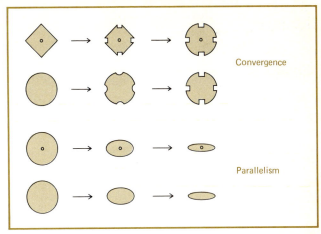

45-1 An idealization of convergence and parallelism.

45.3 Many examples of convergence are found between placentals and marsupials.

The opportunity to burrow underground exists in many parts of the world. But moles (placentals) are not so widespread. In Australia and Africa, their place is taken by other groups of animals.

One of Australia's marsupials (section 30.25) looks remarkable like a mole: It has weak eyes, a short tail, a rotund body, broad digging front feet and fur with no "grain" or direction, so that the animal can go forward or backward in his tunnel without being rubbed the wrong way. Apparently, to live successfully in the manner of a mole, you've got to look like one. Even though we cannot yet say that we could have predicted the evolution of the mole, we know from this spectacular convergence that someday we should be able to make evolutionary predictions. One way to explain this is to hypothesize that evolution is primarily adaptive. If evolution were mostly random and nonadaptive, such convergence would be extremely rare, or perhaps totally nonexistent.

In addition to the marsupial mole, marsupials have evolved to exploit many other environments similar to those of New World and Old World placentals. The Tasmanian wolf is the marsupial version of the dog-family carnivore. He runs on his toes to pursue his prey, as do wolves. There are marsupial "flying squirrels" and marsupial "rats" and "mice" (Figure 45-2).

Perhaps the most startling convergence of marsupials and placentals occurred in some recently extinct animals—the saber-toothed carnivores (Figure 45-3). Sabertooths could open their mouths to a wide enough angle so that their long, sharp saber teeth stuck out. Then, mouths agape, they could plunge their teeth deep into a prey with the aid of unusually powerful neck muscles. This feat obviously took some doing, and there is a long list of special skeletal adaptations that made it possible: long canine teeth, sheathlike extensions of the lower jaw that helped prevent the teeth from breaking, modifications of the skull that permitted the jaws to be held wide open so the teeth protruded, and very robust bones for attachment of the neck muscles.

The placental sabertooths (northern hemisphere) and the marsupial ones (in South America) shared all these adaptations. Imagine how many muscular convergences must have accompanied the skeletal ones. But certain identifying features of the marsupial skull did not need to change; so by looking at their skulls we can easily determine that the two types of sabertooths were descended from ancestors that were totally unlike.

45.4 Arid deserts have engendered still other characteristic adaptations which demonstrate convergence.

Desert plants have evolved as thorny, succulent species especially adapted to conserve moisture. Most lack leaves and carry on photosynthesis in their stems. Leaves have a relatively large surface area, and as they transpire, they lose too much water in the desert. In the New World, cacti have adapted to desert living. They are closely related to roses. In Old World deserts, environmental pressures have caused members of an entirely different family, the Euphorbiacae, to evolve the same set of adaptations as the Cactacea (Figure 45-4). However, their flowers remain like regular euphorbias, and so they can easily be distinguished from cacti.

One of the sophisticated convergences of cacti and other arid-zone plants involves a biochemical

45-2 Some marsupials and their placental ecological equivalents. (From *Life: An Introduction to Biology* by George Gaylord Simpson and William S. Beck, copyright 1965, by Harcourt Brace Jovanovich, Inc., and reproduced with their permission.)

45-3 Skulls of marsupial (A) and placental (B) sabertooths. The marsupial sabertooth is also shown restored (C). (From *And Replenish the Earth.* © 1974 by Michael L. Rosenzweig.)

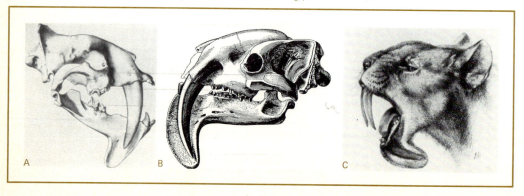

Astrophytum asterias   Euphorbia symmetrica   Cereus validus   Euphorbia heptagona   Cleistocactus buchtienii   Euphorbia morinii

45-4 Cacti and similar-looking *Euphorbia*. (From Rosenzweig, *And Replenish the Earth*.)

system. In order to carry on photosynthesis, plants require CO_2. To get it, their stomates must be open to the atmosphere. But in addition to collecting CO_2, open stomates lose water. (This can be a serious loss of water, for CO_2 is required when there is sunlight, at the time when the desert heat is high. Cacti and Crassulaceae (thick-leaved plants like jade plant and *Kalanchoë*) have solved this problem the same way. Their stomates open at *night* when the desert humidity is high, heat is very low, and water loss is minimal. The CO_2 is stored at this time by reacting with special molecules—probably the

same in both families. When daylight and heat return, the plants close their stomates and use the CO_2 they have stored during the night.

Deserts have influenced animals, as well. The sparsely vegetated deserts of North America harbor the Kangaroo rat (many species); the deserts of Africa and the Near East have jerboas (Figure 45-5). We know these animals to be different because, for one thing, all members of the kangaroo rat's super-family (including pocket gophers and some rather ordinary ratlike rodents) have fur-lined external pockets in their cheeks. Jerboas do not.

Many more cases of convergence could be cited involving birds, insects, fish, and other organisms.

45-5 A bannertail kangaroo rat (*A*) and a jerboa (*B*). (*Dipodomys* courtesy Prof. K. Schmidt-Nielsen. Jerboa from Brehm's *Tierleben*, Bibliographisches Institut, Leipzig and Vienna, 1914.)

45.5 Like convergence, mimicry shows the predictable character of evolution.

Another set of situations is known in which we can say in advance that we know what a fit phenotype will be like. These situations are collectively called **mimicry**. In mimicry, organisms must somehow appear to be like other organisms that have evolved a high degree of fitness.

One kind of mimicry is called **Batesian mimicry**, after the famous nineteenth-century naturalist H. W. Bates, who described it. Batesian mimicry requires at least three, and perhaps four, kinds of organisms. There must be a consumer capable of learning the patterns or colors (or odors, etc.) of his favorite foods, as well as of obnoxious foods. There must be an obnoxious food species (perhaps it tastes bitter or is poisonous), and a tasty species which resembles the distasteful one. (Probably there must also be another tasty food which is dissimilar to the

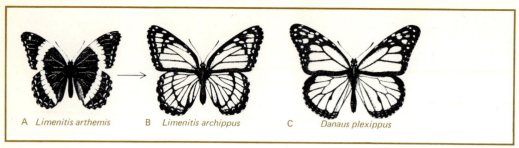

A   *Limenitis arthemis*   B   *Limenitis archippus*   C   *Danaus plexippus*

45-6 A Batesian mimic, model, and the more primitive pattern of the mimic's close relative. The monarch butterfly, *Danaus plexippus* (C) is the model and is distasteful. Its tasty mimic is the viceroy, *Limenitis archippus* (B). *L. arthemis* (A) is thought to resemble the pattern from which *L. archippus* evolved. (From L. P. Brower, *Ecological Chemistry*. Copyright © by Scientific American, Inc. All rights reserved.)

distasteful one.) The distasteful species is called the **model**; the one that resembles it is the **mimic**.

Clearly the tasty mimic relies both on the consumer's ability to remember the distasteful model and on its own resemblance to the model. The advantage to the mimic is that since it resembles an unpalatable form, it is consumed less often. Its death rate is reduced and its fitness improved.

One of the most famous cases of Batesian mimicry occurs in butterflies. The monarch butterfly (Figure 45-6C) feeds on milkweed which contains toxic chemicals. Apparently, having fed on milkweed as a larva, the adult butterfly is rendered ex-

tremely bad tasting. The viceroy butterfly (vice, meaning "instead of"; roy, meaning "the king") looks much like the monarch (Figure 45-6B), but its larvae feed on nontoxic plants. When the viceroy emerges, it is potentially a delectable morsel for any bird that can tell it from the monarch. But by mimicking the monarch, it has made that distinction a formidable problem for the bird. Few humans can do it without training. The bird is forced to look elsewhere for its food.

J. Van Zandt Brower showed the importance of distasteful models to the tasty mimic. She actually fed viceroys to caged, inexperienced scrub jays. The birds ate them readily. When fed monarchs, they sampled a few and then would not eat any more. Once a jay had learned just how obnoxious monarchs are, it refused both monarchs and viceroys—even if, in its naive state, it had feasted on viceroys.

46

The Evolution of Diversity

QUESTIONS TO KEEP IN MIND

What is the alteration of species?

How does speciation by geographical isolation occur?

How does instant speciation by polyploidy occur?

Do speciation and extinction balance?

Why is there such a great variety of living things? In the years since Darwin's time evolutionists have learned how mutations occur, how heterosis maintains much of the variability introduced by mutation, and how the problems of adaptation differ from environment to environment. Yet a satisfactory explanation for the diversity of life is still a long way off. Surely the first step is to understand how new species originate, and much progress has been made here. Curiously, however, even though Darwin's chief work is entitled *The Origin of Species,* this progress is not due to him. What he really described and explained was the alteration of species, not the process whereby two distinct species arise from one.

The second step is to understand how it is that so many species can survive. Here, as you will see, our understanding leaves much room for improvement.

46.1 A species that is spread over a large area will tend to look quite different in the different environments in which it is found.

A common grain such as barley will exhibit **eco-geographical variation,** a change in appearance de-pendent on the nature of the environment in which it lives. In a humid climate, barley often has a stiff stem which resists the tendency to droop to the ground, but in the windy, arid plains of the western Unites States a more flexible stem is fittest because grain tends to be blown off a stiff stem.

Ecogeographical variation occurs because a single phenotype is not equally fit in all environments.

When a species varies from place to place, it usually does so almost imperceptibly. The lizards of one valley are almost the same as those of the next. And those of the next are quite like those of the third, and so on. But when we come to the thirtieth valley and compare its lizards to those of the first, we often see noticeable differences. Such a gradual change of a species' average appearance over space is called a **cline.**

Clines are quite common. Furthermore, in the same species there may be clines of more than one characteristic; body size may be larger at higher latitudes, blackness may intensify with increasing humidity, brownness with increasing temperature, and so on. Thus, a species varies in almost infinite ways. In one characteristic, its members at place A may appear like those at B. But in another characteristic, those at A resemble those at place C.

It takes a huge amount of study to describe such complexity, and as scientific description has progressed and more and more intermediate organisms have been discovered, the dividing lines between species have become harder and harder to draw.

One partial solution has been to assign organisms to **subspecies,** i.e., give them a third name in addition to genus and species. For example, the long-tailed weasel lacks prominent facial markings in the northeastern United States. Here it is called *Mustela frenata noveboracensis*. In Texas, members of the same species have a white "bridle," making them look masked, and these are called *Mustela frenata frenata* (Figure 46-1). The only trouble is that the bridle is not an all-or-nothing thing. Many long-tailed weasels have only part of a bridle. In fact all intermediate gradations can be found. Where is the line to be drawn between the subspecies?

The answer is it must be drawn on a map, more or less arbitrarily. There is most often no really good place to draw the line at all.

Among humans, subspecies are called **races**.

46.2 All biologists recognize that human beings form one variable species.

Generally people are divided into four races: **mongoloid** (including Amerindians), **negroid, caucasoid,** and **australoid** (the black inhabitants of Australia, sometimes known as aborigines). Yet human races fade into each other just as the bridled and unbridled weasels do, and taxonomists could argue for 60 races as well as for 4.

Though races are largely artifacts, who would deny that human gene pools vary geographically? A Swede is more likely to be six feet tall than is an Italian; a resident of Peking has more fat around his eyes, on the average, than one of Window Rock, Arizona; and people from Norway have whiter skin than those in the Congo.

Many characters vary geographically in humans, but since skin color is probably the most important in people's minds, we will choose it as an example. Has skin color evolved? If so, when? and why?

46.3 Many species of animals, including humans, exhibit geographical variation in external color as an adaptation to their environment.

In both humans and animals much of the variation in external color is due to the same kind of pigments—the *melanins*. Also, the variation often tends to be parallel for a wide variety of related species. In other words, if one species in an area is particularly dark, others living in the same area will probably be dark too.

46-1 Long-tailed weasels, with and without the facial "bridle." (From Burt and Grossenheider, *Field Guide to the Mammals,* Houghton-Mifflin.)

46-2 Forms of four species of the genus *Melitaea* in California. Butterflies in each column are members of the species named at the base of the column. In addition to the black being most intense in specimens at the top of their column, red-brown and yellow are also most intense at the top and weaken progressively toward the bottom. (From Rosenzweig, *And Replenish the Earth.*)

William Hovanitz has made a detailed study of butterfly-wing pigments in many California species. He has found parallel evolution of wing color for most of them. The melanic pigments are usually more intense in cooler, more humid environments.

Hovanitz has arranged butterflies of the genus *Melitaea* (the checkerspot) according to the intensity of their color (Figure 46-2). Heavily melanic forms are at the top of each column. These forms all tend to come from the cool, moist northern areas of California. The lightest forms are from the hot, dry

Mojave Desert. Clines of melanic intensity join the two regions for all the species. Hovanitz obtained the same results from the genus *Argynnis*.

Very often investigators can associate phenotypic clines with environmental clines. Hovanitz's work is an excellent example. Specific generalizations about such associations are called ecogeographical rules. There are many such rules. Allen's rule states that protruding parts of birds and mammals tend to be shorter in colder climates. Bergmann's rule maintains that birds and mammals of greater latitudes are likely to be larger than members of their species nearer the equator. Rapoport's rule notes that species of springtails (order Collembola; primitive wingless insects found mostly in the soil) are more often dark-colored in colder climates.

Because ecogeographical variations tend to be exhibited by so many species in parallel, we believe them to be adaptive. Undoubtedly, the most convincing evidence for this comes from certain species which produce two generations per year, each of which differs in appearance, apparently so as to "fit" the different seasons. The mustard-white butterfly, *Pieris napi*, for example, produces a darker generation in the spring, the relatively cool post-rainy season in California, and a lighter generation in summer, when the weather is hot and dry (Figure 46-3). It is clear that the species has adapted to the varying environment. Blacker butterflies must be more fit than light ones in cool, humid climates. Light ones must be more fit in warm, dry climates. Yet, we do not know the reason for the

changes in fitness. To find out would require experiments like the ones Kettlewell performed on peppered moths.

46.4 Gloger's rule, originally proposed to relate the colors of birds to four basic climatic conditions, seems to apply to mammals as well.

C. L. Gloger, a nineteenth-century German biologist, proposed the first ecogeographical rule in 1833. It stated that populations of birds in warm, humid climates tend to be composed of intensely black individuals; in warm, dry climates, of browner ones; in cool, dry climates, of very pale ones; and in cool, humid climates, of less intensely black—but still dark—forms.

	Gloger's rule	
Environment	Cool	Warm
Humid	Weakly black	Intensely black
Arid	Pale	Brown

Since 1833, people have found that not only birds, but also mammals follow Gloger's rule. In fact, people follow it, too. More than 80 percent of the species of birds and mammals that have been examined follow Gloger's rule. Butterflies, however, only partly follow Gloger's rule, and some insects exhibit precisely the reverse of Gloger's rule.

Although we are not sure why the variations described by Gloger's rule benefit fitness, it is interesting that they do. At least, it has enhanced fitness in the past. However, now that humans have so radically altered their environments, the geographical, racial patterns of the past are unlikely to be the optimal ones for the future.

46.5 The first humans may have been black or brown or yellow—but not white.

The wide distribution of *Homo erectus* suggests that our skin-color varieties are even older than the human species. Since our ancestors were mammals long, long before they became human (roughly 100 million years before) and since mammals usually follow Gloger's rule, it is quite likely our prehuman ancestors followed it, too. If there were prehumans living in a cold northern climate, they were prob-

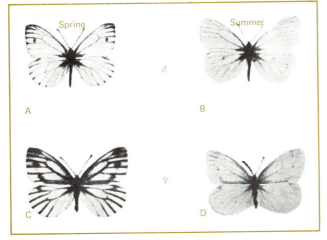

46-3 Adults of *Pieris napi* in two seasons, but from the same place (central California). The darker form is from the spring, the lighter from summer. Males are at the top. (From Rosenzweig, *And Replenish the Earth*.)

ably "white." Those who inhabited tropical rain forests were probably "black."

The pressures that transformed our pelvises, jaws, and brains and thus made us *Homo sapiens* were not skin-color pressures. Perhaps one of these "humanizing" alleles arose in Malaysia, another in Pakistan, another in Guinea, yet a fourth in China. No one knows exactly how many alleles there were, but it is unlikely they all arose in only one population at one place. Instead, each new step arose "never mind where," and then gradually spread by natural selection throughout the vast gene pool of the population of hominids around the world. Each new allele brought our ancestors a bit closer to that exalted and advanced stage of existence where they would have the power to declare nuclear war on themselves. But why change what was already well adapted? Why change skin color?

A few racial varieties may not be very old. In order to live successfully in "white" climates, man probably needed much of the important cultural baggage of the Stone Age: good tools, fire, and knowledge of clothing and sheltering himself. It is possible that prehumanity lived in the cold northland; but we *know* that it inhabited the tropics. It seems likely that the "black race" goes way back to the very origins of humanity, and that only after humans made considerable progress toward civilization would they have been likely to have migrated and successfully colonized colder Europe, subjecting themselves to the climate that caused the evolution of transparent epidermis.

So far we have been talking about the alteration of species or **phyletic evolution,** but have not examined **true speciation,** in which two or more species arise from one. Several radically different mechanisms have been proposed for speciation: At least two are known to be important.

any other living species. It is, in other words, a group of similar phenotypes.

2. The **biospecies** is the group of naturally interbreeding organisms. The virtues of using this concept are many. Taxonomists can actually observe interbreeding in the field, thereby making the classification of organisms into species an objective affair. Members of the same biospecies are an evolutionary unit. Thus, the biospecies is a natural unit of classification.

Most (if not all) other units of classification, such as phyla and classes, are somewhat arbitrary (section 19.3). For instance, birds and crocodiles are quite closely related. Crocodiles, in fact, are more closely related to birds than to turtles. Turtle ancestors separated from the crocodile-bird ancestor some 280 million years ago; birds and crocodiles did not diverge until about 230 million years ago. Yet, biologists agree that it is much more convenient to group crocodiles and turtles as members of the class Reptilia (along with lizards, snakes, and sphenodons) and to create a separate class, Aves, for birds.

With biospecies, such arbitrariness is not allowed. An individual in nature can either breed with another or it cannot. If, through natural interbreeding, the genes of organism A can become part of the genes of B's descendants (and vice versa), then A and B are in the same biospecies. Yet, since sexual reproduction is absent or at least very rare in some organisms, we cannot abandon the concept of phenospecies altogether.

Let us concentrate on the biospecies. How does a sexually reproducing population give rise to another population which is reproductively isolated from the first? There are at least two important answers to this question: geographical isolation and polyploidy.

46.6 Species are defined both in terms of appearance and mating behavior.

We have limped along for many pages without a definition of species, but now that we want to discuss their origin, we will need to know precisely what we are talking about. Unfortunately, there appear to be two necessary definitions of species:

1. The **phenospecies** is the oldest concept of species. It is the population of organisms whose two most dissimilar phenotypes resemble each other more closely than either resembles any member of

46.7 There is evidence that speciation by geographical isolation is the major mechanism of true speciation.

We already know that organisms tend to adapt to their own particular environments. So we might suspect that if some barrier were to separate two groups of a biospecies geographically, each group would begin to evolve in a more or less unique manner. Imagine a squirrel population on two sides of the Grand Canyon. The animals on each side live in a pine forest, but they are separated by the eco-

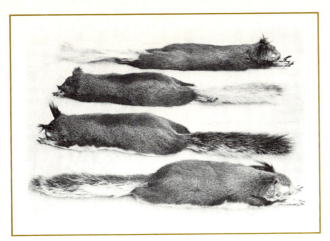

46-4 Tassel-eared squirrels north and south of the Grand Canyon. This is a photograph of four study skins prepared in standard museum fashion for easy storage. (The display skins that one sees in museum exhibits take large amounts of space, so only a small number of specimens can be prepared for display.) The two squirrels at the top of the figure are Kaibab squirrels (from the north rim of the Grand Canyon). Notice their all-white tails and black sides and bellies. (Specimens from the Museum of Southwestern Biology, University of New Mexico.)

logically inappropriate desert of the Colorado River bottom. The altitude of the north rim of the canyon is about a thousand feet higher than the south rim, so the pressures on the two populations may have been somewhat different.

In fact, the squirrel populations on those two rims *are* different. The tassel-eared squirrel of the north rim has an all-white tail and a black belly; the squirrel of the south rim usually has a gray-white belly and a tail with a dark dorsal surface (Figure 46-4). Occasionally one finds a southern-type squirrel in the northern population and vice versa; so we know that selection is maintaining the different phenotypes on the two rims, although we don't know why.

If these populations were to go on evolving for perhaps one or two hundred thousand years, considerable differences between them might arise as each population adapted to its own evolutionary pressures. Perhaps there will be a change in chromosome number in one population, or perhaps a new enzyme system will evolve in the other.

In any case, there is a chance that when we allow them to interbreed (by filling up the Grand Canyon), they will not be able to, or their hybrid offspring (like the mule, offspring of jackass and mare) will be sterile or inviable. If and when this happens, we shall know that two species have evolved from one. Until then, the question of whether they are one or two species is moot and biologically unimportant.

The sequence of events in geographical speciation is then:

1. One biospecies is split into two or more isolated populations by some geographical barrier.
2. Each population evolves independently in its own environment.
3. Rejoining the populations results in the opportunity for resumed interbreeding, but the populations do not or cannot successfully engage in it. They have become separate biospecies.

There is much evidence to suggest that this sequence has been the major mechanism of true speciation.

46.8 Some of the most powerful evidence for geographical speciation has come from studies on archipelagos.

When Darwin visited the Galapagos Islands as a part of his voyage on the *Beagle*, he was intrigued by the unusual varieties of birds he found. Although there were fewer than a dozen different kinds of land birds, there were more than a dozen species of one kind, the *Geospizidae* (Darwin's finches). He could not believe that all these species had been blown across the 600 miles of ocean that separates the archipelago from South America, and he speculated that all had evolved from a single species.

Darwin believed that the original finches had given rise to a population that gradually spread over the five large islands in the archipelago and finally over the smaller, outlying islands as well. Within the small, partially isolated populations on these outer islands, the gene frequencies had probably changed quite substantially due to genetic drift. Eventually the population differed enough from the one on the main islands that interbreeding was not possible. Consequently, when members of outer-island populations returned to the main islands, they formed the nucleus of new populations. They had to compete with existing finches for food, nesting places, and so on, and this pressure forced the new species to diverge still further. They became adapted to eating insects and nectar instead of nectar alone, or seeds and berries and so forth. Descendants of these birds eventually found their way to outer islands and the cycle began again, with genetic drift and natural selection tending to create a new species under the somewhat isolated conditions. This process of speciation and divergence is called **adaptive radiation**, and on the Galapagos, the finches adapted to take advantage of so many ecological niches there is even a finch that performs like a woodpecker (the woodpecker finch). It eats insects dwelling in and under tree bark, but because it lacks the woodpecker's long beak, it breaks off cactus spines or twigs and uses them to dig the insects out.

Six hundred miles from the Galapagos (and 300 from Panama) is Cocos Island, one of the world's most isolated bits of land. Here we find the only other living species of geospizid.

Cocos Island, the size of one of the smaller islands of the Galapagos, is a tropical forest island, and tropical forests usually contain more bird species than any other kind of habitat. So we might expect to find several species on Cocos. But there is only one. Why?

Apparently there is no place for an isolate of the Cocos finch to form. On this tiny island there are no geographical barriers that would permit speciation. And so, though it appears that Cocos could provide for more than one geospizid, it is unlikely that it will ever do so. Here is powerful evidence that geographical barriers may be absolutely essential as the first step in gradual speciation for some kinds of animals.

46.9 The presence of one or more complete extra sets of chromosomes is a condition known as polyploidy; in fertile offspring, polyploidy is a mechanism for instant speciation.

Usually the offspring inheriting an extra chromosome or entire extra set of chromosomes—a condition called **polyploidy**—is inviable or sterile. This is less true for plants than for animals. If the polyploid is fertile, it is a new species, for it cannot interbreed with its parent species, which has fewer chromosomes.

A polyploid may have double, triple, or quadruple sets of chromosomes from its parent species (Figure 46-5) or may inherit a set from each of two different species (Figure 46-6). The first instance is known as **autopolyploidy**, the second as **allopolyploidy**.

Polyploidy requires an anomalous meiosis. The gametes that will form the polyploid are produced

when, by accident, *all* the chromosomes move to one of the daughter cells. The other daughter cell is empty. The full daughter cell goes on to produce two diploid (rather than haploid) gametes by means of a normal cell division.

Sometimes autopolyploids are formed of a diploid gamete and a normal haploid one. These are, of course, triploid; but they are sexually sterile, for

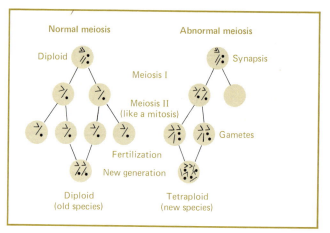

46-5 Speciation by autopolyploidy—doubling of the chromosomes of one species.

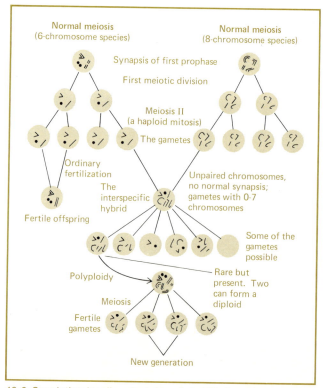

46-6 Speciation by allopolyploidy—hybridization of two species followed by doubling.

their third set of chromosomes has no match in *meiosis I,* so they can't produce normal gametes. Bananas, for instance, are triploid and must propagate vegetatively, because they do not yield viable seeds.

If the new polyploid species is a result of the union of two old species, we say it is an allopolyploid (from *allo,* "different"). Allopolyploidy in plants often results in an unusually large flower or fruit and is therefore desirable, at least to humans. It occurs in a slightly different way from autopolyploidy. The first step is the hybridization of two unlike but normal haploid gametes (Figure 46-6). The product is a haploid zygote—part of its one set of chromosomes comes from one parent, the other part from the second.

Such a hybrid is virtually sterile, because none of its chromosomes can pair when (and if) the hybrid matures and undergoes meiosis. Each of its chromosomes gravitates to one of the daughter cells independent of the others.

Sometimes, purely by chance, all the chromosomes go to only one daughter cell. This yields a functional haploid gamete, a gamete with the same number of chromosomes as the hybrid adult. If two such gametes fuse, a diploid zygote results. When this hybrid matures it can undergo normal meiosis since each of its chromosomes is paired. It is therefore a reproductively competent adult, and if it is viable, represents a new hybrid biospecies.

46.10 Polyploidy has resulted in new species, both naturally and under laboratory conditions.

Polyploidy has apparently yielded many new species of plants: G. L. Stebbins, an eminent botanical evolutionist, estimated that between one-fourth and one-third of the world's plant species were formed by polyploidy. It has also occurred frequently among certain animals.

To discover polyploidy in nature, researchers have examined the chromosomes of organisms and looked for cases of doubling and trebling. For instance, the various species of wheat have chromosome numbers of either 14, 28, or 42. The 28 could be an autopolyploid or an allopolyploid of two 14-chromosome species, and the 42 could be an allopolyploid between a 28 and a 14. Wild cotton species have 26 chromosomes; cultivated ones, 52. The flatworm *Dendrocoelum lacteum* has 16; its

quite similar relative, *D. infernale,* has 32. Many, many other cases could be cited.

46.11 Specialization preserves the variety produced by speciation.

When species have similar ecological roles, one of them may become extinct as a result of interspecific competition. The one that disappears is said to have suffered **competitive exclusion.** Species avoid competitive exclusion if they are each specialized in some different way. Were it not for specialization, only one species at a time could persist at any one place.

There are only a few categories of specialization that can preserve the richness produced by speciation:

1. A species can adapt to a special habitat and stay in it much of the time or even all of the time. Some species of barnacles, for example, occupy the lower (wetter) part of the intertidal zone, whereas other species have adapted to higher (drier) parts.

2. A species can adapt to a certain time of the day or year and become inactive in the off-season. For instance, the desert pocket mouse (hibernates) in the winter, while its competitor, the cactus mouse, aestivates (sleeps) in the hot, dry season.

3. A species can belong to a certain trophic (nutritional) mode. Plants can persist with their consumers, the herbivores. Herbivores and carnivores can share an environment. Their specializations lie in adaptation to each other as well as to their habitat. Only the carnivore is adapted to consuming herbivores. Only herbivores are adapted to consuming plants.

4. A species' fitness may depend on its restricting its feeding behavior. This results in the most complex sort of specialization. It is called resource allocation, because it results in a sharing out of the food resources within a community. A wolf, for example, hunts mostly large-hoofed mammals; it is too big to be fit if it wasted its time on mice or even rabbits. On the other hand, the coyote hunts mostly rabbits and rats; it is too small to hunt moose, elk, and deer. The gray fox, even smaller than the coyote, hunts mostly mice.

Not only can such specialization prevent competitive exclusion of the hunters; it can prevent it in the hunted, as well. The mouse must adapt to the fox, the deer to the wolf.

46.12 The number of species in the world tends to remain constant.

We know from our examination of speciation that both geographical speciation and polyploidy are at work producing new species. At the same time, extinction removes species. (In fact, the fossil record suggests that almost all species that have existed exist no more: Over 99 percent are extinct.)

The available evidence indicates that speciation and extinction tend to balance one another. Figure 46-7 shows the number of families in the fossil record during each epoch of life. It is not a constant record, but except for the marked increase when organisms were first adapting to terrestrial environments and the *apparent* increase in the most recent past, mainly due to our greater knowledge about this time, the data really do suggest that the number of animal families alive at one time hasn't changed much in the ocean for some 600 million years, and hasn't changed on land in about 350 million years. Notice especially that from about 350 million years ago to about 125 million years ago, the number of families seems to have hovered around 325. Since the average number of species per family is probably constant, the number of species appears to have been remarkably stable.

Such long-lasting stability suggests the existence of a steady state. As species increase in number beyond the stable point, extinction rates should get larger than speciation rates; if they decrease, speciation should exceed extinction. There is some evidence that this is true, and the fossil record from North and South America yields an example.

Until several million years ago, South America was a huge island (as Australia is today). In North and South America, many identical environments existed, and many North and South American forms converged. Different species that fill similar roles in different places are called **ecological vicars**. When the North and South American continents were joined by the volcanic activity that completed Central America, there were unprecedented migrations: Yankee species moved south, Latin species north. This migration gave us the opposum and the armadillo, among others. It gave South America the deer and the puma, as well as other animals.

It also brought many ecological vicars into contact and greatly increased the number of species in many places. Competition among vicars for the same territory brought a wave of extinction. This second trend, then, brought the number of species very near what it had been.

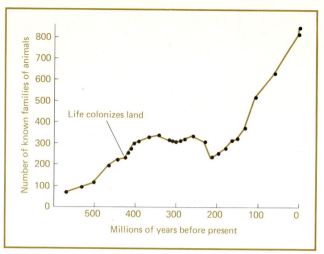

46-7 The number of known families of fossils during different geological epochs. (From Rosenzweig, *And Replenish the Earth.*)

No one hopes to prevent all extinctions. Extinction is a natural process, and it will continue. Moreover, as long as extinctions take their toll, as long as the earth's environments continue to change, and as long as survivors continue to be selected for their increased ability to avoid disease and depredation, so long will phyletic evolution and speciation continue. There is no end in sight. As long as there is life, there will be change. Perhaps Heraclitus was right when he long ago maintained that the only constant in the world is change. In fact, if a single, great unifying factor had to be chosen to characterize life and to discriminate it finally from purely physical and chemical phenomena, it would be evolution directed by natural selection. That which is subject to evolution's pressures is alive. Everything else follows naturally from the environment in which life happens to exist.

SUGGESTED READING

Briggs, D., and S. M. Walters. *Plant Variation and Evolution.* World University Library. New York: McGraw-Hill, 1969.

Colbert, E. H. 1965. *The Age of Reptiles.* New York: Norton, 1965.

Darwin, C. *The Voyage of the "Beagle."* 1845.

Darwin, C. *Autobiography.* Edited by N. Barlow. New York: Harcourt Brace Jovanovich, 1958.

Howells, W. W. *Mankind in the Making.* Garden City, N.Y.: Doubleday, 1959.

Lack, D. L. *Darwin's Finches.* London: Cambridge University Press, 1947.

Lerner, I. M. *Heredity, Evolution and Society.* San Francisco: Freeman, 1968.

Medawar, P. B. *The Future of Man.* New York: Basic Books, 1959.

Pilbeam, D. R. *The Evolution of Man.* Funk & Wagnalls, New York. 216 pp.

Romer, A. S. *The Vertebrate Story.* 4th ed. University of Chicago Press, 1959.

Rosenzweig, M. L. *And Replenish the Earth.* New York: Harper & Row, 1974.

Simpson, G. G. *Horses.* New York: Oxford University Press, 1951.

Smith, H. W. *From Fish to Philosopher.* Brown: Little, Brown, 1953.

Stebbins, G. L. *Processes of Organic Evolution.* Englewood Cliffs, N.J.: Prentice-Hall, 1966.

Volpe, E. P. *Understanding Evolution.* Dubuque, Iowa: Brown, 1967.

Wickler, W. *Mimicry in Plants and Animals.* New York: McGraw-Hill, 1968.

VII ECOLOGY

Earlier parts of this book have presented a picture of
the diversity of organisms and the relationships of
structure, function, and behavior in selected representative organisms. Either the pattern of structure, function, and behavior has been adequate to allow each
species to cope with its environment, or the species
has become extinct. The fossil record tells us that a
fantastic number of species that lived successfully
for a time later became extinct. Many of us who follow
newspaper accounts of the arms race, widespread
industrial pollution, and the population explosion give
occasional thought to the possibility that man himself
may one day suffer the fate of the dinosaurs.

Ecology is the study of the relationships of organisms
to one another and to their changing environment.* It
is important to realize that the earth itself changes
with time, due to both astronomical and physical events
and to the biological activity of its inhabitants. The
ecologist has to understand the natural environment
and how it is subject to change with time. Ecology
deals also with energy fluxes and the cycling of certain
key elements and compounds important to life.

The study of ecology is the arena where biological
science and the social and political affairs of men meet
head on. At this moment in human history, human
population is rising at a rate that can only lead to mass
starvation at the very time that food production increases can no longer be counted on because of
shortages of petroleum and other energy sources. In
the nick of time, the science of ecology is beginning to
develop quantitative approaches to making predictions
about the effects of different courses of human action.

The last chapter in this story cannot yet be written; it
will answer the question Did man learn to utilize his
newly acquired ecological knowledge in time?

* Ecology, to some people, also means "reaction against pollution."
It is used throughout this book in its strict scientific sense.

47

The Planet Earth

What is the scientific meaning of *ecology,* and why has this science assumed social importance recently?

Why do ecologists require a basic knowledge of astronomy, geology, and meteorology?

How do mountain ranges affect the climate in their vicinity?

What factors contribute to the spring and autumn "turnovers" of the water in lakes?

Are estuaries important ecologically, or are they merely wastelands, as many people believe?

<u>Ecology,</u> an old word with new popularity, comes <u>from the Greek word *oikos,* meaning "household."</u> As used today, ecology means the study of nature's household and especially of the interrelationships among the earth's inhabitants and between these inhabitants and the nonliving environment. In the past, the emphasis in ecology has been on learning what these complex interrelationships are: How does nitrogen pass from plant to animal to soil and back to plant again? Or how are sand dunes stabilized by grasses, then shrubbery growth, then trees? Today it is no longer enough to document these processes. Ecologists must use their knowledge to try to predict what *will* happen to these delicate systems if humans strain them. If we are lucky, their predictions will be made in time, and their advice will be heeded.

Because this book will most likely be read only by persons who have more than enough to eat, it may be difficult to convey the urgency of the world's present situation. But for every person who eats well there is another who gets barely enough and a third who is hungry and underfed, day after day after day (Figure 47-1). Yet every hour of these days and nights more than 7,500 additional human beings are added to the population. In a year, this *net* gain amounts to 70 million persons. At this rate the population of the world will double in about 30 years (Figure 47-2). Our food supply will not. And what will happen in the next 30 years? and the next?

As if the problem of producing enough food for the world's expanding millions were not enough, people require other commodities than food for happiness. We value privacy and possessions, blue skies, and forests full of life. We value security and safety from accident and disease. We like quiet walks and bicycle hikes and starry skies. But if populations continue to outstrip food technology, if technology means more pollution than benefits, and if crime and other social disturbances explode with populations—as some experts suggest—where does happiness lie?

It is hoped that some of the understanding on which any effective answer must be based will be found in this and the following chapters.

47.1 Ecology is the study of the interrelationships that exist among living organisms and between these organisms and the nonliving environment.

<u>The thin layer of the earth's surface inhabited by</u> <u>life is known as the **biosphere.**</u> In studying any of its communities of living organisms, from the bio-

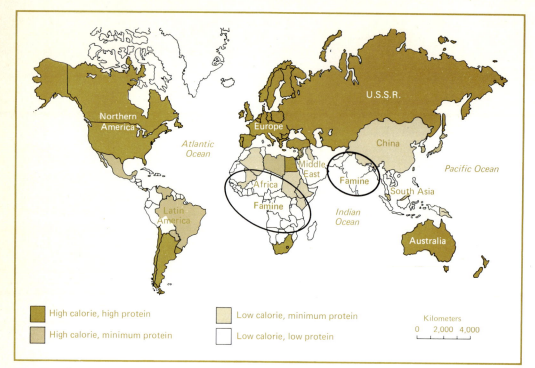

47-1 The global distribution of hunger and of diets rich and poor in protein and calories.

sphere as a whole to the underside of a log, it is clear that there are two sets of components involved —the living and the nonliving. The living or **biotic** components include plants, animals, and micro-

organisms; the nonliving or **abiotic** components include rocks, water, air, carbon, and nitrogen, as well as physical influences such as heat, light, winds, and water currents. The living and the nonliving combine in many ways to form **ecosystems,** which are the functional unit of ecology. **Tundra, desert,** and **marsh** are all ecosystems, and in each,

47-2 The growth of the human population. Man's history would extend the graph at least 40 feet to the left. The sharp rise in population at the right can be attributed to the Industrial Revolution and advances in sanitation and medicine.

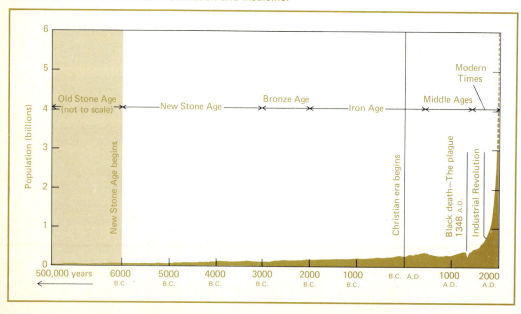

a group of plants and animals depend upon each other and upon the physical environment. Each affects the others.

All ecosystems exhibit certain basic processes. For example, they all have a **flow of energy** that can be followed, often from sun to plants to herbivores to carnivores, and so on. All have a complex of pathways through which substances such as nitrogen and carbon move and in which these and other essential elements are distributed and redistributed throughout the ecosystem.

As the third key component, ecosystems include a changing array of genotypes, which organize or **inform** matter, using available energy to do so. These genotypes, species of plants and animals, move themselves and materials in a generally orderly way in space and time. In addition, all ecosystems change in an orderly manner until they reach a state of dynamic equilibrium, under the influence of prevailing stresses. These stresses may cause ecological regression as well as progression, assuming that we know what ecological "progress" is.

The present chapter concerns itself with the setting for the earth's ecosystems—the abiotic environment. Succeeding chapters will deal with the processes that go on within the systems themselves.

47.2 Our abiotic environment is affected by the earth's position in the solar system.

We live on a lively planet that sweeps around a star we call the sun. The earth also rotates on its own axis, which is tilted relative to the plane of this eccentric path at an angle of $23\frac{1}{2}°$. This tilt (and not the earth's eccentric path around the sun) provides the earth with its seasons. Summer comes to a hemisphere (Northern or Southern, alternately) when that half of the earth is tipped toward the sun and the days grow longer. Day and night result from the spinning of the earth on its axis.

The earth's environment is determined in part by the planet's spherical shape. Because of this, the surface nearest the sun receives sunlight more directly and is warmer than the more polar areas, where light strikes at a glancing angle. The differences in temperature thus produced convert the planet into a giant heat engine that transports heat from the tropics poleward. The energy gradient drives the currents of the atmosphere and oceans, which in turn participate in shaping land masses and ocean basins and in giving them particular climates.

Mass movements of air are shown in Figure 47-3

47-3 Mass movements of air in the earth's atmosphere. See text for explanation.

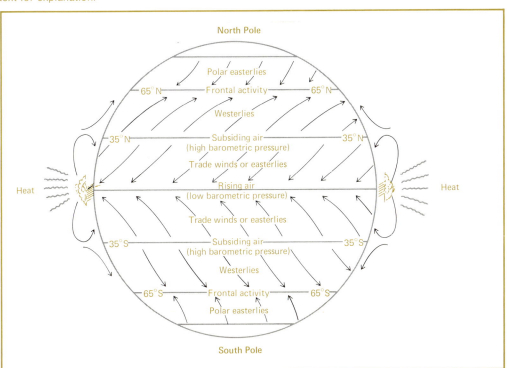

and may be described in a generalized way as follows: Warm surface air at the equator expands and becomes less dense; it is pushed upward by currents of cooler and denser air that are traveling toward the equator near the surface of the earth. The warmer air thus lifted loses its heat in precipitation, and through radiation into the higher atmosphere, and then settles in two bands about 30° on either side of the equator called the *horse latitudes*.

This subsiding air warms through compression (as does air forced into a scuba tank or bike tire), and this increases its ability to absorb and carry away moisture; the great deserts of the world occur in these two belts. Part of the air returns toward the equator at the surface; part travels poleward, but both current systems travel at an angle to the meridians of longitude. In the Northern Hemisphere, north of the horse latitudes and the great desert zones, winds are deflected eastward and appear to the ground observer to be coming out of the southwest. These winds are the *westerlies*. South of the northern horse latitudes, the winds are deflected westwards and in turn are called the *easterlies*. (Note that affairs are reversed south of the equator.) These winds are important in determining the climate of the land and the temperature of the sea.

47.3 The distribution of continents and ocean basins may be largely determined by convection currents in the earth's interior.

In addition to the sun, another important source of heat is the earth's core. For various reasons, including the generation of heat through radioactive decay, the core of the earth is hot and molten. **Convective currents** in this molten material carry heat upward into the earth's mantle. The mantle radiates this heat into the atmosphere. Certain of the earth's surface features appear to be related to these convection systems (Figure 47-4).

The recently expounded theory of sea-floor spreading suggests that these upwelling, diverging convection currents may stretch and thin the earth's crust and produce ocean basins. Converging currents may form deep-sea trenches. Along the backs of these huge convection currents that are created and extinguished in the earth's interior in a slow and unknown rhythm, the surface of the earth rides as if on a conveyor belt. This theory fits well with the concept of **continental drift,** which was sug-

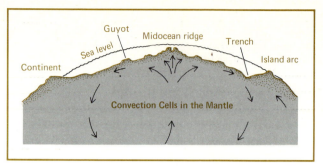

47-4 It is believed that the sea floor spreads by the action of convection cells in the earth's mantle. Such currents are partly responsible for such features of the bottom terrain as the midocean ridge.

gested by the apparent fit of Africa and South America and which states that continents are not stationary but are constantly moving.

47.4 Mountains profoundly affect the nature of the land.

The convection currents in the earth may also play a part in mountain building. More directly, compressive forces, caused by continent-sized pieces of the earth's crust sliding into each other, buckle the crust to form ranges of mountains. Along many of the cracks and faults thus created molten material from the lower crust and upper mantle is squeezed out as lava flows. Some of these flows build up into great piles—volcanoes. Other flows form broad sheets that cover the land for thousands of square miles.

When a range of mountains is formed, it becomes an obstruction to air currents, and other consequences follow. Winds are deflected and some of the air is forced to rise and thus to cool. As air cools, its capacity to carry water vapor lessens. Precipitation then results, especially on the windward slopes of a mountain. The same air, now deprived of its moisture, slides down the leeward slopes and warms as it descends. As it warms, it absorbs moisture from the land and produces a dry zone spoken of as a **rain shadow.** Most mountain systems have moist and dry slopes that are derived in this manner.

In addition, the temperature of the surface of the mountain becomes less from bottom to top. Air temperatures decline about 0.7° C per 100 meters of ascent into the lower 10 kilometers of the atmosphere.

For mountains situated in latitudes above 30° or

so, still another factor comes into play. Remember that as the earth revolves around the sun, its tilted axis has the effect of making the sun appear to rise higher or lower in the sky. Usually it will be low enough to cause extensive shading on north-facing slopes of mountains in the Northern Hemisphere and on south-facing slopes in the Southern Hemisphere. A mountain may thus have a sunny wet quarter, a sunny dry quarter, a shadowy wet quarter, and a shadowy dry quarter (Figure 47-5). Different kinds of organisms will eventually populate different sections according to their needs.

Mountains are not static features. They are built, and they are worn down. Particles of rock and soil eroded from them are carried down streams and deposited in nearby basins, layer upon layer, century after century. The sedimentary deposits thus formed may be thousands of meters thick. Eventually, these sedimentary basins may be compressed, beginning a new cycle of mountain building.

47.5 Other geological factors, such as rock type, affect the environments.

The composition of rocks, in combination with the erosional forces acting upon them, has a large part in determining the environment. Sedimentary rocks (such as sandstone, limestone, and shale) can usually be recognized by their layer-cake–like structure. Cliff faces such as those of the Grand Canyon display beautiful layers of various kinds of sedimentary rock. Though originally deposited in horizontal layers, the strata may be tilted to almost any

47-5 How mountains affect nearby weather conditions.

"Dew point"
Wet rate cooling
Dry rate cooling
Wind
Sunlight

Shady wet   Shady dry
Sunny wet   Sunny dry

47-6 The tilting of a streambed can lead to contrasting habitats along its banks.

degree by later crustal movements. One may thus see strips of different types of soils derived from the weathering of different kinds of sedimentary rock. These sandy (i.e., particles 2 to 0.02 mm in diameter), silty (0.02 to 0.002 mm), or clayey (less than 0.002 mm) soils, along with their chemical characteristics, will support distinctive communities of plants and animals.

When riverbeds form in country underlain by gently tilted sedimentary rocks, one slope of the valley can be steeper than the other (Figure 47-6), because sections of the uphill strata or layers tend to fracture and slide into the stream, whereas those on the downhill side will tend to lie in place. In higher latitudes, freezing and thawing may be of special importance in the erosion of slopes. On the steep slopes of the valley, plants able to maintain a stronghold on the substratum and perhaps even to live on bare rock surfaces will flourish, whereas plants needing deeper soils are more or less restricted to gentler slopes. Thus, geological orientation of sedimentary strata may play an important role in determining the patterns of plants and animals that will inhabit an area.

The tilt of rock strata also plays an important role in the movement of subterranean water. Rainfall readily seeps into a particularly porous layer and may even travel underground to emerge as a spring on the opposite side of the mountain, as you can see in Figure 47-5.

47.6 The nature of the earth's atmosphere affects the biosphere.

The troposphere, that part of the atmosphere in which we live and in which weather and mixing occur, is an envelope of gases some 10 to 14 kilometers thick that covers the earth. It is composed of 79 percent nitrogen, almost 21 percent oxygen, 0.03 percent carbon dioxide, and traces of other gases. The pressure of all these gases weighing upon the earth's inhabitants is 10,332 kg/m² or 14.7 lb/in² at sea level, but only some 30 percent of that amount on a high peak in the Himalayas. In such a rarefied atmosphere, animals and plants must adapt to less oxygen and carbon dioxide per unit volume of air.

Our atmosphere is moderately transparent to wavelengths of light ranging from 4,000 to 7,000 Ångstroms (400 to 700 nanometers). This is the part of the energy spectrum we experience as visible light, and it constitutes about 50 percent of the energy emitted by the sun. Once this light has reached the surface of the earth and is re-emitted as longer wavelength infrared (heat) radiation, atmospheric clouds and carbon dioxide tend to act as a barrier to its escape back into space (Figure 47-7). The absence of clouds allows heat to escape readily into space, and the earth cools sharply. Because of this, a clear, starry night is often followed by a cold

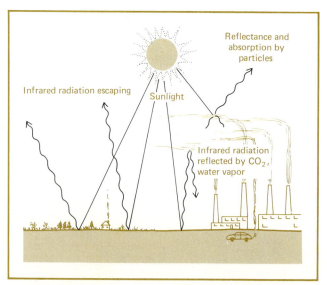

47-7 Clear air in the country is transparent to visible and infrared light, but carbon dioxide and some other pollutants given off by cities trap infrared irradiation, just as the glass of a greenhouse traps the energy of sunlight by not allowing infrared reradiation to escape.

morning. Deserts, with their cloudless skies, often grow dramatically colder at night.

The burning of fossil fuels such as oil and coal, and disturbances of soil, vegetations, and forests that hasten oxidation are contributing to a worldwide increase of carbon dioxide, a gas opaque to infrared or heat radiation. This in turn produces a warming near the earth's surface analogous to that of a greenhouse, whose glass is also opaque to radiant heat. The net effect of atmospheric pollution by all dust and particulate materials and by gas, fluid, and solid suspensions known as **aerosols** is, however, thought to be a cooling rather than a heating. This is because the particles tend to prevent solar radiation from reaching the earth's surface in the first place.

47.7 The atmosphere transports water as rain, snow, and mist.

Annual precipitation ranges from virtually zero over the driest deserts to seven or more meters in the wettest rain forests. Such moisture may come during one season, as it does in the winter months in the Mediterranean or southern California, or it may well be distributed throughout the year, as in New York state. Plants and animals must adapt to these conditions.

Desert rodents and frogs may **aestivate** (go into summer "sleep," section 30.15) during dry periods; desert plants may store water or lose their leaves or die back to underground structures at such times, or they may survive as seeds and germinate and bloom in haste after abundant rains. The life cycles of organisms are thus adjusted to take full advantage of favorable periods and to avoid harsher ones when water is absent or present in excess.

The air and seas around us carry solid particulate material. Such particles play a critical role in weather and climate. Consider the frothy surf of the seacoast. The surf is white because of tiny bubbles of air in the water. After the waves break, the bubbles rise to the surface and explode in a tiny but violent burst and eject a small drop of water into the air. This droplet is traveling so fast that it quickly evaporates, and any particles it carried float away in the air. These particles then act as **condensation nuclei** around which the air's water vapor condenses. The result is **fog** or, if the air is polluted with smoke, **smog**.

If the water droplet that results grows large enough, it may fall as rain. In the absence of con-

densation nuclei, rain may not fall. If there are too many nuclei, so that droplets remain small, precipitation may still not occur. One danger of air pollution is that an excess of condensation nuclei may result in prolonged mists rather than much-needed precipitation.

47.8 About 70 percent of the earth's surface is covered by water, and this fact affects the entire biosphere.

Water is constantly cycling through the biosphere. Water and heat are lost to the atmosphere from living and nonliving surfaces alike, through the process of evaporation. As the resultant water vapor moves with the winds, it loses its heat and condenses on aerosols and other condensation nuclei. It falls as rain, snow, or hail or it recondenses directly on colder surfaces as dew or frost.

Water that has been captured by the land may pass immediately into living tissues, to be stored or reevaporated. Some will percolate through the soil and collect above some impervious layer of shale or clay, but either way, all of it eventually gets back to standing bodies of water. Disregarding subterranean waters, the lakes and oceans on earth cover more than 70 percent of the globe.

Because of water's unique physical properties (section 23.1) which are due to its molecular structure and the hydrogen bonds that tend to form among water molecules, bodies of water have a profound influence on nearby temperatures. Water heats and cools more slowly than the land, and therefore the former exerts a moderating influence on the climate. If you study the autumn foliage around larger lakes, you will notice that trees near the shore change color later in the year than do those farther away. But spring is also delayed along the shore, because it takes more energy to warm the lake than it does the surrounding lands.

Oceans act in an even more pervasive way, and coastal communities generally experience cooler summers and warmer winters than inland areas. Such bodies of water are **thermal buffers**.

47.9 An aquatic environment often exhibits well-defined temperature layers.

A typical lake can be divided into three major zones: an upper one, where temperatures are quite

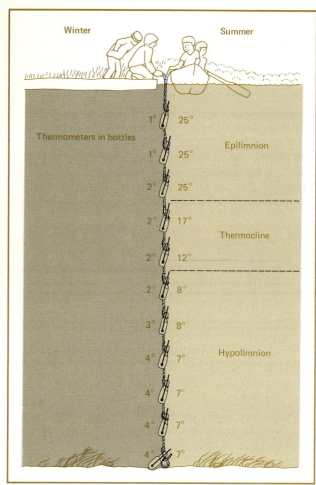

Winter   Summer

Thermometers in bottles

1°   25°
1°   25°   Epilimnion
2°   25°
- - - - - - - - - - - - - - - - - -
2°   17°   Thermocline
2°   12°
- - - - - - - - - - - - - - - - - -
2°   8°
3°   8°
4°   7°   Hypolimnion
4°   7°
4°   7°
4°   7°

47-8 The temperature profile of a lake may be determined by lowering into the water a series of bottles, each carrying a thermometer. The temperature in the bottle remains constant long enough for it to be read at the surface.

uniform; a middle layer, where temperatures change rapidly; and a deeper layer, where temperatures are uniformly cool. In shallow lakes the mixing action of the wind rarely allows this stratification to occur, but in deeper bodies of water this pattern exists (depending on the climate) during all but the colder times of year (Figure 47-8). Such lakes go through an annual cycle. In spring, the surface waters begin to warm. Winds and currents may mix this heat into the topmost several meters of water but not into the deep water, which tends to remain at a relatively constant temperature. As summer arrives, a surface band of warm water, the **epilimnion,** has formed, and beneath it lies a zone of rapid temperature change, the **thermocline.** The lower layer, the **hypolimnion,** changes the least in temperature from summer to winter. With this stratification, there is little mixing between deep and surface waters in warm weather. In autumn, surface

waters begin to cool, winds stir the water, and gradually the temperature throughout the upper layer drops. In late fall or early winter, the temperature of the upper and middle layers becomes equal to that of the bottom, and the waters of the entire lake begin to mix. This brings to the surface nutrients and other materials that have been trapped in the depths of the lake, and it carries well-aerated waters from the upper layers into the deep regions. This period of mixing is called the **fall turnover.**

As winter continues, the lake grows still colder; cold water goes to the bottom, because it is denser than warmer water. When a uniform temperature of 4° C is reached, however, a further fall in temperature results in the still colder water rising to the surface, because fresh water achieves its maximum density at about 4° C, and below that it becomes lighter as it turns into floating ice. With the formation of ice, water stratification ensues.

With the coming of longer days in spring, ice melts. The water warms rather evenly from top to bottom until it reaches a uniform temperature of 4° C throughout its depths. It then undergoes another mixing, the **spring turnover.** A most significant mixing of nutrients and oxygen accompanies this. As spring moves into summer, stratification into upper, middle, and lower layers is again established, and so the cycle continues.

Some lakes do not turn over twice a year, but only once or not at all. If waters of the well-established upper layer have little contact with dissolved nutrients of the bottom, many nutrients become scarce and may finally limit growth of the plankton. Turnover may be accompanied by a great burst of growth (or "bloom") of algae and other organisms. If the lower layer is rich in organic matter, respiration (including decay) will consume available oxygen, and in extreme cases will produce anaerobic conditions. Many organisms cannot live in the lower water under these circumstances. Furthermore, hydrogen sulfide may form and accumulate during anaerobic periods. If fall turnover occurs and the hydrogen sulfide is mixed into the upper layers, the lake may experience a severe mortality of fish and other organisms. These same varieties of conditions exist over large portions of the oceans as well.

47.10 Water pressure is much greater than atmospheric pressure and affects marine animals considerably.

Organisms that live in the water must deal not only with temperature changes, but with water pressure, or **hydrostatic pressure,** as well. For every 10 meters descent in water, the pressure increases by 1 atmosphere (10,332 kg/m²). When a fish descends, the increased pressure causes gases in its air bladder to contract, and it becomes more prone to sink than before. Normally, this increase in pressure does not have serious effects on the fleshy parts of the organism, because body fluids transmit pressures uniformly. It is gas chambers which suffer most, as you may have experienced while diving. Even a dive of a meter can be painful on the eardrums, and successful divers learn to adjust pressures by keeping air flowing freely through the eustachian tubes that lead from the throat to the inner ear. In addition to affecting gas chambers, increasing pressure increases the amount of gas (nitrogen, oxygen, carbon dioxide) that will dissolve in body fluids.

When the pressure is released, these gases must come out of solution. If you have been diving at 25 meters for half an hour or more, bubbles of gas will tend to form in your bloodstream when you return to the surface. They may cause extreme pain, bleeding, and even more serious difficulties. These symptoms were first observed in men building caissons under rivers and were called "caisson disease" or "the bends." Diving mammals appear to overcome the problem mainly by diving with empty lungs and ceasing to breathe while underwater. Their bodies thus do not continue to accumulate gases while they are submerged.

47.11 Tides, currents, and other movements mix the waters, thereby dispersing organisms, distributing nutrients, and affecting the lives of marine organisms in many ways.

Tides, the rhythmic daily or twice-daily rise and fall of sea level, have been observed and commented upon by man for centuries. The relationship between the tides and the moon was grasped very early on.

Tides are caused by a combination of **gravitational attraction** exerted on all parts of the earth (land, water, atmosphere) by the sun and the moon, and of **centrifugal forces**—those that tend to cause the mass of a spinning object to move away from the axis of spin. Together these forces cause the water on the earth to "bulge out," both on the side of the globe facing the moon and on the opposite side. Because these bulges remain in alignment with the moon as the earth rotates, a point in the sea will usually experience two high tides ("bulges") and two lows in each 24-hour-and-50-minute period (the time it takes the earth to rotate relative to the moon).

Tides have dramatic ecological impact upon organisms. Every aspect of the lives of the animal and plant inhabitants of the edge of the sea (or **littoral zone**) is governed by this massive, varied, relentless rise and fall of the sea that occurs no matter what the temperature or relative humidity of the air. Thus, twice a day, they may be hot and dry, wet and cold; their feeding and their growth, their reproduction and their dispersal, their escape and their vulnerability to being eaten are all strictly regulated by overwhelming forces outside their watery medium.

Although lakes are too small to have conspicuous tides, they can have **seiches.** Like water sloshing back and forth from one end to the other in a bathtub, the water in the lake behaves like a pendulum and maintains a constant frequency. The amplitude may vary, however, as it depends upon the climatic conditions of winds, local air pressures, and so forth. You may observe a 25- to 30-cm seiche in the eastern and western ends of Lake Erie. The ecological consequences of this are clear: Animals and plants living in the zones are periodically exposed and submerged, and must adapt.

Water is also moved by the great currents of the world. Driven by winds and by differences in the temperature and salinity of the waters, these currents transport tremendous volumes of water throughout the seas. They affect the climate of the entire globe, in addition to dispersing organisms, transporting nutrients, and so on.

47.12 Special conditions exist where land and water meet, as in an estuary.

The **shoreline** is usually defined as the place where land and water meet. It may a beach, rocky cliff, marsh, lagoon, estuary, or a variation of these. Each presents a different environment. The **estuary,** a semienclosed body of coastal water with a supply of fresh water and access to the sea, is one of the most interesting environments, for it is where land and sea and fresh water meet (Figure 47-9).

Just as waters of different temperatures (and therefore different densities) fail to mix freely, waters with different amounts of dissolved salts delay mixing. Where waters of different salinity meet, as in an estuary, two enormously different ways of life intermingle, and a curious richness may result. Fresh water from a river, in spite of the silt and dissolved materials it carries, is lighter than salty seawater and spreads out over the waters of the sea. The seawater below forms a **salt water wedge** beneath the river runoff. There is actually some mixing between the two layers (Figure 47-9), and some seawater tends to return seaward at the surface, pulled by the fresh water; and some fresh water gets pulled back toward the river by the inflowing seawater. Some organisms use this movement to maintain a favorable position. For others, the interface is a death trap, for few nonestuarine organisms find it possible to adjust successfully to such abrupt changes in the composition of water.

As mixing occurs in estuaries and other shallow areas such as tidal marshes, small, suspended, charged particles (often clay particles) clump together or **flocculate** and settle on the bottom. Because of this, a significant proportion of the nutrients carried by rivers are not washed out to sea, but are trapped in the bottom sediments of a marsh or estuary. **Nutrient traps** are formed and the estuary becomes a perfect nursery ground for many small animals. When we allow persistent pesticides and other toxins to accumulate in estuaries, they become traps of another sort.

Estuaries are among the most fertile places on earth. Geese, swans, and other water birds flock to estuarine waters. Vast hordes of young menhaden, mullet, flounder, channel bass, and other fishes and shrimp hatch there. Salmon, eels, and other species pass through on their ways to or from breeding grounds. Oysters, clams, crabs, and diamond-back terrapins are permanent residents.

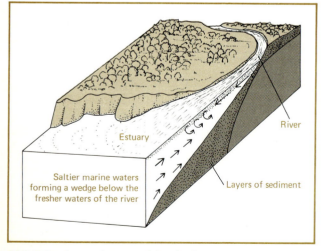

Estuary

River

Saltier marine waters forming a wedge below the fresher waters of the river

Layers of sediment

47-9 Estuaries are bays receiving freshwater runoff from the land. A "salt wedge" forms as the heavier, more saline water slides beneath the fresher water at the surface. The arrows indicate the directions of currents that organisms may use in their movements within the estuary.

Estuaries produce six times as much organic matter per unit as the average wheat land, and perhaps two-thirds of our enormous harvest of marine fishes and shellfish comes from species that are directly dependent upon estuarine food and protection at some time in their lives. Yet we constantly look for new ways to fill or drain coastal marshes as "wasteland" and we foul them all with junk, sewage, and sludge. The accumulation and concentration of DDT and related pesticides is a potentially dangerous result of our activities, but even nontoxic sediment produces turbidity that shuts off light; and when the turbid material settles, it buries the eggs of all species.

47.13 When water falls on the land as precipitation it eventually reaches the water table.

The soil moisture from the atmosphere (rain, mist, snow, etc.) that is not immediately used by plants and animals or reevaporated will percolate through the soil and collect in a saturated zone above some impervious layer of shale or clay. The location of this water is called the **water table.** The underground layer of water (often flowing through coarse sand or gravel) is an aquifer. Ultimately, all water returns to aquifers and standing bodies of water such as lakes or the sea.

Some of this subterranean water, such as the "fossil" water of the Sahara, may be locked underground for tens of thousands of years, but most moves slowly through aquifers to reemerge along river banks or in springs.

The pathways water follows in this **hydrologic cycle** vary from place to place. In deserts where rainfall is rare but often torrential, much of the water travels over the surface, rushes along temporary stream courses, and flows to a catchment basin. If these basins have no outlet, they become salt lakes such as the Dead Sea or the great salt lakes of Iran, eastern Turkey, and Utah; or they may dry seasonally to wonderfully flat **playas** such as those found in Death Valley.

In low- and middle-latitude regions with modest rainfall, the precipitation may be entirely consumed by the vegetation. Little or no water is left to run off directly. Our prairie and Great Plains grasslands typify this situation. Of course, permeability of the substratum is also important and soils having a high clay content may be susceptible to runoff and consequent soil erosion, especially if vegetation is destroyed.

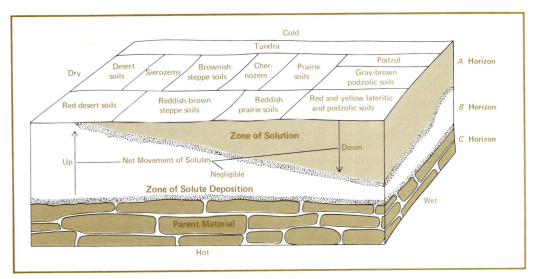

47-10 Soils are the result of the prolonged interaction of rock and sediment with water and organisms at various changing temperatures. The ratio of evaporation to precipitation is especially critical in determining the direction of the net flow of solutes and the establishment of the two upper soil horizons (A and B).

47.14 The ratio of precipitation to evaporation is critical.

In regions where there is an excess of precipitation, water percolates through the soil separating important soil constituents and moving them elsewhere.

The soils remaining contain insoluble residues that are often brightly colored metal oxides. Such soils are the norm in the high rainfall tropics. When a rain forest is cleared, the dark-colored soils rich in organic materials deposited by the former plant cover are oxidized within a few years' time. The yellows and oranges of insoluble iron and aluminum oxides remain.

On the other hand, where evaporation exceeds precipitation, water moves upward through the soil by capillarity, the way oil moves up a lamp wick. It evaporates at the surface and deposits its load of dissolved materials. If this process continues unchecked, the soil becomes so salty that many plants will not grow in it. Extremes of salt deposition may even cause soil to buckle and rupture, forming amazingly rough terrain, as in the famous Devil's Golf Course in Death Valley. If fresh water is drawn excessively from seaboard regions for homes, irrigation, industry, and so on, salt water sometimes rises to replace it. Obviously, the balance of precipitation and evaporation has a great deal to do with the formation of soil and with the use of soil by organisms, including humans.

47.15 Soil evolves in many forms, due to the materials it is derived from, the climate it experiences, and the organisms that live on and in it.

When one considers the play of ecological forces, it becomes clear that not only does the abiotic environment affect life, but that living organisms change their environment. Soil, for example, evolves as a result of a long series of interactions among parent material (rocks and sediment of different kinds), relief, climate (precipitation, temperature), and organic forces (plants and animals). All such factors are **soil-forming factors** (Figure 47-10) and together produce layers of various soils called **horizons.** The horizons constitute a soil's **profile.**

Observe the cross section of a hill exposed in a highway cut. You will see the parent rock at the base of the soil profile. Resting on this are rock and gravel of various sizes whose slow breakdown into small particles contributes to the formation of the mineral part of the soil that lies above.

Obviously, the soil profile can vary strikingly from place to place. It may differ in amount and mineral character. Limestone soils may be thin, because of the rapid dissolution of the base rock and particles and translocation of the residual material by surface erosion or percolation into the cavernous and tunneled strata below. Granite soils, on the other hand, may be deep and well drained, since granite is relatively insoluble and breaks readily into bits that allow water to percolate through easily. Some soils may be water-logged; others may be so well drained that they are leached of all soluble nutrients.

Soil **texture** is due to the relative abundance of clay, silt, and sand. If the percentage of sand (particles 0.02 to 2 mm in diameter) is too high, the soil will be so permeable to water that it cannot retain it and so permeable to air that important organic materials will be oxidized away. If the clay (colloidal particles less than 0.002 mm in diameter) and silt fraction is too high, the soil will be less permeable to the movement of water and gases.

Clay, along with partially decomposed organic material, or **humus,** also plays an important role in holding soil nutrients such as calcium, potassium, phosphorus, and nitrogen (section 23.10). The tiny sizes of clay particles, their crystalline characteristics, and the charged sites on their surfaces provide a vast area capable of holding nutrients in the soil until plants use them. (The surface area of a gram of ordinary clay is about 800 square meters.) In general, soils with high clay content accommodate and hold the nutrients necessary for active plant growth.

Both plants and animals also affect the character of the soil. In upper parts of the soil profile, plant materials are deposited and are subsequently broken down by various soil organisms. Beneath this zone of organic breakdown and accumulation, roots of plants penetrate deeply into the soil. When plant roots die and decay, they leave passageways for air and water to travel. Burrowing organisms, ranging in size from woodchucks and gophers to ants, earthworms, and the tiniest invertebrates, also provide channels through the soil.

Soils in turn affect the rate of erosion, although the process varies more significantly with rainfall, topography, and extent of vegetative cover. The maximum erosion occurs at a rainfall of about 30 cm per year, in the transitional zone between grassland and desert.

Humans have altered the rate of erosion significantly. Ecologists working on watersheds in northern Mississippi, for example, have found that cultivated areas erode several hundred to a thousand times more rapidly than those stabilized by mature forest cover.

48

The Flux of Energy and the Cycling of Materials

QUESTIONS TO KEEP IN MIND

How does energy move through the biosphere?

How does nitrogen pass from the air into living organisms and back into the air again?

Why do elements that exist in solid form, such as phosporus, tend to "leak" out of the cycle of available materials?

How is human civilization upsetting the balance of these important cycles?

The ecosystem is the fundamental unit or natural community of the biosphere and consists of the abiotic environment discussed in the previous chapter; the biotic components—plants, fungi, animals, and microorganisms; and the relationships which bind all parts of the system together. A small or large salt marsh is an ecosystem, a dune along the coast is another; or the marsh, dunes, and impinging ocean may all be studied together as part of a larger ecosystem.

The ecosystems found on our planet vary in wonderous detail from one region to the next. However, although the luxuriant complexities of a tropical rain forest or a coral reef do not seem to have much in common with the stark tundra or open sea, the operation of all ecosystems proceeds along four common lines: (1) Each must have a way of trapping energy and routing it to all its inhabitants, (2) each must cycle essential nutrients, and (3) each develops and then maintains a dynamic equilibrium. Finally, (4) the flux of energy and matter depends upon the **informed substance** of organisms; all of it is useless without organisms and the unique roles each of them plays.

48.1 An ecosystem is held together by a complex web of interdependencies.

Ecosystems are exquisitely organized in all details. Over and over we realize, usually too late, the implications of this fact. We frequently hear that we can just as well do without this or that species. How do we know?

Over a century ago, Charles Darwin related the success of red clover seed production to the number of cats living in the English countryside. He found that red clover develops seed only if cross-pollinated and that the process could only be done by bumblebees. The number of bumblebees, he found, was severely curtailed by field mice, and mice, of course, were controlled by cats.

Due to the complexities of such interdependencies, it seems totally possible that some apparently unfit species or even a demonstrably detrimental species or physical-chemical component may play an unexpectedly important role in the success of another, unrelated species.

From Borneo comes an example of the uncertainties that dog us when we attempt to tamper with natural relationships. There, pesticides were used against mosquitoes in an antimalaria campaign. Roaches, not a target, got sprayed too. They did not die, but merely became somewhat poisonous to small lizards. Because DDT is a nerve poison, these lizards became less agile and less able to escape from the cats that eagerly ate them. The cats were very sensitive to DDT and died. Rats, carrying plague bacillus, invaded the catless houses from nearby forests. Cats were flown in (from an ever-helpful world), but their continued preference for the lizards controlled the lizards' numbers so effectively that there was an outbreak of small caterpillars (the lizards' food) that proceeded to eat the villagers' thatched roofs. This is certainly not the end of the story, and it is unlikely that the widening swirl of disturbances will stop before real damage has been done. Even if the original plan to spray for mosquitoes and reduce malaria finally works, has anyone assured Borneo that the people saved will be provided with food, fiber, fertilizer, tractors, automobiles, roads, airplanes, and hospitals? Put more directly, the question is, Are we using our technology to save people from malaria only to let their expanded populations die of starvation at a later time?

If taking organisms out of an ecosystem can be risky, so is their introduction. Mongooses brought to islands to control snakes have caused the extinction of some native bird species.

The difficulty of assessing the consequences of an act in advance is easy to see. The ability of ecosystems to evolve control mechanisms is also obvious. The lessons are that small changes may snowball and that an ideal predator control must have the capacity to evolve with the predator it is intended to control.

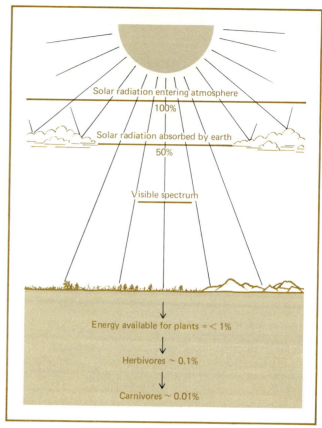

48-1 Of the total solar energy impinging on the earth, less than 1 percent is available to plants. Roughly 90 percent of this energy is lost when plants are consumed by herbivores. When herbivores are consumed by carnivores, another substantial loss occurs.

48.2 Almost all the energy used by ecosystems comes from the sun.

Energy has been defined (section 1.2) as the ability to do work. With the exception of terrestrial nuclear energy, all of it ultimately comes from nuclear transformations in the sun. Of the solar radiation entering the atmosphere (Figure 48-1) almost half is reflected back to space either by the atmosphere or the earth itself. The other half is absorbed by the earth. Except for generalized warming, however, this would do us no good if it were not for the ability of plants to use this incoming solar energy and thus make it available to the rest of the ecosystem.

On the average, each square foot of land receives about 1 kcal of solar radiation each minute. This is roughly equivalent to the heat produced by burning half a wooden match. Only about half of this is even potentially available to plants, the rest not being within the visible spectrum. Of that energy

that does reach the earth's surface, plants absorb some 1 to 5 percent and fix it in the process of photosynthesis to form sugars and other carbohydrates. These plants, the **producers,** use part of this fixed energy for their own respiratory requirements, for they must replace, repair, and reproduce their cellular substance. All herbivorous and carnivorous animals, the **consumers,** depend upon the fixed energy that is left over for their energy and raw materials, as do the **decomposers,** the fungi and microorganisms that break down dead tissues and release nutrients in a form that the producers may use again. Each ecosystem has its producers, consumers, and decomposers.

48.3 In the conversion of energy from plants to animals, about 90 percent is lost.

A cow eating grass illustrates primary conversion from producer to herbivorous consumer. Tiny shrimplike copepods grazing upon microscopic algae of the sea are an aquatic parallel. Cows and copepods can only extract a certain percentage of the energy available in the plant material consumed. The figure varies from situation to situation, but an average of about 10 percent of the consumed producer results in weight gain by the herbivore. This 10 percent represents the conversion efficiency. The other 90 percent is either lost as heat, used to operate cellular and bodily machinery, or is released directly to decomposers in various waste products.

The herbivore is ordinarily eaten by a carnivore, and transformation of herbivore into carnivore occurs with a conversion efficiency of from 10 percent to about 30 percent. This increased conversion efficiency is apparently due to the similarity of tissue composition in animals. One carnivore may eat another, and perhaps yet another may eat the second, to produce, finally, a five-link food chain. Rarely do terrestrial food chains exceed this number of conversion steps, because of the great loss at each conversion, which means that less and less energy is available at each higher food, or trophic, level. If you calculate the amount of energy present at each trophic level (plants, herbivores, primary carnivores, secondary carnivores, ets.), you will see that you have a broad-based energy pyramid with a very narrow peak that represents the top-level carnivore (Figure 48-2). With some interesting exceptions, there tends to be a gradual decrease in energy and in numbers of individuals but an increase in the size of these individuals from the bottom of the pyramid upward.

With conversion efficiency so low between trophic levels, it is clearly an advantage—but a highly vulnerable one—to be as close to the primary producers as possible. This applies to human as well as other consumers, and in overcrowded countries, the largely vegetarian diet makes a depressing kind of good sense. Many more people can be fed on rice from a paddy than could be sustained if the rice were fed to chickens or pigs, and these animals eaten in turn.

A food chain that extends from rice to man or from diatom to copepod to herring to seagull is unrealistically simple. Although there are numerous animals that depend solely on one species for their food, usually each organism at each trophic level has the flexibility of eating a variety of organisms, sometimes ones from its own, sometimes from greatly different trophic levels. The result is a complex food web, which because of its myriad pathways and seasonal transitions, is more stable than a straight food chain, which could be disrupted by the scarcity of any one species.

48.4 Decomposers break down organic wastes and detritus and reconvert them into raw materials for plants.

The amount of energy used and lost at the primary consumer level—when a cow eats grass, for example—is less than the approximately 90 percent already stated if decomposers are taken into account. Then, a bale of hay not only provides energy for a cow (most of which is turned into body heat) and her myriad digestive system symbionts; the hay also, through her excrement and finally her corpse, supplies insects, worms, and millions of microorganisms with the energy they need to run their lives. In the process of meeting their own energy needs, these organisms reconvert materials into forms suitable for reuse by plants.

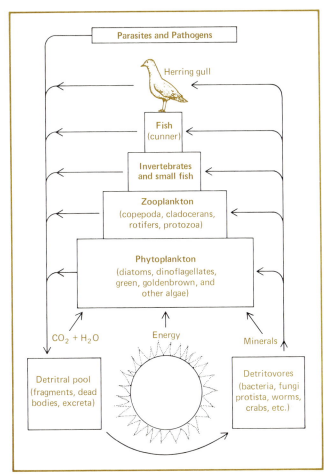

48-2 Energy flow in an oversimplified food chain. The size of the boxes suggests the relative biomasses.

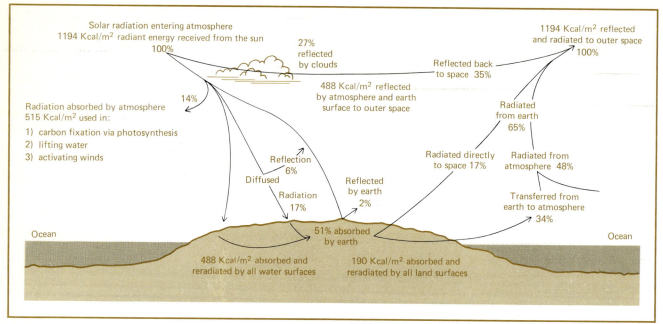

Solar radiation entering atmosphere
1194 Kcal/m² radiant energy received from the sun
100%

27% reflected by clouds

Reflected back to space 35%

488 Kcal/m² reflected by atmosphere and earth surface to outer space

1194 Kcal/m² reflected and radiated to outer space
100%

14%

Radiation absorbed by atmosphere 515 Kcal/m² used in:
1) carbon fixation via photosynthesis
2) lifting water
3) activating winds

Radiated from earth 65%

Reflection 6%

Diffused

Radiation 17%

Reflected by earth 2%

Radiated directly to space 17%

Radiated from atmosphere 48%

Transferred from earth to atmosphere 34%

Ocean

51% absorbed by earth

Ocean

488 Kcal/m² absorbed and reradiated by all water surfaces

190 Kcal/m² absorbed and reradiated by all land surfaces

48-3 The earth's solar energy budget.

48.5 Energy used on earth is finally converted into radiant energy that flows back into space.

The earth's energy budget is a balanced one, as can be seen in Figure 48-3. The biosphere and the rest of the earth act as a gigantic energy converter, receiving energy-rich visible light that is returned to space in the form of degraded and invisible radiation. The biosphere effectively slows down the energy conversion, by temporarily trapping it as chemical energy and using it to work in the interest of life. But except for energy fossilized in coal, petroleum, and related compounds, it is not the fate of energy to remain trapped very long on earth.

When talking about energy flow in an ecological community, it is important to remember that once solar energy has been fixed by green plants, energy always moves through the ecosystem coupled with matter. Consequently, the flow of energy through any ecosystem is inseparable from the flow of matter. If you see a hawk capture a field mouse, you are seeing one route by which energy passes through the ecosystem.

Indirect observations are helpful too, and one of the best techniques of tracing the routes of energy and matter through ecosystems is to incorporate the radioactive isotopes of phosphorus, carbon, potassium, or other elements into certain plants. A later study of the test environment (using a Geiger counter to detect the radioactivity) might indicate that

an ant's nest was especially radioactive. This situation would suggest that ants had harvested the plant product or that they were keeping aphids as "cows," the aphids having tapped sap from the plants. On a later visit, a beetle and bird might show a higher-than-average level of radioactivity, and eventually the radioactivity might show up in rodents and in the young in a nearby hawk nest. We might finally note the dispersion of radioactivity back into the soil and back into primary producers.

48.6 Carbon, like other essential elements, is involved in a complex biogeochemical cycle.

Although coupled, the flows of energy and matter through an ecosystem differ markedly. Whereas energy must constantly be renewed, materials are used over and over again. A certain carbon atom in your body has probably participated in the structure of millions of other organisms during life's history, and the same can be said for any atom likely to be incorporated into living tissues.

Of the more than 90 elements known to exist naturally on earth, about 30 are essential to living organisms. Some are needed in large quantities—carbon, hydrogen, oxygen, and nitrogen—and some in much smaller quantities. All must cycle through the biosphere, however. There are two main types of cycles: One includes gases and solids, the other only solids. Phosphorus is one element that does not flow through the biosphere in gaseous form, although it may enter the atmosphere in particulate form. Carbon is an element that passes as a solid into living organisms and the earth itself and as a gas, CO_2, into the air.

Let us consider the possible fate of a single carbon atom in a molecule of CO_2. This gas may dissolve in the sea ($CO_2 + H_2O \rightarrow H_2CO_3$) to form carbonic acid (H_2CO_3) or its dissociated parts—H^+ ions, bicarbonate (HCO_3^-), or carbonates (CO_3^{2-}). These anions associate with the calcium cation (Ca^{2+}), and in warm water, may precipitate as lime, calcium carbonate ($CaCO_3$). In this form, our carbon atom may become part of the calcareous skeleton of a coral reef.

The coral eventually dies and deposits lime on the sea floor. Thousands of years pass. The lime becomes more deeply buried beneath a blanket of sediments. Some stress in the mantle of the earth ultimately forces a limestone mountain chain into the air. Winds and rains slowly erode the overlying strata, allowing the rock to go into solution in the groundwater of the soil. Interaction of $CaCO_3$ with acid soil ($2H^+ + CaCO_3 \rightarrow Ca^{2+} + CO_2 + H_2O$) releases carbon dioxide, the gas, into the air. From the air, the carbon dioxide can enter an oak leaf.

Photosynthesis causes the CO_2 to be incorporated into carbohydrates. This is oxidized in the synthetic activities of the oak back to carbon dioxide. From the air, the molecule enters the stoma of a nearby oleander leaf and is again reduced to carbohydrate form.

But this time an aphid draws the carbon atom into its stomach and incorporates it into an amino acid. A few hours later a ladybird beetle eats the aphid. Three days later, when the beetle is eaten by a robin, the same carbon atom becomes part of a robin muscle-cell protein. The robin is eaten a week later by a sharp-shinned hawk, which converts the amino acid containing our carbon atom into a protein in one of its feathers. During the hawk's close escape from a wildcat, the feather is torn away and falls into a bog, where it is covered with layer after layer of sphagnum moss and becomes peat. One day this peat is cut and burned in a smoky fire, and the carbon atom is released again as carbon dioxide. A few days later it dissolves in a passing raindrop and reenters the sea off the Azores. Figure 48-4 illustrates a more generalized carbon cycle.

This fanciful account is indeed cyclic. Anywhere along the line, however, another route could have been followed. There is, in fact, no such thing as

48-4 The earth's carbon budget.

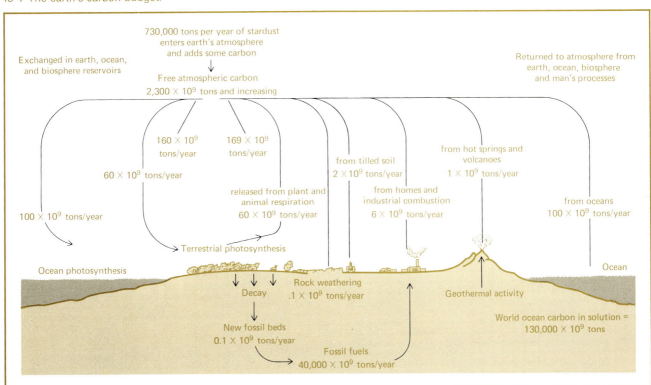

the carbon cycle. Rather, the molecules move, sometimes quickly, sometimes slowly, in thousands of different directions—from land to sea, from sea to sea, from continent to continent, from plant to animal, animal to plant, from organism to atmosphere, and so on. The cycle of every carbon atom is different.

48.7 Nitrogen cycles through the biosphere much as carbon does.

The nitrogen cycle (Figure 48-5) is much the same as that of carbon, except that most green plants are not able to extract nitrogen from the atmosphere. Fortunately, certain **nitrogen-fixing bacteria** (section 23.9) and blue-green algae do have the ability to fix atmospheric nitrogen in a variety of combinations that green plants can use. As a few scientists have pointed out, it is curious that all plants and animals need quantities of nitrogen, yet only a few very small ones can get it from the atmosphere. Instead, plants (and animals) must rely on nitrogen-fixing bacteria and algae to convert atmospheric nitrogen to usable compounds.

Green plants take in nitrogen mainly in the form of nitrate and use it for the synthesis of proteins and nucleic acids (section 23.7). If the plants are eaten, the nitrogen in their amino acids and pro-

teins is incorporated into the consumer's body. Eventually, an organism will release the nitrogen as nitrogenous wastes such as urea, uric acid, or ammonia, or as part of its decaying body. The waste materials are oxidized by several species of nitrifying bacteria and again become available as nitrate or nitrite. In this way the nitrogen atom will usually be used over and over again, but sometimes, in the form of ammonia, nitrite, or nitrate, it is acted upon by **denitrifying bacteria** and released into the atmosphere as nitrogen gas. From the air, nitrogen gas may again be fixed into nitrates by bacteria or by photoelectrical activities in the atmosphere.

48.8 Man's activities have disturbed the balance of the nitrogen cycle in several ways.

Under natural conditions, the nitrogen cycle is balanced, but conditions in some parts of the world have not approached naturalness for close to three-quarters of a century. In the United States, for example, farmers hungry for protein-rich grains soon depleted the soil's natural supply of nitrates. Some began to rotate crops and spread manure on their fields. These practices maintained the organic nitrogen content of the soil and also left the physical structure of the soil undisturbed, so that drainage and aeration (by worms, etc.) continued to be good.

48-5 The earth's nitrogen cycles.

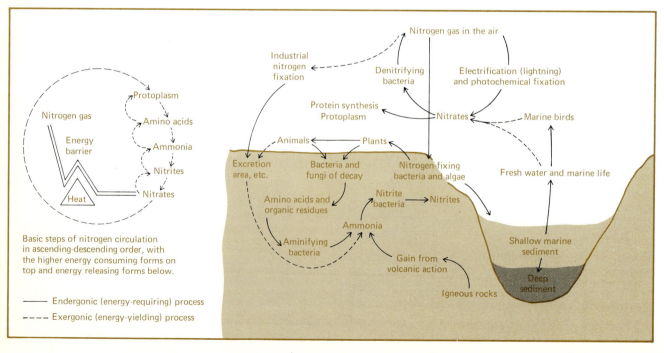

BIOLOGICAL INSIGHTS

What Has the Application of Fertilizers to do With Methemoglobinemia in Children?

American farmers annually apply well over 6 million metric tons of nitrogen to their lands, usually in the form of ammonium nitrate. As a consequence, nitrate levels have been rising in drinking water. The United States Geological Survey reports that 39 different regions of the country are reaching dangerous levels of 8 to 9 parts per million—40 percent of the wells in Missouri are said to be so polluted. Why are these levels dangerous? Nitrate is transformed by bacteria in the digestive tract into nitrite, which combines with the hemoglobin of blood to reduce its effectiveness, causing a kind of anemia or "cyanosis" in children called **methemoglobinemia.**

Spinach, kale, cabbage, cauliflower, radishes, and related plants are capable of storing relatively high concentrations of nitrate nitrogen and become dangerous, especially to babies, when this nitrate is transformed into nitrite under conditions that may occur on the way to the table. Awareness of the complexity of the problem has grown with the realization that the application of common herbicide 2,4-D (usage of which has been severely restricted) may increase not only the uptake of nitrate but also the amount of sodium absorbed. The increased sodium uptake may increase systolic blood pressure and induce electrocardiac abnormalities.

The use of chemical fertilizers instead of nitrogen-containing organic wastes also produced good crops economically but did not maintain a good proportion of organic debris in the soil. The soil therefore became less and less porous. Plants grown in such nonporous soil had insufficient oxygen around their roots and could not use all of the nitrate that was being added to the soil. The remainder was lost through leaching, runoff, or conversion to ammonia, nitrogen gas, and nitrogen oxides.

Thus, in humus-depleted soil, nitrate fertilizer supplies nutrients to the immediate crop, but it also furthers the progression of soil depletion by upsetting the balance of the soil system. Most of the nearly 10 million tons of nitrogen added to American soils winds up in the atmosphere as a gas or in rain, snow, and in lakes and rivers—but not in the soil or in the crops.

This has one particularly hazardous effect, which can be seen, for example, in Lake Erie. There, an overabundance of nitrate wastes from agricultural fertilization coupled with increased levels of phosphorus has caused great algal "blooms." The process producing these blooms is called **eutrophication.** The algae take in the inorganic nitrogen, convert it to organic form for their own growth, grow quickly, die quickly, and pollute the lake with organic material. The main problem is that the algae die as quickly as they bloom, and the bacteria and fungi of decay use up so much oxygen in degrading the dead algae that the waters are left without sufficient oxygen for other organisms. The decomposers

that live on such organic matter and convert it into inorganic salts need oxygen, and if they don't get it the organic matter and the byproducts of anaerobic decomposition build up on the lake floor.

The atmospheric portion of the nitrogen cycle is also in trouble. Each year the engines in factories and motor vehicles of the United States operating at high temperatures, combine atmospheric nitrogen and oxygen into more than 8 million tons of nitrogen oxides. Some of these oxides combine with sunlight and fuel wastes to cause smog, and some are oxidized to nitrates, which, through rain and snow, reach land and water once more, where they may fertilize crops or add to the water eutrophication problem.

The heavy use of artificial fertilizer has been successful in maintaining a temporarily high level of crop production. However, it has seriously jeopardized future crops by causing loss of nitrogen-fixing bacteria and upsetting the balance of the nitrogen cycle. In addition, it has been responsible for a large portion of the algal blooms that disturb the nation's waters.

48.9 Sulfur cycles through the biosphere and as sulfates, ties up a large quantity of free oxygen.

All living organisms need sulfur for certain amino acids. Figure 48-6 illustrates the major pathways

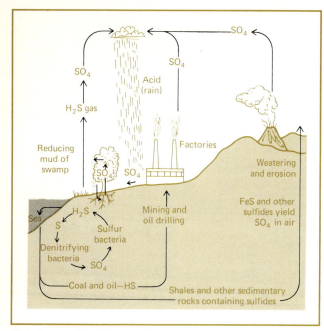

48-6 The earth's sulfur cycle.

their way into the sea. There are plenty of sulfates to replenish the supply, but the sulfates (SO_4^{2-}) that leach away represent part of the earth's oxygen supply, and they are out of the cycle for millenia. There is only one process that works in reverse and liberates the oxygen from sulfates. This process is accomplished by **sulfate-reducing bacteria,** organisms that live in the smelly muds of lakes, swamps, and estuaries. These bacteria are killed by free oxygen and must live in reducing (anoxic) muds. They use sulfur much as other plants use oxygen and convert SO_4^{2-} into hydrogen sulfide (H_2S) and oxygen. Yet in spite of their importance or because their importance is not apparent to most people, modern societies drain marshes and fill swamps. Some scientists believe that to do so even to the extent now underway in this country, may affect our supply of oxygen.

that the element follows. Plants take sulfate ions from the soil and pass them on to animals. Some sulfate washes into the sea, where it is used by aquatic organisms or lost for millenia as sediments. These are eventually compressed into coal- and oil-bearing rock, shales, and so on, and reenter the cycle due to weathering or, through humans' use of oil and coal as a fuel, as a product of combustion.

Like nitrates, more and more sulfates are finding

48.10 Phosphorous and several dozen other minerals cycle through ecosystems in nongaseous forms.

Phosphorus, needed by plants and animals for construction of DNA and the energy-rich compound ATP and by animals for bones and so on, cycles in a somewhat different manner from carbon, nitrogen, and sulfur, because it never exists in gaseous form.

It moves mostly in water or as part of some organism (Figure 48-7) in what is called a **sedimen-**

48-7 The earth's phosphate cycles.

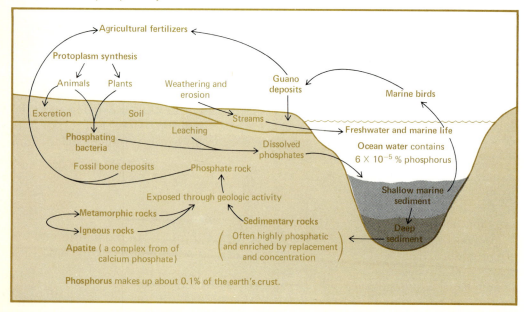

TABLE 14 ENERGY INPUTS (KILOCALORIES) IN CORN PRODUCTION

Input	1945	1950	1954	1959	1964	1970
Labor	12,500	9,800	9,300	7,600	6,000	4,900
Machinery	180,000	250,000	300,000	350,000	420,000	420,000
Gasoline	543,400	615,800	688,300	724,500	760,700	797,000
Nitrogen	58,800	126,000	226,800	344,400	487,200	940,800
Phosphorus	10,600	15,200	18,200	24,300	27,400	47,100
Potassium	5,200	10,500	18,900	36,500	30,400	63,000
Seeds for planting	34,000	40,400	50,400	60,400	68,000	68,000
Irrigation	19,000	23,000	27,000	31,000	34,000	34,000
Insecticides	0	1,100	3,300	7,700	11,000	11,000
Herbicides	0	600	1,100	2,800	4,200	11,000
Drying	10,000	30,000	60,000	100,000	120,000	120,000
Electricity	32,000	54,000	100,000	140,000	203,000	310,000
Transportation	20,000	30,000	45,000	60,000	70,000	70,000
Total inputs	925,500	1,206,400	1,548,300	1,889,200	2,241,900	2,896,800
Corn yield (output)	3,427,200	3,830,400	4,132,800	5,443,200	6,854,400	8,164,800
Kcal return/input kcal	3.70	3.18	2.67	2.88	3.06	2.82

Source: From David Pimentel *et. al.*, *Science* 182, 1973: 443–449. Footnotes to the original table indicate the assumptions underlying all input estimates.

tary cycle. Smaller amounts move as particles in the atmosphere. It is especially susceptible to the pull of gravity and readily concentrates in lakes and seas. In basins it tends to be trapped in sediments, to be returned to the land only through the enormously slow processes of mountain building. For this reason and because phosphorus is proportionally more concentrated in organisms than in the environment, it frequently becomes the limiting factor for organic growth. (That is to say that if an organism has more than enough of all its essential nutrients except one, that one shortage will limit growth.) Because its source is rock and soil, phosphorus is often not readily replenished once removed from a site. An ecosystem, be it forest, lake, bay, or grassland, may be severely retarded in its productivity by want of sufficient phosphorus.

Copper, iron, magnesium, cobalt, zinc, boron, and several dozen more elements are also required by ecosystems. They have cycles similar to the phosphorus cycle, because they do not commonly exist in gases and cannot be transported by the atmosphere. Several of these elements are being added in increasing quantities to farmland, although such usage can disturb the natural texture and balance of the soil, pollute waterways, and so on. The more these disruptions and others like them go on, the more chemical energy—in the form of fertilizer, pesticides, fuel for farm machinery, and so forth—must be invested in a farm to produce the same yield. Table 14 shows how the number of kilocalories of corn that can be produced with one kilocalorie of energy has dropped from 3.70 in 1945 to 2.82 in 1970. The table also shows that the real power for U.S. agriculture comes from fossil fuels (used as fuel and in the production of fertilizer, electricity, etc.), not from manpower or from natural recycling. As fossil fuels become scarce and expensive, food costs rise, yet the need for food is continuing to grow. Is there a better way?

49

Time and Change

If you could watch a sand dune or an abandoned field for 200 or more years, what changes might you see?

How does a forest moderate the physical factors of the environment, and why does it have a greater variety of species than a field?

Time is an ecological factor. Over the years, ecosystems change: The abiotic environment is modified, the number and kinds of organisms living in it change, the routing and efficiency of energy and material transfer vary, and if catastrophic conditions do not occur, the stability of the whole increases. We may thus think of primitive, simpler ecosystems in the past evolving into the more complex, biologically richer, and physically more stable ecosystems of the present. Whatever their characteristics, all ecosystems appear to evolve in the direction of greater stability, which is to say that they move toward a dynamic balance in which all species' needs for food, shelter, waste disposal, regulation of population size, and control of disease are met.

49.1 Ecosystems evolve in a predictable sequence.

The process by which an ecosystem evolves is called **succession. Primary succession** begins in a barren environment that has never been inhabited before—a new volcanic island, a bare rock, a building sand dune. **Secondary succession** proceeds from land that has been colonized but has been disrupted, either by natural means—for instance, fire or storms or climatic shift; or by artificial means—farming, bulldozing, and so on. In both cases a simple community of pioneer organisms develops and gradually provides conditions favorable for a greater and greater variety of inhabitants. The **community,** the populations of plants and animals that live in the same area, will become more complex until it comes into equilibrium with the physical environment. It is then called a **climax community** and unless disturbed will tend to be self-perpetuating and will remain as a tropical forest or beech-and-sugar-maple forest or northern tundra, for example, without further change.

An example of primary succession (Figure 49-1) is seen in an area of outward-building dunes, where the beach represents the new land, inhabited only by a few invertebrates; the grasses and shrubby growth represent slightly older, more established communities; a stand of eastern cottonwoods, an older community still; and on and on, until, several miles from the beach, a beech-maple forest represents the climax community.

To study secondary succession it is usually necessary to watch an abandoned field or burned-over forest develop over many years. In the second instance (assuming we are in the northeastern United States) we would find amid black stumps a few pioneering shrubs and trees. These typically favor sunlight and are not very tolerant of shade. In the shade that they themselves create, however, **shade-tolerant** species spring up. As pioneer vegetation grows, it also improves the soil, both by preventing erosion and by adding organic litter in the form of leaves and dead plants. Some of the

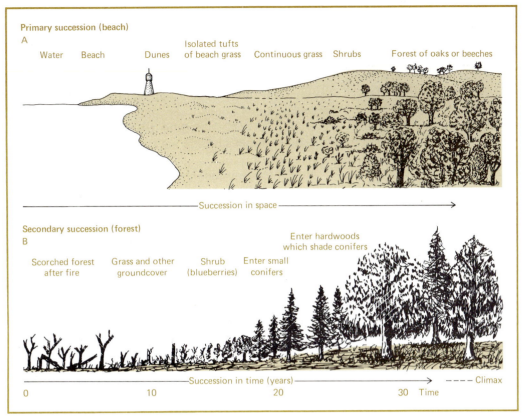

Primary succession (beach)
A

Water   Beach   Dunes   Isolated tufts of beach grass   Continuous grass   Shrubs   Forest of oaks or beeches

——————————— Succession in space ———————————→

Secondary succession (forest)
B

Scorched forest after fire   Grass and other groundcover   Shrub (blueberries)   Enter small conifers   Enter hardwoods which shade conifers

——————————— Succession in time (years) ———————————→   - - - - Climax
0   10   20   30   Time

49-1 Examples of (A) primary and (B) secondary succession. For details see the text.

new species that take advantage of these improved conditions may be the same as those growing in the undisturbed (unburned) portion of the forest.

49.2 Successions commencing in different but neighboring habitats tend to converge in a typical regional climax.

A lake may eventually fill and be covered by the same kind of forest that emerges on nearby sand plains, clay banks, and floodplains. Different soils (where minerals and moisture may differ) and different slopes (where moisture and shade differ) may carry different species for a very long time, however, before all reach a **regional climax.**

Ecologists assign the term **climatic climax** to self-perpetuating distributions of organisms largely dictated by climate. (These climaxes are typically widely distributed, as with polar tundra or northern coniferous forest.) The term **soil climax** indicates distributions influenced by soil. The subject of climax ecosystems is an enormously complex one, and hundreds of variations may result from

the chance domination of an area by pine trees; overuse by deer, cattle, or bison; repeated fires (natural or man-made); glaciation; frequent flooding; erosion; landslides; and so on.

49.3 In the course of evolutionary time, climax communities have advanced, retreated, disappeared, and reappeared in response to changes in climate.

It is one thing to trace the probable succession of events leading to a familiar forest climax. This is essentially a short-term study covering a few hundred, or rarely, a few thousand years. However, it is possible to study succession on a broader scale, that is, in terms of **evolutionary time.** Then the questions asked are ones like, Where did the ancestral pines of central Pennsylvania come from? Canada or the southern United States? Where did oaks, hemlocks, chestnuts, and maples originate? Are these species advancing northward or retreating southward from primitive centers of distribution?

If you were able to see global vegetations in some sort of worldwide time-lapse view, you would observe a most interesting series of advances and retreats from different population centers. Bare earth would be covered, bared again; grasslands would give way to forests; forests would change from coniferous to deciduous.

One of the ways in which the picture of the past is reconstructed is by studying fossilized pollens trapped in sediments. Different sections of cores taken from lake bottoms, bogs, coral reefs, or other areas where the rate of sedimentation is fast enough to trap the pollen that blows in from nearby plants, can be dated in a number of ways. The pollens found in the core then indicate the kinds of vegetation present in the vicinity over hundreds or even thousands of years.

49.4 As succession progresses, the physical environment becomes more moderate, while the biotic environment becomes more demanding.

As pioneering plants and animals move onto a glacial moraine (the piles of earth and stones left by a glacier) or an abandoned field, the factors concerning them most will be physical and chemical. There will be plenty of sunlight (probably too much), an uneven supply of water (for it will run off almost as soon as it arrives), abundant minerals, and no protection from temperature fluctuations or winds.

As the years pass and the community of plants and animals becomes more diverse, the physical environment will become less of a "problem," but the problem of competition from other organisms will increase. This is because the physical factors will have become more moderate. Temperatures, wind, and relative humidity beneath the forest canopy do not fluctuate as widely as they do over bare soil. The difference is comparable to the differences between a parking lot and a deep forest: You can readily feel the difference in either winter or summer. Chemical factors (mainly of the soil) will also have altered, usually toward a steady but rather skimpy supply where demands of producers practically balance supplies turned over by decomposers. On the other hand, there will be more organisms competing for available sunlight, moisture, minerals, foods, and so on. Two sorts of things will have happened. The challenge to both plants and animals will have altered and the organisms themselves will have altered the habitat.

49.5 Succession results in "complexification."

An area does not necessarily lose all or even most of its species when it advances to a new sere, or stage of the successional continuum. Species of earlier seres tend to remain, but in more restricted circumstances. A certain plant requiring mineral soil or a lot of sunlight may prosper even in a climax forest, but only alongside a stream or cliff face, or atop throw mounds produced when large trees are uprooted. These species no longer dominate the scene, but neither do they disappear entirely. Succession thus results in "complexification." Not only are a new species added, but an increasingly complicated distributional pattern emerges.

In a recently burned site, only a few dozen species of plants and animals grow. In an unburned forest there are many, many more. Perhaps only a few species (beech, oak, and maple trees) may be conspicuous in a climax community, but small organisms play a far greater role there, in terms of numbers, species, and complexity of flow of materials and energy, than in pioneer communities.

Naked mineral soil provides few places for life. An organism there may live on the surface or burrow into the ground to one depth or another; but beyond this the habitat has little to offer. Bare soil also fluctuates greatly in temperature, moisture, and wind exposure, conditions that few creatures welcome.

Compare this with a forest of oak, pine, and hickory. In the forest, each tree species provides an array of habitats. Different kinds of creatures specialize to harvest different parts of a single leaf. One attacks the upper tissue layers, another the lower, another the mesophyll cells, others the petiole, and so on. Other organisms are adapted to consume either bark, cambium, phloem, or xylem tissues. Others similarly specialize on roots, flowers, fruits, and seeds.

A single tree may thus provide habitat and food for several hundred different kinds of organisms, ranging in size from the tiniest viruses to squirrels and porcupines. (Larger arboreal animals such as leopards and primates might regularly visit several species of trees in the course of a year, specializing more in terms of prey or fruit than species of tree.) Each species of tree is host to its own assemblage of organisms. The general increase in number of species is paralleled by increased specialization and dependency of each species upon other species, upon the soil, and on microclimates that they themselves have created.

This increased dependency and attendant specialization reflect the organisms' functions or **niches in the community**. The numbers of opportunities to specialize increase with successional stages, hence the increase in number of species in a community. It is well to distinguish habitat from niche: A habitat is an organism's address; its niche is the sum of its activities, its *role in the ecosystem.*

49.6 The productivity of ecosystems tends to increase as they advance toward climax communities.

There are several consequences of the proliferation of smaller organisms and their greater specialization that can be noted in a climax community. First, resources of an environment are used more efficiently. Instead of an organism hunting everything, it specializes; rather than consuming everything, much of which may be indigestible, the specialist eats selectively and evolves greater ability to extract materials and energy from its food.

Second, smaller size is associated with increased metabolic rate. This leads to short generation times, which help a species make relatively fast evolutionary changes, either to adjust to unfavorable conditions or to take advantage of favorable ones. A bacterium's metabolic rate may be some 10,000 times that of a human being's on a per unit rate basis. (A human being with the metabolic rate of a bacterium might need 20,000 eggs, 20,000 pieces of toast, and about 50,000 glasses of milk for breakfast alone; he would also explode from his inability to get rid of excess heat through his comparatively small surface area.)

Because of the small size and fast generation times of many organisms, their total productivity may be underestimated. A community of marine plankton, for example, will produce a lot of living material in a year, although the biochemical machinery is turning over so fast that there is little organic matter for one to see per unit area or per unit volume at any one time; that is, the **standing crop** will be small. (An early marine biologist estimated that the waters of the Baltic Sea were as productive as an equal area of prairie land—certainly not an obvious comparison.)

There is still considerable controversy as to the productivity of different seres in a successional series. There seems to be a general tendency for the productivity of an ecosystem to increase with advances in successional stages. It must be noted that a substantial amount of the increase in productivity (in a forest, for example) is in products such

as wood, bacteria, fungi, and a multitude of small animals. The last three are hard to count and are not considered "products" in terms of economic use. On the other hand, these myriads of small organisms make energy and matter move slowly through the ecosystem. In other words, they conserve.

Cultivation essentially turns soil back to an earlier successional stage. Although this practice produces desirable crops, it also frequently leads to erosion, nutrient loss, depletion of the oxygen of our atmosphere, and other undesirable consequences. Deciding which of the world's communities should be allowed to attain climax status and which can safely be kept cultivated is a challenging job for ecologists of the future.

49.7 As a community passes through its successional stages, it usually develops stabilizing mechanisms of population control.

As has already been noted, the interaction between herbivores and predators tends to keep the population of both groups under control. Complex food webs represent a system of checks and balances, and the numbers of individuals of certain species may be remarkably constant from year to year. The fact that exotic species often become extremely abundant when first introduced to an ill-prepared community attests to the presence of these controls. We are also beginning to understand that many species may have intrinsic means of population control which join with predators and disease to regulate dispersal and birth–death rates. The emergence (or possible presence) of such intrinsic controls for human population is a matter of critical importance.

49.8 Certain communities become increasingly susceptible to certain kinds of catastrophes as they advance in their succession.

The maturing communities of the land and the shallow sea floor stand apart from the surface communities of lakes and oceans in that they invest in supportive structures. The coral reef secretes and deposits a vast amount of calcium carbonate that allows it to expand and diversify but may be broken up in a major storm. A forest invests in the construction of monumental columns of nonliving wood appropriate to support the thin film of sapwood and bark essential to nourish the leafy canopy and the root system. This bulk of supportive material in the forest often proves the undoing of the community, even though it serves the vitally important functions of storage of carbon and release of oxygen. The flammability of woody materials may encourage destructive forest fires that plunge the complex ecosystem back to pioneer stages. The great weight of a forest on a mountain slope may eventually precipitate an avalanche that leaves nothing but bare rock behind. The component organisms, accustomed to the mild physical conditions and multitude of niches fostered by the climax condition, cannot repopulate the disrupted areas. Furthermore, in our era, mature climaxes almost invariably prove attractive to humans and thus are (often unwittingly) destroyed in the name of "progress." We harvest the chemical energy accumulated by a climax community—that is, we cut down a forest for wood—and return the area to an earlier, harsher, and less efficient stage of succession.

49.9 Does the biosphere exhibit evolutionary trends?

Without human intervention, the species array at any place on the surface of the earth tends to change in a predictable pattern that we think of as governed by the principles of succession. Viewed over a much longer period of time, ecosystems have changed in more radical ways as the evolution of species and species groups has occurred. Looking at these changes, can we see any evolutionary trends within ecosystems?

Ecosystems have become richer in variety and stored elements as land plants have evolved from moss-sized plants to lush forests of trees. Soils have become deeper as land plants have evolved soil-binding roots (absent from the earliest land plants). There has also evolved a greater interdependence among species, as between bacteria and flowering plants, ants and plants, or ants and aphids.

Is the ability of the primary producer to capture the energy of sunlight increasing? The evolution of better leaves as land plants have evolved from mosses to horsetails to ferns indicates this is happening. Answers are harder to give for other questions. Are the conversion efficiencies between different trophic levels of modern ecosystems improving to allow less energy to escape? Are new methods being evolved to overcome certain environmental stresses? These are questions about which one can only speculate, but they are exciting and deserve consideration.

50

The Impact of Man

What is the earth's carrying capacity for human population?

Can humans reliably "improve" their environment by building dams, ridding wetlands of insects, and so on, without paying a price?

What is the difference between the commercial price of a sheet of paper and its true (ecological) cost?

What is the relationship of humans to the earth's ecosystems? Are humans contributing, like other organisms, to the increased productivity and stability of the biosphere? Or are they causing confusion and imbalance, through shortsighted intervention? Clearly, they do both.

Humans have been responsible for destroying ecosystems, because we have lowered the ability of the ecosystem to sustain our populations without additives, often costly ones. There are, of course, plenty of instances where the benefits of the ecosystem have been more and more clearly made available to us. In only a few instances, however, can it be claimed that these benefits to humans are secure for long periods of time. We must learn to make our insults to the ecosystem minimal and then begin the challenging and often expensive business of resurrecting destroyed or depleted ecosystems.

50.1 What is the nature of "natural"?

Before examining the human impact on the biosphere, we must clarify our meaning of *natural* and *unnatural*. Since humans are, strictly speaking, a part of nature, should human activities be termed natural or unnatural? In this chapter we will apply *natural* or *wild* to any assemblage of plants and animals in a habitat that has evolved without deliberate ecological planning. Within this conceptual framework, garbage heaps, gardens, and great forests all have some degree of naturalness. There are in each, tendencies toward species enrichment and energy storage over time. Natural places, in contrast to places disrupted by humans, are more self-repairing and persistent. A human hunter, one predator among many, does not disturb the natural balance as much as the human who keeps flocks of sheep or goats. The farmer and the city dweller change both the physical environment and the balance of native species even more.

50.2 If modern man is not familiar with the land in its wild states, will he work to conserve it?

Aldo Leopold wrote in *A Sand County Almanac,* "Your true modern is separated from the land by many middlemen, and by innumerable physical gadgets. He has no vital relation to it; to him it is the space between cities on which crops grow."

Unfortunately, people who ignore the land also ignore their close, although not always obvious, dependence on it. For the urban dweller, water comes from a faucet, and the purity of a river or a lake is not of immediate concern. Milk comes from a cardboard carton bought at a supermarket: How many consumers worry when they hear of insecti-

411

cides and radioactive fallout raining onto a pasture where a herd of dairy cows is grazing?

As we lose the sense of where things come from we also forget where they go when discarded. The sink and the trash basket are not the final repositories for waste materials. The incinerator, the sky, the "sanitary" landfill, the bay, or the river must accept the millions of tons of garbage the people in the United States produce each year.

Conserving pure waters and wild lands costs money. Without a sense of ecological cause and effect we are not likely to pay for these services voluntarily.

50.3 Humans start dramatic chain reactions in the environment, the results of which they can neither predict nor control.

In the 1960s the Egyptians built the High Dam across the Nile near the town of Aswan. It is a mammoth structure that has created one of the largest man-made lakes in the world. The water turns giant turbines that generate millions of kilowatts of electrical power each year and irrigates hundreds of thousands of acres of land. Now, where previously only one crop was grown annually, three or even four crops may be grown each year. The dam will also control the serious difficulties brought about by the Nile's seasonal fluctuations. Navigation will be enhanced, industries will have electric power, and the lake itself is expected to support an important freshwater fishery. These are the positive aspects. The negative ones were given less thought.

One of the first changes the dam effected was a serious decline in the Egyptian sardine fishery that is centered in waters off the Nile Delta. As is well known, the annual flood of the Nile, arising in the highlands of Ethiopia, gathers thousands of tons of rich silt which were, prior to the building of the High Dam, transported all the way to the sea. This nutrient-rich water not only fertilized the banks of the Nile, but the neighboring waters of the Mediterranean as well. Small floating plants of the sea grew rapidly and provided food for small animals which then became food for sardines. With the completion of the dam, however, the fertile sediments suspended in the Nile no longer reach the Mediterranean Sea, so the sardine population dropped. Egyptian fishery authorities approximate the loss at about 18,000 tons annually, or a financial loss of at least $5 million.

The absence of the rich sediments is also felt in the agricultural lands of the lower Nile Valley. An increased application of synthetic fertilizers—at a price, of course—has partially made up for the loss, but like other countries (section 48.7), Egypt will probably discover that commercial fertilizers alone are often unable to maintain high nitrogen levels in the soil.

Parasitologists worry that the dam will cause an increase of flatworm infestations (section 24.17). In particular, they are concerned about the debilitating diseases caused by *Schistosoma mansoni* and *S. haematobium*, which, in their human hosts, inhabit the urinary bladder and lower intestine. In some areas of the Nile Delta where drainage canals provide standing water throughout the year, suitable for the species of snail that supports one stage of the parasites' life cycle, more than 90 percent of the population may carry these painful and debilitating parasites. With the spread of perennial irrigation into the newly developing lands around Aswan, there is fear that incidence of the disease will greatly increase. Public health authorities also fear resurgence of malaria epidemics.

A broader set of secondary problems comes to mind when one considers that the new farmlands and additional power will allow the population of Egypt to grow. This will create an increasing need for water, power, fertilizer, and so on. Egypt's population in 1973 was 36.9 million. Egypt will probably double its present population in just 33 years. How will increased demands be satisfied?

So, building a dam across a river commits many people to a large, expensive, and sometimes dramatic experiment. We do not know all the consequences of the Aswan Dam, and the same is true for almost every dam being built today.

50.4 When humans alter the environment without sufficient information and wisdom, they are often faced with long-term effects that were not predicted and are difficult to control.

In addition to creating relatively large-scale environmental changes, humans also alter the environment in less dramatic but more persistent ways.

One of the most insidious experiments of this sort so far launched is the widespread application of chlorinated hydrocarbons as insecticides. In the 1940s, these poisons—DDT, chlordane, and dieldrin, for example—proved so effective against flies, mosquitoes, lice, fleas, and other potential disease carrying organisms that their use spread throughout the world. Hundreds of thousands of tons were being applied to hundreds of millions of acres of farm forest, and urban land, with only the short-range consequences of insect control in mind.

The long-range consequences, however, were quite severe. Pesticide-resistant insects began to thrive and increase in numbers. The pesticides themselves do not deteriorate and accumulated in the environment and were ingested by all sorts of organisms. Many larger organisms amassed large quantities of pesticides in their tissues, and their fitness was affected. Often the insects' natural enemies suffered great and permanent damage and so were no longer around to cope with the pesticide-resistant insects. With their natural enemies gone, the pests can only be kept under control, if at all, with increasing amounts and kinds of pesticides. Furthermore, like the caterpillars that ate the thatched roofs in Borneo, previously innocuous insects may now become enemies because they are no longer under natural control.

Worst of all, we continue to make the same mistake. As one insecticide becomes useless, we tend to replace it with another, too often under the same policy of complete eradication that failed before. How many kinds of toxic materials can we find to dump on insects? And how many of these will, like the chlorinated hydrocarbons, turn up in foods to poison the very people they were meant to protect? Many pesticides (or their end products) are not very degradable, and in addition they have a great affinity for organisms. The result is **bioaccumulation**. This process may be illustrated thus: In water that contains perhaps 50 parts per trillion of a certain pesticide, microscopic zooplankton may carry a burden 1000 times as great. Shrimp feeding on the zooplankton concentrate the material four times; small fish and yet later fish have still higher amounts of residues; and diving birds that eat the larger fish may carry tremendously high quantities — up to 600 to 800 parts per million. Birds the size of robins have been known to die in trembling heaps. Eagles lay some eggs whose shells are so thin the young do not hatch. And bioaccumulation may account for the fact that coho salmon caught from Lake Michigan sometimes contain so much DDT residue that they are considered by the United States Government to be unfit for food. The same is true of some human mothers' milk.

Bioaccumulation also concentrates the radioactive isotopes of certain metabolically important elements. These isotopes are normally present in nature in biologically insignificant quantities but have been appearing in greater amounts in the environment since the beginning of the "atomic age" (i.e., since 1945) due to the testing or use of atomic weapons and the development of nuclear energy technology. The isotopes that have been introduced include uranium 235 and 238 (^{235}U and ^{238}U), thorium 232 (^{232}Th), radium 226 (^{226}Ra), potassium 40 (^{40}K), carbon 14 (^{14}C), iodine 131 (^{131}I), strontium 90 (^{90}Sr), and cesium 137 (^{137}Ce). A principal danger of these isotopes is that they release ionizing radiation that causes an increase in the rate of genetic mutations. When this occurs in gametes, these mutations are of course heritable.

The most dangerous isotopes to humans are iodine 131, which is accumulated by the thyroid gland; strontium 90, which concentrates in bones; and cesium 137, which accumulates in muscles. When these elements are introduced into a food chain, each trophic level concentrates them more than the previous level. For example, plants may concentrate iodine 131 to some extent, but grazing mammals may accumulate levels of the isotope 500 times higher than those of the plants on which they feed.

While strontium 90 and cesium 137 are not metabolically essential elements, living systems absorb and accumulate them anyway. Perhaps the main consideration causing world leaders to agree to halt the testing of nuclear devices was the growing realization that radioisotopes were being accumulated in the teeth, bones, and genes of children the world over.

50.5 The unique capacity humans have for adaptation has allowed them to penetrate habitats foreign to their physiological nature, and to broaden and diversify their ecological niche.

Because the human brain is capable of storing information and reasoning abstractly, humans can compete with any other organisms for any product or position in the environment. We compete with giant fishes and mammals in the harvest of the sea; we have spread mechanical wings and entered the air. We live in the Antarctic, in jungles and deserts, and in every place in between. It is inevitable that such expansion of habitat and ecological niche trespasses on the "rights" of other kinds of organisms.

The results have been traumatic for wild things. It is estimated that for every 1000 years of human existence, commencing a million or so years ago, one species of organism has been exterminated

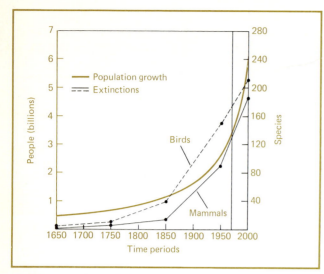

50-1 A comparison of the rates of human population growth and the numbers of extinct species of birds and mammals (projected to the year 2000).

through human action. This is only an educated guess for times past, but the rate of extinction is certainly on the increase. Today, we may be operating at the rate of one extinction every five or ten years (Figure 50-1). Over 1 percent of the known species of mammals and birds have been exterminated, and over 300 species of birds are presently seriously endangered. In the United States alone, at least 28 species have become extinct over the past 200 or so years, and another 100 species are now threatened with extinction.

So what? Can we not rationalize these losses by saying that humans belong to the most important species on earth? Many will agree, but human survival depends upon other species. This is most obvious at the dinner table, but a visit to the drugstore can bring the same point home. How many millions of rats, mice, guinea pigs, sheep, horses, cattle, and other animals and plants give their lives in experiments and for materials to make the drugs we need to maintain our level of health? What if, at some time in the past, we had been able to wipe out the species of rat and mouse that have become our laboratory animals? Who would not, a century ago, willingly have destroyed every fruit fly (*Drosophila*) in existence? Genetics would be less advanced without them. Who would have marched on the White House to save the penicillin mold 50 years ago? What insects or weeds are we eradicating now which might give us a valuable drug? We know relatively little of the great storehouse of species in

the tropics, and yet human population growth is expanding rapidly there and will surely endanger many of the native species. Organisms that have taken several billion years to evolve can be destroyed during one eradication program.

50.6 Humans have substantially modified the environment by building an incredible network of roads.

More than 1 percent of the total surface area of the United States is now covered with roads. No argument needs to be made for the fact that roads and their vehicles are a characteristic phenomenon of our age, yet we have given an astonishingly trivial amount of attention to their consequences in the environment.

Roads dissect land into discrete parts and act as a barrier in a variety of ways. A forest is dissected by a wide swath, from which topsoil is removed and which is routinely sprayed and mowed to keep vegetation under control. In addition, tons of salt are applied each winter; while in the summer, the pavement absorbs the sun's heat and rises to temperatures of 50° C and above. Traffic lays down a wall of sound, ground vibrations, light, air pollutants, and moving steel which few organisms can traverse. The consequences of the interruption of genetic continuity of species is not known.

Roads can also allow species to invade areas they previously could not reach. For example, a bridge may allow a species to cross a river which had been a natural barrier for thousands of years. The armadillo, generally unwilling to ford even smaller streams, seems to be spreading northwards by means of bridges. Because cars and trucks carry dirt on their wheels and undercarriage, they have disseminated seeds of weedy plants and disease-causing organisms. Our roadsides today have a remarkably uniform flora, due in part to the ease with which weed species move along them. Farmers name roadsides as the major source of the weeds infesting their fields.

Roads sometimes disrupt the natural flow of water and air as well as the movement of organisms. A road built across the drainageway of a shallow valley, for example, can convert a rich pasture into an undesired marsh within a few years if the road is not properly drained.

Roads act both as barriers and as routes of dispersal. They alter the primeval situation in ways that ve do not yet fully comprehend. They create new habitats as well as destroy old ones, but most important of all, roads spread human influence. Each year we commit ourselves to thousands of additional miles of roads and additional millions of automobiles and other vehicles. The tonnage of exhaust materials such as hydrocarbons, carbon monoxide, lead compounds, nitrogen oxides, and carbon dioxide being spewed into the air each year is ever increasing.

Can such a runaway situation be controlled? Only one part of the answer is certain. The problems that accompany a country's increasing number of cars and roads, its increasing fuel consumption and air pollution will *not* be solved without a great deal of thought, planning, reeducation, and regulation. Almost no one is going to be happy with the regulations, for they will be inconvenient. Will people learn to use mass transportation instead of insisting on a private car for each person? Will young couples limit the size of their families to reduce the number of potential drivers?

50.7 Cities, like roads, disrupt natural ecosystems.

A city is a site of aggregation. People, buildings, roads, business, industry, amusements, transportation, almost all the tangible qualities of our culture accumulate in urban areas. As cities grow, the amount of flora- and fauna-filled land decreases. The accumulated waste products of a city stress the city environment, since it no longer has the abundance of organisms that break down waste matter. Air, water, soil, and space itself will become more and more burdened with pollution. Unless we change the pattern, the city will remain the major source of air, water, heat, and noise pollution problems.

The thermal properties of a city stand in striking contrast to those of wild places. Because of reduced amounts of living plant tissue, so rich in moisture and active in the release of water to the atmosphere, the city warms up rapidly during the day and cools quickly at night, if it is fortunate enough to be free of a smog blanket.

Cities are centers of noise pollution too. On a noisy city street one's brain must work harder than in a quiet place to sort out the relevant spoken sounds from the bombardment of other sounds that hammer away in the background. As has been pointed out in section 34.3 and Table 10, noise pollution is a very real threat to human health.

50.8 Ecological responsibility must begin with a willingness to pay the true costs of the things we use.

Traditionally, humans have not repaid the environment for the commodities and services they take from it. We pay for the *production* of an object but not for its *renewal* or its safe *disposal*. For example, let us examine the process by which paper is produced and the practices which papermakers have employed in the past. When wood pulp replaced rags and other fibers as the raw material for paper about the middle of the nineteenth century, manufacturers went headlong into the wilderness and cut down trees. Trees cost nothing more than the price of pulping them. Square kilometer after square kilometer was stripped for spruce and pine. No one bothered about the soil or its wildlife. Wasteful processes of cutting led to severe forest fires as well.

Large masses of pulp logs moved down rivers to the mill sites, tearing stabilizing vegetation from river banks. The unwanted bark was left to rot in lakes and rivers, suffocating the water's inhabitants by using large quantities of oxygen for decay. The logs finally entered the pulping process itself, and an insoluble mixture of toxic waste materials, including mercurial compounds used in fungus control, was dumped into the overloaded waterways. The Atlantic salmon disappearef from most of northeastern North America. Other valuable natural resources followed. Stinking, lifeless rivers replaced lovely—and rich—crystalline waterways. The mill furnaces also emitted noxious air pollutants.

The cost of paper does not cover all the expenditures made in harvesting, processing, transporting, and retailing. It does not include the cost of half a century's uncontrolled flood waters that sweep across the land or the cost of lowered soil and forest fertility. Nor does it include the increased cost of recreation as citizens drive farther and farther away to find clean water in which to swim and fish. These are all costs, incurred by the manufacturer of paper and paid for by someone—but not necessarily by the biggest users of paper products.

Matters are improving, in some ways. Larger paper companies, having exhausted the supply of pulp wood, are now attempting to operate on a principle of **sustained optimal yield;** that is, they reforest lumbered areas, because if they do not they will eventually put themselves out of business. Much pulp wood is now being transported by truck and train to allow rivers to recover. Mills still have a long way to go before they stop overburdening both water and air with more pollutants than either can handle, but citizens are beginning to insist that the mills clean up, even though this means charging more for their products. If people become willing to pay the real cost of paper—including its disposal once it has left the home or office, and the restoration of soil fertility in forestlands—we will be able to strike a viable balance with nature. Total tree-harvest methods now emerging pose a special challenge in regard to the maintenance of soil fertility.

Whether we ignore the ecological facts of life or not, the true cost of living is going to catch up with us eventually, and it will be high, because we will have to pay back taxes for about 200 years of neglect. Will we be able to pay this increased cost for all of the essential ingredients of our luxurious Western way of life? Probably not. Our view of what is "essential" may well change. But there is room for some optimism, for there are signs that we are at last turning to **ecological cost accounting.**

50.9 "Natural" ecosystems can teach us a lot about ordering our own human ecosystems.

Undisturbed ecosystems have evolved to states of dynamic equilibrium in which all the inhabitants live in balance with one another. This is not an unreasonable goal for humans to aspire to as well. Either we will learn to live in harmony with the organisms and elements we need, or we will exploit them until the whole structure falls down on our heads. If, then, dynamic equilibrium is our goal what can ecosystems teach us about ordering our own communities?

One of the first lessons concerns the spatial distribution of producers and consumers. For the most part, nature operates a production-decomposition cycle with little lateral transport. A tree rises from the earth and uses its own productivity to enhance its growth. Its leaves are not burned or used for compost a mile away but fall directly over its own roots and eventually return part of their stored nutrients to the ground and to the tree again. The forest loses few nutrients, and it even captures and holds many that happen to be moving through water and air.

A deer or wild pig eating the seeds and nuts of a beech forest may wander over an area of 12 to 15 square km, but its comparatively wide range merely evens out the effects of such a large mammal. An eagle may swoop to steal a fish from an osprey and then fly several miles to its nest. But these are small distances compared to the thousands of miles that we transport coal, steel, wheat, beef, and countless other commodities of our commerce. We might well think about how to reduce the amount of materials transported rather than simply giving all our attention to transporting materials more efficiently. We might also try to counteract the loss of energy and materials to air, soil, sea, and streams, where they become, often only because of their enormous amounts, pollutants.

One result of the natural lack of mobility found in a wild ecosystem is the apportionment of space in a way which we term **territoriality.** In this process, an organism marks out a bit of space for itself and perhaps its mate, and defends it against others of its kind. Small fishes, birds, and many kinds of mammals, including humans, illustrate territorial behavior. Apparently this is one satisfying solution to the problem of adequate and efficient

---

BIOEPICUREAN DELIGHTS

Truite de Montagne Amandine

The incomparable flavor of freshly caught brook trout, *Salmo*, is best preserved by keeping the fish cool on the way to the frying pan. Placing fish on a bed of moist fern leaves in a porous creel as they are caught keeps them several degrees below ambient temperature, due to evaporation.

Prepare chopped almonds by browning slowly in butter in a small skillet, and set aside while the trout are cleaned, dipped first in beaten raw egg, then in cornmeal or flour, and then fried over low heat in a thin layer of bacon fat. As the trout become ready, reheat the almonds in butter and mix in the juice of half a lemon. Remove from the heat immediately, and serve, pouring the almond-lemon-butter over the trout. Garnish with watercress (which is probably available in the stream the trout came from) or with parsley.

apportionment of the requirements of life: food, shelter, building materials, and space.

Each ecosystem has an appropriate working distribution of organisms in the space available. We call this optimal capacity of an environment to support life its **carrying capacity**. A certain grassland in southern Texas may support one head of beef cattle per 15 hectares, on the average. A certain forest in Vermont may carry four healthy deer per square kilometer. Each habitat has its distinctive carrying capacity for each of its resident species. This clearly applies to humans, too.

In 1973 the world's population was about 3.86×10^9 people, each requiring about 10^6 kcal of food per year. The population as a whole thus requires approximately 3.86×10^{15} kcal per year. If crops could give us in an edible form some 1 percent of the energy received from the sun (and many cannot) we would have a total energy budget of 7.3×10^{17} kcal (area of arable land in the world times 1 percent of the average solar energy received by these lands). Much of the energy fixed by plants is now being "lost" to domestic animals. Fishery resources gathered at the present rate (1971) can add another 6.9×10^{13} kcal, which brings the world's approximate total energy budget for man to 7.31×10^{16} kcal. This means that *if* everyone in the world were willing largely to forego meat, *if* production levels could be maintained, and *if* food were shared equally, the world would have a carrying capacity of 7.31×10^{10} human beings. At the present rate of growth (Figure 47-2) this level of population will be reached soon after A.D. 2100.

Still another lesson that we can learn from ecosystems is that nature produces few toxic substances that are not biodegradeable. The protective toxins produced by so many thousands of animals and plants are neither spread indiscriminately over the earth nor do they remain toxic for long in the areas where they are used.

Humans, on the other hand, spread poisons that kill on contact all over the world, and these biocides do not degrade into nontoxic substances. Some, in fact, react with things in the environment and become more toxic. Continued application may bring the amounts of these materials in soil and water to a serious level. Since organisms absorb and accumulate these materials along the food chain, it is entirely possible that there is already enough DDT in the biosphere to cause a great deal of trouble.

Nature does produce a few nonbiodegradable carbon compounds. However, in the long eons of the geological past, such substances accumulated, at the end of a long trail of decomposers, deep underground, where they were quite safely out of the way of living things. Some of these super-resistant carbon compounds make up part of the petroleum deposits of the earth. When we use petroleum constituents and mix them with inorganic materials (which, likewise, nature tended to take out of circulation and store underground), we release a deadly arsenal. In fact, once back in circulation, these substances, unless biodegradable, do not sift slowly down through pathways of decomposition and into earth's basement. Instead, they have a curious affinity for living tissue and actually migrate upward in the ecological pyramid.

In addition to toxins, we also manufacture articles made out of hundreds of substances which cannot be broken down by other organisms. Plastics are a prime example. How unpleasant it is to walk the shores of a bay where rafts of plastic toothbrushes, dismembered dolls, boots, and bottles have been cast up. Nature, normally omnivorous, is being fed an increasing volume of indigestible items.

Thus the three main lessons that humans could learn from balanced ecosystems are: (1) Limit lateral transport; (2) do not exceed the earth's carrying capacity; and (3) limit the production of items, both toxic and nontoxic, that cannot be naturally decomposed or artificially recycled.

50.10 Natural systems tend to stabilize. Do human systems?

As an ecosystem grows more complex, its parts become more interdependent, a system of checks and balances develops, and it becomes more able to recover from disturbances. We cannot say the same for our own society.

The more complex our technology, the more complex and destructive become the consequences of its misuse and the more helpless we tend to feel about correcting the ensuing problems. We have unlocked nuclear power and have created an international political fabric of such fragility that almost anything can happen tomorrow. This is far from stability. World population is increasing at an ever-increasing rate, while, in general, the environment's capacity to support man is declining due to abuse. This is not stability. Our renewable resources such as forests, fishery stocks, and grasslands are degenerating. This is not stability.

Accelerated abuse of our environment is probably the prime symptom of our ecological failure.

Having removed ourselves from an ecosystem shared with many other helpful species, we now live alone. Our domesticated farms assume almost all of the responsibility for maintaining an ecosystem friendly to us. Our contribution, so far, has largely been to make our environment unfriendly. Oxygen overuse, resource depletion, soil erosion, forest destruction, and carbon dioxide pollution are only a few of our activities that interfere with natural stabilizing mechanisms.

The first step in trying to reverse this trend is to make people aware of the existing predicaments. Once they are willing to acknowledge the problems, they may decide either to resist the destructive trends or to give up.

The rapidity and efficiency with which we are disrupting the very mechanisms on which our lives depend make the problems seem so far beyond our control that we despair of ever reversing the trend. "What can I do?" That cry of despair that comes at some time to each person who honestly faces a seemingly insoluble problem can too easily preface a long list of "reasons why I am not responsible." That you cannot stop the soft drink or liquor industries from using nonreturnable containers does not relieve you of your own responsibility in this matter. You can choose to drink beverages that come in returnable bottles or recyclable cans. You can choose to be responsible for your share of the mess and for voting for public servants who take responsible positions on ecological issues. To the degree that you accept such responsibility you will escape the sense of despair that so paralyzes vast numbers of people. Inactivity threatens, more than DDT, more than air or water pollution, to signal a premature and inglorious end to what still may be nature's most glorious experiment—*Homo sapiens.*

SUGGESTED READING

Allee, W. C., et al. *Principles of Animal Ecology.* Philadelphia: Saunders, 1949.

Bates, M. *The Forest and the Sea.* New York: Random House, 1960.

Blanchard, D. C. *From Raindrops to Volcanoes.* Garden City, N.Y.: Doubleday, 1967.

Blanchard, D. C., and L. Syzdek. "Mechanism for the Water-to-Air Transfer and Concentration of Bacteria." *Science* 170 (1970): 626–628.

Borgstrom, G. *Too Many: A Study of Earth's Biological Limitations.* New York: Macmillan, 1969.

Bormann, F. H., and G. E. Likens. "The Nutrient Cycles of an Ecosystem." *Scientific American,* October 1970, pp. 92–101.

Bryson, R. A. "A Perspective on Climatic Change." *Science* 184 (1974): 753–760.

Bryson, R. A., and D. A. Barreis. "Possibilities of Major Climate Modification and Their Implications: Northwest India, A Case Study." *Bulletin of the American Meteorological Society,* March 1967, pp. 136–142. (See also *Saturday Review,* 1 April 1967, pp. 52–55.)

Burdick, G. E., et al. "The Accumulation of DDT in Lake Trout and the Effect on Reproduction." *Transactions of the American Fisheries Society* 93 (1964): 127–136.

Carlquist, S. *Island Biology.* New York: Columbia University Press, 1974.

Clark, C. W. "The Economics of Overexploitation." *Science* 181 (1973): 630–634.

Cloud, P. "Evolution of Ecosystems." *American Scientist,* January–February 1974, pp. 54–66.

Cole, L. C. "Can the World be Saved?" *BioScience* 18 (1968): 679–684.

Darling, F. F. *Wilderness and Plenty.* Boston: Houghton Mifflin, 1970.

Dorst, J. *Before Nature Dies.* Boston: Houghton Mifflin, 1970.

Ehrenfeld, D. W. *Biological Conservation.* New York: Holt, Rinehart and Winston, 1970.

Ehrlich, P. R. *Biochemical Coevolution.* Corvallis: Oregon State Press, 1970.

Ehrlich, P. R., and A. H. Ehrlich. *Population, Resources, Environment: Issues in Human Ecology.* San Francisco: Freeman, 1972.

Elton, C. S. *Animal Ecology.* New York: Macmillan, 1927.

Elton, C. S. *The Ecology of Invasions by Animals and Plants.* London: Methuen, 1958.

Elton, C. S. *The Pattern of Animal Communities.* New York: Wiley, 1966.

Farvar, M. Taghi, and J. P. Milton, eds. *The Careless Technology; Ecology and International Development.* Garden City, N.Y.: Natural History Press, 1972.

Fisher, J., N. Simon, and J. Vincent. *Wildlife in Danger.* New York: Viking Press, 1969.

Gates, D. M. *Man and His Environment: Climate.* New York: Harper & Row, 1972.

Geiger, R. *The Climate Near the Ground.* Cambridge: Harvard University Press, 1965.

George, C. J. "The Role of the Aswan High Dam in Changing the Fisheries of the Southeastern Mediterranean." In *The Careless Technology,* edited by M. T. Farvar and J. P. Milton, pp. 159–178. Garden City, N.Y.: Natural History Press, 1972.

George, C. J., and D. McKinley. *Urban Ecology: In Search of an Asphalt Rose.* New York: McGraw-Hill, 1974.

Hardin, G. *Birth Control.* New York: Pegasus, 1970.

Hardin, G. *Exploring New Ethics for Survival.* New York: Viking Press, 1972.

Harte, J., and R. H. Socolow. *Patient Earth.* New York: Holt, Rinehart and Winston, 1971.

Heller, A., ed. *The California Tomorrow Plan.* Los Altos, Calif.: William Kaufman, 1971.

Henry, S. M., ed. *Symbiosis.* New York: Academic Press, 1966.

Jackson, J. H., and E. D. Evans. *Spaceship Earth: Earth Science.* Boston: Houghton Mifflin, 1973.

Janzen, D. H. "Tropical Agroecosystems." *Science* 182 (1973): 1212–1219.

Jeffries, D. J., and M. C. French. "Lead Concentration in Small Mammals Trapped on Roadside Verges and Field Sites." *Environmental Pollution* 3 (1972): 147–156.

Kormondy, E. J. *Concepts of Ecology.* Englewood Cliffs, N.J.: Prentice-Hall, 1969.

Landsberg, H. E. *Weather and Health: An Introduction to Biometeorology.* Garden City, N.Y.: Doubleday, 1969.

Leopold, A. *A Sand County Almanac with Other Essays on Conservation from Round River.* New York: Oxford University Press, 1966.

Likens, G. E., and F. H. Bormann. "Acid Rain: A Serious Regional Environmental Problem." *Science* 184 (1974): 1176–1179.

Marx, W. *Man and His Environment: Waste.* New York: Harper & Row, 1971.

Murdoch, W. W., ed. *Environment, Resources, Pollution and Society.* Stamford, Conn.: Sinauer Associates, 1971.

Murphy, E. F. *Man and His Environment: Law.* New York: Harper & Row, 1971.

Neill, W. T. *The Geography of Life.* New York: Columbia University Press, 1969.

Odum, E. P. *Fundamentals of Ecology.* 3rd ed. Philadelphia: Saunders, 1971.

Owen, D. F. *Man in Tropical Africa.* New York: Oxford University Press, 1973.

Phillipson, J. *Ecological Energetics.* New York: St. Martin's Press, 1966.

Pianka, E. R. *Ecology: An Evolutionary Approach.* New York: Harper & Row, 1974.

Pimentel, E., et al. "Food Production and the Energy Crisis." *Science* 182 (1973): 443–449.

Richards, P. W. "Tropical Rain Forest." *Scientific American,* December 1973, pp. 58–67.

Rudd, R. L. *Pesticides and the Living Landscape.* Madison: University of Wisconsin Press, 1964.

Ryther, J. H. "Photosynthesis and Fish Production in Sea." *Science* 166 (1969): 72–76.

Schaefer, V. J. "The Inadvertent Modification of the Atmosphere by Air Pollution." *American Meteorological Society Bulletin* 50 (1969): 199–206.

Schindler, D. W. "Eutrophication and Recovery in Experimental Lakes: Implications for Lake Management." *Science* 184 (1974): 897–899.

Schlee, S. *The Edge of a Unfamiliar World: A History of Oceanography.* New York: Dutton, 1973.

Shelford, V. E. *The Ecology of North America.* Urbana: University of Illinois Press, 1963.

Shepard, P., and D. McKinley, eds. *The Subversive Science, Essays Toward an Ecology of Man.* Boston: Houghton Mifflin, 1969.

Shepard, P., and D. McKinley, eds. *Environ/Mental, Essays on the Planet as a Home.* Boston: Houghton Mifflin, 1971.

Sondheimer, E., and J. B. Simeone, eds. *Chemical Ecology.* New York: Academic Press, 1970.

Southward, A. J. *Life on the Sea-Shore.* London: Heineman, 1965.

Thomas, W. L., Jr., ed. *Man's Role in Changing the Face of the Earth.* Chicago: University of Chicago Press, 1956.

Van den Bosch, R., and P. S. Messenger. *Biological Control.* New York: Intext Publishing Co., 1973.

Walter, H. *Vegetation of the Earth in Relation to Climate and the Eco-Physiological Conditions.* New York: Springer-Verlag, 1973.

Watt, K. E. F. *Principles of Environmental Science.* New York: McGraw-Hill, 1973.

Watt, K. E. F. *The Titanic Effect: Planning for the Unthinkable.* Stamford, Conn.: Sinauer Associates, 1974.

Whittaker, R. H., and G. E. Likens, eds. "The Primary Production of the Biosphere." In *Human Ecology* 1 (1973): 299–369.

Whittaker, R. H., and G. M. Woodwell. "Evolution of Natural Communities." In *Ecosystem Structure and Function,* edited by J. A. Wiens, pp. 137–156. 35th Biology Colloquium. Eugene: Oregon State University Press, 1972.

Wickler, W. *Mimicry in Plants and Animals.* New York: McGraw-Hill, 1968.

Woodwell, G. M. "Success, Succession, and Adam Smith." *BioScience* 24 (1974): 81–87.

Woodwell, G. M., and E. V. Pecan, eds. *Carbon and the Biosphere.* Brookhaven Symposia, no. 30. Washington, D.C.: Office of Information Services, U.S. Atomic Energy Commission, 1973.

Woodwell, G. M., and H. H. Smith, eds. *Diversity and Stability in Ecological Systems.* Brookhaven Symposia in Biology, no. 22. Washington, D.C.: Office of Information Services, U.S. Atomic Energy Commission, 1969.

Zaret, T. M., and R. T. Paine. "Species Introduction in a Tropical Lake." *Science* 182 (1973): 449–455.

Glossary

abiotic   nonliving

abscission   the normal falling away of leaves, flowers, or fruit from plants

acellular   pertaining to living material not compartmentalized into cells

acetylcholine   one of the chemicals that function as transmitters of nerve impulses in synapses in nervous systems

acid   any substance that releases protons in solution

acoelomate   any animal lacking a coelom or body cavity

acrosome   reactive tip of a sperm

acrosomal reaction   the reaction of the acrosome in response to contact with the egg surface or fluid which contains substances dissolved from the egg surface

acrosomal filament   a filament (in some cases consisting of actin) that forms during the acrosome reaction and that penetrates the egg surface

actin   a "contractile" protein found in muscle and many other cells

action potential   a nerve impulse consisting of a progressive change in electrical polarity, a change due to ion migration that passes along an axon or dendrite of a nerve

activation energy   energy required to initiate a chemical reaction

active site   the region of an enzyme (or other protein) to which a substrate (or other molecule) attaches in order to react

active transport   the movement of molecules through a membrane against a concentration gradient and requiring the expenditure of metabolic energy

adaptive radiation   the diversity that evolves among related species as each adapts to different environmental circumstances

adenine   a nitrogenous base, a constituent of DNA, RNA, ATP, ADP, and AMP

adenosine triphosphate (ATP)   a trinucleotide, the breakdown of which (to form ADP) releases energy that can be used for nearly all life processes

adrenalin   a hormone produced by the adrenal medulla that prepares an animal for "fight or flight"

adrenal gland (cortex, medulla)   a gland located just anterior to the kidney of vertebrates. The cortex (outer portion) and medulla (inner portion) produce corticosteroids and adrenalin, respectively

aerobic   requiring oxygen

aestivation   summer dormancy

alimentary tract   gut, digestive tract

alkaloid   a class of alkaline plant products often with strong medicinal biological effects

alkaptonuria   a heritable disorder of phenylalanine metabolism

allele   one of two or more particular forms of a gene

allopolyploidy   possession by a polyploid individual of sets of chromosomes originating from two or more species

alpha cells   cells of the endocrine portion of the pancreas that secrete glucagon

alpha-helix   one of the possible configurations of a polypeptide chain of a protein

alpha motor neuron   nerve cells in the spinal cord that cause muscles to contract when stimulated

alternation of generations   the alternate succession of haploid and diploid phases of a sexually reproducing organism

alevolar ducts   smallest branches of the bronchioles leading to the alveoli of the lung

alveoli   tiny air sacs in the lungs

amino acids   about 20 small molecules that constitute the building blocks of proteins and polypeptides. All have an amino group (—NH$_2$) and an organic acid group (—COOH)

amniocentesis   sampling of the amniotic fluid by insertion of a hypodermic needle through the maternal abdominal and uterine walls into the amniotic cavity of the embryo

amnion   one of the extraembryonic membranes of the vertebrate embryo

421

amnionic cavity the fluid-filled cavity which contains the embryo and which is surrounded by the amnion

amoeba acellular organism (protist) that moves by pseudopods; free-living or parasitic

amoebocyte cell from a multicellular organism that resembles an amoeba

amoeboid movement movement by means of pseudopodia

ampullae ampule-like part of the water-vascular system of echinoderms

amyloplasts starch grains in plant cells

anaerobic not requiring oxygen

anaphase phase of mitosis during which chromosomes move to the poles of the mitotic spindle

anemia deficiency in either blood cells or hemoglobin

angiosperm flowering plant

animal hemisphere hemisphere of an animal egg that gives rise to ectodermal parts of an embryo (*opposite:* vegetal hemisphere)

annual rings concentric circles in a cross section of a woody stem that indicate such a plant's age

anther male sex organ of a flower

antheridium (*pl.* antheridia) male sex organ of plants other than flowering plants

antibody substance produced by an organism in response to introduction of an antigen or foreign protein

anticodon a triplet of three nucleotides in transfer RNA that can pair with a complementary triplet in the codon of messenger RNA

antigen any foreign protein that stimulates an immune response

antitoxin an antibody to a toxin

antrum fluid-filled chamber of an ovarian follicle

anus exit opening of the alimentary tract

aorta main artery from the heart to the body

apical cell cell at the apex of a shoot

apical meristem regions of cell growth and differentiation in the root and shoot of plants

aqueous humor a fluid-filled cavity in the eye

archegonium (*pl.* archegonia) femal sex organ of nonflowering plants

archenteron primitive gut in embryonic stages

arteriole a small branch leading from an artery

articular cartilage cartilage in joints

articular capsule ligaments and other connective tissue surrounding a joint

ascospore one of eight haploid cells, products of meiosis, contained in the ascus

ascus a zygote found at the tips of certain fungal hyphae (in Ascomycota)

asexual reproduction reproduction without sex

association neuron a neuron in the spinal cord intermediate between the sensory and motor neurons of a reflex arc

aster part of the mitotic apparatus of dividing animal cells

auditory nerve sensory nerve from the ear involved in hearing

autonomic nervous system that portion of the nervous system responsible for control over the viscera and circulatory system

autosome a nonsex chromosome

auxin plant growth hormone

atria smaller receiving chambers of the heart

atrio-ventricular node specialized conductile tissue of the heart

axon portion of a neuron that carries impulses away from the cell body toward the synapse with the next neuron

axoneme the complex of $9 + 2$ microtubules and associate structures in a cilium or flagellum

backcross (*also* testcross) mating of F_1 products with the homozygous recessive strain

bacteriophage bacterial viruses

Barr body sex chromatin body found in nuclei of female human cells; the presence of two is used as a diagnostic feature for sex chromosome anomalies

basal body a centriolar derivative at the base of every cilium or flagellum

Batesian mimicry mimicry in which a relatively harmless organism has adopted features of a dangerous organism, thereby gaining protection

B-cells lymphocytes involved in humoral immunity

B-cell effector plasma cells that produce antibodies against foreign proteins

benthic pertaining to the deep sea

beta cells cells of the endocrine portion of the pancreas that secrete insulin

binomial nomenclature the system of naming an organism by its genus and species

bioaccumulation the concentration of certain poisons in organisms at the top of the food chain

biogenesis the origin of life

biological clock endogenous rhythms that continue to be expressed in the absence of normal patterns of light, temperature, and so on.

biomass the total weight of organisms living in a given area

biosynthetic pathway a chain of reactions leading to the synthesis of substances required for cellular metabolism

biosystematics taxonomy; the science of classification of organisms living or once alive

biotic pertaining to life

bipedal having two legs

bivalent a tetrad; paired homologous chromosomes and their division products

blastocoel the fluid-filled cavity in a blastula-stage embryo

blastocyst an early blastulalike stage of a mammalian embryo

blastodisc blastulalike stage of a bird embryo

blastomere a cell of an early embryo

blastula early stage in embryonic development in which the cells of the embryo form a single layer around a fluid-filled cavity

Bowman's capsule a cup-shaped structure at one end

of a kidney tubule, into which the renal filtrate is filtered from the blood

bryophyte   fern

budding   a form of asexual reproduction in which a bud appears as a part of the parent organism and develops into a new individual

buffer   a combination of a weak acid and its salt that serves to maintain the acidity (or pH) of a solution at a relatively constant level

caecum   a blind sac extending from the beginning of the large intestine

calorie   a unit of heat; the amount of heat required to raise the temperature of 1 gram of water 1 degree centigrade; a Calorie (capitalized) is 1000 times greater, and is also called a kilogram calorie

Calvin cycle   the cyclic dark reactions of photosynthesis

cambium   the meristematic layers in the stem of a plant

carbon fixation   the addition of carbon from carbon dioxide to organic molecules in the cell

cardiac muscle   heart muscle

carnivore   meat-eating animal (or plant)

carrying capacity   the largest population (or biomass) that a given environment or region can support over an extended period

Casparian strip   a thick, waxy strip that prevents the diffusion of dissolved material into the vascular tissue of the root

catalysis   the acceleration of a chemical reaction by a catalyst, e.g., an enzyme

cell plate   the first structure separating the daughter cells of a dividing plant cell; gives rise to the middle lamella

cell theory   theory proposed in 1838 to 1839 by Schleiden and Schwann suggesting that organisms are composed of similar units called cells

cellular immunity   the defense mechanism of an organism against foreign cells, such as bacteria and grafted tissues from a donor organism

cellulose   a polysaccharide; a straight-chain polymer of glucose found in plant cell walls

cell wall   the relatively rigid container secreted by many plant cells; it limits their swelling in hypoosmotic (too diluted) solutions and is partly responsible for turgor

central nervous system   that portion of the nervous system that contains most of the cell bodies and exercises control over the rest of the nervous system

centrifugal force   the product of the centrifugal acceleration and the reduced weight of a particle in a centrifugal field

centriole   a cytoplasmic organelle that serves as an important microtubular organizing center

cephalization   in evolution, a trend toward the enlargement of the anterior portion of the nervous system to form a control center in the head

cerebellum   portion of the brain concerned with balance, equilibrium, and muscular coordination

cerebral ganglion   a collection of nerve-cell bodies in the head region

cerebrospinal fluid   the fluid contained in the vesicles of the brain and in the spinal cord

cerebrovascular accident (stroke or embolism)   any interruption of the blood supply to the cerebrum (by hemorrhage or by a clot or embolus)

cerebrum   portion of the brain responsible for voluntary action and mental activity

cervix   posterior portion of the uterus where it joins the vagina

chemical synapse   a junction between neurons at which interneuronal communication is effected by chemical transmitters

chemoreceptor   a sensory neuron specialized for responding to chemical stimuli

chitin   a polysaccharide of which arthropod exoskeleton is composed; found also in some fungi

chiton   a type of mollusk

chlorophyll   green pigments found in plants, blue-green algae, and some bacteria that serve as energy transducers in photosynthesis

chloroplast   a chlorophyll-containing organelle that serves as the intracellular site of photosynthesis in plant cells

choanocytes   flagellate collar cells in sponges

chromatic adaptation   the automatic alteration of the color of an alga depending on its depth in the ocean; depends on the amounts of accessory pigments

chromatid   one of a pair of sister chromosomes together from the time of molecular duplication until separation at anaphase of division

chromatid tetrad   a group of four chromatids from homologous pairs of chromosomes and their duplication products

chromatin   the nucleoprotein substance of chromosomes

chromosome mutation   a heritable genetic change caused by a rearrangement of, or loss of, genes

chyme   the fluid contents of the small intestine

cilia   cellular organelles composed of axonemes surrounded by an extension of the cell membrane; cilia beat like tiny oars to accomplish cellular locomotion or to propel fluids past the cell

cistron   a sequence of nucleotides in RNA or DNA that specifies the sequence of amino acids in a polypeptide or protein

cleavage   the division of an egg

climax community   a population of organisms that has reached an equilibrium with its environment

cline   minor genetic variation within a species resulting from adaptations to slightly different environments

clitoris   part of the female external genitalia; a source of pleasurable stimuli during sexual excitation

club fungi   common name for basidiomycota

cnidoblast (*also* nematocysts)   stinging cells of cnidarians (or coelenterata)

cochlea   the inner ear

codon the sequence of three nucleotides in DNA that specifies a particular amino acid at the time of protein synthesis

coelom a true body cavity, lined with mesoderm

coelomate an animal possessing a true body cavity

coenzyme a small organic molecule that acts as an accessory in an enzymatic reaction

coitus sexual intercourse

collagen a fibrous protein found in bone and connective tissue

collar cell (choanocyte) a flagellated cell found in sponges that causes water to circulate

collenchyma a supporting tissue of plants composed of elongated cells with thick walls

colon large intestine

colony a group of similar cells or organisms living together

community interacting populations in a common environment

companion cell a narrow cell associated with the sieve elements in the phloem of flowering plants

competitive exclusion the principle that, with at least one resource in limited supply, not more than one species can occupy a given constant environment indefinitely

complement factor any protein that assists in the formation of an antigen-antibody complex formation

compound eye a type of eye found typically in arthropods; consists of many segments or ommatidia, each of which forms a miniature image

conception initiation of egg development by fertilization; term usually implies initiation of human development

condom a protective sheath, usually of thin rubber, that may be drawn over the penis; used both to prevent conception and to avoid the spread of venereal disease

conductive tissue refers to both the phloem and xylem of plants

cone receptor cell light receptors of the eye responsible for color vision

conformation the shape of a molecule (e.g., alpha-helical or random coil conformations of a polypeptide or protein)

conidia haploid spores of sac fungi (Ascomycota)

conifer cone-bearing trees or shrubs; gymnosperms

conjugation the exchange of nuclei between different mating strains of protists or algae

constitutive enzyme an enzyme normally present at all times in a cell

continental drift very gradual movements of continents due to convection currents in the earth's mantle

contractile vacuole (or expulsion vesicle) organelle in many protists and in cells of freshwater metazoans that pumps out excess water that enters by osmosis

convergence independent evolution of similar structures in unrelated organisms

copulation sexual union in which sperm is transferred to the reproductive tract of the female

corepressor a low-molecular-weight compound that combines with a protein (the aporepressor) to form a complete, active repressor

cork cambium a meristematic layer that forms the bark of woody plants

cornea the "window" of the eye, consisting of a layer of transparent flattened cells

coronary artery the artery supplying the heart muscle with blood

coronary vein the vein draining the heart muscle of deoxygenated blood

corpus luteum the "yellow body" formed from follicle cells of the ovary of mammals after the egg has been released; secretes the hormone progesterone

cortex outer portion (e.g., of cell, adrenal gland, brain, etc.)

corticosteroids steroid compounds synthesized by the adrenal cortex

cotyledon the leaf developed by an embryo plant while still in the seed, which unfolds during sprouting

covalent bond a chemical bond resulting from the sharing of an electron by adjacent atoms

Cowper's gland a gland in the male reproductive system that contributes substances to the semen

crop a muscular storage organ of some digestive systems

crossing-over the reciprocal exchange of corresponding segments between homologous chromosomes

cyclosis rotational cytoplasmic streaming in some plant cells and protists

cytochromes cell pigments that serve in oxidative reactions

cytokinesis division of the cytoplasm; in eggs, called cleavage

cytology the study of cells; now subsumed under cell biology

cytoplasm the portion of a cell outside the nucleus

cytoplasmic differentiation the formation of specialized cytoplasmic structures

cytoplasmic streaming (cytoplasmic transport) the bulk flow of cytoplasm within a cell

cytosine one of the nitrogen bases in nucleic acids

cytosome the cytoplasm taken as a whole, as contrasted with the nucleus

decibel a logarithmic unit of sound intensity

decomposer an organism that obtains energy by degrading organic matter in the environment (e.g., bacteria, fungi)

dehydration synthesis the synthesis of a larger molecule by the removal of water from two smaller molecules as they are joined

denaturation the loss of the chemical and biological properties of a molecule, especially a protein

dendrite the extensions of a neuron that carry nerve impulses toward the cell body

denitrifying bacteria bacteria that release nitrogen from the soil

deoxyribonucleic acid (DNA) the hereditary material

of all living organisms except certain plant viruses. Stored in nucleus or nucleoid

deuterostomes   a broad group of animal phyla including the echinoderms and chordates

diabetes mellitus   sugar diabetes, caused by insufficient insulin secretion

diapause   dormancy during development, especially in insects

dicot (short for dicotyledonous)   plants in which the embryo develops two leaves before germination

differentiation   process by which different cells or cytoplasmic regions follow different developmental pathways

dioecious organisms   organisms in which eggs and sperm are produced in different (male and female) individuals

diploid   having two homologous sets of chromosomes

dominant trait   a hereditary trait that is expressed in the heterozygote

Down's syndrome   also called mongolism, a genetic abnormality in man caused by failure of the offspring to receive a normal, complete complement of chromosomes

duodenum   the portion of the small intestine adjacent to the stomach

ecology   the scientific study of the interactions of organisms with their physical and biological environment

ecosystem   the organisms of a selected habitat, together with their physical environment

ectoderm   outermost embryonic layer

ectoplasm   outermost cytoplasmic layer

egestion   the casting out of undigested food from the mouth (e.g., *Hydra*) or anus

egg   female germ cell

ejaculation   forcible expulsion of semen during sexual excitement

electronic synapse   a junction between neurons that is not bridged by a chemical transmitter

electron transport system   the system of coenzymes, flavoproteins, cytochrome pigments, and enzymes that accomplishes the oxidation of cell substrates

Embden-Meyerhoff common pathway   the metabolic pathway for the breakdown of glucose to form pyruvic acid

embryonic induction   the process whereby one embryonic tissue influences the development of a nearby tissue

endergonic process   a process requiring the input of energy

endogenous rhythm ("biological clock") *or* circadian rhythm   a roughly 24-hour activity rhythm that continues in the absence of all environmental cues

endoskeleton   an "inside" skeleton, e.g., of bones in vertebrates

endosymbionts   harmless or beneficial organisms living inside another organism, for example, zoochlorellae living inside *Paramecium bursaria*

endocrine organ   an organ that secretes hormones into the blood

endocytosis   cell eating and drinking (phagocytosis and pinocytosis)

endoderm   the innermost embryonic layer

endoplasm   the innermost region of the cytoplasm

endoplasmic reticulum   a system of intracellular membranes that serves as a site for macromolecular synthesis

endosperm   the nutritive tissue in the seeds of flowering plants

energy   the capacity to perform work

energy transducer   a machine or molecule capable of transforming energy from one form to another

entropy   a measure both of the randomness of a system and of its energy that is unavailable to do work

enzyme   a biological catalyst; a protein capable of hastening a chemical reaction to its equilibrium

epiblast   the upper layer of a chick embryo at the blastula stage

epidermis   outer layer of skin

epididymis   accessory male sex organ

epiglottis   a valve in the pharynx that closes the trachea during swallowing

epilimnion   the upper layer of a lake

epithelial cells   cells lining various cavities such as the gut and its glands

esophagus   tube leading from the throat to the stomach

essential amino acids   amino acids that cannot be synthesized in the human body from other amino acids

estrogen   female hormone primarily responsible for the development of female secondary sexual characteristics

estuary   the region where a great river meets the sea, and where tides alter the level of the river water

eucaryotic   having a "true" nucleus, with a nuclear membrane

eustachian tube   tube leading from the pharynx to the middle ear; responds to atmospheric pressure changes

eutrophication   the overgrowth, death, and sedimentation of plants due to excessive nutrients

evagination   outpocketing of a cell layer in embryogenesis

exergonic   spontaneous reaction or process from the point of view of energy requirement

exoskeleton   outer skeleton (e.g., in lobster or crayfish)

expiration   breathing out

fallopian tube   tube through which the ovum reaches the uterus

fall turnover   the turnover of water in a lake when the temperature of the water falls and the density becomes higher than underlying water

fecal matter (feces)   solid intestinal wastes

femur   the long bone of the thigh

fertilization   union of an egg and sperm

F_1 (first filial) **generation**   the progeny of two genetically different parents

fiber tract   a bundle of nerve fibers in the central nervous system

fission   division of a cell

fitness   a measure of the reproductive success of a given genotype

flagellum   a whiplike structure found in bacteria, protists, metazoans, and sperm

flame cell   a cell specialized for excretion in flatworms and rotifers

follicle   an ovum surrounded by a vesicle containing follicle cells in the ovary

food web   a diagram showing the food links between organisms in a community or ecosystem

foreplay   amorous behavior leading to sexual intercourse

formed elements (of the blood)   blood cells (red and white) and platelets

fovea   region of the retina containing the densest accumulation of receptors, and the region on which the center of the visual field is focused

free energy   chemical energy available to do work

fruiting body   sporangium, e.g. of a slime mold

functional group   the reactive portion of a molecule

galactose   a monosaccharide, a component of lactose, or milk sugar

gallbladder   organ that stores bile and releases it when required

gametes   eggs and sperm

gametogenesis   the production of gametes in the male and female gonads

gametophyte   the plant generation giving rise to gametes

gamma motor neurons   nerve cells which on stimulation result in the relaxation of muscles and the prevention of movement

ganglion (*pl.* ganglia)   a collection of nerve-cell bodies

gastric gland   a gland secreting into the stomach

gastropods   members of the class of mollusks including snails and whelks

gastrula   an embryonic stage characteristic of most animals in which three germ layers are present: the ectoderm, endoderm, and mesoderm

gastrulation   the formation of the endoderm in an embryo

gemmule   an overwintering asexual reproductive structure in sponges

gene   a unit of heredity

gene mutation   heritable change in a gene

genetic drift   evolution, or change in gene proportion, by chance processes alone

genetic engineering   modification of the genetic constitution of an individual

genome   an individual organism's collection of genes

genotype   the genetic constitution of an individual with respect to a particular trait

genus (*pl.* genera)   a group of related species

geotropism   plant growth or movement in response to gravity

germ cell   a cell that will give rise to gametes

germinate   the initiation of growth or development from a spore or seed

germ layers   the ectoderm, endoderm, and mesoderm of an animal embryo

gibberellin   a plant growth regulator involved in stem elongation, flowering, and seed germination of some plants

gill   a respiratory structure of aquatic animals

gizzard   a digestive organ specialized for grinding food (e.g., in birds)

Gloger's rule   the generalization that populations of birds in warm, humid climates tend to be darker than those in cooler, less humid climates

glomerulus   a cluster of capillaries enclosed in each of the Bowman's capsules of the kidney

glucagon   a pancreatic hormone that stimulates the liver to release glucose into the blood stream

glucose   a monosaccharide that serves as the most important energy source for living organisms; also a product of photosynthesis

glycogen   a polymer of glucose found as a storage product in animal cells

glycolysis   the breakdown of glucose or glycogen to form lactic or pyruvic acid

Golgi body or apparatus   an organelle concerned with synthesis of macromolecules for export from the cell

gonad   sex organ (ovary, testis)

grana   green flecks visible by the light microscope in chloroplasts; consists of stacks of flattened membrane sacs called thylakoids

gray crescent   a primary organizing region of the amphibian embryo that gives rise to the dorsal lip of the blastopore

gray matter   the portion of the brain and spinal cord consisting of neuron cell bodies

ground cytoplasm   the cytoplasm minus its organelles

growth   an increase in size or mass

guanine   one of the nitrogenous bases of nucleic acids

guard cell   one of a pair of epidermal cells that surround and open and close the stomates of leaves

gymnosperm   a member of the class of seed plants in which seeds are borne on cone scales and not in an enclosed ovary (e.g., conifers)

haploid   having a single set of chromosomes

Hardy-Weinberg law   the law stating that the frequencies of alleles in a gene pool remain constant unless acted upon by natural selection

Haversian system   system of concentric microcylinders of bone, each surrounding a blood vessel

heme   an organic ring compound containing iron, found in hemoglobin

hemodialysis   the dialysis of blood to remove impurities or wastes, e.g., urine, an "artificial kidney" machine

hemoglobin   the chromoprotein found in red blood cells

hemophilia   a hereditary disease in which affected individuals bleed because of a defect in the clotting mechanism

hepatic cell   liver cell

hepatic duct   duct conducting bile from the liver toward the gallbladder and small intestine

hepatic portal system   veins draining the small intestine, carrying nutrients to the liver

herbivore   consumer of plants

hereditary symbiosis   a mutually beneficial relationship between two organisms, the mechanism for which is regulated by genes

hermaphrodite   individual having organs of both sexes, a condition that may be normal or abnormal to the organism

Herring-Breuer reflex   a neural mechanism that regulates breathing, mediated by stretch receptors in the lung

heterokaryon   a cell (normally in the mycelium of a fungus) in which two genetically different nuclei reside

heterosis   hybrid vigor or heterozygote superiority

heterotroph   organism requiring complex organic foods and incapable of synthesizing its own from inorganic materials

heterozygote   an organism having two different alleles with respect to the same trait

hibernation   winter dormancy of animals

hominid   a manlike primate

homologous   alike in structure (e.g., homologus chromosomes)

homosexuality   sexual orientation toward members of the same sex

homozygote   organism possessing two similar alleles for a given trait

hormone   a substance secreted by cells that has a physiological, behavioral, or developmental effect on cells at another location in the organism

horny layer   tough outer layer of the skin

humoral immunity   immunity against virus particles

humus   the organic component of soil resulting from the decomposition of plant material

hydrogen bonds   weak chemical bonds between hydrogen atoms and nearby atoms of oxygen or nitrogen

hydrolysis   splitting by water, e.g., chemical reactions in which water is added to break a polymer into its constituent monomers

hydrophilic   water-loving

hydrophobic   water-hating

hypoblast   the lower layer of a chick blastula

hypolimnion   the deeper layer of a lake or pond

hypothalamus   the portion of the brain controlling the internal environment

immunization   the conferral of immunity by artificial means

inducible enzyme   an enzyme the synthesis of which is initiated only when the enzyme's substrate is present

infertility   apparent inability to have offspring

inhalation   the taking of air into the lungs

innate behavior   inborn or instinctive behavior

inner ear   the cochlea, or organ of hearing (in man)

instinct   inborn or innate behavior

integument   outer covering, skin

interferon   a substance synthesized by tissue cells that controls virus reproduction

interphase   the "resting stage" between divisions of a cell

intervertebral disk   cartilaginous cushion separating vertebrae of the spinal column

intrauterine device (IUD)   a coil, loop, chain, or other device inserted in the female's cervix that prevents attachment of early human embryos to the uterine wall

invagination   impocketing of an embryonic tissue layer; a common event in development

invertebrate   any animal without a backbone

in vitro ("in glass")   under artificial, experimental conditions (*opposite:* in vivo)

in vivo ("in life")   under living conditions (or with an intact organism or cell)

islets of Langerhans   endocrine portion of the pancreas where insulin and glucagon are produced

iso-   prefix meaning "equal"

isotonic   refers to solutions with the same osmotic strength relative to the normal environment of a particular cell

kelp   a bladelike brown alga found in shallow sea water

kinetic energy   energy in an active form (*opposite:* potential energy)

kinetochore   mitotic-spindle-fiber attachment region of a chromosome

Krebs-Henseleit cycle   chemical reaction cycle in which urea is produced

Krebs tricarboxylic acid cycle   chemical reaction cycle in which acetyl coenzyme A is oxidized to carbon dioxide and water

labor   physiological state of pregnant mammals in which repeated uterine contractions occur, leading to birth of young

lacteal   one of the lymph vessels in small intestine transporting lipid droplets to blood

Lamarckism   view, now discredited, that organisms inherit acquired characteristics

lamella (*pl.* lamellae)   a sheet or membrane

lamellar frets   lamellae extending from grana into the stoma of chloroplasts

larva (*pl.* larvae)   motile, feeding embryos (esp. of invertebrates)

larynx   voice box of mammals

leukocyte   any of several types of white blood cells

ligament   soft tissue holding bones together at joints

light reactions   photosynthetic events utilzing light energy and producing reducing substances and ATP for the dark reactions

linkage   occurrence of several genes (linkage group) on same chromosome

linkage map   map showing "genetic distance" of genes in a linkage group (i.e., on same chromosome)

lipase   enzyme that catalyzes the breakdown of lipids

littoral zone   zone near shore of a sea or large body of fresh water

liver   organ that performs most biochemical operations in vertebrates

loop of Henle   tubular loop in nephrons of the kidney; site of active transport processes

lymph   fluid derived from the leakage of blood plasma out of the vascular system in vertebrates

lymph node   a rounded mass of lymphatic tissue that helps to filter tissue exudates of foreign material

lymphocyte   cell of the lymphatic system

lysosome   cellular organelle containing degradative enzymes used in cellular digestion

macrogamete   the larger of two gametes; egg

macromolecules   large molecules, esp. proteins, nucleic acids, and polysaccharides

macronucleus   the larger of two types of nuclei found in many ciliates, such as *Paramecium*

mammary gland   milk gland of mammals; human breast

mantle   tissue lining the gill cavity of mollusks which gives rise to the shell

marsupials   members of an order of mammals found in Australia that carry young in immature developmental stage, e.g., kangaroo

masturbation   erotic stimulation of sex organs to produce orgasm

mating type   one of several "sexes" that are morphologically indistinguishable

medulla   inner portion of kidney, adrenal, or brain

meiosis   special cell divisions in gametogenesis that reduce the genetic information in gametes to half that of regular cells

melanin   a dark pigment in many cells

menses   monthly period of human females; menstruation

meristem, meristematic tissue   undifferentiated, rapidly dividing plant tissue which gives rise to differential portions

mesenchyme   undifferentiated "third-layer" tissue found in flatworms and higher metazoans

mesoderm   the "middle layer" of developing animal embryos

mesoglea   a gelatinous middle layer in cnidarians

messenger RNA (mRNA)   molecules of ribonucleic acid that carry a transcribed coded message from DNA for the synthesis of one polypeptide

metabolism   the sum total of chemical reactions in a cell or organism

metabolites   chemical intermediates in metabolism

metamorphosis   change of form; may be radical, as in metamorphosis of insect pupa to adult

metaphase   a phase of cell division in which the chromosomes line up preparatory to being segregated into two equal groups

microfilaments   filaments about 7 nm in diameter and of indeterminate length found in almost all cells; apparently most are composed of F-actin

microgamete   small gamete; sperm

micronuclei   smaller of two types of nuclei in ciliate protists

micronutrient   nutrients required in minute quantities

microtubule   tubular structures 25 nm in diameter and of indeterminate length found in nearly all eucaryotic cells; composed of tubulin and associated with other proteins

microvillus (*pl.*, microvilli)   minute, fingerlike projection on cell surface

middle ear   cavity inside tympanic membrane in higher vertebrates

mimicry   resemblance of one organism to another harmful organism; an adaptive feature

mitochondrion (*pl.*, mitochondria)   organelle in which oxidative phosphorylation occurs

mitosis   nuclear division in all eucaryotic cells; involves equal apportionment of chromosomes to daughter cells

mitotic apparatus   in plants, the spindle alone; in animals, the spindle and asters

mitotic spindle   organelle that self-assembles only for mitosis; composed primarily of microtubules

Monera   kingdom of all procaryotic organisms

monocots   those seed plants possessing only a single cotyledon

monomer   small molecule capable of being assembled into a polymer

morphogenesis   change in form during development

motor cortex   portion of cerebral cortex controlling voluntary movements

motor neuron   neuron with its cell body in the spinal cord acting on effectors, such as muscle fibers

motor end-plate   neuromuscular junction

mucosa   inner lining of intestinal organs

multiple alleles   several possible genes that may occupy one of two homologous loci on a chromosomal pair

multiple factor inheritance   inheritance governed by more than a single genetic locus

mutation   stable change in a gene

muton   unit of mutation in a gene

mycelium   cellular extension in fungi

myelin sheath   a membranous wrapping applied to developing neurons by Schwann cells

myocardial infarction (coronary thrombosis)   heart attack caused by blockage of coronary artery or one of its branches in the heart

myofibrils   contractile fibrils in muscle fibers

myosin   the mechanochemical transducer molecule of muscle and some other motile cells

natural selection   the process in which the fittest organisms survive

negroid   pertaining to the black people who originated in Africa

nematocyst   one of several types of stinging cells found in cnidarians

nephridia   excretory tubules found in annelids

nephrons   functional units of the kidney

nerve net   nervous system in cnidarians

neural plate   region of an early embryo that will give rise to the neural tube, the forerunner of the vertebrate nerve cord

neuroblast differentiating neuron

neuron nerve cell

neurotransmitter substance secreted at a nerve ending that affects another neuron

nitrogen fixation conversion of atmospheric nitrogen to organic nitrogen by certain bacteria and blue-green algae

nondisjunction failure of the kinetochores to separate in the first meiotic division

notochord skeletal element in primitive and embryonic chordates

nuclear envelope structure separating the nucleus from the cytoplasm of all interphase eucaryotic cells

nucleoid region of a procaryotic cell that contains the chromosome

nucleolus component of the eucaryotic nucleus; concerned with ribosome synthesis

nucleotide intermediates in nucleic acid and energy metabolism

occipital lobe posterior dorsal portion of cerebral cortex

olfactory receptor cell specialized for receiving smelled chemical stimuli

olfactory lobe portion of the brain receiving olfactory stimuli

ommatidia segments of a compound arthropod eye

oocyte developing egg cell

operator gene a gene that controls the operation of other genes

optic vesicle forerunner of the embryonic eye

organelle functional particle in a cell

organism living plant, animal, fungus, or microbe (microorganism)

organogenesis development of organs

osmosis the diffusion of water through a selectively permeable membrane in response to a difference in water (or solute) concentrations on either side of the membrane

ovary female sex organ

oviparous producing eggs

ovulation the release of one or more eggs from the ovary

ovule egg cell of a plant

ovum egg cell of an animal

oxidation the removal of electrons, hydrogen, or both from a substrate

oxidative phosphorylation the coupled reactions that generate ATP from ADP and phosphate, powered by oxidative electron transport in mitochondria

oxygen debt an animal's requirement of extra oxygen after vigorous exercise to oxidize accumulated lactate in muscle

parasympathetic the craniosacral portion of the autonomic nervous system

parathyroid gland located in the neck; controls calcium metabolism

parenchyma plant tissue that is active in photosynthesis and storage; found especially in leaves

parthenogenesis development of an egg without activation by or union with sperm (may be natural or artificial)

pedigree analysis genetic analysis of the history of various recognizable traits in a family

pelagic of the open sea, ocean, or lake

penis male organ of copulation

peptide bond bond between carbon and nitrogen connecting amino acids in polypeptides and proteins

periosteum sensitive outer coating of bones

peripheral nervous system the spinal nerves, affectors, and effectors; in contrast to the central nervous system

peritoneum cellular lining of the body cavity in a vertebrate

phage transduction the carrying of one gene from one bacterial cell to another by viral infection

phagocytosis cell eating, part of endocytosis, along with pinocytosis

pharyngeal gill slit slits between the gill arches in primitive and embryonic chordates (a phylum characteristic of the chordata)

pharynx throat region

phenotype the genetic appearance of an organism

pheromone chemical released by an organism that modifies the behavior of other members of the same species

phloem plant tissue that conducts nutrients from the leaves to stems and roots

phospholipid lipid containing a nitrogen base and phosphoric acid

photoautotrophic needing only CO_2 as a source of carbon and light energy to feed itself

photoperiodism the property of responding to light cycles, e.g., induced flowering

photophosphorylation the generation of ATP by the light reactions of photosynthesis

photosynthesis the synthesis of carbohydrate from CO_2 and water, requiring light energy

phyla the broadest categories into which organisms of one kingdom can be sorted taxonomically

phyto- prefix meaning plant

pinocytosis the drinking of fluids into vesicles; especially in amoebae and mammalian cells in culture

pistil female sex organ of flowering plants which carries ovules

pith unspecialized plant tissue found at the center of a vascular cylinder

pituitary gland master endocrine gland of man and other vertebrates

placenta an organ formed of embryonic and maternal tissue, through which the embryo (and later the fetus) is fed

planarian free-living flatworm

plankton organisms that can be collected from fresh or sea water

plasma cell cell derived from lymphocyte; found in inflammations

plasma membrane   cell membrane; permeability barrier of the cell

plasmodesmata   bridges between cells through which materials may pass

plasmodium   multinucleate stage of an acellular slime mold

plasmolysis   cell shrinkage, a reaction of plant cells to high osmotic pressure

platelet   smallest formed element in the blood; plays a role in clotting

pleura   membrane lining the pleural cavity surrounding the lungs

point mutation   mutation of a gene (as opposed to chromosomal mutation)

polar nuclei   nuclei present in plant ovules

pollen   cells of flowering plants specialized for delivery of the male gamete nuclei to the ovule

pollination   delivery of pollen from a stamen of one flower to the pistil of another

polygenic   having many genes

polymer   a macromolecule composed of many monomers

polyp   sessile asexual stage of some jellyfish species

polyploidy   having many (haploid) sets of genes

polyribosome (polysome)   a molecule of $_mRNA$ with its attached ribosomes

potential energy   energy that is stored, not used or in use

primary succession   changes in plant life in a given area as the soil develops from erosion and so on

procaryotic   having a nucleoid, with no nuclear envelope or mitosis

progesterone   hormone secreted by the *corpus luteum* in the ovary; its most important function is to forestall menses

prostate gland   an accessory male sex gland; secretes a major portion of the semen

proteins   a class of macromolecules that are polymers of amino acids

prothoracic gland   an endocrine gland of insects

protist   a one-cell or acellular eucaryotic organism

Protista   kingdom of the protists

proton   positively charged hydrogen ion (H⁺)

protoplasm   outdated generalized concept of the contents of cells. Not used in this text

protozoan (*pl.* protozoa)   animal-like protist

pseudocoelomates   animals containing a "false" body cavity

pulmonary vein   the only vein that carries oxygenated blood toward the heart

Punnett square   checkerboard square useful in working out the results of genetic crosses

radicle   primary root of a plant

radula   rasping digestive organ of mollusks, especially snails

rain shadow   dry side of a mountain range

recessive trait   a trait controlled by a gene that must be present in homozygous form to be expressed visibly

recombination   result of independent segregation and assortment of chromosomes in meiosis

recon   subdivision of a gene; smallest unit of recombination

red tide   a "bloom" of poisonous dinoflagellates

refractory period   period after stimulation before an excitable cell can react a second time

regulator gene   gene specifying the synthesis of a repressor or other regulatory macromolecule

repressible enzyme   an enzyme, the synthesis of which can be held back

repressor protein   a protein that represses gene function

respiration   gas exchange; oxygen uptake, carbon dioxide production

respiratory center   a region of the hypothalamus that controls breathing

resting potential   the voltage difference between the inside and outside of an unstimulated cell

retina   layer of light receptors in the eye upon which images are formed

Rh factor   "Rhesus factor"; an inherited antigen, the inheritance of which involves risks to human newborn

rhizoid   root or rootlike structure

rhizome   a horizontal underground stem

ribonucleic acid (RNA)   a nucleic acid polymer consisting of the sugar ribose, phosphoric acid, and four nitrogen bases

ribosome   organelle which synthesizes polypeptide in cooperation with mRNA

rod   red-sensitive receptor of the retina; responsible for crepuscular vision

root cap   loose cap of cells on root tip

root hair   hairlike lateral extensions of root cells

rough ER   endoplasmic reticulum covered with ribosomes and mRNA

sarcolemma   membrane plus connective tissue around muscle fibers

sarcomere   repeating unit of function in a myofibril

sarcoplasm   fluid in a muscle fiber

sarcoplasmic reticulum   calcium-sequestering membrane system in vertebrate striated muscle

scenescence   aging

Schwann cell   a cell that forms myelin around some kinds of neurons

sclerenchyma   a supporting tissue of plants with thick-walled cells

scrotum   saclike extension of the body cavity containing the testes in mammals

selective permeability   the unique property of the plasma membranes of cells that they permit some substances to pass but not others

semen   sperm-containing fluid formed by the testes and accessory organs

semicircular canal   the organ of orientation in the inner ear

seminal vesicle   organ for storage of semen

sensory neuron   nerve cell that receives stimuli from the environment and transmits impulses to ganglia or the brain

sessile   nonmotile (e.g., polyps)

sex linkage   linkage of traits on the X chromosome

sieve plate   horizontal, perforated septa in sieve tubes of plant phloem

smooth ER   endoplasmic reticulum lacking attached ribosomes

smooth muscle   muscle characteristic of viscera

sodium pump   refers to the ability of the membranes of most cells to exclude sodium ions, using the energy of ATP

soma   body (in any context)

somatic cell   body cell (as opposed to germ, or sex, cells)

somites   characteristic blocks of embryonic bone, muscle, and dermis that appear beside the developing nervous system

species   a group of organisms usually defined as a breeding population

sperm   male germ cells

spermatogenesis   the production of spermatids by mitosis and meiosis

sphincter   a ring of muscle that controls the flow through a continuous tube

spinal cord   the nerve cord in the spinal column (of vertebrates)

sporangium   spore case

spore   asexual reproductive cell of lower plants and fungi

sporophyte   asexual generation of plants producing spores

spring turnover   the inversion of water in a lake that occurs as the top and bottom densities change with temperature

stamen   pollen-bearing male organ of flowering plants

starch   a glucose polymer synthesized by plants

steady state   a "dynamic equilibrium"

stolon   a horizontal stem that gives rise to a new plant at its tip

stoma (*pl.* stomata)   opening between guard cells into parenchyma of a leaf

striated muscle   skeletal muscle, usually under voluntary control, composed of multinucleate fibers (cells)

structural gene   a segment of DNA specifying an enzymatic or structural protein

substrate   substance acted upon by an enzyme

swim bladder   buoyancy organ of bony fish (teleosts)

sympathetic nervous system   the adrenergic portion of the autonomic nervous system

synapse   the junction of two neurons where communication occurs. (*See also* chemical synapse; electronic synapse)

synapsis   the pairing of homologous chromosomes during meiotic prophase

syncytial   having many nuclei not separated by compartments

synovial fluid   fluid bathing human joints

systole   the power stroke of the heart; the cause of systolic blood pressure

target cell   terminology used in endocrinology to designate the cell or cell type on which a hormone acts

taste bud   one of four types of special chemoreceptors on the tongue

teleost   bony fish

telophase   terminal phase of mitosis or meiosis, when the daughter nuclei are reforming

template   mold or model for synthesis

tendon   soft fibrous tissue connecting a muscle to a bone

tension-cohesion theory   the currently accepted theory explaining the ascent of water in trees by tension from transpiration and the cohesion of water molecules

territoriality   the tendency of higher animals (probably including man) to stake out a territory and defend it

testcross   a cross of the members of the first filial generation to a homozygous recessive

testis   primary sex organ of male animals

tetrad   *See* chromatid tetrad

thermodynamics   the science dealing with energy and its interconversions

threshold   the minimum stimulus intensity that elicits a response

thymine   a nitrogen base in DNA

tonicity   osmotic strength relative to normal (isotonic) conditions

trachea   air passage to the bronchi and lungs

tracheid   a conducting element of the xylem

transamination   transfer of an amino group from one molecule to another

transcription   the rewriting of a DNA-coded message in the RNA language

transduction   the introduction of foreign genetic material into the genome of a cell by bacterial infection

transfer RNA (tRNA)   RNA that escorts amino acids to the site of protein synthesis

translation   the synthesis of a protein from the instructions transcribed in mRNA

translocation   the displacement of a gene along a chromosome during meiosis

transmitter substances   several substances that are released at synapses by one nerve cell to stimulate its neighbor

transpiration   the passage of water out of a plant by evaporation from the insides of leaves through stomata

trichinosis   a serious parasitic disease, caused by ingesting undercooked pork containing trichina worms

triplet code (the genetic code)   system of three-nucleotide "letters" that specify amino acids to be added to polypeptide chains

trochophore larva   basic larval type for annelids and mollusks

tropism   a slow response of a lower organism to an environmental stimulus or gradient

tubal ligation   surgical interruption of the oviduct (fallopian tube) that prevents conception

tube feet   organs of locomotion of echinoderms; part of the water vascular system

tuber   enlarged portion of an underground stem (stolon)

tubulin   monomeric, dimeric, or polymeric form of a protein that self-assembles into microtubules

turgor pressure   inward pressure exerted on plant cells by the cell walls acting against osmotic pressure

tympanic membrane   eardrum

umbilical cord   nutritive cord connecting fetus to its placenta

unit membrane   double line seen in fixed membranes of all kinds with an electron microscope

urea   nitrogen waste product of many animals

ureter   tube connecting the kidneys with the bladder

urethra   tube connecting the kidneys with the outside

uterine cycle   uterine changes during the menstrual cycle of the female

uterus   organ in which young are nourished before birth

vacuole   membrane-bounded sac; often holds water or watery fluid

vagina   a pouchlike female organ that receives the penis during copulation

valve   device that prevents back-flow, especially in circulatory systems

vascular cambium   meristematic tissue that differentiates into vascular tissue

vascular plants   higher plants, containing vascular tissue (xylem, phloem)

vas deferens   duct leading from the testis and male accessory organs to the urethra that carries seminal fluid to be discharged

vasectomy   operation that closes off the vas deferens, preventing the escape of sperm and rendering the individual sterile

vegetal hemisphere   the hemisphere of an egg (e.g., amphibians) containing the yolk

vegetative propagation   asexual propagation in plants that will give rise to endodermal parts of the embryo

venous blood   blood returning to the heart

ventricle   thick-walled pumping chamber of the heart

venule   tiny vein

vesicle   membrane-bounded watery body in a cell

vessel element   conducting relic of a xylem cell

villi   digitlike extensions of a surface, such as the lining of the intestine

virus   submicroscopic, infectious, noncellular particle

vitreous humor   clear fluid in the main cavity of the eye

viviparous   bearing young alive

water table   groundwater level

white matter   portion of the brain composed mainly of bundles of myelinated axons and dendrites

xylem   conductive tissue of higher plants that carries water and nutrients upward toward the leaves

yolk   nutritive storage product in animal eggs and embryos

yolk sac   one of the extraembryonic membranes of vertebrate embryos

Z lines   line that separate adjacent sarcomeres in striated muscle fibers

zone of elongation   zone of a root tip in which cells are elongating

zoospore   motile spore (*general term*)

zygote   fertilized ovum

Index

76 77 78 79 80 9 8 7 6 5 4 3 2 1